Studies in Surface Science and Catalysis 51

NEW SOLID ACIDS AND BASES
their catalytic properties

Studies in Surface Science and Catalysis
Advisory Editors: B. Delmon and J. T. Yates
Vol. 51

NEW SOLID ACIDS AND BASES

THEIR CATALYTIC PROPERTIES

by

Kozo TANABE — Professor, *Department of Chemistry, Faculty of Science, Hokkaido University, Sapporo, Japan*

Makoto MISONO — Professor, *Department of Synthetic Chemistry, Faculty of Engineering, The University of Tokyo, Tokyo, Japan*

Yoshio ONO — Professor, *Department of Chemical Engineering, Faculty of Engineering, Tokyo Institute of Technology, Tokyo, Japan*

Hideshi HATTORI — Associate Professor, *Department of Chemistry, Faculty of Science, Hokkaido University, Sapporo, Japan*

KODANSHA
Tokyo

1989

ELSEVIER
Amsterdam - Oxford - New York - Tokyo

Copublished by
KODANSHA LTD., Tokyo
and
ELSEVIER SCIENCE PUBLISHERS B.V., Amsterdam

exclusive sales rights in Japan
KODANSHA LTD.
12-21, Otowa 2-chome, Bunkyo-ku, Tokyo 112, Japan

for the U.S.A. and Canada
ELSEVIER SCIENCE PUBLISHING COMPANY, INC.
655 Avenue of the Americas, New York, N.Y. 10010, U.S.A.

for the rest of the world
ELSEVIER SCIENCE PUBLISHERS B.V.
25 Sara Burgerhartstraat, P.O. Box 211, 1000 AE Amsterdam, The Netherlands

Library of Congress Cataloging-in-Publication Data

New solid acids and bases : their catalytic properties / by Kozo Tanabe ... [et al.].
p. cm. -- (Studies in surface science and catalysis ; 51)
Includes bibliographical references.
ISBN 0-444-98800-9
1. Acids. 2. Bases (Chemistry) 3. Catalysts. I. Tanabe, Kōzō.
II. Series.
QD477.N49 1989
546'.24--dc20 89-23475
CIP

ISBN 0-444-98800-9 (Vol.51)
ISBN 0-444-41801-6 (Series)
ISBN 4-06-204394-7 (Japan)

PRINTED IN JAPAN

Preface

Nineteen years have passed since the monograph "Solid Acids and Bases" was published in 1970. During this period many new kinds of solid acids and bases have been found and synthesized. The surface properties (in particular, acidic and basic properties) and the structures of the new solids have been clarified by newly developed measurement methods using modern instruments and techniques. The characterized solid acids and bases have been applied as catalysts for diversified reactions, many good correlations being obtained between the acid–base properties and the catalytic activities or selectivities. Recently, acid–base bifunctional catalysis on solid surfaces is becoming an ever more important and intriguing field of study.

It has been recognized that the acidic and basic properties of catalysts and catalyst supports play an important role even in oxidation, reduction, hydrogenation, hydrocracking, etc. The effect of the preparation method and the pretreatment condition of solid acids and bases on the acidic and basic properties, the nature of acidic and basic sites and the mechanism regarding the generation of acidity and basicity have been elucidated experimentally and theoretically. On the basis of the accumulated knowledge of solid acids and bases, it is now possible to design and develop highly active and selective solid acid and base catalysts for particular reations.

Moreover, the chemistry of solid acids and bases is being related to and utilized in numerous areas including adsorbents, sensors, cosmetics, fuel cells, sensitized pressed papers, and others.

In the present volume, the great progress in solid acids and bases made over the past two decades is summarized and reviewed with emphasis on fundamental aspects and chemical principles.

We wish to express our gratitude to Ms. Cecilia M. Hamagami and Mr. I. Ohta of Kodansha Scientific Ltd. for their invaluable assistance of the preparation of the English manuscripts which comprise this book.

Summer 1989

Kozo TANABE
Makoto MISONO
Yoshio ONO
Hideshi HATTORI

Preface

Contents

1
Definition and Classification of Solid Acids and Bases

In general terms, a solid acid may be understood to be a solid on which the color of a basic indicator changes or a solid on which a base is chemically adsorbed. More strictly, following both the Brønsted and Lewis definitions, a solid acid shows a tendency to donate a proton or to accept an electron pair, whereas a solid base tends to accept a proton or to donate an electron pair. These definitions are adequate for an understanding of the acid-base phenomena shown by various solids, and are convenient for a clear description of solid acid and base catalysis.

TABLE 1.1 Solid Acids

1.	Natural clay minerals: kaolinite, bentonite, attapulgite, montmorillonite, clarit, fuller's earth, zeolites (X, Y, A, H-ZSM etc), cation exchanged zeolites and clays
2.	Mounted acids: H_2SO_4, H_3PO_4, $CH_2(COOH)_2$ mounted on silica, quartz sand, alumina or diatomaceous earth
3.	Cation exchange resins
4.	Charcoal heat-treated at 573 K
5.	Metal oxides and sulfides : ZnO, CdO, Al_2O_3, CeO_2, ThO_2, TiO_2, ZrO_2, SnO_2, PbO, As_2O_3, Bi_2O_3, Sb_2O_5, V_2O_5, Cr_2O_3, MoO_3, WO_3, CdS, ZnS
6.	Metal salts : $MgSO_4$, $CaSO_4$, $SrSO_4$, $BaSO_4$, $CuSO_4$, $ZnSO_4$, $CdSO_4$, $Al_2(SO_4)_3$, $FeSO_4$, $Fe_2(SO_4)_3$, $CoSO_4$, $NiSO_4$, $Cr_2(SO_4)_3$, $KHSO_4$, K_2SO_4, $(NH_4)_2SO_4$, $Zn(NO_3)_2$, $Ca(NO_3)_2$, $Bi(NO_3)_3$, $Fe(NO_3)_3$, $CaCO_3$, BPO_4, $AlPO_4$, $CrPO_4$, $FePO_4$, $Cu_3(PO_4)_2$, $Zn_3(PO_4)_2$, $Mg_3(PO_4)_2$, $Ti_3(PO_4)_4$, $Zr_3(PO_4)_4$, $Ni_3(PO_4)_2$, AgCl, CuCl, $CaCl_2$, $AlCl_3$, $TiCl_3$, $SnCl_2$, CaF_2, BaF_2, $AgClO_4$, $Mg(ClO_4)_2$
7.	Mixed oxides : SiO_2-Al_2O_3, SiO_2-TiO_2, SiO_2-SnO_2, SiO_2-ZrO_2, SiO_2-BeO, SiO_2-MgO, SiO_2-CaO, SiO_2-SrO, SiO_2-ZnO, SiO_2-Ga_2O_3, SiO_2-Y_2O_3, SiO_2-La_2O_3, SiO_2-MoO_3, SiO_2-WO_3, SiO_2-V_2O_5, SiO_2-ThO_2, Al_2O_3-MgO, Al_2O_3-ZnO, Al_2O_3-CdO, Al_2O_3-B_2O_3, Al_2O_3-ThO_2, Al_2O_3-TiO_2, Al_2O_3-ZrO_2, Al_2O_3-V_2O_5, Al_2O_3-MoO_3, Al_2O_3-WO_3, Al_2O_3-Cr_2O_3, Al_2O_3-Mn_2O_3, Al_2O_3-Fe_2O_3, Al_2O_3-Co_3O_4, Al_2O_3-NiO, TiO_2-CuO, TiO_2-MgO, TiO_2-ZnO, TiO_2-CdO, TiO_2-ZrO_2, TiO_2-SnO_2, TiO_2-Bi_2O_3, TiO_2-Sb_2O_5, TiO_2-V_2O_5, TiO_2-Cr_2O_3, TiO_2-MoO_3, TiO_2-WO_3, TiO_2-Mn_2O_3, TiO_2-Fe_2O_3, TiO_2-Co_3O_4, TiO_2-NiO, ZrO_2-CdO, ZnO-MgO, ZnO-Fe_2O_3, MoO_3-CoO-Al_2O_3, MoO_3-NiO-Al_2O_3, TiO_2-SiO_2-MgO, MoO_3-Al_2O_3-MgO, heteropoly acids

TABLE 1.2 Solid Bases

1.	Mounted bases: NaOH, KOH mounted on silica or alumina; Alkali metal and alkaline earth metal dispersed on silica, alumina, carbon, K_2CO_3 or in oil; NR_3, NH_3, KNH_2 on alumina; Li_2CO_3 on silica; *t*-BuOK on xonotolite
2.	Anion exchange resins
3.	Charcoal heat-treated at 1173 K or activated with N_2O, NH_3 or $ZnCl_2$-NH_4Cl-CO_2
4.	Metal oxides: BeO, MgO, CaO, SrO, BaO, ZnO, Al_2O_3, Y_2O_3, La_2O_3, CeO_2, ThO_2, TiO_2, ZrO_2, SnO_2, Na_2O, K_2O
5.	Metal salts : Na_2CO_3, K_2CO_3, $KHCO_3$, $KNaCO_3$, $CaCO_3$, $SrCO_3$, $BaCO_3$, $(NH_4)_2CO_3$, $Na_2WO_4 \cdot 2H_2O$, KCN
6.	Mixed oxides: SiO_2-MgO, SiO_2-CaO, SiO_2-SrO, SiO_2-BaO, SiO_2-ZnO, SiO_2-Al_2O_3, SiO_2-ThO_2, SiO_2-TiO_2, SiO_2-ZrO_2, SiO_2-MoO_3, SiO_2-WO_3, Al_2O_3-MgO, Al_2O_3-ThO_2, Al_2O_3-TiO_2, Al_2O_3-ZrO_2, Al_2O_3-MoO_3, Al_2O_3-WO_3, ZrO_2-ZnO, ZrO_2-TiO_2, TiO_2-MgO, ZrO_2-SnO_2
7.	Various kinds of zeolites exchaged with alkali metal or alkaline earth metal

TABLE 1.3 Solid Superacids

Group	Acid	Support
1a	SbF_5	SiO_2-Al_2O_3, SiO_2-TiO_2, SiO_2-ZrO_2, TiO_2-ZrO_2
1b	SbF_5	Al_2O_3-B_2O_3, SiO_2, SiO_2-WO_3, HF-Al_2O_3
2	SbF_5, TaF_5	Al_2O_3, MoO_3, ThO_2, Cr_2O_3, Al_2O_3-WB
3	SbF_5, BF_3	graphite, Pt-graphite
4	BF_3, $AlCl_3$, $AlBr_3$	ion exchange resin, sulfate, chloride
5	SbF_5-HF SbF_5-FSO_3H	metal (Pt, Al), alloy (Pt-Au, Ni-Mo, Al-Mg), polyethylene, SbF_3, AlF_3, porous substance (SiO_2-Al_2O_3, kaolin, active carbon, graphite)
6	SbF_5-CF_3SO_3H	F-Al_2O_3, $AlPO_4$, charcoal
7	Nafion (polymeric perfluororesin sulfonic acid)	
8	TiO_2-SO_4^{2-}, ZrO_2-SO_4^{2-}, Fe_2O_3-SO_4^{2-}	
9	H-ZSM-5 zeolite	

In accordance with the above definitions, a summarized list of solid acids and bases is given in Tables 1.1 and 1.2. The first group of solid acids in Table 1.1 includes naturally occurring clay minerals. The main constituents are silica and alumina. Various types of synthetic zeolites such as zeolites X,Y,A, ZMS-5, ZSM-11, etc. have been reported to show characteristic catalytic activities and selectivities. The well-known solid acid, synthetic silica-alumina, is listed in the seventh group, which also includes the many oxide mixtures which have recently been found to display acidic properties and catalytic activity. In the fifth and sixth groups are included many inorganic chemicals such as metal oxides, sulfides, sulfates, nitrates, phosphates and halides. Many have been found to show characteristic selectivities as catalysts.

Of the solid bases listed in Table 1.2, special mention should be made of the alkaline earth metal oxides in the fourth group and mixed metal oxides in the sixth group, whose basic properties and catalytic action have been recently found to be striking and interesting. A solid superacid is defined as a solid whose acid strength is higher than the acid strength of 100% sulfuric acid. Since the acid strength of 100% sulfuric acid expressed by the Hammett acidity function, Ho, is -11.9, a solid of Ho < -11.9 is called a solid superacid. The kinds of solid superacids are shown in Table 1.3. The groups 1 through 6 include acids supported on various solids.

On the other hand, a solid superbase is defined as a solid whose base strength expressed by the basicity function, H_-, is higher than $+26$. The basis of the definition has been described in the literature.[1)] The kinds of solid superbases are shown in Table 1.4 together with their preparation method and pretreatment temperature.

TABLE 1.4 Solid Superbases

	Starting material, Preparation method	Pretreatment temp. K	H_-
CaO	$CaCO_3$	1173	26.5
SrO	$Sr(OH)_2$	1123	26.5
MgO-NaOH	(NaOH impregnated)	823	26.5
MgO-Na	(Na vaporized)	923	35
Al_2O_3-Na	(Na vaporized)	823	35
Al_2O_3-NaOH-Na	(NaOH, Na impregnated)	773	37

REFERENCES

1. K. Tanabe, *in*: *Catalysis by Acids and Bases*, (eds. B. Imelik, C. Naccache, G. Coudurier, Y. Ben Taarit, J. C. Vedrine) Elsevier, Amsterdam, 1985, p.1.

2
Determination of Acidic and Basic Properties on Solid Surfaces

A complete description of acidic and basic properties on solid surfaces requires the determination of the acid and base strength, and of the amount and nature (Brønsted or Lewis type) of the acidic and basic sites.

2.1 ACIDIC PROPERTY

2.1.1 Strength and Amount of Solid Acid

When measuring the strength of a solid acid or base, it should be recognized that activity coefficients for species on the solid are unknown. Therefore, acidity and basicity functions for the solid are not properly defined thermodynamically. Nevertheless, the acidity and basicity functions are clearly valuable in a relative sense, while the absolute values are also useful provided the above limitations are recognized and numerical accuracy is not overstated.

The acid strength of a solid is defined as the ability of the surface to convert an adsorbed neutral base into its conjugate acid. If the reaction proceeds by means of proton transfer from the surface to the adsorbate, the acid strength is expressed by the Hammett acidity function H_0,[1)]

$$H_0 = pK_a + \log [B] / [BH^+], \tag{1}$$

where [B] and $[BH^+]$ are, respectively, the concentrations of the neutral base (basic indicator) and its conjugate acid and pK_a is pK_{BH^+}. If the reaction takes place by means of electron pair transfer from the adsorbate to the surface, H_0 is expressed by

$$H_0 = pK_a + \log [B]/[AB], \tag{2}$$

where [AB] is the concentration of the neutral base which reacted with the Lewis acid or electron pair acceptor, A.

The amount of acid on a solid is usually expressed as the number or mmol of acid sites per unit weight or per unit surface area of the solid, and is obtained by measuring the amount of a base which reacts with the solid acid. This is also sometimes loosely called "acidity".

For the determination of strength and amount of a solid acid, there are two main methods: an amine titration method using indicators and a gaseous base adsorption method.

A. Amine Titration Method Using Indicators

The color of suitable indicators adsorbed on a surface will give a measure of its acid strength: if the color is that of the acid form of the indicator, then the value of the H_0 function of the surface is equal to or lower than the pK_a of the conjugate acid of the indicator. Lower values of H_0 correspond to greater acid strength. Thus, for indicators undergoing color changes in this way, the lower the pK_a, the greater the acid strength of the solid. For example, a solid which gives a yellow coloration with benzalacetophenone (p$K_a = -5.6$), but is colorless with anthraquinone (p$K_a = -8.2$), has an acid strength H_0 which lies between -5.6 and -8.2. A solid having $H_0 \leq -16.04$ will change all indicators in Table 2.1 from the basic to the acidic colors, whereas one which changed none of them will have an acid strength of $H_0 > +6.8$.

The experimental details of the acid strength determination are described in earlier publications.[2,3] The acid strength of a solid superacid which is very sensitive to moisture can be determined by observing the color change of an indicator whose vapor

TABLE 2.1 Basic indicators used for the measurement of acid strength

Indicators	Color		pK_a †1	$[H_2SO_4]$†2/%
	Base-form	Acid-form	pKa	
Neutral red	yellow	red	+ 6.8	8×10^{-8}
Methyl red	yellow	red	+ 4.8	—
Phenylazonaphthylamine	yellow	red	+ 4.0	5×10^{-5}
p-Dimethylaminoazobenzene	yellow	red	+ 3.3	3×10^{-4}
2-Amino-5-azotoluene	yellow	red	+ 2.0	5×10^{-3}
Benzeneazodiphenylamine	yellow	purple	+ 1.5	2×10^{-2}
Crystal violet	blue	yellow	+ 0.8	0.1
p-Nitrobenzeneazo-(*p*′-nitro-diphenylamine)	orange	purple	+ 0.43	—
Dicinnamalacetone	yellow	red	− 3.0	48
Benzalacetophenone	colorless	yellow	− 5.6	71
Anthraquinone	colorless	yellow	− 8.2	90
2,4,6-Trinitroaniline	colorless	yellow	−10.10	98
p-Nitrotoluene	colorless	yellow	−11.35	†3
m-Nitrotoluene	colorless	yellow	−11.99	†3
p-Nitrofluorobenzene	colorless	yellow	−12.44	†3
p-Nitrochlorobenzene	colorless	yellow	−12.70	†3
m-Nitrochlorobenzene	colorless	yellow	−13.16	†3
2,4-Dinitrotoluene	colorless	yellow	−13.75	†3
2,4-Dinitrofluorobenzene	colorless	yellow	−14.52	†3
1,3,5-Trinitrotoluene	colorless	yellow	−16.04	†3

†1 pK_a of the conjugate acid, BH^+, of indicator, B, (=pK_{BH^+})

†2 wt. percent of H_2SO_4 in sulfuric acid solution which has the acid strength corresponding to the respective pK_a

†3 The indicator is liquid at room temperature and acid strengh corresponding to the indicator is higher than the acid strength of 100 percent H_2SO_4.

is adsorbed on a solid sample through a breakable seal in a vacuum system at room temperature.[4] The indicators used for the determination are included in Table 2.1.

The amount of acid sites on a solid surface can be measured by amine titration immediately after determination of acid strength by the above method. The method consists of titrating a solid acid suspended in benzene with *n*-butylamine, using an indicator. The use of various indicators with different pK_a values (see Table 2.1) enables us a determination of the amount of acid *at various acid strengths* by amine titration. The experimental details such as the effects of titration time, volume of added indicator, pore size, and moisture on measured acid amount are given in Reference 2.

As an example, the acid strength and amount of $ZnO-Al_2O_3$ having different compositions as well as those of ZnO and Al_2O_3, when calcined at 773K in air, are shown in Fig. 2.1.[5] The maximum acid amounts were observed when the content of ZnO was 10 mol% at any acid strength. Many examples of good correlations between acid amount and catalytic activity have been reported. An example is shown in Fig. 2.2, where the catalytic activity of various binary oxides increases linearly with increasing acid amount at acid strength $H_0 \leq -3$ of the catalysts.[6]

The amine titration method gives the sum of the amounts of both Brønsted and Lewis acid, since both proton donors and electron pair acceptors on the surface will react with either the electron pair ($-\ddot{N}=$) of the indicator or that of amine ($\equiv N$:) to form a coordination bond. This method is rarely applied to colored or dark samples where the usual color change is difficult to observe. However, the difficulty can be minimized by mixing a white substance of known acidity with the sample or by employing the spectrophotometric method.[2,7] Calorimetric titration of a solid acid with amine is also available for the estimation of the acid amount of a colored or dark sample.[2,7-9] Recently, Hashimoto *et al.* developed a method to measure the acid strength

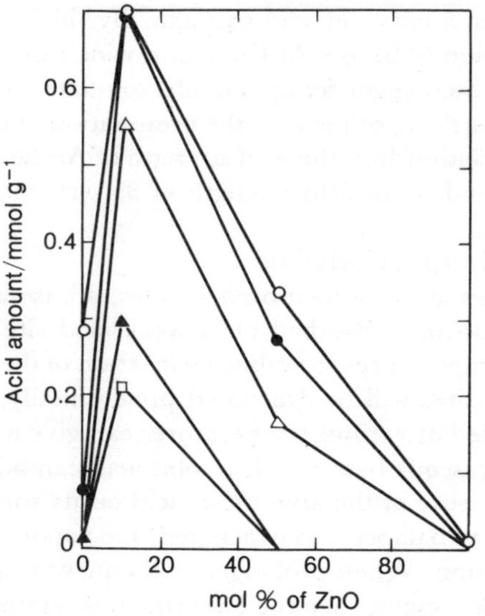

Fig. 2.1 Acid amounts at various acid strengths of $ZnO-Al_2O_3$ *vs.* mol % of ZnO. —○—; $H_0 \leq 4.8$, —●—; $H_0 \leq 3.3$, —△—; $H_0 \leq 1.5$, —▲—; $H_0 \leq -3.0$, —□—; $H_0 \leq -5.6$

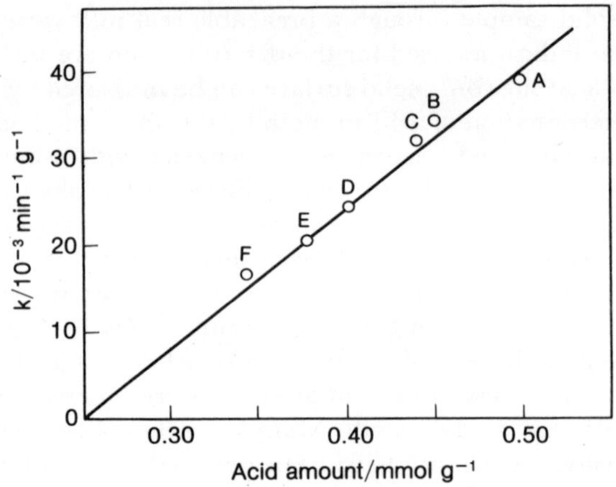

Fig. 2.2 First order rate constant of depolymerization of paraldehyde over various mixed catalysts *vs.* acid amount at $H_0 \leq -3$ of the catalysts.
A; SiO_2-MoO_3, B; $Al_2O_3-MoO_3$, C; SiO_2-WO_3,
D; $Al_2O_3-WO_3$, E; $SiO_2-V_2O_5$, F; $Al_2O_3-V_2O_5$
(Reproduced with permission by S. P. Wavekar *et al.*, *J. Res. Inst. Catal.*, **20**, 223(1972)).

distribution on a solid surface by utilizing the chemisorption isotherms of a series of Hammett indicators on a solid suspended in a nonpolar solvent such as benzene or cyclohexane.[10)] The fraction of acid sites covered by the indicator was expressed by a Langmuir type equation, which includes both acid strength and the indicator concentration. The chemisorption isotherms of Hammett indicators can be converted to a cumulative distribution curve of acid strength. By this method, they measured the acid strength distribution of $SiO_2-Al_2O_3$ over a wide range of acid strength ($-15 \leq H_0 \leq -3$).[10)] Rys and Steinegger set up a model to relate the sorption of Hammett indicators onto proton-carrying solids with the protonation of these indicators in acid solutions.[11)] From the relationship, the acid strength of Amberlyst-15 dispersed in water was found to correspond to an acid strength of 35 wt% aqueous sulfuric acid.

B. Gaseous Base Adsorption Method

When gaseous bases are adsorbed on acid sites, a base adsorbed on a strong acid site is more stable than one adsorbed on a weak acid site, and is more difficult to desorb. As elevated temperatures stimulate evacuation of the adsorbed bases from acid sites, those at weaker sites will be evacuated preferentially. Thus, the proportion of adsorbed base evacuated at various temperatures can give a measure of acid strength.

The amount of a gaseous base which a solid acid can adsorb chemically from the gaseous phase is a measure of the amount of acid on its surface. After a solid sample is put in a quartz spring balance and evacuated, the vapor of an organic base may be introduced for adsorption. When prolonged subsequent evacuation produces no further decrease in sample weight, then the base which is retained upon the sample is understood to be chemically adsorbed.[12)]

Recently, temperature programmed desorption (TPD) of basic molecules such as ammonia, pyridine, *n*-butylamine, etc. is frequently used to characterize the acid

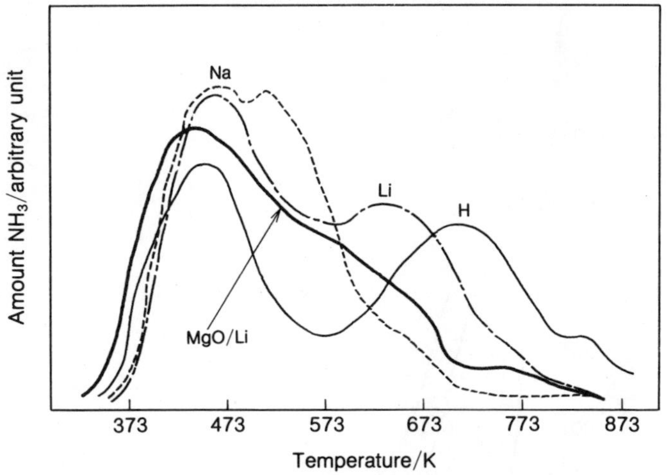

Fig. 2.3 TPD spectra of NH_3 on cation-exchanged ZSM-5 zeolites (Si/Al=44).

strength as well as acid amount on a solid surface. Fig. 2.3 shows a diagram of TPD of ammonia adsorbed on cation-exchanged ZSM-5.[13)] Two distinct peaks were observed for H–ZSM-5, indicating the existence of strong (a peak at 723K) and weak (a peak at 463K) acid sites. The acid strength of cation-exchanged ZMS-5 follows the order H > Li > MgO/Li > Na, the acid amount being shown by respective peak intensity. The details of the experimental procedure of the TPD method is given in the literature.[14)] A method which calculates a density distribution function of activation energy for desorption of ammonia by utilizing the TPD spectrum of ammonia is also presented.[15)]

The heat of adsorption of various bases is also clearly a measure of the acid strength on a solid surface.[2)] The differential heat of ammonia adsorption on $SiO_2-Al_2O_3$ and SiO_2 plotted against the surface coverage is shown in Fig. 2.4.[16)] Heat of adsorption corresponding to acid strength increases with increasing alumina content in $SiO_2-Al_2O_3$. Differential thermal analysis (DTA) and thermogravimetry (TG) of desorption of basic molecules is available for the estimation of the acid amount together with the acid strength of a solid.[2)]

Ammonia, *n*-butylamine, and pyridine are used extensively as gaseous bases for the determination of strength and amount of a solid acid.[2)] However, ammonia and *n*-butylamine in which hydrogen atoms are attached to the nitrogen atom have a tendency to dissociate (e.g., $NH_3 \rightleftharpoons NH_2^- + H^+$) and adsorb on both acidic and basic sites depending on the kinds of solids and the adsorption condition. In this regard, triethylamine, which is much more difficult to dissociate, is recommended for use as an adsorbate.[17)] Care must also be taken in the use of pyridine, because pyridine has recently been found to adsorb on strong basic sites to form an anion radical of pyridine.[18)] It appears necessary to check the adsorbed states of basic molecules by IR or ESR spectroscopy.

Methods utilizing adsorption and desorption of gaseous bases have the advantage that the acid amount for a solid at high temperatures (several hundred degrees centigrade), or under its actual working conditions as a catalyst, can be determined. The

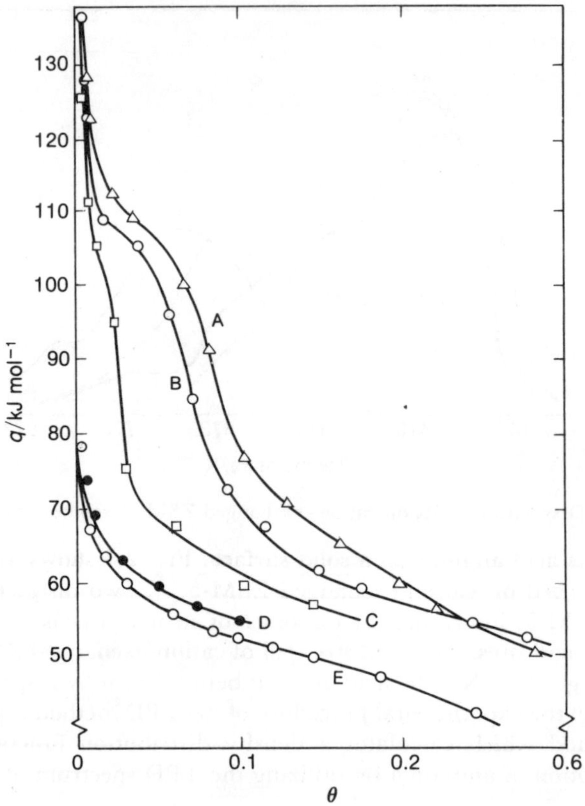

Fig. 2.4 Heat of adsorption of NH_3 on SiO_2–Al_2O_3 and SiO_2 at 298 K. A; SiO_2–Al_2O_3 (Al_2O_3 : 28 wt %), B; SiO_2–Al_2O_3 (13 %), C; SiO_2–Al_2O_3 (0.7 %), D; NH_3-preadsorbed SiO_2–Al_2O_3 (13%), E; SiO_2

methods apply even to colored samples. They suffer from the disadvantage that it is difficult to distinguish between chemical and physical adsorption and to differentiate between the amounts of acid at various acid strengths.

C. Other Methods

Catalytic activity has been used as a measure of acidity and acid strength.[2)] Recently, the activity for the dehydration of isopropyl alcohol or the isomerization of butene in the presence of an excess of air has been reported to be a good measure of acidity of some oxidation catalysts whose surface areas are so small that a gas adsorption method is difficult for the determination of acidity.[19–22)] In the case of the isomerization of 1-butene, it is necessary to poison basic sites with the products formed by the oxidation of a mixture of air and butene, which is passed through the catalyst at 470 – 570 K before the activity measurement, since the isomerization is catalyzed not only by acidic sites, but also by basic sites.[23)] Fairly good correlations are found between the acidity measured by adsorption of ammonia or pyridine and the activity for dehydration of isopropyl alcohol and isomerization of 1-butene.[21)]

Kinetics of the coisomerization of *cis*-2-butene-d_0/d_8 gives an insight as to whether the active sites on SiO_2-MoO_3[24], MoO_3[24], and WO_3[25] are acidic or basic.

Besides the above reactions, any kind of acid-catalyzed reactions such as cracking of cumene, alkylation of benzene with propene, hydration of olefins, isomerization of cyclopropane, esterification of acetic acid with ethanol, etc. can be used for the estimation of the acidic property of solid acids. Skeletal isomerization of *n*-butane to *i*-butane is used to check whether a solid acid has superacidity,[26] since the isomerization is known not to be catalyzed even by 100% sulfuric acid. However, it should be noticed that the differentiation between acid strength and acid amount is not easy from the measurement of catalytic activity for an acid-catalyzed reaction. Characterization of acid catalysts by use of model reactions has been reviewed recently by Guisnet.[27]

The measurement of acid strength and amount of surface hydroxyl group by infrared transmittance spectroscopy established by Peri is well known.[28,29] Infrared diffuse reflectance spectroscopy has been shown to be effective for characterization of acidic properties of hydroxyl groups.[30] The nature of different Lewis acid sites in dehydroxylated zeolites and oxides was also characterized by utilizing the shift of the fundamental stretching vibration of hydrogen molecule adsorbed at low temperature (77 K).[31-33] Recently, progress in nuclear magnetic resonance (NMR) with cross polar (CP), magic angle spin (MAS), or multipulse has made quantitative studies of acidic property possible. Freude *et al.* gave a useful information about Brønsted acidity and structure defect in zeolites by means of ^1H MAS NMR spectra.[34] The acid strength of bridging OH groups was shown to increase with the Si/Al ratio from 1.4 to 7 but to remain constant above Si/Al = 10. ^{15}N NMR high resolution measurements of adsorbed pyridine and acetonitrile were used for the detection of both Brønsted and Lewis acid sites on HY zeolite.[35] The acid strength and amount and the distance among OH groups on Al_2O_3, SiO_2, and $SiO_2-Al_2O_3$ were also estimated by ^1H-NMR study.[36,37]

The acidic property on a solid surface in aqueous solution can be measured by a potentiometric acid-base titration method,[38] where the amounts of protonated hydroxyl group $M-OH_2^+$ and of ionized hydroxyl group $M-O^-$ are measured as a function of solution pH. The acid amounts of various oxides such as Al_2O_3, SiO_2, TiO_2, ZrO_2, etc. measured by this method were found to be consistent with those obtained by a *n*-butylamine titration method or a gas adsorption method, though the hydrated surface is measured in aqueous solution.

2.1.2 Brønsted and Lewis Acid Sites

The methods for determining the strength and amount of acid described in the foregoing sections (2.1.1.A, B, and C) do not distinguish between Brønsted acid sites and Lewis acid sites. The acid amount which is measured is the sum of the amounts of Brønsted acid and Lewis acid at a certain acid strength. In order to elucidate the catalytic actions of solid acids, it is often necessary to distinguish between Brønsted acids and Lewis acids.

Infrared spectroscopic studies of ammonia and pyridine adsorbed on solid surfaces make it possible to distinguish between Brønsted and Lewis acids and to assess their amounts independently. Basila and Kantner showed that the modes in which ammonia is adsorbed on $SiO_2-Al_2O_2$ are as physically adsorbed NH_3, as coordinately bonded NH_3, and as NH_4^+, each of which can be detected by means of their absorption bands.[39] Their investigations of the relative intensities of the corresponding bands showed a ratio of Lewis to Brønsted acid sites of 4/1. The spectrum of pyridine coordinatively bonded to the surface is very different from that of the pyridinium ion, as shown in Table 2.2[40] This fact permits differentiation between acid types on the surface of a solid acid. Fig. 2.5 shows the infrared spectra of pyridine adsorbed on SiO_2-ZnO of various compositions which has been calcined at 773 K for 3 h in air.[41] The bands at 1,450, 1,490, and 1,610 cm^{-1} which are observed on all the mixed oxides are characteristic bands of pyridine coordinatively bonded to Lewis acid sites. No bonds were detected at 1,540 cm^{-1} on any sample; this is due to pyridinium ion

TABLE 2.2 Infrared bands of pyridine on solid acids in the 1,400–1,700 cm^{-1} region[†1]

Hydrogen bonded pyridine	Coordinately bonded pyridine	Pyridinium ion
1,400–1,447 (vs)	1,447–1,460 (vs)	
1,485–1,490 (w)	1,488–1,503 (v)	1,485–1,500 (vs)
		1,540 (s)
1,580–1,600 (s)	to 1,580 (v)	
	1,600–1,633 (s)	to 1,620 (s)
		to 1,640(s)

[†1] Band intensities: vs, very strong; s, strong; w, weak; v, variable (Reproduced with permission by E. P. Parry, *J. Catal.*, **2**, 374 (1963)).

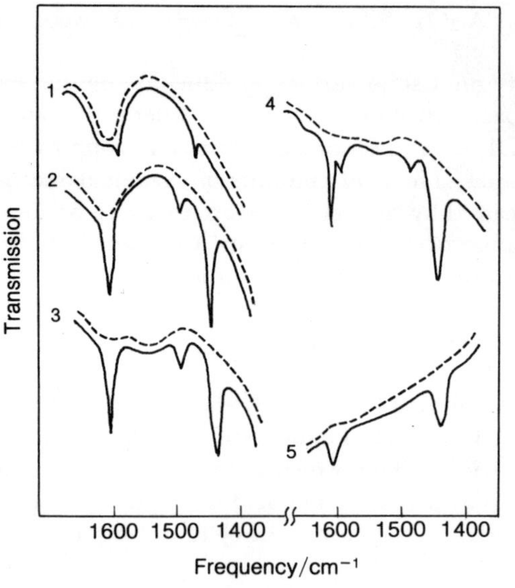

Fig. 2.5 Infrared spectra of pyridine adsorbed on $ZnO-SiO_2$. 1; SiO_2, 2; $ZnO-SiO_2$ (1/9), 3; $ZnO-SiO_2$ (3/7), 4; $ZnO-SiO_2$ (9/1), 5; ZnO, Broken lines : backgrounds

formed by the adsorption on Brønsted acid sites. Therefore, the kind of acid sites on the mixed oxides is Lewis type. The Brønsted and Lewis acidity of nickel sulfate heat-treated at various temperatures, as derived from the infrared spectra of adsorbed pyridine is shown in Fig. 2.6.[42] A maximum of Brønsted acidity appears when the sulfate is heat-treated at 523 K, while that of Lewis acidity appears at higher temperatures (576 − 670 K). The sum of the two acidity curves gives the total acid amount which can be measured by the amine titration method (cf. 2.1.1.A).

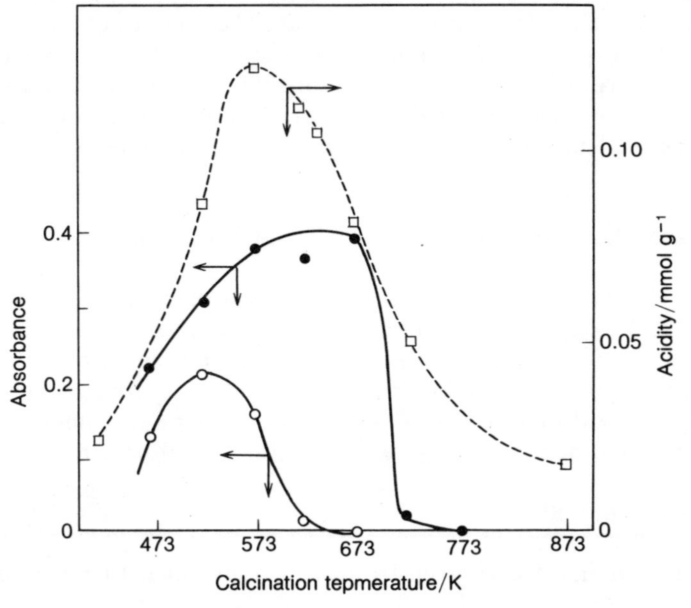

Fig. 2.6 **Variation of absorbances at 1425 cm^{-1} (●)and at 1520 cm^{-1} (○)with calcination** temprature of $NiSO_4$. Dotted line shows acid amount at acid strength $H_0 \leq 3.3$ measured by n−butylamine tiration.

Infrared spectroscopic method using pyridine as an adsorbate is extensively used and considered to be a most reliable method, though there are many other methods to distinguish between Brønsted and Lewis acid.[2] It was reported recently that both Brønsted and Lewis acid sites can be measured by ^{13}C-NMR and ^{15}N-NMR study of adsorbed pyridine.[36] Kazansky *et al.* showed that IR diffuse reflectance spectroscopy of adsorbed hydrogen molecule provided information about the nature of different Lewis acid sites in dehydroxylated H-forms of zeolites.[43]

A reliable reaction which can be used to measure Lewis acidity alone of a solid is hydrolysis of methylene chloride at 573 − 623 K.[44] It may be worthwhile to mention the determination of the hard (soft) property of a solid acid on the basis of the concept of "Hard and Soft Acids and Bases (HSAB)" proposed by Pearson[45] and extended by Klopman.[46] According to the concept , *o*-xylene and *p*-xylene are considered to be formed on hard and soft sites on solid acids, respectively, in the methylation of toluene. From the ratio of *o*-xylene/*p*-xylene, Wendland and Bremer determined the order of hardness of various aluminosilicates to be $Ca_{0.15}Na_{0.7}-Y>H-M>SiO_2-Al_2O_3>H-Y>H_{0.7}Na_{0.3}-Y>H-ZSM$-5.[47)]

2.2 BASIC PROPERTY

The basic strength of a solid surface is defined as the ability of the surface to convert an adsorbed electrically neutral acid to its conjugate base, i.e. the ability of the surface to donate an electron pair to an adsorbed acid. The amount of base (basic sites) on a solid is usually expressed as the number (or mmol) of basic sites per unit weight or per unit surface area of the solid. It is also sometimes more loosely called "basicity." There are two main methods for the measurement of strength and amount of basic sites: benzoic acid titration method using indicators and geseous acid adsorption method.

2.2.1 Benzoic Acid Titration Method Using Indicators

When an electrically neutral acid indicator is adsorbed on a solid base from a non-polar solution, the color of the acid indicator is changed to that of its conjugate base, provided that the solid has the necessary basic strength to impart electron pairs to the acid. Thus, it is generally possible to determine the basic strength by observing the color changes of acid indicators over a range of $pK_a = pK_{BH}$ values.

For the reaction of an acid indicator BH with a solid base $\overline{B}$,

$$BH+\overline{B} \leftrightarrows B^- + \overline{B}H^+ \tag{3}$$

the basic strength H_- of $\overline{B}$ is given by an equation similar to equation (1),

$$H_- = pK_a(=pK_{BH}) + \log[B^-]/[BH] \tag{4}$$

where [BH] is the concentration of the acidic form of the indicator and $[B^-]$ the concentration of the basic form.

The first perceptible change in the color of an acid indicator occurs when about 10 percent of the adsorbed layer of indicator is in the basic form, i.e. when the ratio $[B^-]/[BH]$ reaches 0.1/0.9 (= 0.11). Further increase in the intensity of the color is only perceptible to the naked eye when about 90 percent of the indicator is in the basic form, i.e.$[B^-]/[BH]$ = 0.9/0.1 (= 9). Thus the initial color change and the subsequent change in intensity are observed at values of H_- equal to $pK_{BH} - 1$ and $pK_{BH} + 1$ respectively. If we assume that the intermediate color appears when the basic form reaches 50 percent, i.e. when $[B^-]/[BH]$ = 1, we have $H_- = pK_{BH}$.

According to this assumption, the approximate value of the basic strength on the surface is given by the pK_{BH} value of the adsorbed indicator at which the intermediate color appears.[49)] Indicators which lend themselves to this method are listed in Table 2.3.[49-51)] Non-polar solvents such as benzene and isooctane are used for the indicators.

The amount of basic sites can be measured by titrating a suspension in benzene

TABLE 2.3 Indicators used for the measurement of basic properties

Indicators	Color		pK_a[†1]
	Acid-form	Base-form	
Bromothymol blue	yellow	green	7.2
Phenolphthalein	colorless	red	9.3
2,4,6-Trinitroaniline	yellow	reddish-orange	12.2
2,4-Dinitroaniline	yellow	violet	15.0
4-Chloro-2-nitroaniline	yellow	orange	17.2
4-Nitroaniline	yellow	orange	18.4
4-Chloroaniline	colorless	pink[†2]	26.5[†3]
Diphenylmethane	colorless	yellowish-orange	35.0
Cumene	colorless	pink	37.0

[†1] pK_a of indicator, BH, $(=pK_{BH})$

[†2] The color disappears with the addition of benzoic acid

[†3] This value was estimated from the data of Stewart, R. and Dolman, D. : *Can. J. Chem.*, **45**, 925 (1967).

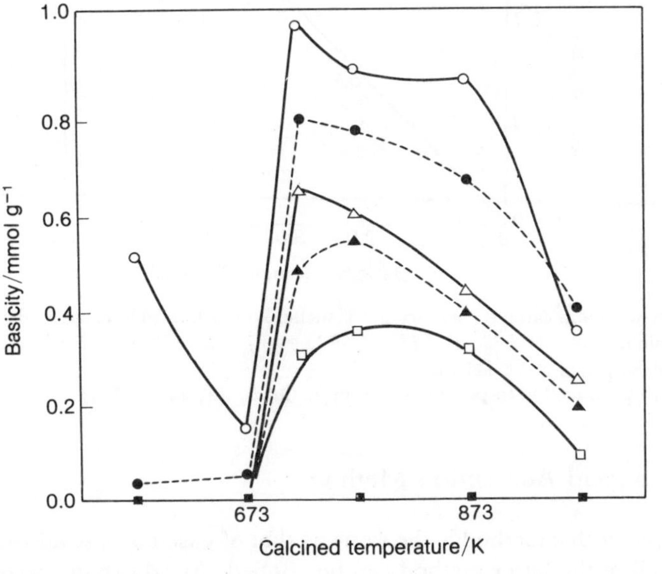

Fig. 2.7 Basicities at various basic strengths of CaO calcined at various temperatures in air. ○; $H_- \geq 7.1$, ●; $H_- \geq 12.2$, △; $H_- \geq 15.0$, ▲; $H_- \geq 17.2$ □; $H_- \geq 18.4$, ■; $H_- \geq 26.5$

of a solid on which an indicator has been adsorbed in its conjugate basic form, with benzoic acid dissolved in benzene. The benzoic acid titers are a measure of the amount of basic sites (in mmol g^{-1} or mmol m^{-2}) having a basic strength corresponding to the pK_{BH} value of the indicator used.[2)]

The base amounts (basicity) at different base strengths of CaO calcined in air at various temperatures which were measured by the benzoic acid titration method are shown in Fig. 2.7.[52] As calcination temperature is raised, the basicities at basic strengths of $pK_{BH} = 7.1 - 18.4$ increase rapidly and attain maximum values and then decrease. A very good correlation was reported between the basicity at $pK_{BH} = 7.1$ per unit surface area of CaO obtained by calcining $Ca(OH)_2$ at $573 - 1073$K and the catalytic activity for the conversion of benzaldehyde into benzyl benzoate as shown in Fig. 2.8.[53] Calcium oxide obtained by thermal decomposition of $CaCO_3$ at 1173K showed high activity, though CaO obtained by calcining $Ca(OH)_2$ at 1173K showed little activity. The measurement of basicity by using Hammet indicators will be described in 2.3.1.

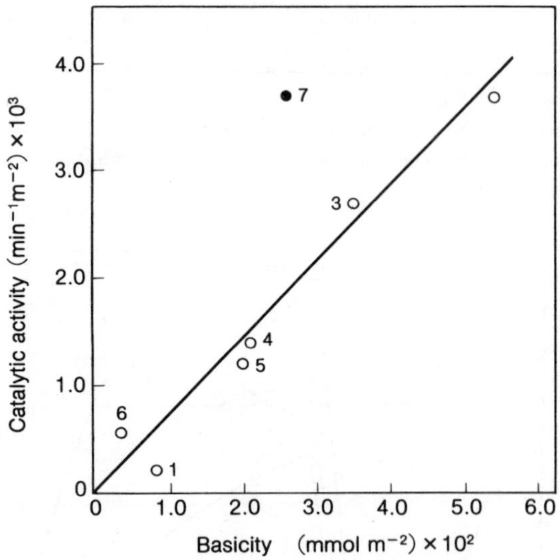

Fig. 2.8 Basicity and catalytic activity for Tishchenko reaction of benzaldehyde of $Ca(OH)_2$ calcined at : 1; 573, 2; 673, 3; 773, 4; 873, 5; 973, 6; 1073 K and of $CaCO_3$ decomposed at 7; 1173 K.
(Reproduced with permission from *J. Catal.*, **35**, 250(1974)).

2.2.2 Gaseous Acid Adsorption Method

The principle of this method is the same as that of gaseous base adsorption method (2.1.1.B) and all of the latter method can be applied. As adsorbates, acidic molecules such as carbon dioxide, nitric oxide and phenol vapor have been used. The adsorption of phenol[54] is not necessarily good for the measurement of basic property, because phenol is easily dissociated to adsorb on both acidic and basic sites[55,56] and hence acidic property affects the adsorption of phenol. Nitric oxide is used for the measurement of unusually strong basic sites.[57] The amount of carbon dioxide irreversibly adsorbed is a good measure of the amount of basic sites on solid surfaces. The TPD profiles of carbon dioxide desorbed from alkaline earth oxides are shown in Fig. 2.9.[58] Since acidic carbon dioxide desorbs at higher temperature from stronger base sites, the base strength is estimated to be in the order BaO > SrO > CaO > MgO. In the case of CaO, carbon dioxide is reported to adsorb on the basic site as a unidentate complex when the pressure of carbon dioxide is relatively high, but on both acidic and basic sites as a bidentate complex when the pressure is low (cf. Fig. 2.10).[59] However, only a unidentate complex of carbon dioxide is formed over ZrO_2 regardless of the pressure of carbon dioxide. The measurement of differential heat of CO_2 adsorption was applied to characterize the basic properties of MgO, SiO_2, Al_2O_3, and zeolites.[60] Ai has recently found a good correlation between the basicity of $Co_3O_4-K_2O$ measured by carbon dioxide adsorption and the oxidation activity for *n*-hexane, phenol, and methanol.[61]

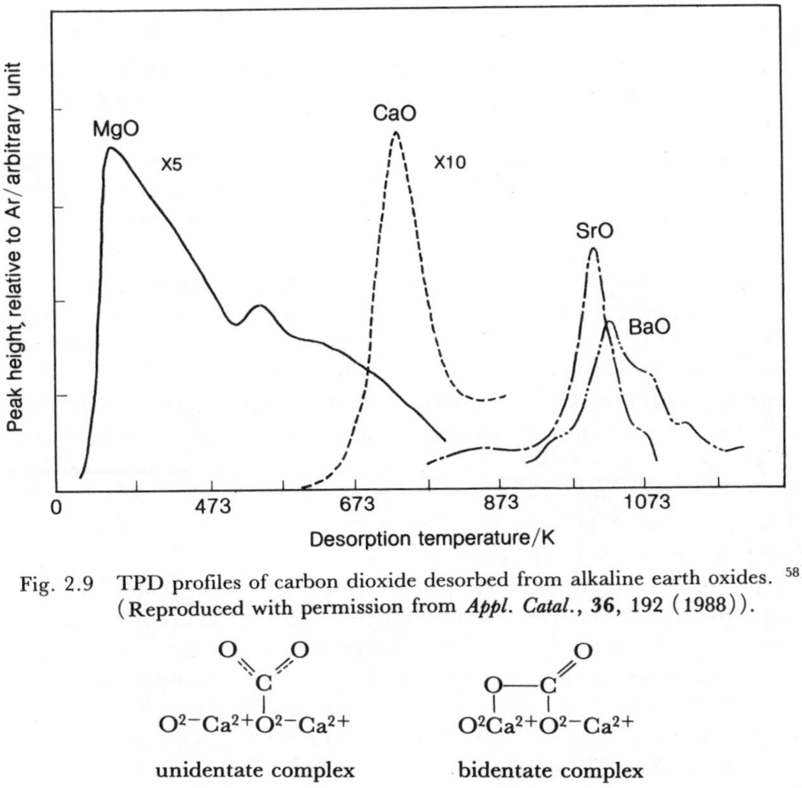

Fig. 2.9 TPD profiles of carbon dioxide desorbed from alkaline earth oxides.[58] (Reproduced with permission from *Appl. Catal.*, **36**, 192 (1988)).

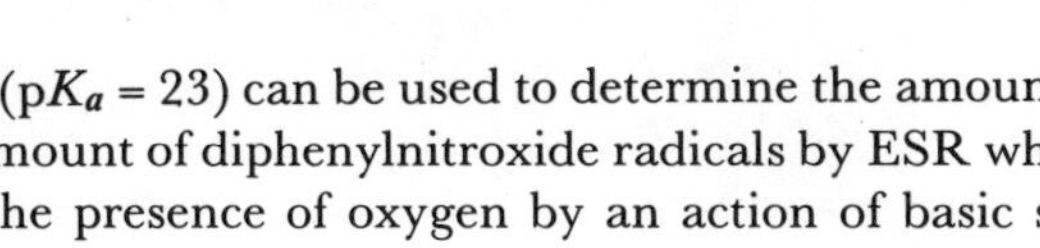

Fig. 2.10 Adsorbed states of CO_2 on CaO.

Diphenylamine ($pK_a = 23$) can be used to determine the amount of strong base sites by measuring the amount of diphenylnitroxide radicals by ESR which are formed from diphenylamine in the presence of oxygen by an action of basic sites.[62]

2.2.3 Other Methods

As mentioned in 2.1.1.C, the catalytic activity for dehydration of isopropyl alcohol to propylene (r_p) is proportional to the acidity of a catalyst.

$$r_p = k \cdot \text{acidity} \tag{5}$$

On the other hand, the activity for dehydrogenation of isopropyl alcohol to acetone (r_a) is assumed to be proportional to the acidity and basicity of a catalyst, since the dehydrogenation is considered to proceed by a concerted mechanism, for example:[20,22)]

$>$C$-$H$^-$···acidic site
O$-$H$^+$···basic site

$$r_a = k' \cdot \text{acidity} \cdot \text{basicity} \tag{6}$$

From equations (5) and (6), the following equation is derived,

$$\text{basicity} = k'' \cdot r_a / r_p, \tag{7}$$

where k, k', and k'' are constants.

Thus, r_a/r_p can be used as a measure of the basicity of a catalyst. In fact, a good correlation is found between r_a/r_p and the amount of carbon dioxide irreversibly adsorbed.[21,22)]

This method can be applied well to the basicity measurement of some oxidation catalysts such as $V_2O_5-K_2SO_4-H_2SO_4$ whose surface area is so small (about 0.7 m^2 g^{-1}) that the accurate measurement of the amount of carbon dioxide irreversible adsorbed is not easy.[20)]

The other reactions which can be used to estimate the basic property of a solid are the decomposition of 4-hydroxyl$-$4-methyl$-$2-pentanone (diacetone alcohol)[63)] and the isomerizaiton of 1-butene.[63)] In the latter reaction, use of isotope tracer gives information regarding the activity of basic sites.

Calorimetric titration with trichloroacetic acid[49)] and potentiometric acid-base titration[38)] are also applicable to basicity measurement. The amount of surface basic hydroxyl group in aqueous solution can be measured by exchanging the hydroxyl group with fluorine ion.[64)] The basic hydroxyl group on Al_2O_3, $SiO_2-Al_2O_3$, SiO_2-MgO, Al_2O_3-MgO, etc. was found to play an important role for controlling the amount of effectively mounted MoO_3.

The O_{1s} binding energy of metal oxides, which can be measured by x-ray photoelectron spectroscopy (XPS), is also a measure of basic strength of metal oxides, since the electron pair donating ability of oxides is assumed to be expressed by the O_{1s} binding energy. The order of basic strength determined by this method is as follows:[65)] La_2O_3 (529.0 eV) $>$ Sm_2O_3 (529.2) $>$ CeO_2 (529.4) $=$ Dy_2O_3 (529.4) $>$ Y_2O_3 (529.5) $>$ Fe_2O_3 (530.3) $>$ Al_2O_3 (531.8) $>$ GeO_2 (532.4) $>$ P_2O_5 (532.4) $>$ SiO_2 (533.1). The metal oxides whose binding energy is less than 529.5 eV are reported to be catalytically active for the selective formation of 1-olefin from secondary alcohol.[65)] Infrared and NMR spectroscopy can be applied also to basicity measurement similarly as in 2.1.1.B.

2.3 ACID-BASE PROPERTY

2.3.1 Representative Parameter, $H_{0,max}$, of Acid-Base Property

As described in 2.1.1.A and 2.2.1, acid strength (H_0) is expressed by the pK_a values of the conjugate acids of basic indicators, while base strength (H_-) is expressed by the pK_a values of acidic indicators. Since the indicators used for the basicity measurement are different from those used for acidity measurement (cf. Tables 2.1 and 2.3) it was impossible to determine the acid-base strength distribution on a common scale. Recently, a new method which determines the basicity at various base strengths of solid samples by using a series of Hammett indicators as shown in Table 2.1 has been presented.[66)] By this method, both acidic and basic property can be determined on a common H_0 scale, where the strength of basic sites is expressed by the H_0 of the conjugate acidic sites. It was found that the strongest H_0 value of the acidic sites was approximately equal to the strongest H_0 value of the basic sites.[67)] The equal strongest H_0 was termed "$H_{0,\max}$" which is a practical parameter to represent acid-base property on solid surfaces.

Before discussing the significance and usefulness of $H_{0,\max}$, we shall study the principle of the method of expressing basic property by an H_0 scale.

A. Basic Property Expressed by H_0 Scale

The acidity and acid strength of a solid can be determined by the amine titration method using a series of Hammett basic indicators, B, listed in Table 2.1, as mentioned in 2.1.1 A. When a solid has no acid sites of $H_0 \leq pK_{BH^+}$, the color of the basic indicator does not change. In this case, if a standard solution of Brønsted acid in benzene is added gradually, the color of the basic indicator on the surface will change to the color of its conjugate acid. The color change is taken as the end-point of the titration. At the end-point, the acid strength H_0 of the resultant solid, which was formed by the addition of Brønsted acid to the original solid, is equal to the pK_{BH^+} of the indicator used. As basic sites are neutralized by Brønsted acid at the end-point, the titers of Brønsted acid required for the neutralizaiton should give a measure of the number of basic sites (basicity) on the surface. During the titration, stronger basic sites are neutralized earlier and weaker ones later and weaker basic sites require stronger acids for the neutralization. Therefore, it can be assumed that the weakest basic sites have been finally neutralized by an acid having an acid strength of $H_0 = pK_{BH^+}$.

The proton donating ability of the solid at the end-point of titration is considered to be either due to the conjugate acids which were formed by the proton transfer from Brønsted acid solution to the original solid or due to the Brønsted acid which was physically adsorbed on the surface during the titration. The proton donating ability of both the conjugate acid and the Brønsted acid used for titration is assumed to be equal. Since the weakest basic sites form the strongest conjugate acids, the acid strength, H_0, of the conjugate acid of the weakest basic sites should be equal to or greater than the pK_{BH^+} of the indicator used.

Thus, "basic strength H_0" of basic sites is defined as the acid strength, H_0, of the conjugate acids of the basic sites. We shall express the function H_0 used previously by "acid strength H_0" in cases where it is necessary to distinguish between this and "basic strength H_0." As the basicity at "basic strength H_0" = pK_{BH^+} is easily determined by using a series of basic indicators as described above, the distribution of basic strength of a solid as well as that of acid strength can be expressed by a common scale of acid-base strength. The use of the function H_0 for basic strength is neither surprising nor curious, because the basic strengths of the organic compounds in homogenous solution are usually expressed by pK_a's of the conjugate acids. It should be noted that the measurement of the basicity when the basic strength H_0 is equal to or greater than a pK_{BH^+} value is possible only when there are no acid sties whose acid strength is equal to or less than the same pK_{BH^+} value.

Figure 2.1.1 shows the results of acid-base strength distribution on a common H_0 scale of some solids,[66] where the acidity at various acid strengths was measured by the method described in 2.1.1 A, while the basicity at various basic strengths by titrating the solid suspended in benzene with a 0.1 N solution of trichloroacetic acid in benzene using the same indicators as those used for acidity measurement.[66] The acidity at an H_0 value shows the number of acid sites whose acid strength is equal to or less than the H_0 value and the basicity at an H_0 value shows the number of basic sites whose basic strength is equal to or greater than the H_0 value.

Titanium oxide exhibited high basicity at basic strength $H_0 > 1.5$, but low acidity at acid strength $H_0 \leq 6.8$, while $MgSO_4$ showed high acidity at acid strength $H_0 \leq 6.8$ but low basicity at basic strength $H_0 \geq 1.5$. Acidic and basic sites of equal strength do not coexist on the same solid surface. Therefore, the measurement is to determine a significant acid−base strength distribution of a given solid in the full range of the H_0 scale.

B. Significance and Usefulness of $H_{0,max}$

As seen in Fig. 2.11, the acid−base strength distribution curves intersect at a point on the abscissa where acidity = basicity = 0. Hence, the strongest H_0 value of the acid sites is equal to the strongest H_0 value of the basic sites. $H_{0,max}$ is defined as the H_0 value at a point of intersection, which expresses the equal strongest H_0 value of both acidic and basic sites. Each $H_{0,max}$ value, which was determined from a point of intersection of each acid-base strength distribution curve and the abscissa, is given in Table 2.4. A unique $H_{0,max}$ is found for every solid. The $H_{0,max}$ value changes on calcination.

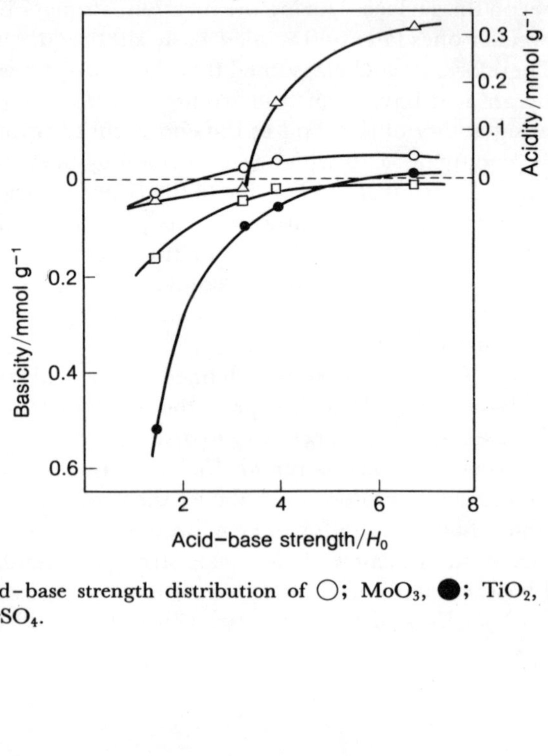

Fig. 2.11 Acid−base strength distribution of ○; MoO_3, ●; TiO_2, □; V_2O_5, and △; $MgSO_4$.

TABLE 2.4 Acidities, basicities, $H_{0,\max}$[†1][67)]

Solids	Basicity/mmol g^{-1}				Acidity/mmol g^{-1}					$H_{0,\max}$
	$H_0 \geq 1.5$	$H_0 \geq 3.3$	$H_0 \geq 4.0$	$H_0 \geq 6.8$	$H_0 \leq -3$	$H_0 \leq 1.5$	$H_0 \leq 3.3$	$H_0 \leq 4.0$	$H_0 \leq 6.8$	
activated Al_2O_3	0.30	0.10	0.06	0.01	0	0	0	0	0	8.0
γ-Al_2O_3	0.43	0.22	0.16	0.01	0	0	0	0	0	7.2
ZrO_2	0.08	0.03	0.02	0.01	0	0	0	0	0	9.5
TiO_2	0.52	0.10	0.06	0	0	0	0	0	0.01	5.5
B_2O_3	0.27	0.04	0.02	0	0	0	0	0	0.01	8.0
ZnO		0.07	0.05	0	0	0	0	0	0.005	6.4
BaO		0.09	0.06	0.05	0	0	0	0	0	15
MoO_3	0.03	0	0	0		0	0.02	0.04	0.05	2.1
$MgSO_4$[†1]	0.03	0.02	0		0	0	0	0.16	0.32	3.4
$MgWO_4$	0.23	0.03	0		0	0	0	0	0.05	4.0
Ta_2O_5	0				0	0.02	0.05	0.06	0.07	2.0
WO_3	0				0	0.01	0.06	0.04	0.14	1.3
Bi_2O_3	0.05	0.07	0.03	0	0	0	0	0	0.003	6.6
V_2O_5	0.16	0.05	0.02	0.01	0	0	0	0	0	8.5
SrO	0.14	0.04	0.02	0.01	0	0	0	0	0	9
$ZnSO_4 \cdot H_2O$	0				0	0.07	0.07	0.20	0.30	1.5
$CuSO_4$					0	0.11	0.22	0.22	0.28	0
$CaSO_4 \cdot 0.5H_2O$	0.03	0.01	0.01	0	0	0	0	0	0.003	6.0
$MnSO_4$[†2]	0				0	0.14	0.28	0.30	0.30	0.2
$Al_2(SO_4)_3$	0				0	0.13	0.20	0.43	0.67	−1.0
$AlPO_4$	0.61	0.08	0.07	0	0	0	0	0	0.01	6.0
$Zn_3(PO_4)_2 \cdot 4H_2O$	0.64	0.04	0.04	0	0	0	0	0	0.02	5.2
$CaWO_4$	0.07	0.01	0.02	0	0	0	0	0	0.01	5.0
$Na_2WO_4 \cdot 2H_2O$	0.50	0.03	0.06	0.03	0	0	0	0	0	12
$CaCO_3$	0.14	0.03	0.02	0	0	0	0	0	0.01	6.0
$Ba(OH)_2$	0.13	0.03	0.02	0.01	0	0	0	0	0	9.0
$Mg(OH)_2$	0.09	0.03	0.02	0	0	0	0	0	0	6.8
$NiSO_4$[†3]	0.46	0			0	0	0.44	0.43		2.0

†1 $MgSO_4 \cdot 7H_2O$ was calcined at 673 K, 3 h.
†2 $MnSO_4$ was calcined at 523 K, 4h.
†3 $NiSO_4 \cdot 7H_2O$ was calcined at 573 K, 4 h.

For example, the $H_{0,\max}$ values of $MgSO_4 \cdot 7H_2O$ calcined at 573, 673, 793, and 943 K are 3.0, 3.4, 3.3, and 3.5, respectively. Since $MgSO_4 \cdot 7H_2O$ without calcination shows an $H_{0,\max}$ of 6.0, the solid calcined at 573 K has the minimum $H_{0,\max}$.[66)] On the other hand, $H_{0,\max}$ of TiO_2 does not much change on calcination and the variances were less than 1.0 unit of H_0 scale.[66)]

$H_{0,\max}$ can be regarded as a practical parameter to represent an acid-base property on solids which is sensitive to the surface structure. A solid with a large positive $H_{0,\max}$ has strong basic sites and weak acidic sites. Thus, basic sites play an important role. On the other hand, a solid with a large negative $H_{0,\max}$ has strong acidic sites and weak basic sites. In this case, acid sites often become important.

A good linear relation was reported between $H_{0,max}$ of $Al_2O_3-SiO_2$ treated with fluorine and the catalytic activity for the synthesis of β-ethylpyridine from acrylaldehyde (Fig. 2.12).[68] The activity increases with decreasing $H_{0,max}$, but is not correlated with simple H_0. In the case of dehydration of isopropyl alcohol, the catalytic activity of $F-Al_2O_3$ and $Na-Al_2O_3$ showed a maximum at $H_{0,max} = +4$ as seen in Fig. 2.13,[69] suggesting the necessity of coexistence of both acidic and basic sites each having appropriate strength and acid-base bifunctional catalysis.

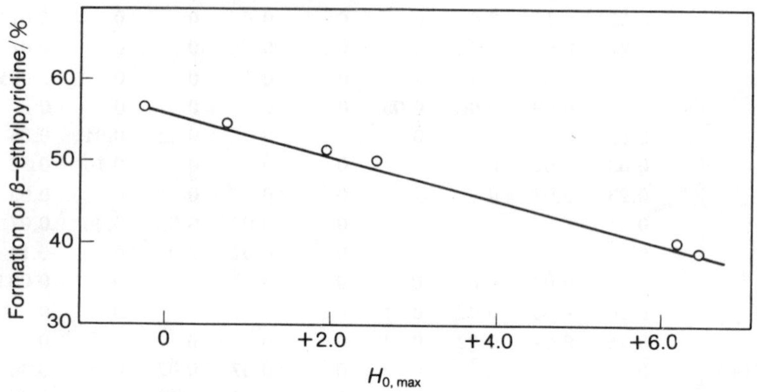

Fig. 2.12 Activity of $F-Al_2O_3$ for formation of β-ethylpyridine from acrylaldehyde *vs*. $H_{0,\,max}$.

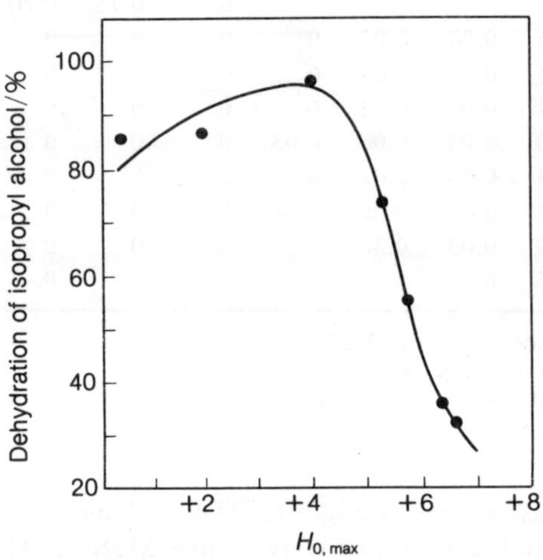

Fig. 2.13 Activity of $F-Al_2O_3$ and $Na-Al_2O_3$ for dehydration of isopropyl alcohol *vs*. $H_{0,\,max}$.

2.3.2 Acid-Base Pair Sites

Even in reactions which have been recognized to be catalyzed only by acid sites on a catalyst surface, basic sites also act more or less as active sites in cooperation with acid sites. The catalysts having suitable acid-base pair sites sometimes show pronounced activity, even if the acid-base strength of a bifunctional catalyst is much weaker than the acid or base strength of simple acid or base. For example, ZrO_2 which is weakly acidic and weakly basic shows higher activity for C—H bond cleavage than highly acidic $SiO_2-Al_2O_3$ or highly basic MgO.[70] The cooperation of acid sites with basic sites is surprisingly powerful for particular reactions and causes highly selective reactions. This kind of reaction is often seen in enzyme catalysis. Thus, it becomes sometimes necessary to know not only the strengths of the acidic and basic sites but also the orientation of acid-base pair site (distance between acidic and basic sites, sizes of acidic and basic sites, etc.). To characterize the nature of an acid-base pair site, the TPD method using phenol is useful.

Phenol is known to adsorb on both acidic $SiO_2-Al_2O_3$ and basic MgO, as shown in Fig. 2.14.[55] It was found recently that phenol also adsorbs on ZrO_2 and the desorption temperature of phenol adsorbed on ZrO_2 is higher than those of phenol adsorbed on MgO and $SiO_2-Al_2O_3$ as shown in Fig. 2.15.[56] Namely, phenol adsorption is

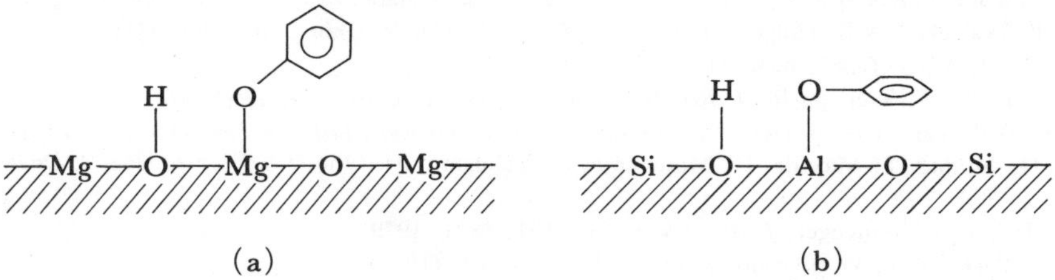

Fig. 2.14 Adsorbed states of phenol on MgO(a) and $SiO_2-Al_2O_3$(b).

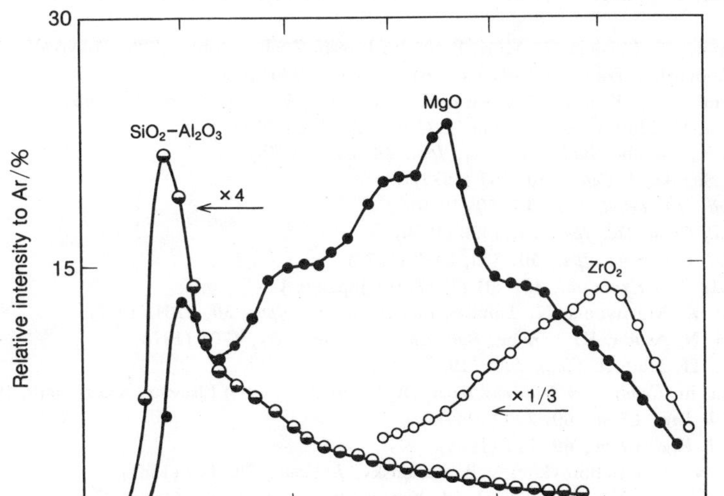

Fig. 2.15 Temperature-programmed desorption profiles of phenol. ○; ZrO_2, ●; MgO, ◒; $SiO_2-Al_2O_3$.[56]
(Reproduced with permission from *Materials Chem. and Phys.*, **19**, 293 (1988)).

strongest on ZrO_2 and weakest on $SiO_2-Al_2O_3$, the adsorption strength of phenol on MgO being intermediate between that on ZrO_2 and that on $SiO_2-Al_2O_3$. This supports a characteristic acid—base bifunctional catalysis of ZrO_2. It was also found that ZrO_2 showed higher activity and selectivity than $SiO_2-Al_2O_3$ and MgO for formation of nitriles from alkylamines,[71)] which can be interpreted by the bifunctional catalysis of ZrO_2.

REFERENCES

1. L. P. Hammett, A. J. Deyrup, *J. Am Chem. Soc.*, **54**, 2721 (1932).
2. K. Tanabe, *Solid Acids and Bases*, Kodansha, Tokyo and Academic Press, New York London, 1970.
3. K. Tanabe, in *Catalysis: Science and Technology*, (eds. J.R. Anderson and M. Boudart) Springer-Verlag, Berlin, 1981, Vol.2, Chapt. 5.
4. K. Tanabe, H. Hattori, *Chem. Lett.*, **1976**, 625.
5. K. Shibata, T. Kiyoura, J. Kitagawa, T. Sumiyoshi, K. Tanabe, *Bull. Chem. Soc. Jpn.*, **46**, 2985 (1973).
6. S. P. Walvekar, A.B. Halgeri, *J. Res. Inst. Catal.*, Hokkaido Univ., **20**, 219 (1972).
7. L. Forni, *Advan. Catal.*, **8**, 65 (1974).
8. K. Tanabe, T. Yamaguchi, *J. Res. Inst. Catal.*, Hokkaido Univ., **14**, 93 (1966).
9. S.P. Walvekar, A.B. Halgeri, S. Ramanna, T.N. Srinivasan, *Fertilizer Tech.*, **13**, 241 (1976).
10. K. Hashimoto, T. Masuda, H. Motoyama, H. Yakushiji, M. Ono, *I & EC Prod. Res. & Develop.*, **25**, 243 (1986).
11. P. Rys, W.J. Steinegger, *J. Am. Chem. Soc.*, **101**, 4801 (1979).
12. R.L. Richardson, W. Benson, *J. Phys. Chem.*, **61**, 405 (1957).
13. H. Sato, N. Ishii, K. Hirose, S. Nakamura, Proc. 7th Intern. Zeolite Conf., 1986, Tokyo (Y. Murakami *et al.*, eds.) Kodansha, Tokyo and Elsevier, Amsterdam, 1986, p.755.
14. For example, G. Wang, H. Itoh, H. Hattori, K. Tanabe, *J. Chem. Soc., Faraday Trans.* 1, **79**, 1373 (1983).
15. K. Hashimoto, T. Masuda, T. Mori, Proc. 7th Intern. Zeolite Conf., 1986, Tokyo (Y. Murakami, *et al.*, eds) Kodansha, Tokyo and Elsevier, Amsterdam, 1986, p.503.
16. K. Tsutsumi, H.Q. Koh, S. Hagiwara, H. Takahashi, *Bull. Chem. Soc. Jpn.*, **48**, 3576 (1975).
17. K. Shimazu, H. Hattori, K. Tanabe, *J. Catal.*, **48**, 302 (1977).
18. T. Iizuka, K. Tanabe, *Bull. Chem. Soc. Jpn.*, **48**, 2527 (1975).
19. M. Ai, S. Suzuki, *J. Catal.*, **30**, 362 (1973).
20. M. Ai, *Bull. Jpn. Petrol. Inst.*, **18**, 50 (1976).
21. M. Ai, *Bull. Chem. Soc. Jpn.*, **49**, 1328 (1976).
22. M. Ai, *Bull. Chem. Soc. Jpn.*, **50**, 355, 2579 (1977).
23. M. Ai, *Yuki Gosei Kyokai-shi*, **35**, 201 (1977) (in Japanese).
24. H. Hattori, K. Maruyama, K. Tanabe, *Bull. Chem. Soc. Jpn.*, **50**, 2181 (1977).
25. H. Hattori, N. Asada, K. Tanabe, *Bull. Chem. Soc. Jpn.*, **51**, 1704 (1978).
26. K. Tanabe, H. Hattori, *Chem. Lett.*, **1976**, 625.
27. M. Guisnet, in: *Catalysis by Acids and Bases*, (B. Imelit *et al.*, eds.) Elsevier, Amsterdam, 1985, p. 283.
28. J.B. Peri, *J. Phys. Chem.*, **69**, 220 (1965).
29. J.B. Peri, *J. Phys. Chem.*, **69**, 211 (1965).
30. L.M. Kustov, V. Yu. Borovkov, V.B. Kazansky, *J. Catal.*, **72**, 149 (1981).
31. V.B. Kazansky, V. Yu. Borovkov, L.M. Kustov, Proc. 8th Intern. Congr. Catal., Verlag Chemie, 1984, Vol. 3, p. 3.
32. V.B. Kazansky, L.M. Kustov, V. Yu. Borovkov, *Zeolites*, **3**, 77 (1983).
33. V.B. Kazansky, in: *Structure and Reactivity of Modified Zeolites*, Elsevier, 1984, p. 61.
34. D. Freude, M. Hunger, H. Pfeifer, W. Schwieger, *Chem. Phys. Lett.*, **128**, 62 (1986).
35. D. Freude, *Advan. Colloid and Interface Sci.*, **23**, 21 (1985).
36. C.E. Bronnimann, R.C. Zeigler, G.E. Maciel, *J. Am. Chem. Soc.*, **110**, 2023 (1988).

37. G.E. Maciel, J.F. Haw, I-S. Chuang, B.L. Hawkins, T.A. Early, D.R. McKay, L. Petrakis, *J. Am. Chem. Soc.*, **105**, 5529 (1983).
38. H. Kita, N. Henmi, K. Shimazu, H. Hattori, K. Tanabe, *J. Chem. Soc., Faraday Trans.* 1, **77**, 2451 (1981).
39. M.R. Basila, T.R. Kantner, *J. Phys. Chem.*, **71**, 467 (1967).
40. E.P. Parry, *J. Catal.*, **2**, 371 (1963).
41. T. Sumiyoshi, K. Tanabe, H. Hattori, *Bull. Jpn. Petrol. Inst.*, **17**, 65 (1975).
42. H. Hattori, S. Miyashita, K. Tanabe, *Bull. Chem Soc. Jpn.*, **44**, 893 (1971).
43. V.B. Kazansky, V. Yu. Borovkov, L.M. Kustov, Proc. 8th Intern. Congr. Catalysis, Verlag Chemie, 1984, Vol. 3, p. 3.
44. T. Yamaguchi, K. Tanabe, Proc. 4th Intern. Congr. Catal., 1969, p. 410.
45. R.G. Pearson, *J. Chem. Education*, **45**, 581 (1968).
46. G. Klopman, Chemical Reactivity and Reaction Path, MIR, Moscow, p. 91 (1977).
47. K.P. Wendlandt, H. Bremer, Proc. 8th Intern. Congr. Catalysis, Verlag Chemie, 1984, Vol. 4, p. 507.
48. S. Malinowski, S. Szczepanska, *J. Catal.*, **2**, 310 (1963).
49. K. Tanabe, T. Yamaguchi, *J. Res. Inst. Catal., Hokkaido Univ.*, **11**, 179 (1964).
50. J. Take, N. Kikuchi, Y. Yoneda, *J. Catal.*, **21**, 164 (1971).
51. G. Suzukamo, M. Fukao, M. Minobe, *Chem. Lett.*, **1987**, 585.
52. H. Hattori, N. Yoshii, K. Tanabe, Proc. 5th Intern. Congr. Catalysis, 1972, 10-233.
53. K. Tanabe, K. Saito, *J. Catal.*, **35**, 247 (1974).
54. O.V. Krylov, E.A. Fokina, *Problemy Kinet. Katal., Acad. Nauk, USSR*, **8**, 248 (1955).
55. K. Tanabe, T. Nishizaki, Proc. 6th Intern. Congr. Catal., **2**, 863 (1977).
56. B. Xu, T. Yamaguchi, K. Tanabe, *Materials Chem. and Phys.*, **19**, 291 (1988).
57. T. Iizuka, Y. Endo, H. Hattori, K. Tanabe, *Chem. Lett.*, **1976**, 803.
58. G. Zhang, H. Hattori, K. Tanabe, *Appl. Catal.*, **36**, 189 (1988).
59. Y. Fukuda, K. Tanabe, *Bull. Chem. Soc. Jpn.*, **46**, 1616 (1973).
60. A. Auroux and J.C. Vedrine, *Catalysis by Acids and Bases*, (B. Imelik *et al.*, eds.) Elsevier, Amsterdam, 1985, p. 311.
61. M. Ai, *J. Catal.*, **54**, 223 (1978).
62. Y. Nakano, T. Iizuka, H. Hattori, K. Tanabe, *J. Catal.*, **57**, 1 (1979).
63. Y. Fukuda, H. Hattori, K. Tanabe, *Bull. Chem. Soc. Jpn.*, **51**, 3150 (1978).
64. N. Yamagata, Y. Owada, S. Okazaki, K. Tabane, *J. Catal.*, **47**, 358 (1977).
65. H. Vinek *et al.*, *J. Chem. Soc. Faraday Trans. 1*, **1977**, 734.
66. T. Yamanaka, K. Tanabe, *J. Phys. Chem.*, **79**, 2409 (1975).
67. T. Yamanaka, K. Tanabe, *J. Phys. Chem.*, **80**, 1723 (1976).
68. C-B Wang, Y-R. Li, *J. Catal.* (Cuiha Xuebao), **3**, 187 (1982).
69. Z-F. Qian, Z. Xu, Q. Tang, *Shi Yu Hua Gong*, **15**, 567 (1986) (in Chinese).
70. T. Yamaguchi, Y. Nakano, T. Iizuka, K. Tanabe, *Chem. Lett.*, **1976**, 677.
71. B. Xu, T. Yamaguchi, K. Tanabe, *Chem. Lett.*, **1988**, 281.

3
Acid and Base Centers: Structure and Acid-Base Property

3.1 METAL OXIDES

3.1.1 Li_2O, Na_2O, K_2O, Rb_2O, Cs_2O

The observation that sodium deposited on alumina acts as an effective catalyst for the isomerization of olefins triggered much research in the field of solid base catalysis.[1] However, the catalytic properties of alkali oxides themselves have not been studied so extensively. Many studies have been done on alkali metals doped on or intercalated in other materials. Some of them are known as superbases and described in section 3.10.

The basicities of some of the oxides were measured by benzoic acid titration using indicators as shown in Fig. 3.1.[2] Basic strengths of Rb_2O outgassed at 643 K and Cs_2O outgassed at 573 K exceed $H_- = 26$, which is the critical strength for superbases.

The single component oxides are recognized as base catalysts by their catalytic features for butene isomerization. The reaction network in butene isomerization over

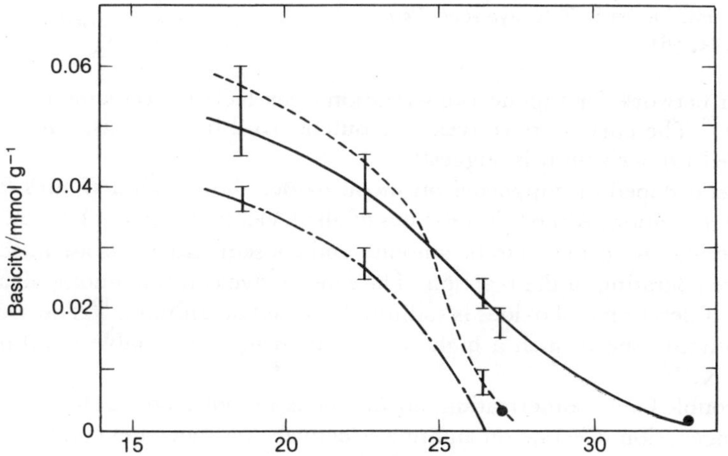

Fig. 3.1 Basicity *vs.* basic strength.—— ;Rb_2O outgassed at 643 K , —·— ;Rb_2O outgassed at 573 K , - - - - - ; Cs_2 outgassed at 573 K

Rb_2O is shown in Fig. 3.2.[2,3] In the base-catalyzed isomerization of butene, the intermediates are primarily the *cis* form of allylic anions, because the *cis* allylic anion is more stable than the *trans* allylic anion. A curve convex to 1-butene – *cis*-2-butene axis is caused by the intermediate being *cis* form of allylic anion and characteristic of base-catalyzed butene isomerization. Essentially the same curves are observed for the other alkali metal oxides such as Li_2O, Na_2O, K_2O, and Cs_2O.

Graphite reacts with alkali metals to give lamellar compounds in which alkali metals are present in the form of monolayers separated by one or more carbon layers. The basicities measured by benzoic acid titration are shown in Fig. 3.3.[4] The strongest basic sites are $H_- = 18$ both for potassium and cesium intercalated compounds.

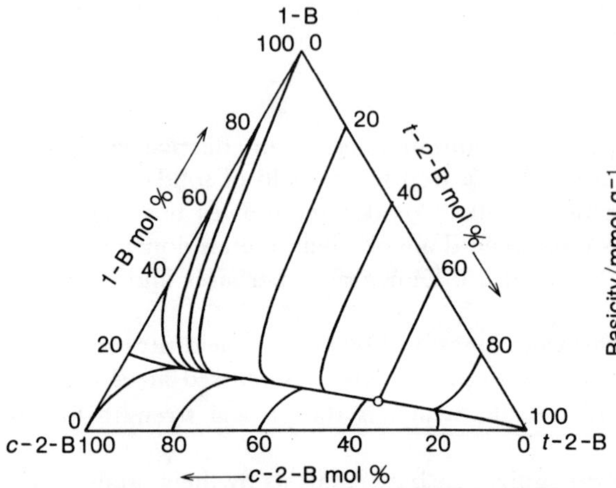

Fig. 3.2 Reaction profile of isomerization of butenes over Rb_2O at 413 K. (Reproduced with permission by S. Tsuchiya *et al.*, *Chem. Lett.* **1984**, 664)

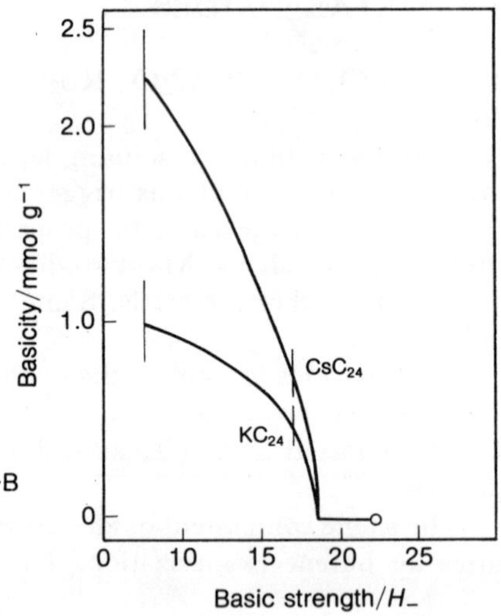

Fig. 3.3 Basicity distributions of KC_{24} and CsC_{24}.

The reaction network for butene isomerization over KC_8 is also similar to those of alkali oxides.[5] The curves are convex to 1-butene–*cis*-2-butene axis, and therefore, base-catalyzed isomerizaiton is suggested.

Alkali metal doped or supported on metal oxides show high activities for alkene double bond migration. Although the states of alkali elements are not known, the reaction intermediates are believed to be anionic, and consequently, it is assumed that the basic sites are operating in the reaction. The most active catalyst among alkali metals dispersed on different metal oxides is sodium dispersed on alumina.[6] The sodium dispersed on alumina shows such a high activity as to proceed double bond migration even at 213 K.[7]

Besides double bond isomerization, alkali metals supported on metal oxides are active for hydrogenation. Cesium on alumina selectively hydrogenates conjugated dienes to monoolefins.[8] Ethylene is more easily hydrogenated over alkali oxides; in particular, Na and Li supported on alumina are active to promote ethylene hydrogenation below room temperature.[9] Benzene is also hydrogenated over Cs and K supported on alumina. The types of supporting oxide are crucial to reveal the activity of alkali oxides for hydrogenation. For benzene hydrogenation, Cs, or K supported on SiO_2, $SiO_2-Al_2O_3$, and MgO show no activity.[10)]

Side-chain alkenylation of alkyl benzenes with conjugated dienes is catalyzed by alkali metals supported on CaO. Among alkali metals, K supported on CaO shows the best results. Alkylbenzenes such as toluene, *o*-xylene, *p*-xylene, ethylbenzene, and 4,5-tetraethylbenzene react with butadiene below 382 K. The mechanism involves the formation of a benzyl anion which successively adds to the conjugated dienes.[11)]

REFERENCES

1. H. Pines, J.A. Vesely, V.N. Ipatieff, *J. Am. Chem. Soc.*, **77**, 347 (1955).
2. S. Tsuchiya, S. Takase, H. Imamura, *Chem. Lett.*, **1984**, 661.
3. H. Noumi, T. Misumi, S. Tsuchiya, *Chem. Lett.*, **1978**, 429.
4. S. Tsuchiya, A. Fukui, H. Imamura, 55th Catalysis Soc. Japan Meeting, AS8 (1985).
5. S. Tsuchiya, T. Misumi, N. Ohuye, H. Imamura, *Bull. Chem., Soc. Jpn.*, **55** 3089 (1982).
6. W.O. Haag, H. Pines, *J. Am Chem. Soc.*, **82**, 387 (1960).
7. T.M. O'Gray, R.M. Alm, M. C. Hoff, Preprint, Meeting, Am. Chem. Soc. (Pet. Div.), 136th. Atlantic City,4, B65 (1959).
8. A.J. Hubert, *J. Chem. Soc.* [C] 2419 (1967).
9. S.E. Voltz, *J. Phys. Chem.*, **61**, 756 (1957).
10. L.H. Slaugh, *Tetrahedron*, **24**, 4525 (1968)
11. G.G. Eberhardt, H.J. Peterson, *J. Org. Chem.*, **30**, 82 (1965).

3.1.2 BeO, MgO, CaO, SrO, BaO, RaO, $Ba(OH)_2$

Magnesium oxide, CaO, SrO, and BaO are typical solid base catalysts. In particular, MgO is a representative one and positioned as a sort of reference catalyst among solid base catalysts like $SiO_2-Al_2O_3$ among solid acid catalysts. In contrast, very little investigation of BeO and RaO as catalysts has been done because of toxicity and radioactivity, respectively.

In addition to these alkaline earth oxides, application of $Ba(OH)_2$ as a solid base catalyst in organic reactions has been developed in recent years. A discussion of $Ba(OH)_2$ is included in this section.

Magnesium oxide, CaO and BaO were once regarded as catalytically inert materials, but at present they are known as very active catalysts for certain base-catalyzed reactions if properly activated. High temperature heat treatment is required to obtain highly active catalysts.

A. Preparation and Activation

The catalysts are prepared from hydroxides or carbonates by thermal decomposition. Equilibrium pressures for decomposition of carbonates and peroxides are shown in Fig. 3.4.[1)] To obtain oxides from hydroxides or carbonates, high temperature pretreatment is required. During pretreatment, evolution of H_2O, CO_2, and O_2 occurs. Evolution of H_2O begins at about 673 K as $Mg(OH)_2$, $Ca(OH)_2$, and commercially available BaO are heat-treated *in vacuo*.[2,3)] Carbon dioxide starts to evolve at a temperature slightly higher than that for H_2O evolution. From commercially available

BaO, O_2 evolves at about 823 K.[3] As oxide surfaces are revealed by removal of H_2O, CO_2, and O_2, the basic properties appear.

Surface areas change depending upon heat treatment conditions. Outgassing results in high surface area as compared with calcination under atmosphere.[4] Presence of water vapor facilitates sintering arising from the forward and backward reactions, as shown by the following equation.

$$MgO + H_2O \rightleftharpoons Mg(OH)_2$$

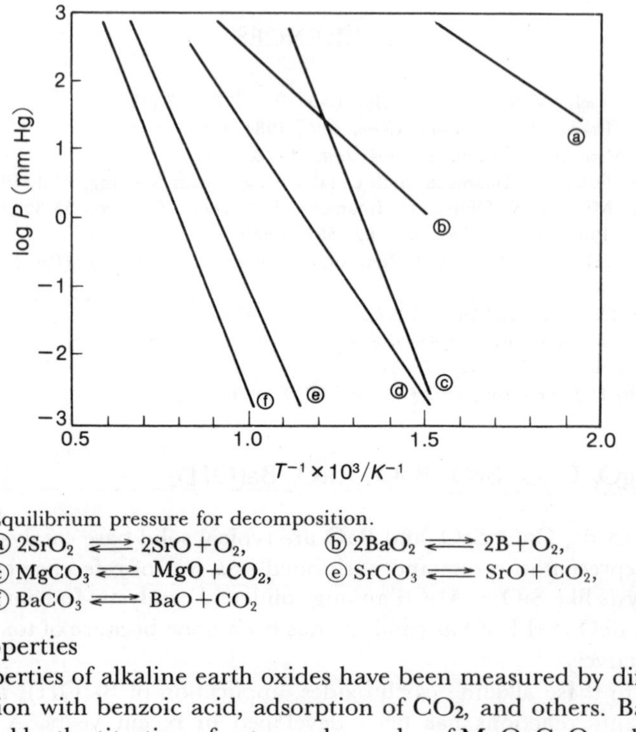

Fig. 3.4 Equilibrium pressure for decomposition.
ⓐ $2SrO_2 \rightleftharpoons 2SrO + O_2$, ⓑ $2BaO_2 \rightleftharpoons 2B + O_2$,
ⓒ $MgCO_3 \rightleftharpoons MgO + CO_2$, ⓔ $SrCO_3 \rightleftharpoons SrO + CO_2$,
ⓕ $BaCO_3 \rightleftharpoons BaO + CO_2$

B. Basic Properties

Basic properties of alkaline earth oxides have been measured by different methods such as titration with benzoic acid, adsorption of CO_2, and others. Basicity distributions measured by the titration of outgassed samples of MgO, CaO and SrO are shown in Fig. 3.5.[5] Magnesium oxide and CaO possess base sites stronger than $H\_=26$.

Variations of basic properties of MgO with calcination temperature are shown in Fig. 3.6.[2] Base sites appear by heat treatment above 673 K at which surface oxide is revealed by removal of H_2O and CO_2. However, base sites stronger than $H\_=26$ do not appear for samples calcined in air. Similar variation was observed for CaO.[2,6]

Besides basic properties, alkaline earth oxides exhibit electron donating properties, which can be measured by observing the anion radicals formed on the surfaces as probe molecules are adsorbed. Tench and Nelson first observed the formation of nitrobenzene anion radicals on MgO surfaces.[7] The formation of nitrobenzene anion radicals was also reported for CaO.[4] The heat treatment temperatures to generate maximum amounts of electron donating sites on CaO are 773 K for outgassed samples, and 973 K for calcined samples. The electron donating sites are much more numerous for samples outgassed than for those calcined in air, and far fewer than the basic sites measured by the titration method. The electron donating sites on CaO have been sug-

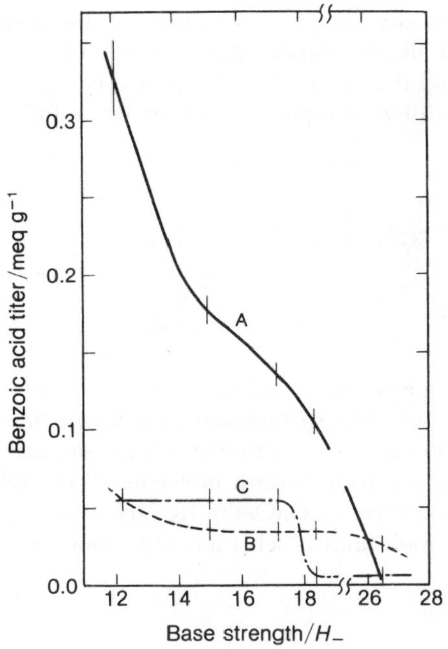

Fig. 3.5 Benzoic acid titer *vs.* base strength of (A) MgO, (B) CaO, and (C) SrO. (Reproduced with permission by J. Take *et al.*, *J. Catal.*, **21**, 167 (1971)).

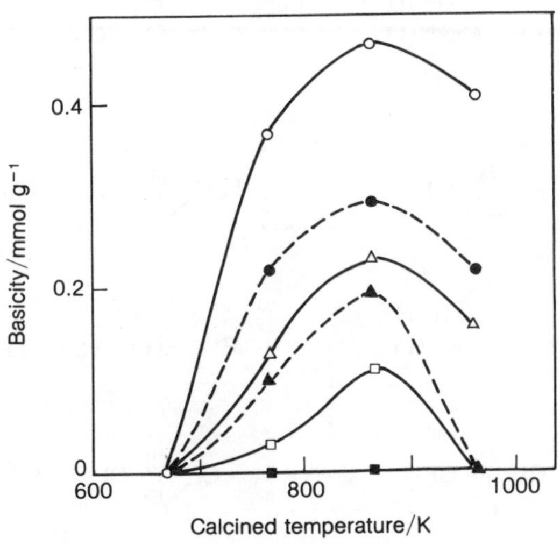

Fig. 3.6 Variation of basicities at different strengths for MgO calcined at various temperatures in air.
○; $H_- \geq 7.1$, ●; $H_- \geq 12.2$, △; $H_- \geq 15.0$, ▲; $H_- \geq 17.2$, □; $H_- \geq 18.4$, ■; $H_- \geq 26.5$

gested to be different from the base sites measured by the titration method.[4)]

The basic character of alkaline earth oxides was demonstrated by the detection of anionic species adsorbed on the surfaces by IR spectroscopy. On adsorption of CO on MgO, anionic species of different types were observed as follows.[8)]

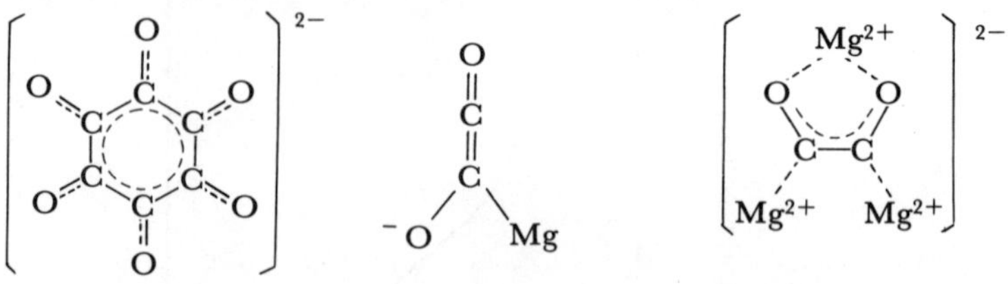

The formaiton of anionic species indicates the existence of an electron or electron pair donating sites on the surface. The surface acts as a Lewis base toward CO.

The existence of the sites acting as a Brønsted base was also demonstrated by observing dissociation of an H^+ from certain molecules.[4)] On adsorption of benzaldehyde, 2-propanol, and chloroform on CaO, the IR OH stretching band is intensified.[4)] The dissociation of these molecules is schematically drawn as follows.

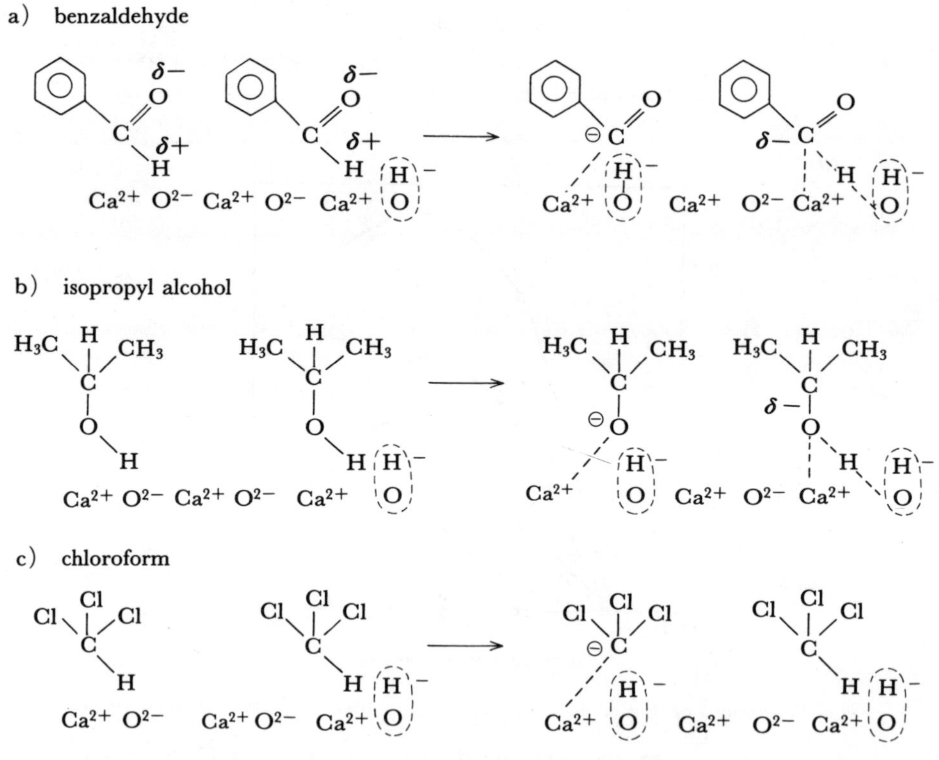

Model of adsorption state of benzaldehyde, isopropyl alcohol and chloroform on CaO : OH^- in broken oval denotes surface hydroxide ion.

The anions formed by dissociation of an H^+ are stabilized by surface metal cations.

Carbon dioxide is adsorbed on the surfaces of alkaline earth oxides in different forms depending on the adsorption condition. Carbon dioxide is adsorbed on MgO in bidentate form at low coverage and in unidentate form at high coverage, while on CaO CO_2 is adsorbed as a bidentate regardless of the coverage.[9)]

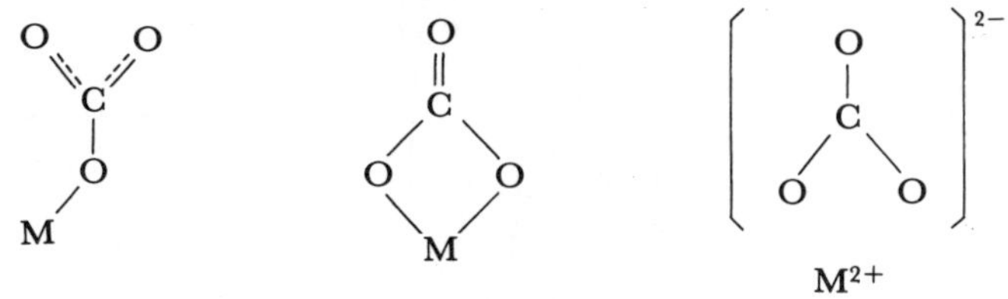

unidentate bidentate carbonate

It should be noted that in bidentate form, not only surface oxide ions but metal cations are also involved in the adsorption sites.

Evans and Whately reported IR spectroscopic measurement of the adsorption of CO_2 on MgO.[10)] In addition to unidentates, bidentates, and carbonates, bicarbonate species were also detected. This suggests that hydroxy groups on MgO also act as a base toward CO_2.

Basic properties were also measured by TPD of probe molecules such as CO_2, CO and H_2. Although CO_2 appears to be the proper probe molecule because of its acidic nature, TPD profile of CO_2 varies depending on the adsorption condition of CO_2. Only a broad desorption peak appeared if too much CO_2 was adsorbed. The alkaline earth oxide surfaces react with CO_2 to form different surface structures depending on the adsorption time and temperature. TPD profiles of adsorbed CO_2 on MgO, CaO, SrO, and BaO measured under controlled adsorption conditions are shown in Fig.2.9.[11)]

TPD profiles of CO adsorbed on MgO are shown in Fig. 3.7.[12)] Appearance of three peaks at different temperatures indicates the existence of different sites on MgO. Relative quantities of these peaks vary with the pretreatment temperature of MgO. The adsorbed species giving peak I at about 400 K are $(CO)_6^{2-}$ and those for peak III are $(CO)_2^{2-}$, which were identified by IR measurement of these species.

TPD profiles of H_2 adsorbed on MgO are shown in Fig. 3.8.[13,14)] Hydrogen is heterolytically dissociated on the surface to form H^+ and H^-, which are adsorbed on surface O^{2-} ion and Mg^{2+} ion, respectively. Appearance of peaks at different temperatures indicates that several types of ion pairs with different coordination numbers exist on the surface of MgO. The number of hydrogen adsorption sites on MgO pretreated at different temperatures and the coordination numbers of each ion pair are also summarized in Table 3.1.[13-15)] The corresponding surface structure is shown on p.39 (Fig. 3.12).

Heterolytic dissociation of hydrogen on MgO surface is demonstrated by IR spectroscopy.[16,17)] IR bands of both O-H and Mg-H stretching are observed as shown in Fig. 3.9.

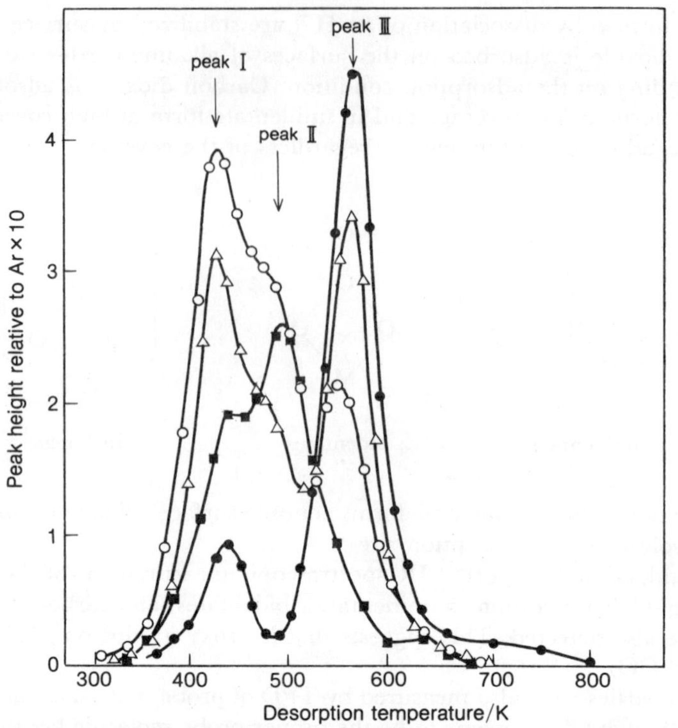

Fig. 3.7 TPD profiles for CO adsorbed on MgO pretreated at the following temperatures : ■; 773 K, ○; 873 K, △; 973 K, ●; 1273 K.
(Reproduced with permission by G. Wang, *et al.*, *J. Chem Soc. Faraday Trans.*, **79**, 1375 (1983)).

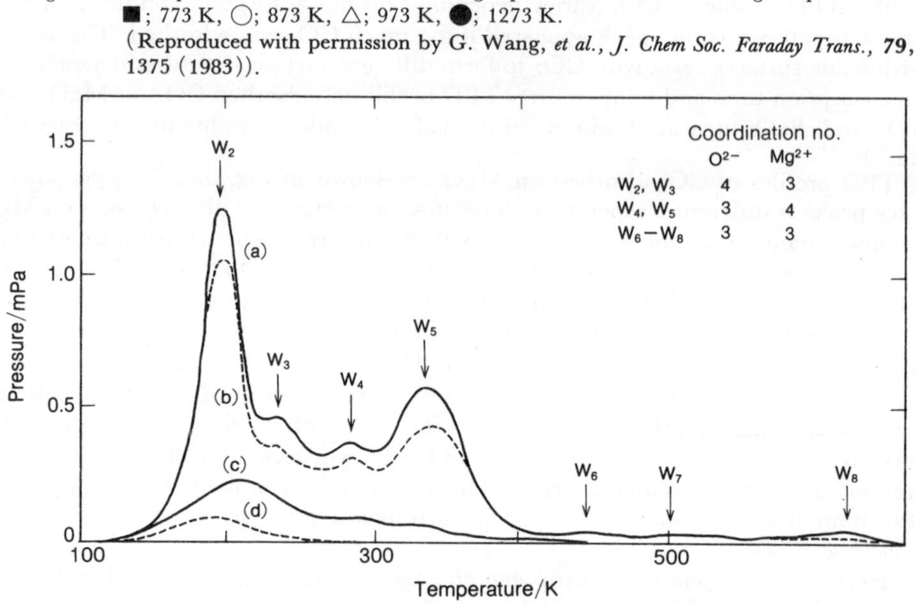

Fig. 3.8 TPD plots for hydrogen adsorbed on MgO.
TPD was run from 100 K. Pretreatment temperature/K; (a) 1123, (b) 973, (c) 823, (d) 673.

TABLE 3.1 Coordination numbers of active sites on MgO and their concentration obtained from TPD for hydrogen adsorbed.

Active site	Coordination no.		Number of sites/10^{15} m^{-2}			
			Pretreatment temperature/K			
	O_{LC}	Mg_{LC}	673	823	973	1123
W_2 and W_3	4	3	4.0	11.6	29.3	32.4
W_4 and W_5	3	4	0.0	4.9	22.1	26.5
W_6 and W_7	3	3	0.0 (W6–W8)	0.3 (W6–W8)	1.3	4.1
W_8	3	3			1.2	4.2

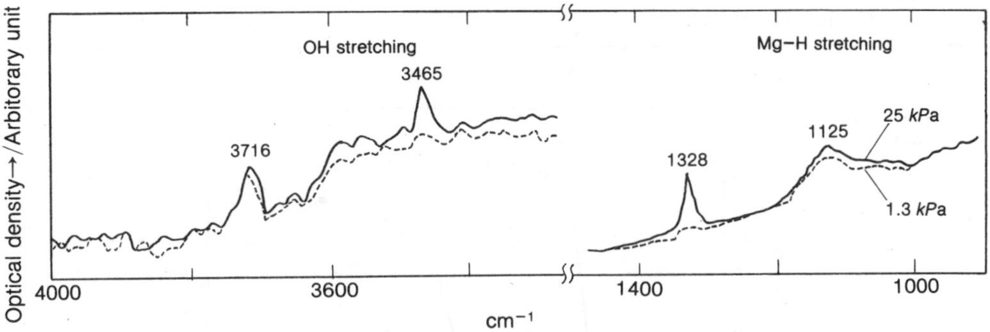

Fig. 3.9 Infrared spectra of hydrogen on MgO pretreated at 1103 K.

C. Catalytic Activities

Base-catalyzed reactions occurring over alkaline earth oxides are listed in Table 3.2. All reactions are initiated by abstraction of an H^+ from the reactants to form anionic intermediates. The surface O^{2-} ions abstract an H^+ and the metal cations stabilize the anionic intermediates.

Butene isomerization over alkaline earth oxides has been studied extensively. The activity and selectivity variations in 1-butene isomerization as a function of pretreatment temperature are shown in Fig. 3.10 for CaO.[2)] Similar variations have been observed for MgO,[2,18)] SrO[19)] and BaO.[3)] The activities appear as H_2O and CO_2 are removed from the surfaces. A high *cis* to *trans* ratio is characteristic of base-catalyzed 1-butene isomerization. This is caused by the high stability of *cis* allylic anions as compared with *trans* allylic anion as described further on.

In most cases the *cis*-to-*trans* ratios become low as the activities become high by changing the pretreatment temperature. This is caused by the generation of a second type of sites which are highly active on heat treatment at certain temperatures: around 873 K for MgO, CaO and BaO. The reaction products consist of two parts, one produced on the highly active sites and the other on the normal basic sites. The products from the highly active sites become close to an equilibrium mixture of butene isomers (1:*cis*:*trans*=3:17:80 at 273 K). Therefore, the sum of the products consists of a low *cis*-to-*trans* ratio. This is evidenced by the coisomerization of butene d_0/d_8.[20)] Iso-

TABLE 3.2 Reaction types catalyzed by solid bases

1	Isomerization of double bond (H migration)
	Olefins, Alkynes, Allenes, Unsaturated compounds containing hetero-atoms
2	Addition
	Hydrogenation, Amination, Aldol addition
3	Decomposition
	Alcohols, Amines, Halogen substituted alkanes
4	Alkylation
	Phenol, Aniline
5	Esterification
	Aldehydes
6	Exchange
	Olefins-D_2, H_2-D_2

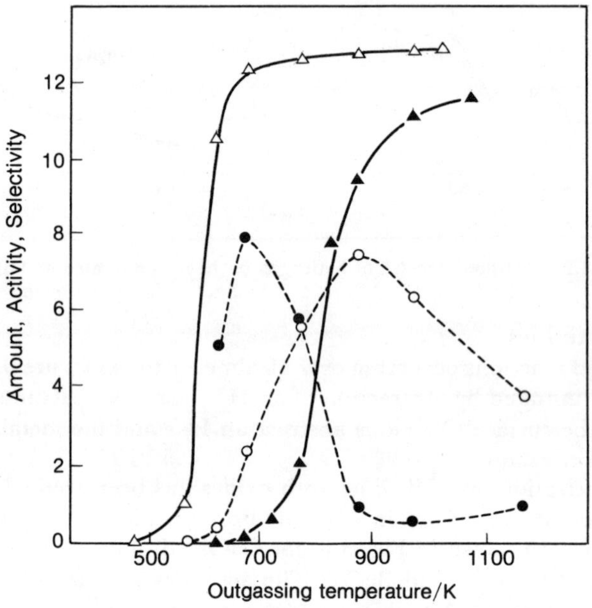

Fig. 3.10 Evolution of water and carbon dioxide from $Ca(OH)_2$ on outgassing at different temperatures and the catalytic activity and selectivity of the resulting CaO for 1-butene isomerization.
▲; number of $CO_2/20^{-1}$ mmol g^{-1}, △; number of H_2O/mmol g^{-1},
○; Activity/10^2 mmHg min^{-1} g^{-1}, ●; Selectivity (*cis/trans*)

topic distributions in *trans*-2-butene resulted from coisomerization of *cis*-2-butene d_0/d_8 over BaO pretreated at 823 K and at 1073 K are different from each other. Over BaO pretreated at 1073 K, isotopic butene is divided into two parts: one a non-exchanged part and the other a "binomial" part. The butene isomers of the "binomial" part are in equilibrium ratio, indicating occurrence of extended isomerization on the highly active sties. The highly active sites are rapidly poisoned by O_2. Similar phenomena were observed for CaO. These results indicate the existence of different active sites on the surfaces.

Benzaldehyde esterification is catalyzed by CaO. The variations of the activity and basicity of CaO catalysts parallel each other as the pretreatment temperature of catalyst changes, indicating that the base sites are the active sites.[6] The reaction is of the Cannizzaro type as shown below, and the slow step involves H^- transfer from (I) to (II).

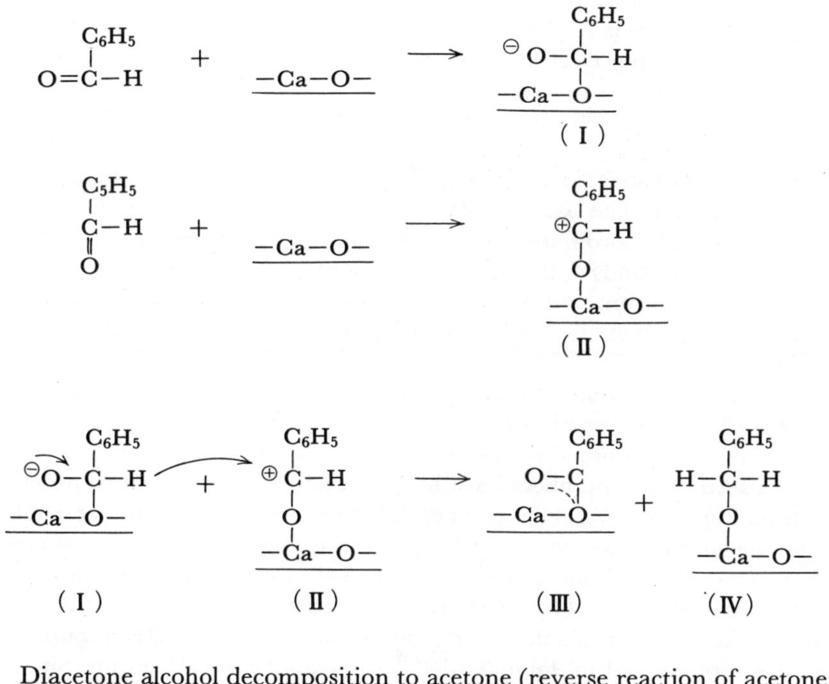

(1) (2) (3)

Diacetone alcohol decomposition to acetone (reverse reaction of acetone aldol condensation) proceeds over alkaline earth catalysts.[21] The active sites are poisoned by CO_2. The slope of the activity decrease with increasing amount of adsorbed CO_2 represents the activity per unit base site. The activities per unit site are in order BaO>SrO>CaO>MgO. The order coincides with the base strength order; the stronger the base strength the more effective the active sites.

D. Structure of Active Sites

Although the appearance of basic sites requires removal of H_2O and CO_2 from the surfaces, the activity variations as a function of pretreatment temperature are not the same for different reactions. The activity variations of MgO for different reactions are plotted against outgassing temperature in Fig. 3.11.[22)] Increasing the pretreatment temperature, activities for butene isomerization appeares at relatively low pretreatment temperature followed by activities for exchange. Hydrogeneration activities appear at high pretreatment temperature and reach maxima around 1273 K. This tendency is also seen for CaO, SrO and BaO, though appearance of activity maxima for different reactions against pretreatment temperature is not so distinct as observed for MgO. Three activity maxima for different reactions indicate that at least three types of sites exist on the surfaces of alkaline earth oxides.

Surface structure of alkaline earth oxides was investigated using UV absorption[23)] and luminescence spectroscopies.[16,24)] High surface area MgO absorbs UV light and emits luminescence, which is not observed with MgO single crystal. UV absorption corresponds to the following electron transfer process involving surface ion pairs.

$$Mg^{2+}O^{2-} + h\nu \longrightarrow Mg^{+}O^{-}$$

Absorption bands at 230 and 274 nm are of lower frequency than the band at 160 nm caused by bulk ion pairs. The bands at 230 and 274 nm are considered to be due to the surface O^{2-} ions of coordination numbers 4 and 3, respectively.

Luminescence corresponds to the reverse process of UV absorption, and the shape of the luminescence spectrum varies with the excitation light frequency and with adsorption of certain molecules. Luminescence involves surface ion pairs of low coordination numbers.

Ion pairs of low coordination numbers responsible for UV absorption and luminescence exist at corner, edge, or high Miller index surface of (100) plane as shown in Fig. 3.12.[16,17)] Effects of adsorbed molecules on luminescence spectra indicate that ion pairs of lower coordination numbers have higher reactivities toward adsorption.

As seen from Fig. 3.12, several ion pairs of different coordination number combinations exist on the surface. Existence of different sites is suggested by different activity maxima as shown in Fig. 3.11 and by the appearance of peaks at different temperatures in TPD profiles of adsorbed CO and H_2 (Figs. 3.7 and 3.8.).

Four active site types on alkaline earth oxides are proposed: OH groups, Sites I – III, and are summarized in Table 3.3.[22,25)] Appearance of OH group, Site I, II, and III with increasing pretreatment temperature is schematically illustrated in Fig. 3.13. Correspondence of these sites to the surface ion pairs in the model structure is also included in Table 3.3.

Quantum mechanical calculations were done to reveal the effect of surface structure on the basic strength of O^{2-} ions.[26,27)] The main factors generating stronger basicity are: i) fewer Mg atoms coordinated to the central oxygen atom in the basic site and ii) more O atoms coordinated to the Mg atoms adjacent to the central oxygen

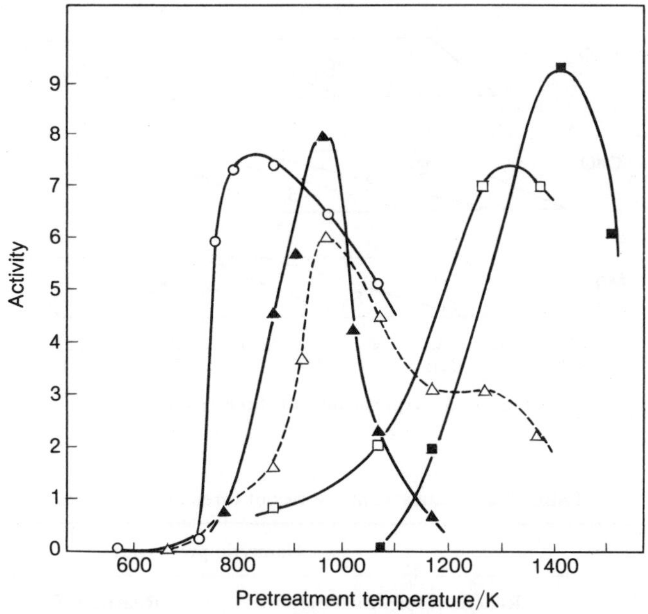

Fig. 3.11 Variations of activities of MgO for different types of reactions as a function of pretreatment temperature
○, 1-butene isomerization ($/3.5\times10^{3}$ mmHg min^{-1} g^{-1})303 K;
▲, CH_4-D_2 exchange ($/4.3\times10^{3}$ % s^{-1} g^{-1}) 673 K;
△, amination of 1,3-butadiene with dimethylamine ($/5\times10^{17}$ molucules min^{-1} g^{-1}) 273 K;
□, 1,3-butadiene hydrogenation ($/2.5\times10^{-1}$ % min^{-1} g^{-1}) 273 K;
■, ethylene hydrogenation ($/0.3$ % min^{-1} g^{-1}) 523 K

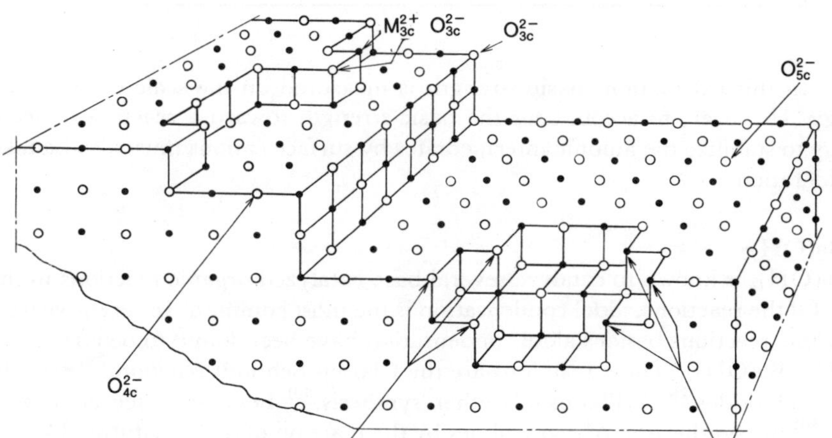

Fig. 3.12 Ions in low coordination on the surface of MgO.
(Reproduced with permission by S. Coluccia, A. J. Tench, Proc. 7th Intern Congr. Catal., Kodansha, 1981, p.1160)

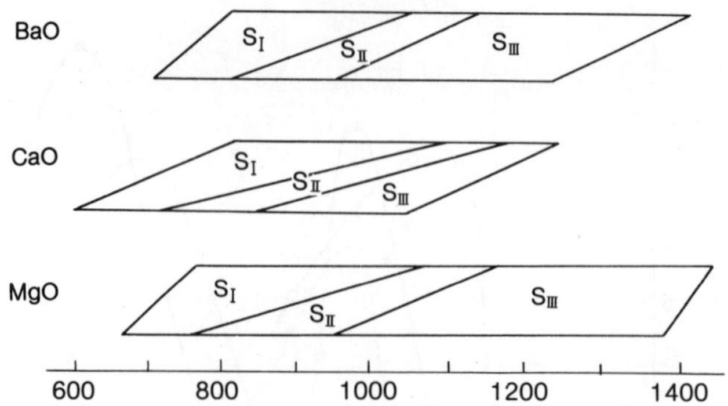

Fig. 3.13 Appearance of three types of sites.

TABLE 3.3 Catalytic properties of three of active sites

Type of sites	Catalytic properties	
	Reactions for which the sites are active	Reactions for which the sites are inactive
S_I	Isomerization (olefins, allyl amines, allyl ethers)	H-D exchange (CH_4-D_2, H_2-D_2, among butenes), Hydrogenation
S_{II}	Isomerization, H-D exchange Amination	Hydrogenation
S_{III}	Hydrogenation Isomerization (slow)	H-D exchange

atom. In this calculation, basic strength is measured on the scale of H^+ stabilizing energy. However, to account for the basic strength toward a reacting molecule, the energy to stabilize the anionic intermediates by surface cations must also be taken into consideration.

E. $Ba(OH)_2$

$Ba(OH)_2$ is known to catalyze several base-catalyzed organic reactions in the solid form. Of the reactions, aldol condensation is the most common. In recent years, several organic reactions besides aldol condensation have been found to be effectively catalyzed by $Ba(OH)_2$. These reactions are the Claisen-Schmidt reaction,[28)] esterification of acid chlorides,[29)] Williamson's ether synthesis,[30)] benzil-benzilic acid rearrangement,[30)] the synthesis of Δ^2-pyrazolines by the reaction of α,β-unsaturated ketone with $PhNHNH_2$[31)] Wittig-Horner reaction,[32,33)] and Michael addition.[34,35)] For these reactions, the $Ba(OH)_2$ catalyst prepared from $Ba(OH)_2 \cdot 8H_2O$ by calcination at 473 K shows the highest activity.

REFERENCES

1. Data cited from Landolt-Bornstein, Zahlenwerte und Functioner, II Band, 2 teil, Springer (1960).
2. H. Hattori, N. Yoshii, K. Tanabe, Proc. 5th Intern. Congr. Catal., 1972, Miami Beach, 10–233.
3. H. Hattori, K. Maruyama, K. Tanabe, *J. Catal.*, **44**, 50 (1976).
4. T. Iizuka, H. Hattori, Y. Ohno, J. Sohma, K. Tanabe, *J. Catal.*, **22**, 130 (1971).
5. J. Take, N. Kikuchi, Y. Yoneda, *J. Catal.*, **21**, 164 (1971).
6. K. Tanabe, K. Saito, *J. Catal.*, **35**, 247 (1974).
7. A.J. Tench, R.L. Nelson, *Trans. Faraday Soc.*, **63**, 2254 (1967).
8. E. Guglielminotti, S. Collucia, E. Garrone, L. Cerruti, A. Zecchina, *J. Chem. Soc., Faraday Trans. 1*, **75**, 96 (1979).
9. Y. Fukuda, K. Tanabe, *Bull. Chem. Soc., Jpn.*, **46** 1616 (1973).
10. J.V. Evans, T.L. Whateley, *Trans. Faraday Soc.*, **63**, 2769 (1967).
11. G. Zhang, H. Hattori, K. Tanabe, *Appl. Catal.*, **36**, 189 (1988).
12. G. Wang, H. Hattori, *J. Chem. Soc., Faraday Trans., 1*, **80**, 1039 (1984).
13. T. Ito, M. Kuramoto, M. Yoshida, T. Tokuda, *J. Phys. Chem.*, **87**, 4411 (1983).
14. T. Ito, T. Murakami, T. Tokuda, *J. Chem. Soc., Trans. Faraday 1*, **79**, 913 (1983).
15. T. Ito, T. Sekino, N. Moriai, T. Tokuda, *J. Chem. Soc., Trans. Faraday 1*, **77**, 2181 (1981).
16. S. Coluccia, A.J. Tench, Proc. 7th Intern. Congr. Catal., Kodansha, Tokyo, 1980, p. 1154.
17. S. Coluccia, F. Bozzuzzi, G. Ghiotti, C. Morterra, *J. Chem. Soc., Faraday Trans.*, **78**, 2111 (1982).
18. H. Hattori, K. Shimazu, N. Yoshii, K. Tanabe, *Bull Chem. Soc. Jpn.*, **49**, 96 (1976).
19. M. Mohri, K. Tanabe, H. Hattori, *J. Catal.*, **32**, 144 (1974).
20. A. Satoh, H. Hattori, *J. Catal.*, **45**, 32 (1976).
21. Y. Fukuda, K. Tanabe, S. Okazaki, *Nippon Kagakukaishi*, 513 (1972) (in Japanese).
22. H. Hattori, in: *Adsorption and Catalysis on Oxide Surfaces*, (eds. M. Che and G.C. Bond), Elsevier, Amsterdam, 1985, p. 319.
23. A. Zecchina, M.G. Lofthouse, F.S. Stone, *J. Chem. Soc., Faraday Trans., 1*, **71**, 1476 (1975).
24. S. Coluccia, A.M. Deane, A.J. Tench, *J. Chem. Soc., Faraday Trans., 1*, **74**, 2913 (1978).
25. H. Hattori, *Materials Chem. Phys.*, **18**, 533 (1988).
26. H. Kawakami, S. Yoshida, T. Yonezawa, *Shokubai* (Catalyst), **25**, 160 (1983) (in Japanese).
27. H. Kawakami, S. Yoshida, *J. Chem. Soc. Faraday Trans., 2*, **80**, 921 (1984).
28. J.V. Sinisterra, A. Garcia-Raso, J.A. Cabello, J.M. Marinas, *Synthesis*, 502 (1984).
29. A. Garcia-Raso, J.V. Sinisterra, J.M. Marinas, *Polish J. Chem.*, **56**, 1435 (1982).
30. A. Garcia-Raso, J.V. Sinisterra, J.M. Marinas., *React. Kinet. Catal. Lett.*, **19**, 145 (1982).
31. J.V. Sinisterra, *React. Kinet. Catal. Lett.*, **30**, 93, (1986).
32. J.V. Sinisterra, Z. Mouloungui, M. Delmas, A. Gaset, *Synthesis*, 1097 (1985).
33. J.V. Sinisterra, A.R. Alcantara, J.M. Marinas, *J. Coll. Interface Sci.*, **115**, 520 (1987).
34. A. Garcia-Raso, J.A. Garcia-Raso, B. Campaner, R. Mestres, J.V. Sinsterra, *Synthesis*, 1037 (1982).
35. M. Iglesias, J.M. Marinas, J.V. Sinisterra, *Tetrahedron*, 2335 (1987).

3.1.3 Oxides of Rare Earth Elements (Sc, Y, La, Ce, Pr, Nd, Pm, Sm, Eu, Gd, Tb, Dy, Ho, Er, Tm, Yb, Lu), Actinide Oxides (ThO_2, UO_2)

A. Oxides of Rare Earth Elements

Among rare earth elements, La and Ce are rather common catalyst components. These two elements are used as exchangeable cations in zeolite to prepare cracking catalysts, and also as one of the components in oxidation catalysts. Elements other than La and Ce are rarely used as catalyst components. However, recent developments in separation techniques have rendered rare earth elements available in purity high enough to warrant fundamental investigation. As a result, studies on catalytic properties of rare earth oxides are being extended. Promethium have never been used as a catalyst in any form because all its isotopes are radioactive

a. Preparation and Activation

Rare earth oxides are prepared from the hydroxides by calcination in air or by outgassing at high temperatures. The hydroxides are obtained from aqueous nitrates by hydrolysis with aqueous ammonia. Except for those of three elements (Ce, Pr, and Tb), rare earth oxides thus prepared are stable in sesquioxide (M_2O_3) stoichiometry. The oxides of the three exceptions are stable in the nominal compositions, CeO_2, Pr_6O_{11}, and Tb_4O_7.

As a typical thermogram for rare earth oxides, a thermogram of La_2O_3, starting from $La(OH)_3$, is shown in Fig. 3.14.[1] Following an initial small weight loss at 373 to 473 K (a − b) due to removal of adsorbed water and/or crystallization water, the first true stage of $La(OH)_3$ decomposition occurs in the temperature range 523 to 623 K, and results in the formation of a well-defined, hexagonal LaOOH intermediate, represented by the break in the integral TG curve at point c. Subsequent dehydration of the oxyhydroxide to La_2O_3 occurs at 523 to 693 K (c − d) and is completed at the later temperature.

The final broad weight loss that occurs in the temperature range 723 to 1073 K

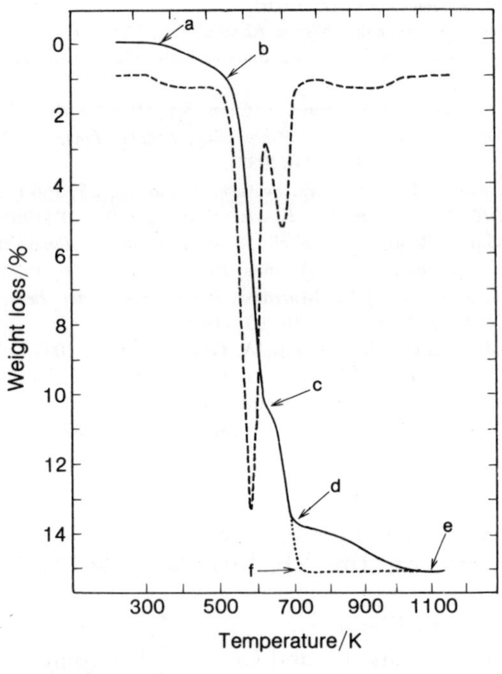

Fig. 3.14 Thermogram of prepared $La(OH)_3$ obtained at 2 ℃/min in vacuum.
— —; Integral weight loss curve, — — —; Time/temperature derivative,
· · ·; Pathway followed by rehydrated La_2O_3 sample during second stage of dehydration.
(Reproduced with permission by M. P. Rosynek *et. al.*, *J. Catal.*, **46**, 407(1997)).

(d − e) is due to decomposition of a surface layer of a unidentate carbonate species that is invariably present on the oxide as a result of interaction of the highly basic trihydroxide precursor with atmospheric carbon dioxide during preparation and handling.

Surface areas of the rare earth oxides prepared by decomposition of hydroxides depend on the decomposition conditions, temperature, atmosphere, etc. The oxides prepared by decomposition of hydroxides at 873 K in a vacuum, for example, have specific surface areas in the range 10 to 50 $m^2\ g^{-1}$.[2)]

b. Basic Property

In the form of single component oxides, one important feature is their basic property. Basic property was measured by benzoic acid titration. Basicity variation of La_2O_3 at $H_- = 12.2$ as a function of calcination temperature was measured.[3)] The maximum basicity upon unit weight basis was observed with pretreatment at 773 K. The measured basicity correlates well with the catalytic activity for diacetone alcohol decomposition.[3)]

IR measurement during activation of La_2O_3 indicates that carbon dioxide is strongly adsorbed and retained on the surface following outgassing at 773 K.[4)] The IR band almost diminishes following outgassing at 923 K. Strong basic sites seem to exist in maximum number at the outgassing temperature of 923 K.

c. Catalytic Activity

The reactions for which basic properties are reported to be relevant are hydrogenation of olefins,[2,7−9)] isomerization of olefins,[1−6,8)] aldol addition of ketones,[2,10,11)] and dehydration of alcohol.[12,13)]

For ethylene hydrogenation, a series of rare earth oxides shows high activity as the reaction proceeds at 195 K.[5)] In particular, La_2O_3 and Nd_2O_3 have activities comparable to those of transition metal oxides such as Cr_2O_3. Oxides such as CeO_2 and Pr_6O_{11} with high oxidation states, however, show low activities. Although the participation of basic sites in the reaction mechanism is not pictured, an importance of basic sites for the reaction is suggested.[5)]

1-Butene isomerization, 1,3-butadiene hydrogenation, and acetone aldol addition are catalyzed by rare earth oxides.[2)] The activity sequences of a series of rare earth oxides for the reactions are shown in Fig. 3.15. The activity sequence is the same for 1-butene isomerization and 1,3-butadiene hydrogenation, which is different from that of aldol addition. For the former reaction group, one characteristic feature is that the oxides of sesquioxide stoichiometry show the activity while the oxides with metal cations of higher oxidation states are entirely inactive. The situation is different in acetone aldol addition. Three oxides with high oxidation state, CeO_2, Pr_6O_{11} and Tb_4O_7, showed considerable activity. Pr_6O_{11} in particular showed activity close to the highest activity among rare earth oxides. Although acetone aldol addition is catalyzed by relatively weak basic sites, stable oxides with metal cations of oxidation state higher than 3 possess weak basic sites which are not strong enough to catalyze hydrogenation and isomerization.

Certain rare earth oxides show characteristic selectivity in dehydration of alcohols. 2-Alcohols undergo dehydration to form 1-alkenes. The formation of thermodynamically unstable 1-olefins contrasts with the formation of stable 2-olefins observed in the dehydration over acidic catalysts. The results of dehydration of 4-methyl-2-pentanol

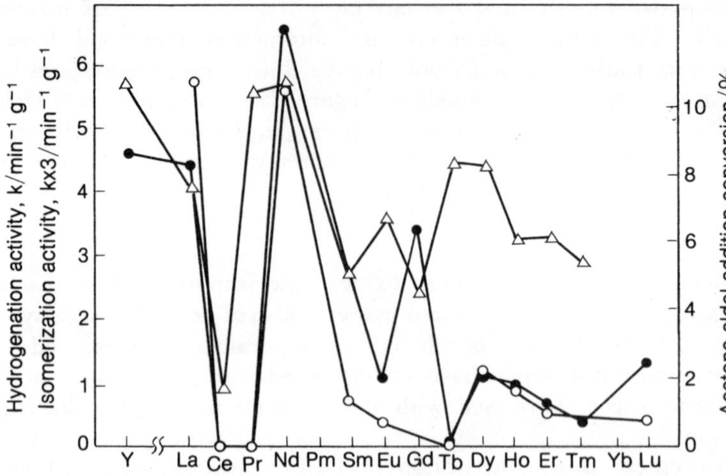

Fig. 3.15 Catalytic activities of rare earth oxides for 1-butene isomerization (○), 1, 3-butadiene hydrogenation (●), and acetone aldol addition (△).

are given in Table 3.4.[14] Selective formation of 1-olefins from 2-alcohols is observed for all catalysts listed in the table. The composition of 1-alkenes exceeds 81% for all catalysts. The selective formation of 1-olefins from 2-alcohols is due to the anionic character of the intermediates.

Variation of the activities of La_2O_3 as a function of the pretreatment temperature is shown in Fig. 3.16 for 1-butene isomerization,[3] 1,3-butadiene hydrogenation,[9] and methane $-$ D_2 exchange.[12] Pretreatment at 923 K results in maximum activity for all reactions. Essentially the same variation is observed for Nd_2O_3 and Sm_2O_3 in that maximum activity is obtained at the pretreatment temperature of 923 K. This is the temperature required to remove all CO_2 from the surface as detected by IR, indicating that strong basic sites are associated with the sesquioxide stoichimetry after complete removal of CO_2 and H_2O from the surfaces. On the other hand, weak basic sites, probably surface OH groups, exist on partially hydrated surfaces whenever the oxidation states of the metal cations are high or low.

B. Actinide Oxides (ThO_2, UO_2)

Thorium oxide and UO_2 are rather basic catalysts, though acidic sites seem to participate in the base-catalyzed reactions. Although basicity has not been measured by usual methods, catalytic selectivities and poisoning experiments suggest the existence of basic sites or acid $-$ base pair sites on the surfaces.

The acidic and basic properties are dependent upon the preparation and activation methods. Thorium oxide is usually prepared from aqueous solution of thorium nitrate or chloride by precipitation with aqueous ammonia followed by washing, and calcining. In some cases, ThO_2 is prepared by thermal decomposition of thorium nitrate or thorium oxalate. The acidic and basic properties of the ThO_2 prepared from $ThCl_4$ are distinctly different from the other ThO_2. The catalytic activities of ThO_2 prepared by different methods for 1-butene isomerizaton and 2-butanol dehydration are summarized in Table 3.5.[15] The ThO_2 prepared from cholride completely lacks the measur-

TABLE 3.4 Dehydration of 4-methyl-2-pentanol and other oxides

Oxide	Temp. K	HLSV	Conversion to olefin, %	Olefin products, %	
				1-Alkene	2-Alkene
Sc_2O_3†1	680	28	5	95	5
Y_2O_3†1	685	26	63	96	4
La_2O_3†2	687	55	39	96	4
CeO_2†2	623	50	14	86	14
Pr_6O_{11}†2	697	55	11	92	8
Nd_2O_3†2	691	55	25	94	6
Sm_2O_3†2	688	52	24	94	6
Eu_2O_3†1	700	46	10	95	5
Gd_2O_3†2	696	48	23	94	6
Tb_4O_7†1	700	16	2	90	10
Dy_2O_3†1	677	45	6	97	3
Ho_2O_3†2	684	52	47	97	3
Er_2O_3†1	676	45	6	97	3
Tm_2O_3†1	676	60	6	95	5
Yb_2O_3†2	676	50	42	97	3
UO_2†3	690	44	49	81	19

†1 Oxides from Michigan Chemical were used as received. †2 Oxalates from K & K Chemical were calcined 16 h at 673 K. †3 Oxalate prepared from nitrate was calcined 16 h at 673 K.

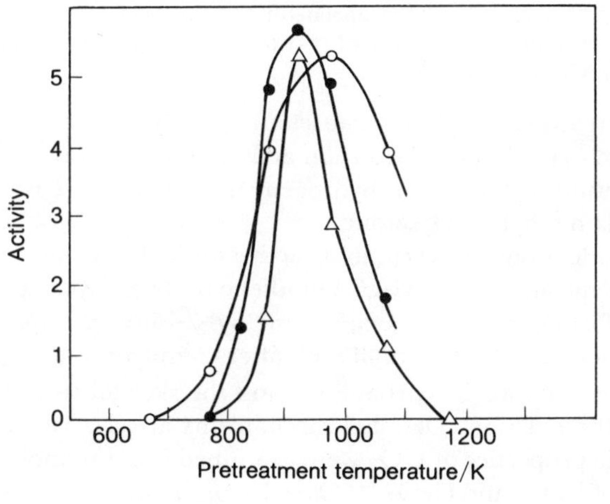

Fig. 3.16 Activity variation of La_2O_3 with pretreatment temperature.
○ ; 1-Butene isomerization at 303 K (1 unit : 6.4×10^{20} molecules·min^{-1} g^{-1}),
△ ; CH_4-D_2 exchange at 573 K (1 unit : 10^{-2} %·s^{-1} g^{-1}),
● ; 1,3-Butadiene hydrogenation at 273 K (1 unit : 1.2×10^{20} molecules·min^{-1} g^{-1})

able activity for 1-butene isomerization, while the other ThO_2 catalysts show base-catalyzed isomerization activity; the *cis/trans* ratio in 2-butene is high. For 2-butanol dehydration, the ThO_2 prepared from chloride shows low selectivity for 1-butene formation. The other catalysts show high selectivity for 1-butene formation. Butene isomerization is retarded by the introduction of both CO_2 and NH_3, indicating that not only basic sites but acidic sites are also participating in the reaction.[15)]

The high selectivity for 1-butene in decomposition of 2-butanol[16)] is explained by E1cB mechanisms operating in the reaction.[17)] An H^+ is abstracted by base sites on ThO_2 to form anionic intermediates.

TABLE 3.5 Catalytic activity and selectivity of ThO_2 for isomerization of butenes and dehydration of *s*-butyl alcohol

Catalyst	Surface area (m^2/g)	1-Butene Activity^{†1}	1-Butene Ratio of *cis* to *trans*	2-Butanol Activity^{†1}	2-Butanol composition of butenes^{†2}: 1-butene	*trans*-2-butene	*cis*-2-butene
ThO_2 (oxalate)	59.1	29.7	3.4	1.87×10^{-1}	84.2	9.3	6.5
ThO_2 (nitrate)	62.3	22.9	3.1	2.36×10^{-1}	82.4	10.6	7.0
ThO_2 (nitrate)	41.7	10.8 (1.1)^{†1}	3.2 (3.3)^{†4}	0.74×10^{-1}	76.6	14.8	8.6
ThO_2 (chloride)	46.3	0.0	—	2.22×10^{-1}	39.8	26.8	33.4

^{†1} Initial activity; $\%^{-1}min^{-1}$. ^{†2} Extrapolated to 0 conversion.
^{†3} Reactant; *cis*-2-butene. ^{†4} Ratio of 1- to *trans*-. Reaction temperature: 353 K for isomerization; 373 K for dehydration

Hydrogen treatment of ThO_2 causes drastic change in the catalytic property of ThO_2. Thorium oxide loses dehydration activity for reaction of alcohols, and becomes a dehydrogenation catalyst on hydrogen treatment.[18)] 2-Btutanol decomposition primarily yields methyl ethyl ketone.

Thorium oxide is one of the catalysts active for hydrogenation by anionic intermediates. 1,3-Butadiene and 2-methyl-1,3-butadiene undergo hydrogenation by 1,4 addition of H atoms to form *trans*-2-butene and 2-methyl-2-butene, respectively.[19–21)] The formation of alkanes is negligibly small even after complete consumption of the reactants. The intermediates are allylic carbanions, and the skeletal structures of carbon are retained during the reaction. Detailed mechanisms are described in section 4.15.

The catalytic properties of UO_2 were examined in 2-butanol dehydration and compared to those of ThO_2 and CeO_2.[17)] Over UO_2, the E2 mechanism is operating, while E1 and E1cB mechanisms operate over CeO_2, and ThO_2, respectively. The O ions in UO_2 have much lower electron density than in CeO_2 and ThO_2. The basic property of UO_2 is weaker than that of CeO_2 and ThO_2.

REFERENCES

1. M.P. Rosynek, D.T. Magnuson, *J. Catal*, **46**, 402 (1977).
2. H. Hattori, H. Kumai, K. Tanaka, G. Zhang, K. Tanabe, Proc. 8th National Symp. Catal. India, Sindri, 1987, p. 243.
3. Y. Fukuda, H. Hattori, K. Tanabe, *Bull. Chem. Soc. Jpn.*, **51**, 3150 (1978).
4. (a) M.P. Rosynek, D.T. Magnuson, *J. Catal.*, **46**, 147 (1977);
   (b) M.P. Rosynek, J.S. Fox, *J. Catal.*, **49**, 285 (1977);
   (c) Y.J. Goldwasser, W.K. Hall, *J. Catal.*, **63**, 520 (1980).
5. Y.J. Goldwasser, W.K. Hall, *J. Catal.*, **71**, 53 (1980).
6. M.P. Rosynek, J.S. Fox, J.L. Jensen, *J. Catal.*, **71**, 64 (1981).
7. Y. Imizu, K. Sato, H. Hattori, *J. Catal.*, **76**, 65 (1982).
8. K.M. Minachev, D.A. Kondratev, G.V. Antosin, *Kinet. Katal.*, **8**, 131 (1967).
9. K.M. Minachev, Y.S. Kondrakov, V.S. Nakhshunov, *J. Catal.*, **49**, 207 (1977).
10. G. Zhang, H. Hattori, K. Tanabe, *Appl. Catal.*, **36**, 189 (1988).
11. G. Zhang, H. Hattori, K. Tanabe, *Appl. Catal.*, **40**, 183 (1988).
12. M. Utiyama, H. Hattori, K. Tanabe, *J. Catal.*, **44**, 237 (1978).
13. A.J. Lundeen, R. van Hoozen, *J. Am. Chem. Soc.*, **85**, 2180 (1963).
14. A.J. Lendeen and R. van Hoozen, *J. Org. Chem.*, **32**, 3386 (1967).
15. Y. Imizu, T. Yamaguchi, H. Hattori, K. Tanabe, *Bull Chem. Soc., Jpn.*, **50**, 1040 (1977).
16. T. Tomatsu, T. Yoneda, H. Ohtsuka, *Yukagaku*, **17**, 236 (1968) (in Japanese).
17. K. Thomke, Proc, 6th Intern. Congr. Catal., 1976, London, p. 303.
18. B.H. Davis and S. Brey, Jr., *J. Catal.*, **25**, 81 (1972).
19. Y. Imizu, H. Hattori, and K. Tanabe, *J. Chem. Commun.*, 1091 (1978).
20. Y. Imizu, H. Hattori, K. Tanabe, *J. Catal.*, **57**, 35 (1979).
21. Y. Tanaka, Y. Imizu, H. Hattori, K. Tanabe, Proc. 7th Intern. Congr. Catal., 1980, Tokyo, p.1254.

3.1.4 TiO_2, ZrO_2

Titania (TiO_2) and zirconia (ZrO_2) have attracted attention as interesting supports for metal catalysts such as Pt and Pd, since strong interaction was found between the oxides and the metals. On the other hand, TiO_2 and ZrO_2 were found to exhibit super acidity when combined with a small amount of SO_4^{2-}. Titania itself is recognized as an acidic oxide, but becomes basic on reduction and ZrO_2 itself has both weakly acidic and weakly basic properties which sometimes show intriguing acid-base bifunctional catalysis. Here, the surface properties of TiO_2 and ZrO_2 are summarized together with typical examples of their catalytic behavior.

A. TiO_2

a. Surface property

The surface area and acidic property of TiO_2, prepared by hydrolysis of $TiCl_4$ with aqueous ammonia followed by washing the precipitates with distilled water until chloride ion was not detected in the washing with $AgNO_3$, drying at 380 K for 8 h, and calcining at 573 − 973 for 3 h are shown in Table 3.6.[1] The highest acid strength of TiO_2 thus prepared was $H_0 \leq -3$, but the acid amount was very small. The acid amount of TiO_2 calcined at 773 was 0.11 mmol/g at $H_0 \leq +4.0$. No basic property was observed on TiO_2 calcined at any temperature.[1] Differential thermal analysis, showed an endothermic peak at 343 K. The structure of TiO_2 dried at 383 K was only of the anatase type, while that of TiO_2 calcined at 573 – 773 K was a mixture of anatase and rutile types according to X-ray diffraction.[1] However, TiO_2 obtained by calcining the precipitates of titanic acid at 623 and 773 K which were prepared similarly as above and aged at 373 K for 1 h gave surface areas of 169 and 85 m^2/g which are considerably larger than the values in Table 3.6.[2] The acid amount of TiO_2 calcined at 773 K was 0.06 mmol/g at $H_0 \leq +1.5$ which was much larger than the value in Table 3.6, but zero at $H_0 \leq -3$.[2] The effect of aging of precipitates is fairly large. The structure of the latter TiO_2 was only anatase even when calcined at 773 K and the crystallization began at 623 K. According to other reports, the surface area and acidic property of TiO_2 prepared from $TiCl_4$ and calcined at 773 K were 38.5 m^2/g and about 0.06 mmol/g at $H_0 \leq +3.3$, respectively, no acid sites stronger than $H_0 = +1.5$ being observed[3] and those of TiO_2 prepared from $TiCl_4$ in the presence of $(NH_4)_2SO_4$ and calcined at 773 K were 80 m^2/g and 0.058 mmol/g at $H_0 \leq +3.3$, 0.032 mmol/g at $H_0 \leq +1.5$, and 0 mmol/g at $H_0 \leq -3$.[4] The acid amount at $H_0 \leq +1.5$ is eight times the value in Table 3.6. Ammonium sulfate which was used to prevent peptization of the precipitates remained in the precipitates even after thorough repeated washing and the small amount of sulfate ion is considered to cause larger surface area and acid amount.

TABLE 3.6 Surface area and acidic property of TiO_2

Calcination temp. K	Surface area m^2/g	Acid amount, mmol/g					
		$pK_a = pK_{BH^+}$					
		+6.8	+4.8	+4.0	+3.3	+1.5	−3.0
573	129	0.15	0.021	0.021	0.004	0.004	0.004
673	85.6						
773	44.4	0.23	0.11	0.11	0.004	0.004	0.004
873	28.9						
973	10.8	0.026	0.015	0.015	0.005	0.004	0.005

(Reproduced with permission from *J. Catal.*, **53**, 4 (1978)).

It can be said from the above results that the surface property of TiO_2 changes depending on the preparation method, but, in general, TiO_2 is classified into a weakly acidic metal oxide. The acid sites of TiO_2 are of the Brønsted type when calcined at low temperatures and of the Lewis type when calcined at higher temperatures.[4] It should be noted, however, that TiO_2 shows super acidity when it contains an appropriate amount of sulfate ions as described in section 3.9.

As for basicity, a commercially available TiO_2 calcined in dry nitrogen showed basicity which decreased with increase of calcination temperature (Fig. 3.17, and see section 2.2 for the measurement method).[5,6] Recently, basicity as well as acidity of TiO_2 prepared from $TiCl_4$ was measured in aqueous solution by a potentiometric acid-base titration method. The acidity and basicity of TiO_2 having a surface area of 124 m^2/g

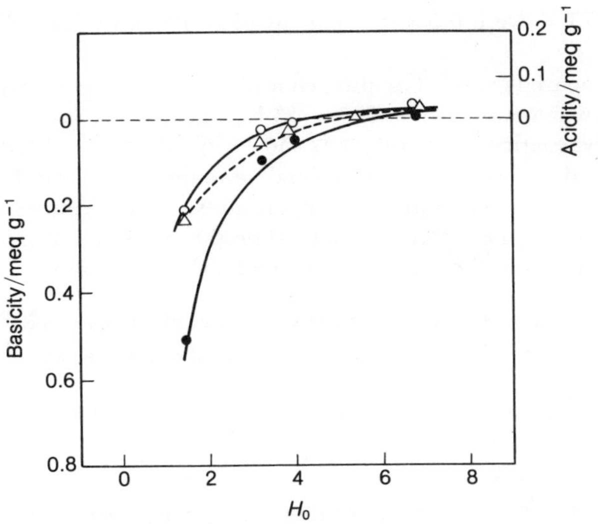

Fig. 3.17 Acid–base strength distributions of TiO_2 without calcination (●), calcined for 2 h at 573 K (○) and 773 K (△). $H_{0,\,max}$'s are 5.5, 4.5, and 5.0, respectively.

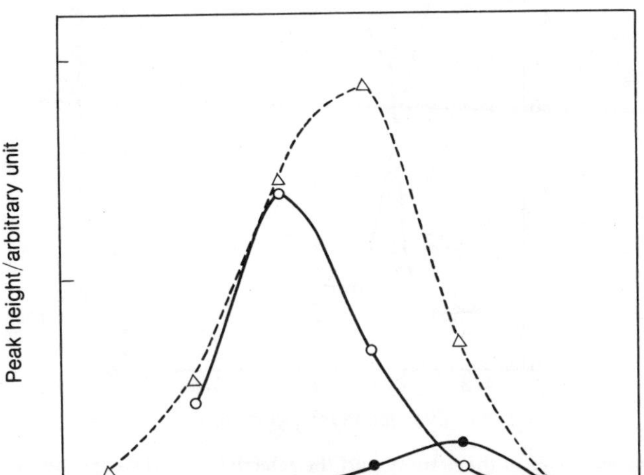

Fig. 3.18 Amounts of nitrobenzene anion radicals and Ti^{3+} of TiO_2 evacuated at various temperatures. (△···) nitrobenzene anion radical, Ti^{3+} before (○—) and after (●—) exposure to nitrobenzene.
(Reproduced with permission from *J. Catal.*, **38**, 176(1975)).

calcined at 773 K were $0.5 \times 10^{14}/cm^2$ at pH = 10 and $0.2 \times 10^{14}/cm^2$ at pH = 3.75, respectively.[7)]

Titania prepared from $TiCl_4$ also generates a reducing property (an electron donating property) when evacuated at 673 – 773 K. The dependence of the amount of reducing sites on evacuation temperature is shown in Fig. 3.18.[8)] According to ESR study, TiO_2 is reduced to form Ti^{3+} and a small amount of Ti^{2+}, and the amount of Ti^{3+} decreases on evacuation at higher temperatures. When a nitrobenzene which has a tendency to accept an electron is adsorbed on the surface of TiO_2, it reacts with Ti^{3+} to form an anion radical of nitrobenzene and Ti^{4+}. This causes the decrease of Ti^{3+}.

$$\text{nitrobenzene} + Ti^{3+} \longrightarrow \text{anion radical of nitrobenzene} + Ti^{4+}$$
$$\text{nitrobenzene} + Ti^{2+} \longrightarrow \text{anion radical of nitrobenzene} + Ti^{3+}$$

The reducing sites act as basic sites for particular molecules (see below).

b. Catalytic activity

The catalytic activity and selectivity of TiO_2 for isomerization of 1-butene change with change in evacuation temperature, as shown in Fig. 3.19[8)] The activity is high

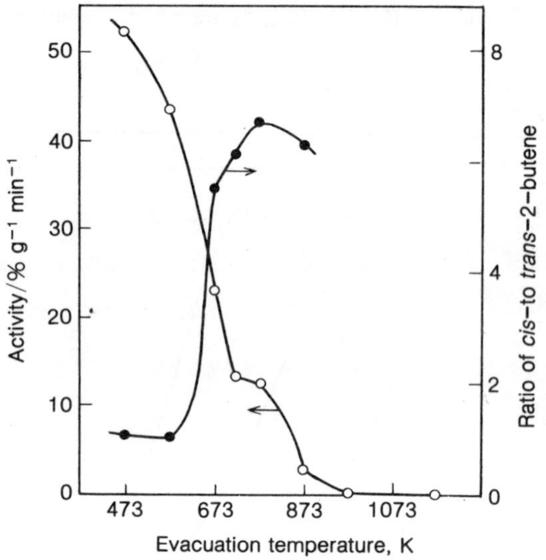

Fig. 3.19 Dependencies of the activity and the selectivity on the evacuation temperature in the isomerization of 1-butene over TiO_2 (I).
(Reproduced with permission from *J. Catal.*, **38**, 174(1975)).

on evacuation at low temperatures, but decrease with rise in evacuation temperature, while the selectivity (the ratio of *cis*-2-butene/*trans*-2-butene) is low (about 1) on evacuation at low temperatures, but high (about 5–6) on evacuation at high temperatures. The active sites on TiO_2 are considered to be Brønsted acid sites or basic sites (Ti^{3+}) depending on whether evacuation temperature is low or high. A tracer study revealed that the isomerization proceeeds by an intermolecular hydrogen transfer mechanism via carbenium ion over Brønsted acid sites and by an intramolecular hydrogen transfer mechanism via carbanion over basic sites (Ti^{3+}).[9,10] A weakly acidic TiO_2 is used as a catalyst for the manufacture of camphene from α-pinene. For this reaction, the use of strongly acidic catalysts causes the formation of by-products such as menthadienes, tricyclene, and limonene.[11]

Titania is also used as a good catalyst support. Table 3.7 shows the activities of molybdenum catalysts supported on various supports for the reduction of N_2O with H_2.[12] $TiO_2(\beta)$ exhibited an extremely high activity. The differences in activity between $TiO_2(\alpha)$ and $TiO_2(\beta)$ is due to the difference in acidity, since $TiO_2(\beta)$ contains a larger amount of $SO^2{}_4{}^-$ than $TiO_2(\alpha)$.

TABLE 3.7 Support effect of molybdenum catalysts on activity for $N_2O + H_2 \longrightarrow N_2 + H_2O$ at 523 K

Catalyst	Conversion, %
MoO_3/TiO_2 (β)	100
MoO_3/active carbon	7
MoO_3/SiO_2	4
MoO_3/MgO	2
MoO_3/TiO_2 (α)	2
MoO_3/Al_2O_3	2

B. ZrO_2

a. Surface property

Most of the ZrO_2 dealt with in this sections was prepared by calcining zirconium hydroxide at various temperatures in air for several hours and evacuating at the calcination temperatures for 2 h before use. The zirconium hydroxide was obtained by hydrolysis of zirconium oxychloride with 28% aqueous ammonia, followed by washing with deionized water until no chloride ion was detected in the filtrate and drying at 373 K for 24 h.[13,14] In some cases, zirconium oxynitrate was used instead of oxychloride.[15] The other preparations are specified at the appropirate places.

Specific surface areas of ZrO_2 pretreated at different temperatures are given in Table 3.8.[14] The surface areas decreased progressively with rise in pretreatment temperature.

The highest acid strength of ZrO_2 calcined in air at 773 K for 3 h is $H_0 = +1.5$ and the acid amounts are 0.06 and 0.280 mmol g^{-1} at $H_0 \leq +1.5$ and $H_0 \leq +4.0$, respectively.[16] Sometimes the highest acid strength is $H_0 = +3.3$. Thus, ZrO_2 is a weakly acidic oxide. The acid is mainly Lewis acid and partly Brønsted acid.[14]

The amount of pyridine irreversibly adsorbed at 373 K on a unit surface area basis in shown in Fig. 3.20 as a function of pretreatment temperature of ZrO_2. The maximum value of 3.9×10^{-8} mol m^{-2} (2.4×10^{16} molecules m^{-2}) was observed when ZrO_2 was pretreated at 673 K.[14]

The highest base strength of ZrO_2 evacuated at 773 K which was measured in an *in situ* cell is $H_- = 18.4$[17], though ZrO_2 calcined in air at 773 K does not show any basic property with the indicator method.[18] The amount of CO_2 irreversibly adsorbed at 373 K on ZrO_2 pretreated at various temperatures are shown in Fig. 3.21.[14] The basicity measured by CO_2 adsorption does not change much with the pretreatment temperature of ZrO_2.

TABLE 3.8 Specific surface areas of ZrO_2 pretreated at various temperatures

Pretreatment temperature (K)	Surface area ($m^2\ g^{-1}$)
573	175.5
673	109.0
773	64.5
873	32.1
973	21.4
1073	10.8
1173	9.9

(Reproduced with permission from *J. Catal.*, **57**, 3(1978)).

The amount of phenylnitroxide radicals formed on the surface of ZrO_2 pretreated at various temperatures when diphenylamine was adsorbed from a vapor phase at 453 K is also shown in Fig. 3.21. The maximum number of radicals formed on ZrO_2 which had been pretreated at 973 K, the number of radicals being 1.7×10^{-7} mol m^{-2} (1.1×10^{17} radicals m^{-2}).[14] Since the adsorbed diphenylamine is converted to phenylnitroxide radical when O_2 is admitted according to the following scheme,

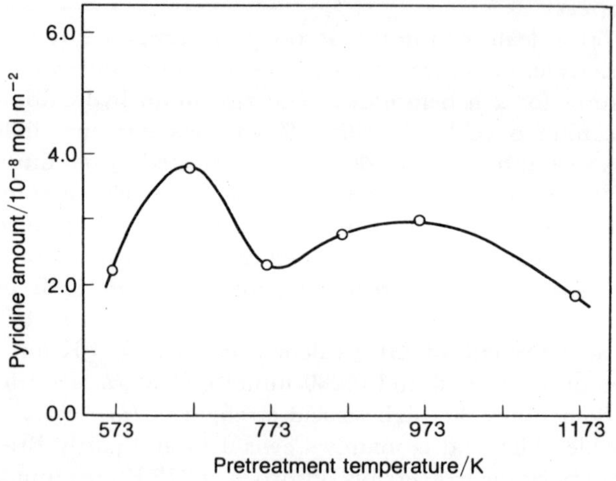

Fig. 3.20 Amount of pyridine molecules irreversibly adsorbed at 373 K on ZrO_2 pretreated at various temperatures.
(Reproduced with permission from *J. Catal.*, **57**, 4(1978)).

$$(C_6H_5)_2NH + B \longrightarrow (C_6H_5)_2N^-(ads) + H^+B \xrightarrow{O_2} (C_6H_5)_2NO_2^-(ads) + H^+$$
$$\longrightarrow (C_6H_5)_2NO^{\cdot}(ads) + {\cdot}OH(ads) + B, \ [B : \text{basic site}]$$

the amounts of the radicals give the basicity. As seen in Fig. 3.21, the basicity is much smaller than that measured by CO_2 adsorption.

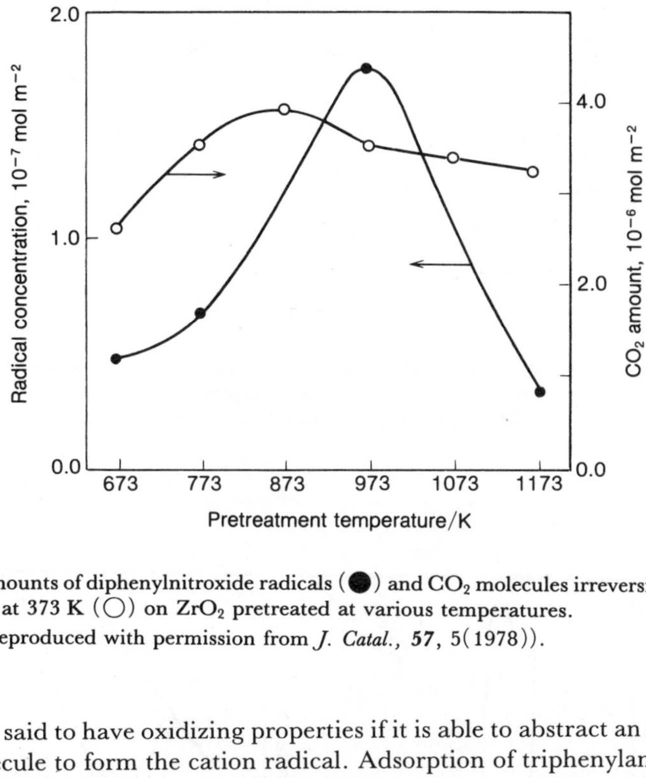

Fig. 3.21 **Amounts of diphenylnitroxide radicals (●) and CO_2 molecules irreversibly adsorbed at 373 K (○) on ZrO_2 pretreated at various temperatures.**
(Reproduced with permission from *J. Catal.*, 57, 5(1978)).

A surface is said to have oxidizing properties if it is able to abstract an electron from a suitable molecule to form the cation radical. Adsorption of triphenylamine on ZrO_2 does not give any ESR signal, though the ZrO_2 surface develops a light blue color. Subsequent addition of O_2 causes an immediate change in surface color to greenish gray, and a triplet signal with $g=2.005$ is observed by ESR, and is assigned to the cation radical of triphenylamine. The amplitude of the signal is independent of oxygen pressure. The number of cation radicals (oxidizing sites on the surface) as a function of pretreatment temperature is shown in Fig. 3.22. The maximum radical concentration was observed on ZrO_2 pretreated at 973 K and its value was 1.5×10^{-7} mol m^{-2} (9.3×10^{16} radicals m^{-2}).[14)]

A surface is said to have reducing properties if it is capable of donating an electron to a suitable molecule to form the anion radical. Nitrobenzen is a suitable molecule to measure the reducing property.[19)] The amounts of nitrobenzene anion radicals formed on ZrO_2 pretreated at various temperatures are shown in Fig. 3.22.[14)]

The maximum value was observed when ZrO_2 was pretreated at 773 K, the concentration being 4.3×10^{-8} mol m^{-2} (2.6×10^{16} radicals m^{-2}).

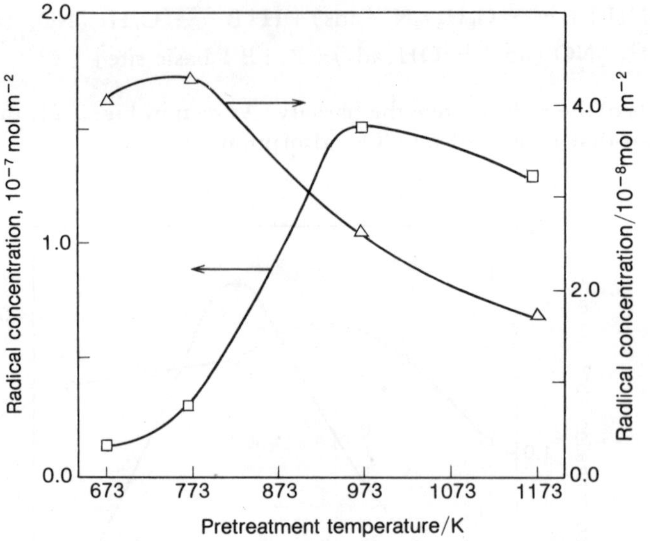

Fig. 3.22 Amounts of triphenylamine cation redicals (□) and nitrobenzene anion radicals (△) on ZrO_2 pretreated at various temperatures.
(Reproduced with permission from *J. Catal.*, **57**, 3(1978)).

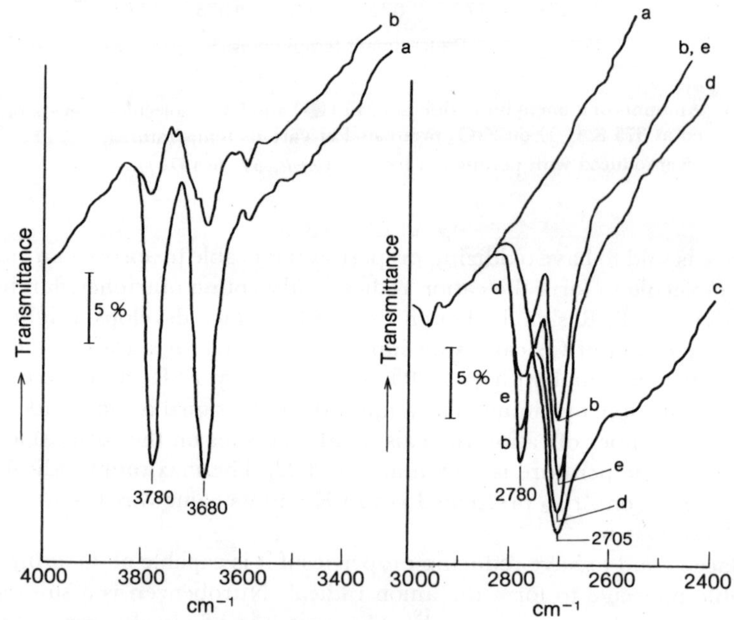

Fig. 3.23 Exchange of hydroxyl groups with D_2O.
(a) After evacuation at 733 K for 5 h. (b) Adsorption of 8 mmHg of D_2O at room temperature followed by evacuation at 773 K for 3 h. (c) Adsorption of 8 mmHg of D_2O at room temperature followed by evacuation at room temperature for 1 h. (d) Evacuation at 473 K for 1 h. (e) Evacuation at 573 K for 1 h.

Two sharp IR absorption bands, 3780 and 3680 cm^{-1}, are observed on the surface of ZrO_2 evacuated at 773 K, as shown in Fig. 3.23.[20] The hydrogen of these hydroxyl groups reacts with D_2O, CD_3COCD_3 or $CD_3CDODCD_3$ at room temperature, but not with $CDCl_3$. Hydroxyls showing a 3780 cm^{-1} band are selectively and irreversibly chlorinated by $CDCl_3$. The hydroxyl group of 3780 cm^{-1} is more reactive than the hydroxyl group of 3680 cm^{-1}.

A FT-IR study of hydrogen adsorbed on ZrO_2 evacuated at 973 K revealed recently that H_2 split heterolytically to form ZrOH (1780 and 3668 cm^{-1}), ZrH (1562 cm^{-1}), and ZrHZr (1371 cm^{-1}).[21] The OH groups showing at 3780 cm^{-1} are reported to be more reactive with CO than those showing at 3668 cm^{-1}. In the case of CO adsorption, formate is formed even in the absence of hydrogen, and the formate is reduced to methoxide in the presence of hydrogen, while, in the case of CO_2 adsorption, bicarbonate is mainly formed and, in the presence of sufficient hydrogen, it is converted into formate.[22,23] He and Ekerdt proposed by infrared spectoscopy that oxymethylene, $H_2CO_2^-$, is formed when H_2CO is adsorbed on ZrO_2.[23]

It is important to note that the temperature programmed desorption (TPD) profiles of CO adsorbed on ZrO_2 prepared by directly evacuating $Zr(OH)_4$ at 773 and 1073 K do not give any desorption peak, but the TPD profiles of CO adsorbed on ZrO_2 prepared by calcining $Zr(OH)_4$ in air at 773 and 1073 K and then evacuating at the same temperatures give two desorption peaks. This indicates that the preparation method of ZrO_2 strongly affects the surface properties.

b. Catalytic behavior

Hydrogen exchange of a methyl group The H–D exchange reaction of a methyl group of adsorbed isopropyl alcohol-d_8 with a surface OH group was found to occur at room temperature over ZrO_2 pretreated at 773 K. However, the exchange reaction was not catalyzed by strongly acidic $SiO_2-Al_2O_3$ and Al_2O_3 or strongly basic MgO and CaO under the same reaction conditions.[24] Therefore, ZrO_2, which is less acidic and less basic but has both acidic and basic sites, is considered to act as an acid-base bifunctional catalyst to activate the methyl group.

Synthesis of α-olefin from sec-alcohol A ZrO_2 catalyst is highly selective for the formation of 1-butene from sec-butanol compared with an Al_2O_3 catalyst as shown in Table 3.9.[25] The poisoning effects with *n*-butylamine and carbon dioxide indicate that both

TABLE 3.9 Selective formation of 1-butene from sec-butanol over ZrO_2

Catalyst	Selectivity (%)		
	1-butene	*cis*-2-butene	*trans*-2-butene
ZrO_2	90.2	7.4	2.4
Al_2O_3	26.9	62.2	10.9

acidic and basic sites on ZrO_2 surface participate in the reaction as active sites. The specific character of ZrO_2 which activates the methyl group of the alcohol mentioned above is capable of abstracting simultaneously both OH^- and H^+ of a terminal methyl group to form 1-olefins from 2-alkanols. The strongly acidic Al_2O_3 which abstracts OH^- first from sec-alcohol to give a carbenium ion mainly produces thermodynamically stable β-olefin.

Isomerization of 1-butene The catalytic activity of ZrO_2 for the isomerization of 1-butene is more than twice that of alumina. The selectively (the ratio of formed *cis*-2-butene to *trans*-2-butene) for the isomerization is 7.3 for ZrO_2 and 3.0 for Al_2O_3.[25] The activity and the selectivity suggest that the basic sites on ZrO_2, which are stronger than those on Al_2O_3, act as active sites for the isomerization reaction.

Formation of 1-butene and ammonia from butanamine Zirconium oxide catalyzes the elimination of ammonia from 2-butanamine to yield 1-butene as the major product. The catalytic activity of ZrO_2 is the highest among ThO_2, La_2O_3, ZnO, and MgO.[26] The selective formation of 1-butene from 2-butanamine is considered to proceed by a carbanion mechanism as shown below.

$$CH_3{-}CH_2{-}\underset{\displaystyle NH_2}{\underset{|}{CH}}{-}CH_3 \longrightarrow CH_3{-}CH_2{-}\underset{\displaystyle NH_2}{\underset{|}{CH}}{-}\bar{C}H_2 + H^+ \longrightarrow CH_3{-}CH_2{-}CH{=}CH_2 + NH_3$$

The reaction is initiated by abstraction of H^+ from carbon atom 1 by basic sites on the catalyst surface. The acidic sites having appreciable acid strength also seem to be necessary to stabilize the carbanion.

Hydrogenation of butadiene, etc. We have noted a few examples in which the cooperation of acid sites with basic sites results in surprisingly high catalytic activity and selectivity. Not only the acid and base strengths but the orientation of acid and base sites are also important for the catalytic activity and selectivity. The example of ZrO_2 pretreated at various temperatures is shown in Fig. 3.24.[13,14] The ZrO_2 catalyst pretreated at 873 K shows maximum activities for the hydrogenation of 1,3-butadiene with H_2 and the exchange between H_2 and D_2, whereas the ZrO_2 catalyst pretreated at 1073 K gives maximum activities for the hydrogenation of 1,3-butadiene with cyclohexadiene and the isomerization of 1-butene. Since the activity changes do not correlate with any surface properties described in the foregoing section, bifunctional catalysis seems to operate. The appearance of two maximum activities is considered to be due to the difference in distance between acid site (Zr^{4+}) and base site (O^{2-}). In fact, an X-ray diffraction study revealed that ZrO_2 pretreated below 973 K is mainly amorphous, small amounts of metastable tetragonal and monoclinic phase being contained, while ZrO_2 pretreated above 973 K is monoclinic.[17] Thus, the lattice constant of ZrO_2 pretreated at 873 K will be considerably different from that of ZrO_2 pretreated at 1073 K.

Syntheses of methanol and iso-butene Iso-butene is produced from $CO + H_2$ over ZrO_2 pretreated at 773 K under moderate conditions (0.5−21 atm, 573−723 K), the selectivity of butenes among hydrocarbons and that of iso-butene among C_4 hydrocarbons being 81.7 and 97.1 mol%, respectively, at 623 K and 0.68 atm.[15] At lower reaction temperatures, methanol is selectively formed [15,27] as an example is shown in Table 3.10.[15]

He and Ekerdt proposed a mechanism of CO and CO_2 hydrogenation over ZrO_2 as shown in Fig. 3.25 on the basis of the results by TPD and IR of adsorbed CO, CO_2, H_2, CH_3OH, HCOOH, H_2CO, and $HCOOCH_3$.[23]

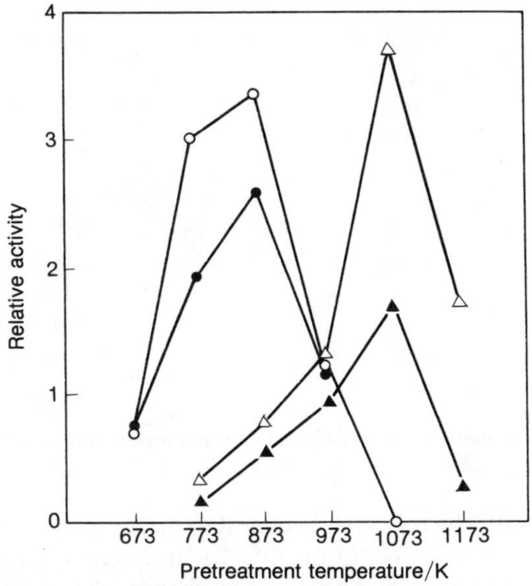

Fig. 3.24 Catalytic activities of ZrO_2 pretreated at different temperatures. ○, hydrogenation of 1,3-butadiene with H_2; ●, H_2-D_2 exchange; ▲, hydrogenation of 1,3-butadiene with cyclohexadiene.
(Reproduced with permission from *J. Catal.*, **80**, 307(1983)).

TABLE 3.10 Product distribution (mol %) from reaction of $CO+H_2$ on ZrO_2[†1]

T(K)	P(atm)	CO(%conv.)	CO_2	MeOH	MeOMe	Hydrocarbons
473[†2]	0.68	0.4	4.3	81.8	13.4	0.5
523[†2]	0.68	1.9	17.4	65.2	16.7	0.6
573[†2]	0.68	4.8	49.7	13.3	34.8	2.2
625[†2]	0.68	10	83.8	0.9	1.6	13.6
673[†2]	0.68	18	80.8	0.0	0.0	19.2
723[†2]	0.68	21	76.6	0.0	0.0	23.4
673[†3]	10	5.3	83.6	2.4	1.6	12.3
673[†3]	21	33.8	81.6	3.4	5.5	9.6

[†1] ZrO_2: 1.5 g, $H_2/CO=3$. [†2] The catalyst was evacuated at 973 K for 3 h befor $CO+H_2$ reaction with gas-circulation system (470 ml). The products were collected at liquid nitrogen temperature for initial 25 h. The surface area of the catalyst was *ca.* 50 $m^2\ g^{-1}$. [†3] The reactor was washed with N_2 flow at 673 K for 2 h before $CO+H_2$ reaction.

Recently, Abe *et al.* proposed a similar mechanism by means of the FT-IR method, but they insist that methoxide species are converted into formate ion, while the conversion of formate ion into methoxide is accelerated when methoxide and formate species are coadsorbed on the ZrO_2 surface at 523 K.[28)]
Hattori and Wang suggested on the basis of TPD profiles and IR spectra of the surface species that CO adsorbed on metal oxides of basic character reacts with H_2 to form

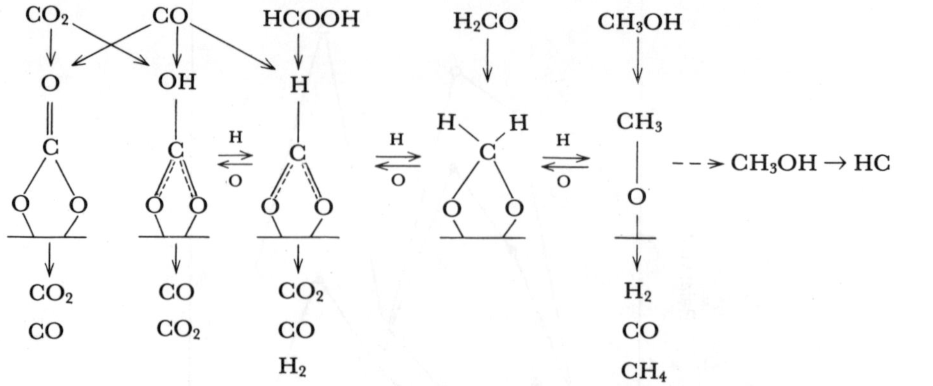

Fig. 3.25 Proposed mechanism of CO and CO_2 hydrogenation over ZrO_2.
(Reproduced with permission by M. He, J. G. Ekerdt, *J. Catal.*, **90**, 21(1984)).

a formyl group (HCO) which is adsorbed on the surface oxygen ion to make a formate ion.[27] The surface formate (HCO on O^{2-}) is different from the formate which is formed on the surface when HCOOH is adsorbed as shown below.

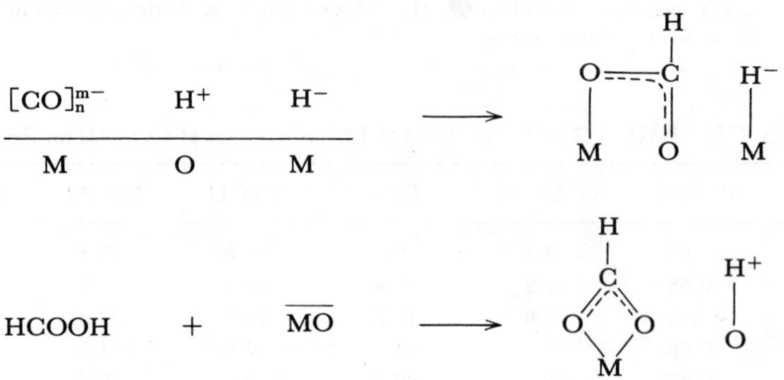

(M; metal cation, O: oxygen anion)

ZrO_2 as Catalyst Support The MoO_3 supported on ZrO_2 shows the highest catalytic activity for the reduction of NO with H_2 at 553 K as shown in Table 3.11.[29] Detailed study of adsorbed species of NO by ESR, IR, and UV revealed that the active site for the reaction of NO with H_2 is Mo^{5+} $(NO)_2$.[30]

The Rh supported on ZrO_2 exhibits higher catalytic activity for the hydrogenation of CO and CO_2 compared with that supported on Al_2O_3, SiO_2, *etc.*[31,32] In particular, the Rh/ZrO_2 catalyst shows the highest activity for the hydrogenation of CO_2, as shown in Fig. 3.26.[31] For the hydrogenation of CO, it was second following Rh/Nb_2O_5. Upon adsorption on CO on reduced Rh/ZrO_2, the carbonate bands due to the reaction, $2CO \rightarrow C + CO_2$, together with the bands of twin, linear and bridge CO species were observed at moderate temperatures, whereas Rh/MgO gave no appreciable formation of CO_2 even at higher temperatures (>373 K) and Rh/Al_2O_3

showed intermediate behavior.[33] On the adsorption of CO_2, the linear CO band is formed at a lower frequency than that on CO adsorption. The linear CO species formed from CO_2 shows higher reactivity toward hydrogen compared with that from CO adsorption. The reducing properties of ZrO_2 seem to play an important role in the support effects.

Recently, ZrO_2 was found to be a surprisingly good support of a $La_{0.8}Sr_{0.2}CoO_3$ catalyst for complete oxidation of propane.[34] The catalyst has been confirmed to be highly dispersed on ZrO_2.

TABLE 3.11 **Support effect of molybdenum catalyst on reduction of NO with H_2 at 553 K**

Catalyst	NO conversion (%)
MoO_3-ZrO_2	90
MoO_3-ZrO_2-TiO_2	71
MoO_3-active carbon	51
MoO_3-TiO_2 (β)	50
MoO_3-TiO_2 (α)	48
MoO_3-TiO_2-SiO_2	38
MoO_3-Al_2O_3	32
MoO_3-SnO_2	29
MoO_3-MgO	25
MoO_3-SiO_2	23

(Reproduced with permission from *Reat. Kinet. Catal. Lett.*, **11**, 151 (1979)).

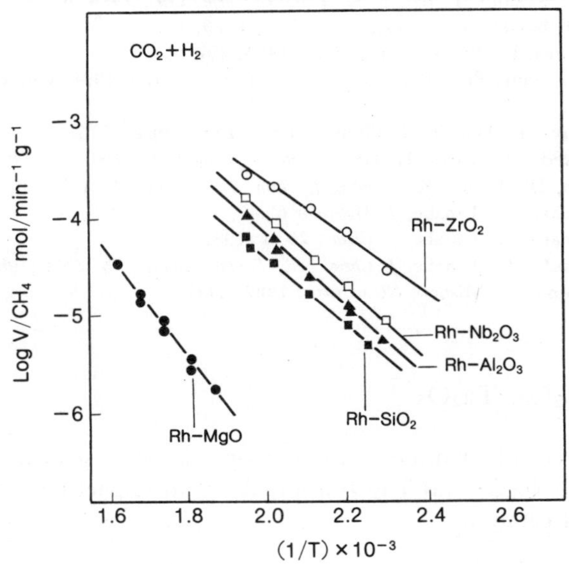

Fig. 3.26 **Arrhenius plots of the reaction of CO_2 and H_2. Effect of catalyst support.** (Reproduced with permission from *J. Molecular Catal.*, **17**, 383(1982)).

REFERENCES

1. K. Tanabe, H. Hattori, T. Sumiyoshi, K. Tamaru, T. Kondo, *J. Catal.*, **53**, 1 (1978).
2. M. Itoh, H. Hattori, K. Tanabe, *J. Catal.*, **35**, 225 (1974).
3. K. Shibata, T. Kiyoura, J. Kitagawa, T. Sumiyoshi, K. Tanabe, *Bull. Chem. Soc. Jpn.*, **46**, 2985 (1973).
4. K. Tanabe, C. Ishiya, I. Matsuzaki, I. Ichikawa, H. Hattori, *Bull. Chem. Soc. Jpn.*, **45**, 47 (1972).
5. T. Yamanaka, K. Tanabe, *J. Phys. Chem.*, **80**, 1723 (1976).
6. T. Yamanaka, K. Tanabe, *J. Phys. Chem.*, **79**, 2409 (1975).
7. H. Kita, N. Henmi, K. Shimazu, H. Hattori, K. Tanabe, *J. Chem. Soc., Faraday Trans. 1*, **77**, 2451 (1981).
8. H. Hattori, M. Itoh, K. Tanabe, *J. Catal.*, **38**, 172 (1975).
9. H. Hattori, M. Itoh, K. Tanabe, *J. Catal.*, **41**, 46 (1976).
10. K. Morishige, H. Hattori, K. Tanabe, *Bull. Chem. Soc. Jpn.*, **48**, 3088 (1975).
11. R. Ohnishi, K. Tanabe, S. Morikawa, T. Nishizaki, *Bull. Chem. Soc. Jpn.*, **47**, 571 (1974).
12. S. Okazaki, N. Ohsuka, T. Iizuka, K. Tanabe, *J. Chem. Soc. Chem. Commun.*, **1976**, 654.
13. Y. Nakano, T. Yamaguchi, K. Tanabe, *J. Catal.*, **80**, 307 (1983).
14. Y. Nakano, T. Iizuka, H. Hattori, K. Tanabe, *J. Catal.*, **57**, 1 (1978).
15. T. Maehashi, K. Maruya, K. Domen, K. Aika, T. Onishi, *Chem. Lett.*, **1984**, 747.
16. K. Shibata, T. Kiyoura, J. Kitagawa, T. Sumiyoshi, K. Tanabe, *Bull. Chem. Soc. Jpn.*, **46**, 2985 (1973).
17. Y. Nakano, "Surface and Catalytic Properties of ZrO_2", Thesis for the degree of Doctor of Science at Hokkaido University, 1982.
18. K. Tanabe, I. Ichikawa, H. Ikeda, H. Hattori, *J. Res. Inst. Catalysis, Hokkaido Univ.*, **19**, 185 (1972).
19. T. Iizuka, H. Hattori, Y. Ohno, J. Sohma, K. Tanabe, *J. Catal.*, **22**, 130 (1974).
20. T. Yamaguchi, Y. Nakano, K. Tanabe, *Bull. Chem. Soc. Jpn.*, **51**, 2482 (1978).
21. T. Ohnishi, H. Abe, K. Maruya, K. Domen, *J. Chem. Soc., Chem. Commun.*, **1985**, 617.
22. M. He, J. G. Ekerdt, *J. Catal.*, **87**, 381 (1984).
23. M. He, J. G. Ekerdt, *J. Catal.*, **90**, 17 (1984).
24. T. Yamaguchi, Y. Nakano, T. Iizuka, K. Tanabe, *Chem. Lett.*, **1976**, 677.
25. T. Yamaguchi, H. Sasaki, K. Tanabe, *Chem. Lett.*, **1973**, 1017.
26. A. Satoh, H. Hattori, K. Tanabe, *Chem. Lett.*, **1983**, 497.
27. H. Hattori and G. Wang, Proc. 8th Intern. Congr. Catal., Berlin, 1984, Verlag Chemie, Weinheim, vol. 3, p. 219.
28. H. Abe, K. Maruya, K. Domen, T. Ohnishi, *Chem. Lett.*, **1984**, 1875.
29. K. Tanabe, H. Ikeda, T. Iizuka, H. Hattori, *React. Kinet. Catal. Lett.*, **11**, 149 (1979).
30. T. Iizuka, M. Itoh, H. Hattori, K. Tanabe, *J. Chem. Soc., Faraday Trans., 1*, **78**, 501 (1982).
31. T. Iizuka, Y. Tanaka, K. Tanabe, *J. Molecular Catal.*, **17**, 381 (1982).
32. T. Iizuka, Y. Tanaka, K. Tanabe, *J. Catal.*, **76**, 1 (1982).
33. Y. Tanaka, T. Iizuka, K. Tanabe, *J. Chem. Soc., Faraday Trans.*, **78**, 2215 (1982)
34. H. Fujii, N. Mizuno, M. Misono, *Chem. Lett.*, **1987**, 2147.

3.1.5 V_2O_5, Nb_2O_5, Ta_2O_5

It is only very recently that the acid-base properties of these metal oxides have been studied. Hydrated Nb_2O_5 and Ta_2O_5 are making an impact for their application as unusual solid acid catalysts.

A. V_2O_5

A commercially available V_2O_5 of reagent grade which was dried in a desiccator for a few days was reported to be acidic, the acid strength being $H_0 \leq +3.3$.[1] However, another V_2O_5 of guaranteed reagent without any treatment did not show any acidity, but showed considerable basicity (0.16 mmol/g at $H_0 \geq +1.5$)[2] when measured by titration with trichloroacetic acid using a Hammett indicator described in Section 2.2. The discrepancy seems to be due to the effect of moisture. The $H_{0,max}$ value of the latter ZrO_2 is 8.5, indicating that V_2O_5 is more basic than Ta_2O_5 ($H_{0,max} = 2.0$), MoO_3 (2.1), TiO_2 (5.5) and γ-Al_2O_3 (7.2) and more acidic than ZrO_2 (9.5) and BaO (15).[2]

B. Nb_2O_5

Hydrated niobium pentoxide ($Nb_2O_5 \cdot nH_2O$), which is usually called niobic acid, was found to exhibit a high acid strength ($H_0 = -5.6$) corresponding to the acid strength of 70% H_2SO_4 when calcined at relatively low temperatures (373 − 573 K), though the surface of niobic acid calcined at 773 K was almost neutral.[3] Since any kind of acidic metal oxide shows acidity on calcination at about 773 K and the acidity is lost or decreased by absorbing water, niobic acid which shows high acid strength on the surface in spite of its containing water is an unusual solid acid. The unusual solid acid is expected to show stable catalytic activity for acid-catalyzed reactions in which water molecules participate or are liberated. In fact, it showed excellent stability as a catalyst for esterification, hydrolysis, and hydration reactions.

a. Surface property

The surface areas of niobic acid were 164, 126, and 42 m^2/g after evacuation at 373, 573, and 773 K, respectively.[4] The ion exchange experiment showed that only 1.2% of the protons of $H_8Nb_6O_{19}$ could be exchanged with sodium ion. The exchange process of protons was very slow.[5]

Acidic property of niobic acid measured by *n*-butylamine titration using Hammett indicators is shown in Fig. 3.27.[4] Considerable acid amounts at $H_0 = -5.6$ were obserbed for niobic acids pretreated at 393 − 573 K, though a niobic acid calcined at

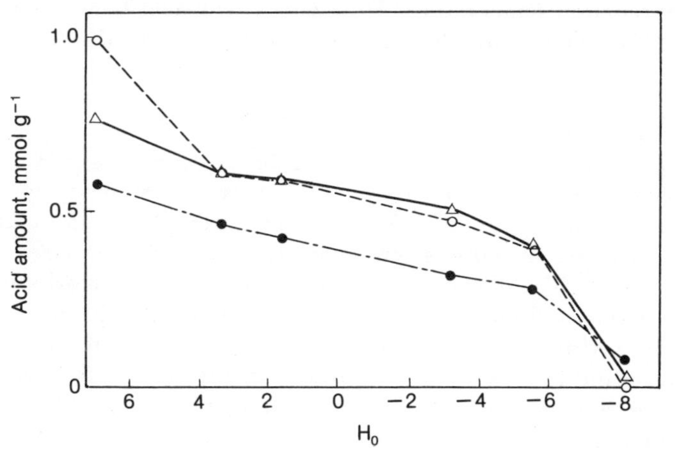

Fig. 3.27 Acid amount *vs.* acid strength of niobic acid. ○; Calcined at 393 K, △; 473 K, ●; 573 K

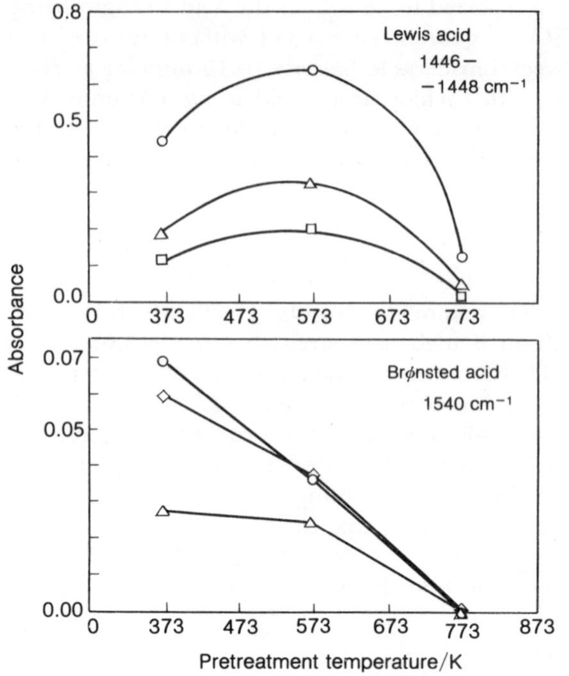

Fig. 3.28 Acidity change of niobic acid with pretreatment temperature. ◇; Evacuated at room temperature after adsorption of pyridine, ○; at 373 K, △; at 473 K, □; at 573 K

773−873 K did not show any acidic property.

According to infrared spectra of pyrindine adsorbed on niobic acid, the Brønsted acid band intensity was strongest on the sample evacuated at 373 K and decreased with increase of evacuation temperature. However, Lewis acid band intensity showed a maximum on the niobic acid which had been evacuated at 573 K as shown in Fig. 3.28.[4)]

b. Catalytic behavior

Isomerization of 1-butene, Dehydration of 2-butanol, and Polymerization of propylene These reactions which are known to be catalyzed by acids were studied over niobic acid to characterize the acidic nature of niobic acid. The catalytic activity and selectivity of niobic acid evacuated at various temperatures for isomerizations of 1-butene are shown in Fig. 3.29.[4)] The niobic acid evacuated at 373 K for 2 h exhibited the highest activity. The selectivity indicates that Brønsted acid is acting as the active sites. The activity decreases with increase of evacuation temperature and the selectivity becomes almost 2, suggesting that Lewis acid is also acting as active sites. On evacuation at 773−873 K, the activity almost disappeared. An interesting finding is that the activity of niobic acid evacuated at 573 K followed by exposure to water vapor and then evacuated at 373 K becomes almost the same as that evacuated at 373 K, whereas the niobic acid once evacuated at 773 K does not increase in activity even if exposed to water

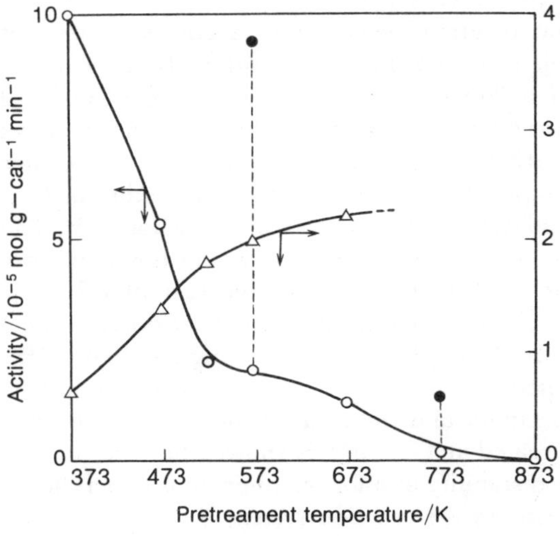

Fig. 3.29 Activity(○) and selectivity(△) of 1-butene isomerization ●; H_2O added after pretreatment and evacuated at 373 K.

vapor. This indicates that the transformation of amorphous to T.T. phase[4] interferes with the regeneration of Brønsted acid by water addition.

In the case of dehydration of 2-butanol, the activity of niobic acid was high and competed with that of $SiO_2-Al_2O_3$ when evacuated at 423 K, but markedly decreased when evacuated at 573 K.[4] Thus, the active sites are considered to be Brønsted acid from the comparison with the data in Fig. 3.28. For polymerization of propylene, niobic acid showed a high activity on evacuation at 373 K and the activity decreased on evacuation at 423−473 K and increased on evacuation at 523−573 K, finally disappearing on evacuation at 773 K.[4] This activity change can be interpreted by taking into account the fact that the main active sites are Brønsted acid in the case of low temperature evacuation, but Lewis acid in the case of high temperature evacuation (*cf.* Fig. 3.28).

Hydration, Esterification and Hydrolysis For hydration of ethylene, the activity of niobic acid was considerably lower in the early stage of the reaction, but increased gradually as the reaction proceeded and reached a steady state in 6 h. The deactivation of the catalyst was not observed when the run was repeated.[6] The steady state activity of niobic acid calcined at 573 K was higher than that of solid phosphoric acid which is widely used in industry. When a niobic acid was calcined at a high temperature of 773 K, the activity was low and did not increase even in the later stage of the reaction. It is interesting and important that niobic acid calcined at relatively low temperature showed high activity and long life. The selectivity for the formation of ethyl alcohol at 473 K over the niobic acid catalyst was more than 97%, a small amount of the other product being acetaldehyde. A niobic acid treated with phosphoric acid was recently found to show higher activity than a simple niobic acid.[7] Treatment with phosphoric acid was effective for maintaining a large surface area and a large amount of strong acid sites and for preventing niobic acid from crystallizing even after heat treatment at higher temperature above 873 K.

For esterification of ethyl alcohol with acetic acid, the catalytic activities and selectivities of niobic acid and the other solid acids are shown in Table 3.12.[3] The niobic acid showed higher activity than resin, $ZrO_2-SO_4^{2-}$, $Fe_2O_3-SO_4^{2-}$, and $SiO_2-Al_2O_3$. The selectivity for the formation of ethyl acetate was 100%. In the case of resin, the selectivity was high, but the resin turned black after 1 h reaction so that repeated use was impossible under the reaction condition. On the other hand, the activity of niobic acid did not change even after use for 60 h. The $TiO_2-SO_4^{2-}$, one of the solid super acids, showed high activity, but the activity rapidly decreased and became much lower than the activity of niobic acid after 2 h reaction. The HZSM-5 catalyst also exhibited high activity, but formed considerable amounts of diethyl ether and ethylene as by-products, the selectivity being less than 92%. It is concluded that niobic acids pretreated at relatively low temperatures are highly active for esterification with 100% selectivity and the catalyst life is long enough.

Niobic acid pretreated at 473 – 673 K showed high activity and selectively (100%) and remarkably good stability also for the hydrolysis of acrylic ester, for which a large amount of water exists in the reaction system.[8]

C. Ta_2O_5

Following niobic acid, hydrated Ta_2O_5 was recently reported to be a strongly acidic oxide. The acid strength of Ta_2O_5 calcined at 473 – 673 K is $H_0 \leq -8.2$[9], which is stronger than that of niobic acid. Even when calcined at 773 and 873 K, it shows high acid strengths of $-8.2 < H_0 \leq -5.6$ and $-5.6 < H_0 \leq -3.0$, though the acid strengths decrease on calcination at 1073 K. Thus, the regeneration of deactivated Ta_2O_5 by

TABLE 3.12 Activities and selectivities of $Nb_2O_5 \cdot nH_2O$ and the other solid acids for esterification of ethyl alcohol with acetic acid.

Catalyst	Reaction temp. K	C_2H_5OH basis		Byproducts
		Conversion/%	Ester selectivity/%	
$Nb_2O_5 \cdot nH_2O$†1	393	72	100	
	413	86	100	
resin†2	393	38	<99	
	413	50	<98	
$ZrO_2-SO_4^{2-}$†3	413	56	90	$(C_2H_5)_2O$
$Fe_2O_3-SO_4^{2-}$†3	413	13	93	$(C_2H_5)_2O$
$TiO_2-SO_4^{2-}$†3	393	95(54)†4	99(98)†4	
	413	100	95	C_2H_4
$SiO_2-Al_2O_3$†3	393	4	99	$(C_2H_5)_2O$
	413	14	98	$+C_2H_4$
HZSM-5†3	393	82	92	$(C_2H_5)_2O$
	413	99	72	$+C_2H_4$

†1 Calcined at 473 K, †2 Calcined at 393 K, †3 Calcined at 773 K, †4 After 2 h reaction time.

Catalyst weight; 1 g, Volume ratio of acetic acid to ethyl alcohol=1, Reaction time; 1 h.

calcination at 773 or 873 K is possible. The difference in calcination temperature dependence of acidity between $Nb_2O_5 \cdot nH_2O$ and $Ta_2O_5 \cdot nH_2O$ is considered due to the difference in temperature of crystallization (860 K for Nb_2O_5[4] and 1003 K for Ta_2O_5[9]).

The catalytic activity of $Ta_2O_5 \cdot nH_2O_5$ for esterification of acrylic acid with metahnol was found to be higher than that of $Nb_2O_5 \cdot nH_2O$, and its stability was also better,[9] indicating it to be a promising solid acid catalyst.

REFERENCES

1. K. Nishimura, *Nippon Kagaku Zasshi*, **81**, 1680 (1960) (in Japanese).
2. T. Yamanaka, K. Tanabe, *J. Phys. Chem.*, **80**, 1723 (1976).
3. Z. Chen, T. Iizuka, K. Tanabe, *Chem. Lett.*, **1984**, 1085.
4. T. Iizuka, K. Ogasawara, K. Tanabe, *Bull. Chem. Soc. Jpn.*, **56** 2927 (1983).
5. B. K. Sen, A. V. Saha, N. Chatterjee, *Mat. Res. Bull.*, **16**, 923 (1981).
6. K. Ogasawara, T. Iizuka, K. Tanabe, *Chem. Lett.*, **1984**, 645.
7. S. Okazaki, M. Kurimata, T. Iizuka, K. Tanabe, *Bull. Chem. Soc. Jpn.*, **60**, 37 (1987).
8. T. Iizuka, S. Fujie, T. Ushikubo, Z. Chen, K. Tanabe, *Appl. Catal.*, **28**, 1 (1986).
9. Mitsubishi Chem. Co., Japan Patent Kokai, 60-082915 (1985).

3.1.6 Oxides of Cr, Mo, W

A. General Remarks

Oxides of Cr, Mo, and W are usually used for catalysts as mixed oxides with other oxides such as alumina and silica which are prepared by coprecipitation, impregnation, etc. They are seldom put to practical use as simple oxides. Principal reactions catalyzed by these oxides, unlike those observed for silica-alumina or zeolites, often involve redox-type reaction steps, and during these steps reaction intermediates having covalent carbon-metal bonds are formed. Examples of those reactions are dehydrogeneration, hydrogenation and skeletal isomerization of hydrocarbons, and polymerization of olefins, as well as metathesis of olefins and hydrodesulfurization. Therefore, acid-base properties of catalysts usually play secondary roles in catalysts.

Cr_2O_3 gels are prepared by decomposition of salts of Cr such as ammonium bichromate and chromic hydroxide. Chromic hydroxide can be prepared by neutralization of an aqueous solution of chromic nitrate with ammonia or urea, followed by washing and drying of the precipitate.[1,2] Surface area is usually $1-10\ m^2g^{-1}$, but it varies depending on the heat treatment. Heat treatment at $600-700$ K causes transformation of chromia gel to α-Cr_2O_3. Molybdenum and tungsten oxides are prepared similarly. Preparation of Cr, Mo, and W oxides which are supported, impregnated and fixed on oxide surfaces may be referred to in the literature.[3]

B. Cr_2O_3

Chemisorption and catalysis on chromia has been discussed in general.[4]

a. Acidic properties

Based on IR of NH_3 adsorbed on Cr_2O_3 it has been reported that the surface of

Cr_2O_3 has only Lewis acid sites.[5] The adsorption of pyridine, O_2, H_2O, and CO on α-Cr_2O_3 has been investigated in detail by means of IR and showed that Cr(III) ions, which differ in the number of coordinated oxide and hydroxide ions, are present on the surface.[6] According to these studies, these Cr(III) ions react with pyridine molecule (py) as follows. On dehydroxylated and oxygen-uncovered surfaces, strong absorption takes places on Cr(III) ion through a coordination bond (eq. 1). The IR band is typical for pyridine bound to a Lewis acid site.

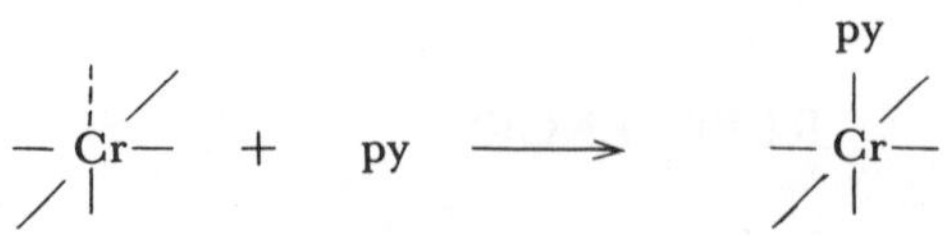

(1)

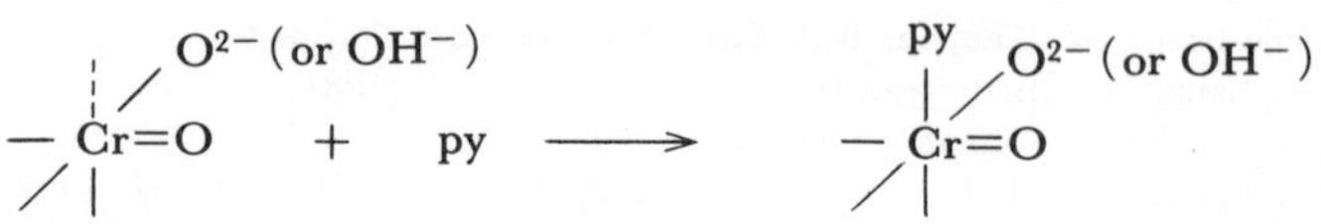

(2)

The coordination sites are not fully occupied by oxygens which are dissociatively adsorbed. Hence, the Lewis acidity due to Cr(III) can be observed by the pyridine adsorption (eq. 2), even after the above surface is exposed to oxygen. Adsorption of CO prohibits the pyridine adsorption,[5] and butene blocks the adsorption of CO.[6]

On the hydrated surface, in addition to weak physical adsorption on surface OH or H_2O group through hydrogen bonding, medium to strong chemical adsorption on Cr(III) ion (Lewis acid site) takes place by eq. 3.

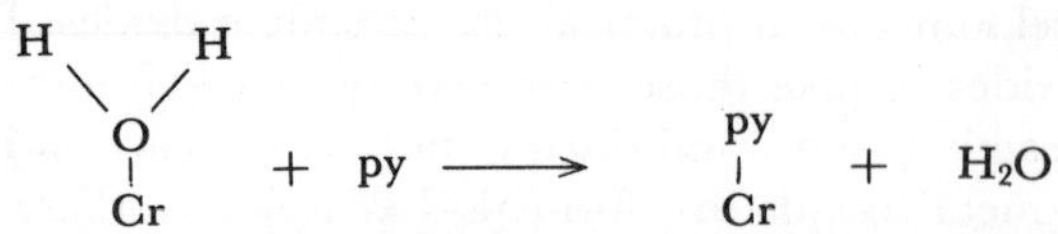

(3)

Pyridine molecules adsorbed following eqs. 1 – 3 were not distinguishable by IR. The presence of Brønsted sites was not indicated by these studies.

b. Catalysis

Based on the changes of the selectivity and the rate of 1-butene isomerization it was proposed that butene isomerizes via a carbenium ion over low temperature-treated chromia and via an allylic-type intermediate over chromia outgassed at higher temperature.[1] Surface hydroxyl groups are responsible for the former mechanism (acid catalysis) and surface sites produced by the removal of water from two adjacent OH groups for the latter. The presence of two allylic intermediates (anionic and cationic) was indicated from the difference in the selectivity between He- and H_2-treated α-Cr_2O_3.[7]

Active sites for oligomerization and polymerization of olefins over chromia supported on silica-alumina are believed to be Cr(II) and/or Cr(III).[8] Dehydrogenation of alcohols proceeds on chromia.[9,10] Formate ion detected by IR has been suggested to be the reaction intermediate for conversion of methanol to H_2, CO and CO_2.[10]

These reactions are probably assisted by the basicity of oxide ions and the redox properties of Cr, although quantitative discussion has not been attempted.

C. MoO_3

a. Acid-base properties

There are only a few studies reported for the acidic or basic properties of simple oxide of molybdenum. The acidity increases when MoO_3 is supported on or mixed with SiO_2, Al_2O_3 or TiO_2, for which Brønsted and Lewis acidities have been shown by IR studies of pyridine adsorption[3,11-14] (see Section 3.2). MoO_3-TiO_2 and $-Al_2O_3$ have significant amounts of strong acid sites at high temperatures, but in the case of MoO_3-SiO_2, Brønsted sites decreased rapidly by heat treatment and Lewis acidity due to Mo ion increased.[14] Only Brønsted sites were indicated by IR of NH_3 adsorbed on MoO_3.[12] But a recent report on IR study of NH_3 adsorption[15] indicated the presence of only Lewis sites for highly dispersed MoO_3-SiO_2. According to an amine titration, very small amount of weak acidity was observed (*ca.* 0.01 mmole g^{-1} for $H_0 \leq +4.8$).[16] Acid strength distributions measured by an amine titration were reported for Mo–, and Mo,Co$-Al_2O_3$.[17] The results are shown in Table 3.13.

Coordinative unsaturation of Mo ion on the surface of $MoO_3-Al_2O_3$, which may be regarded as Lewis acidity, has been investigated extensively by the adsorption of NO and O_2.[15,16,18]

b. Catalysis

Isomerization of butene was examined over MoO_3.[16] Contrasting poisoning

TABLE 3.13 Surface acid strength distribution in the Co-Mo-Al_2O_3 system

Sample	Differences in *n*-butylamine titer values[†1] for indicators of various pK_a values				
	$4.8>pK_a>3.3$	$3.3>pK_a>2$	$2>pK_a>-3$	$-3>pK_a>-5.6$	$pK_a<-5.6$
Al_2O_3	0.8	0	0	3.4	0
Al_2O_3-Na	0	0	0	0.9	0
Mo-Al_2O_3-Na	1.5	0	0	3.6	6.0
Co-Al_2O_3-Na	0.8	0	0	6.2	0
Co-Mo-Al_2O_3-Na	1.0	0	0.8	3.6	7.0
Mo-Co-Al_2O_3-Na	0.7	0	0	3.9	5.0
CoMo-Al_2O_3-Na	2.2	0	0	2.4	6.5
Mo-Al_2O_3	3.4	0	0	5.0	5.0
CoMo-Al_2O_3	1.0	0	5.0	1.6	9.4
Co-Al_2O_3	0.1	0	0	6.6	0
Mo-Co-Al_2O_3	2.7	0	0	5.1	4.5
MoO_3	18.1	0	22.1	20.1	64.2

†1 Number of *n*-butylamine molecules consumed by 1000Å^2 of the surface of the samples; the values are reproducible to about ±0.8.

(Reproduced with permission by P. Ratnasamy *et al.*, *J. Phys. Chem.*, **78**, 2069 (1974)).

effects of CO_2 and NH_3 suggested that active sites of MoO_3 were different from those over MoO_3-SiO_2. The secondary isomerization of olefins produced by dehydration of alcohols was reported to be absent.[19)] But the methyl migration during dehydration tended to occur in some cases, for which an *E*1-like mechanism was proposed based on the olefin distributions. Dehydration of N-ethyl formamide to nitrile was catalyzed by MoO_3-BiPO_4 better than MoO_3 and $BiPO_4$.[20)]

Catalysis of molybdenum oxides mixed with other oxides is described in Section 3.2 in more detail.

D. WO_3

Brønsted as well as Lewis acidity were reported for WO_3-SiO_2, TiO_2, Al_2O_3 by IR of adsorbed pyridine.[14)] Structure of $WO_3-Al_2O_3$ have been studied by ISS.[21)]

Isomerization of butene and dehydration of 2-propanol have been investigated in relation to the acidity of tungsten oxide.[22)] As for the double bond isomerization of 1-butene, the presence of poisoning effect of NH_3 and its absence for H_2O and CO_2 together with the tracer study indicated that the isomerization proceeded via a carbenium ion and that hydrogen atoms of polymerized butene produced on Lewis sites were the catalytically active sites. Tungsten oxides are efficient catalysts for hydration of olefins. Hydrocracking and isomerization of heptane over tungsten oxide reduced to varying degrees were studied.[23)] Presence of acidity was suggestged for WO_3 reduced at about 800 K (WO_2) by the retarding effect of pyridine and alkali. It was reported that hydrocracking over reduced $WO_3-Al_2O_3$ proceeded by a carbenium ion mechanism.[24)] Metathesis and polymerization of olefins proceed probably on lower valent tungsten ions.[25,26)]

REFERENCES

1. N. E. Cross, H. F. Leach, *J. Catal*, **21**, 239 (1971).
2. S. R. Dyne, J. B. Butt, G. L. Haller, *J. Catal*, **25**, 378 (1972); A. Zecchina, S. Coluccia, E. Guglielminotti, G. Ghiotti, *J. Phys. Chem.*, **75**, 2774 (1971).
3. F. E. Massoth, *Advan. Catal. Relat. Subj.*, **27**, 265 (1977); Y. Iwasawa (ed.), *Tailored Metal Catalysts*, D. Reidel, (1985).
4. R. L. Burwell, Jr., G. L. Haller, K. C. Taylor, J. F. Read, *Advan. Catal. Relat. Subj.*, **20**, 1 (1969).
5. Yu. V. Belokopytov, V. A. Kuznetsov, K. M. Kholyavenko, S. V. Gerei, *J. Catal.*, **44**, 1 (1976).
6. A. Zecchina, E. Guglielminotti, L. Cerruti, S. Coluccia, *J. Phys. Chem.*, **76**, 571 (1972).
7. G. L. Haller, J. Saint-Just, Proc. 6th Intern. Congr. Catal., 1976, The Chemical Society, London, p. 235.
8. D. L Myers, J. H. Lunsford, *J. Catal.*, **99**, 140 (1986) and references therein.
9. L. Nondek, D. Mihajlova, A. Andreev, A. Palazov, M. Kraus, D. Shopov, *J. Catal.*, **40**, 46 (1975).
10. K. Yamashita, S. Naito, K. Tamura, *J. Catal.*, **94**, 353 (1985).
11. F. E. Kiviat, L. Petrakis, *J. Phys. Chem.*, **77**, 1232 (1973).
12. Yu. V. Belokopytov, K. M. Kholyavenko, S. V. Gerei, *J. Catal.*, **60**, 1 (1979).
13. T. Fransen, O. van der Meer, P. Mars, *J. Phys. Chem.*, **80**, 2103, (1976).
14. T. Komaya, J. Take, Y. Yoneda, Prepr. 48th Symp. Catal., Catal. Soc. Jpn., 1981, 2W16.
15. E. Alsdorf, W. Hanke, K-H. Schrabel, E. Schreier, *J. Catal.*, **98**, 82 (1986).
16. K. Maruyama, H. Hattori, K. Tanabe, *Bull. Chem. Soc. Jpn.*, **50**, pp.86, 2181 (1977).
17. P. Ratnasamy, D. K. Sharma, L. D. Sharma, *J. Phys. Chem.*, **78**, 2069 (1974).
18. K. Segawa, W. S. Millman, *J. Catal.*, **77**, 221 (1982); **101**, 218 (1986).

19. B. H. Davis, *J. Catal.*, **79**, 58 (1983).
20. P. B. Delmon, *J. Catal.*, **97**, 287, 300, 312 (1986).
21. P. B. Delmon, *J. Catal.*, **100**, 500 (1986).
22. H. Hattori, N. Asada, K. Tanabe, *Bull. Chem. Soc. Jpn.*, **51**, 1704 (1978).
23. Y. Kamiya, E. Ogata, Proc. 5th Intern. Congr. Catal., 1972, pp.93, North-Holland, 1973.
24. E. Ogata, Y. Kamiya, N. Ohta, Kogyo Kagaku Zasshi, **73**, 2582, 2587 (1970) (in Japanese).
25. L. F. Heckelsberg, R. L. Banks, G. C. Bailey, *Ind. Eng. Chem. Prod. Res. Develop.*, **7**, 29 (1968); **14**, 33 (1975).
26. A. Kobayashi, E. Echigoya, *Nippon Kagaku Kaishi*, **1973**, 908, 1057 (in Japanese).

3.1.7 Oxides of Mn, Re

Mn oxides are typical Bertholide compounds. For example, the stoichimetry, x, of MnO_{2-x} varies from 0 to 0.3, and several structures are known. These MnO_{2-x}'s are prepared in various ways, e.g. chemical or electrochemical oxidation, hydrolysis and thermal decomposition of Mn salts.

MnO_{2-x} are generally active catalysts for oxidation reactions and also active electrode materials for dry cells. Probably due to the high oxidation activity, very few studies have been reported for acid-base properties and catalysis of Mn oxides. MnO_2 has been claimed to be efficient for the hydration of nitriles to amides,[1] in which the acidic character of the catalyst seems to play some role. In the case of hydration of acrylonitrile in aqueous solution, a relationship between the equi acid-base point (EABP) of metal oxides and catalytic function has been suggested.[2] MnO_2 (EABP = 3.7) was active and selective for the formation of amide in acidic solution, and MnO (EABP = 11.2) was active in basic solution, but caused more hydration of the C=C bond.

The oxidation states of Re in oxides vary from 2 to 7 and are sensitive to the atmosphere. Among them Re(IV) and Re(VII) are stable. Re_2O_7 is prepared by oxidation of Re metal or low valent oxides by nitric acid followed by evaporation to dryness.

Re_2O_7 supported on metal oxides such as alumina are efficient catalysts for metathesis of olefins, and quite a few studies have been devoted to the subject. The reaction is believed to proceed via a carbene-like intermediate having a direct metal to carbon covalent bond. Hence, the bond is not a typical acid-base interaction and the acidic character of catalysts plays auxiliary roles. A good correlation has been reported[3] between the promoting effect of metal oxide additives on the catalytic activity and the electronegativity of the metal ion of the oxide, M, in the case of metathesis of 2-pentene over $Re_2O_7/Me_xO_y-Al_2O_3$. As for the metathesis of 1,5-cyclopentadiene over Re_2O_7 supported on metal oxides, it has been indicated that the acidic property did not directly control the catalytic activity, although $Re_2O_7-Al_2O_3$ exhibited strong acid sites ($pK_a= -5.6$) in significant amounts.[4]

Isomerization of 1-butene was studied over crystalline ReO_3 and Re_2O_7 which were prepared in several ways, and the presence of two different sites were indicated; one was sensitively promoted by the presence of H_2 and the other (possibly Lewis sites) not.[5]

REFERENCES

1. L. R. Haefele, H. J. Young, *Ind. Eng. Chem., Prod. Res. Rev.*, **11**, 364, (1972); H. Miura, K. Sugiyama, S. Kawakami, T. Aoyama, T. Matsuda, *Chem. Lett.*, **1982**, 183.
2. K. Sugiyama, H. Miura, H. Sekiwa, T. Matsuda, Catal. Symp. Catal. Soc. Jpn., 3S08, Kanazawa, 1985.
3. R. Nakamura, E. Echigoya, Proc. 2nd Intern. Congr. Metathesis, 1977.
4. K. Saito, T. Yamaguchi, K. Tanabe, T. Ogura, M. Yagi, *Bull. Chem. Soc. Jpn.*, **52**, 3192 (1979).
5. T. Yamaguchi, N. Tsuda, K. Tanabe, *Bull. Chem. Soc. Jpn.*, **53**, 539 (1980).

3.1.8 Oxides of Fe, Co, and Ni

A. Fe Oxides

α- and γ-Fe_2O_3, Fe_3O_4 and FeO are ordinary oxides of iron. Iron oxide catalysts are practically used for oxidation or dehydrogenation reactions as single oxides promoted by alkalis or as mixed oxides. Typical examples are dehydrogenation of ethyl benzene (Fe_2O_3-K), water-gas shift reaction ($Fe_2O_3-Cr_2O_3$), ammoxidation of propylene ($Fe_2O_3-Sb_2O_5$), and dehydrogenation of methanol ($Fe_2O_3-MoO_3$).

Usually Fe_2O_3 is regarded to be weakly acidic and basic. It catalyzes the dehydrogenation of ethanol,[1] but only dehydration has been reported for 2-butanol[2] and isopropanol.[3] It has been reported that γ-Fe_2O_3 which had been partially reduced was very active for the isomerization of 1-butene via a cationic allyl intermediate.[4] IR study of adsorption and desorption of NH_3 showed the presence of strong Lewis acid sites, but formation of nitrogen oxides from adsorbed NH_3 was observed at higher temperatures.[5] Interaction of the surface with acetic acid produced acetone.[6]

Tanabe and co-workers[2] have investigated the acid catalysis over several iron oxides prepared by different procedures from different starting materials. They found that Fe_2O_3 (II) prepared by homogeneous precipitation of iron alum ($Fe_2(SO_4)_3 \cdot (NH_4)_2SO_4 \cdot 2H_2O$) using urea was very acidic and active for several reactions. The results are summarized in Table 3.14. Based on IR data of the iron oxide they presumed that strong acid sites were generated by the interaction of the surface with sulfate ion. The acidic iron oxide was also active for hydrogenolysis of diphenyl methane.[7] The acidity of iron oxide containing sulfate ion is described in detail in Section 3.9. They detected no basicity by CO_2 adsorption at 303 K; Fe_2O_3 (II) did not adsorb CO_2, and Fe_2O_3 (III) showed only reversible adsorption. However, the presence of basicity as well as acidity was reported by Ai.[3] In his study, the basic sites determined by irreversible adsorption of CO_2 at 303 K amounted to $1-1.5$ μmole m^{-2} and a similar amount of acid sites was found from irreversible adsorption of NH_3.

The acidic properties of iron ion or small particles of iron oxide dispersed on the surface of SiO_2, MgO, Al_2O_3 and TiO_2 have been studied by means of adsorption of pyridine and Mossbauer spectroscopy.[8] New Lewis acid sites which were associated with coordinatively unsaturated iron cations were generated on SiO_2 upon the addition of Fe. TPD of adsorbed pyridine indicated that the acid strength decreases in the following order: $Fe^{3+} > Fe^{2+} \gg SiO_2$. This order of acid strength was consistent with the IR band shift of adsorbed pyridine. Presence of Brønsted acid sites was suggested

TABLE 3.14 Physical and acidic properties and catalytic activities of iron oxides prepared by different procedures

Catalyst†1	Surface area (m^2g^{-1})	Phase by XRD	SO_4^{2-} content (wt %)	Amounts of NH_3 irreversibly adsorbed at 303 K	Isomerization of 1-butene at 373 K	Dehydration of 2-butanol at 473 K
				mmol g^{-1}	Rate†2	Rate†2
Fe_2O_3 (I)	18.0	α-Fe_2O_3	0	0.033	31	0.4
Fe_2O_3 (II)	53.5	Amorphous	2	0.136	189	40
Fe_2O_3 (III)	13.0	α-Fe_2O_3	0	0.052	9.5	0.2
Fe_2O_3 (IV)	11.3	α-Fe_2O_3	0	0.029	90	0.1
$SiO_2 \cdot Al_2O_3$	350	—	0	0.279	88	11

†1 Preparation procedure: Fe_2O_3 (I) from nitrate and urea, (II) from alum and urea, (III) from nitrate and ammonia, and (IV) from alum and ammonia.
†2 $10^{-1}\mu$mol m^{-2} min^{-1}

for Fe/SiO_2 prepared by coprecipitation,[9] but not in the above case. Acidity was not generated for Fe/MgO, a small amount of acid sites was noted for Fe/Al_2O_3, and increase of acid strength was found for Fe/TiO_2.[10]

B. Co Oxides

CoO and Co_3O_4 are the ordinary oxides of cobalt, the latter being more stable at low temperatures and under high partial pressures of oxygen. Co_3O_4 is one of the most active single metal oxides for oxidation reactions.

Heat treatment of Co_3O_4 in vacuum at 700 – 800 K reveals a high catalytic activity for activation of H_2 at low temperatures[11] and for hydrogenation and isomerization of olefins.[12,13] Poisoning effect of CO indicates that the number of active sites is $1-2 \times 10^{12}$ site cm^{-2}. Treatment with hydrogen brings about similar activation.[14] These sites are plausibly coordinatively unsaturated Co ion and the latter reactions proceed via alkyl intermediate formed from Co-H and olefin, so they may be regarded to be soft Lewis acid sites.

The relationship between the degree of coordinative unsaturation and the catalytic activity has been discussed by Siegel[15] and Tanaka.[16] Dimerization of olefin has been reported for Co_3O_4 supported on carbon.[17]

C. Ni Oxides

Black to gray nickel oxides having excess oxygen, NiO_{1+x}, are known. Nearly stoichiometric NiO, which is prepared by calcination at a high temperature, is green to grayish green. NiO_{1+x} has high catalytic activity for deep oxidation, but when it is evacuated at 700 – 800 K, it shows high activity for H_2-D_2 exchange, hydrogenation and isomerization of olefins.[11-13] Coordinatively unsaturated Ni ion, particularly of lower valence, is likely the active site.

NiO mixed with solid acids such as SiO_2,[18] $SiO_2-Al_2O_3$[19] and $NiSO_4$[20] are active for dimerization of olefins. The active site has been suggested to be a combination of a low valent nickel and an acid site.[18] In the case of $NiO-SiO_2-Al_2O_3$,[19] the activity increases with the acid amount; and the number of Lewis acid sites increases upon the addition of NiO to $SiO_2-Al_2O_3$, while the number of Brønsted sites decreases.[21] Adsorption of CO_2 was studied by IR.[21] Isomerization of butene over $NiO-SiO_2$ proceeds via a butyl cation on a Brønsted site which is induced from a Lewis site and butene.[22]

REFERENCES

1. G. Blyholder, *J. Phys. Chem.*, **66**, 2597 (1962); **68**, 3882 (1964); P. Mars, in: *The Mechanism of Heterogeneous Catalysis* (de Boer, Ed.) Elsevier, 1960.
2. A. Kayo, T. Yamaguchi, K. Tanabe, *J. Catal.*, **83**, 99 (1983).
3. M. Ai, *J. Catal.*, **60**, 306 (1979).
4. M. Misono, Y. Nozawa, Y. Yoneda, Proc. 6th Intern. Congr. Catal., London, 1976, The Chemical Society, London, p.386, 1976.
5. Yu. U. Belokopytov, K. M. Kholyavenko, S. V. Gerei, *J. Catal.*, **60**, 1 (1979).
6. J. C. Kuriacose, S. S. Jewur, *J. Catal.*, **50**, 330 (1977).
7. K. Tanabe, H. Hattori, Y. Yamaguchi, "Studies on Utilization of Coal through Conversion," October 1978, SPEY 16.
8. G. Connell, J. A. Dumesic, *J. Catal.*, **101**, 303 (1986).
9. T. Iizuka, H. Tasumi, K. Tanabe, *Aust. J. Chem.*, **35**, 919 (1982).
10. G. Connell, J. A. Dumesic, *J. Catal.*, **102**, 216 (1986).
11. D. A. Dowden, N. Mackenzie, B. M. W. Trapnell, *Proc. Roy. Soc.*, **A237**, 245 (1956).
12. K. Tanaka, H. Nihira, A. Ozaki, *J. Phys. Chem.*, **74**, 4510 (1970).
13. D. Harrison, D. Nicholls, H. Steiner, *J. Catal.*, **7**, 359 (1967).
14. T. Fukushima, A. Ozaki, *J. Catal.*, **41**, 82 (1976).
15. S. Siegel, *J. Catal.*, **30**, 139 (1973).
16. K. Tanaka, T. Okuhara, *J. Catal.*, **65**, 1 (1980).
17. R. G. Schultz, R. M. Engelbrecht, R. N. Moore, L. T. Wolford, *J. Catal.*, **6**, 385 (1966); **7**, 286 (1967).
18. K. Kimura, H. Ai, A. Ozaki, *J. Catal.*, **18**, 271 (1970).
19. H. Uchida, H. Imai, *Bull. Chem. Soc. Jpn.*, **35**, 995 (1962).
20. K. Maruya, A. Ozaki, *Bull. Chem. Soc. Jpn.*, **46**, 351 (1973).
21. M. Sano, T. Yotsuyanagi, K. Aomura, *Kogyo Kogaku Zasshi*, **74**, 1563 (1971) (in Japanese).
22. A. Ozaki, K. Kimura, *J. Catal.*, **3**, 395 (1964).

3.1.9 Oxides of Cu, Ag, Au

Copper oxides CuO and Cu_2O are semiconductors and effective for redox-type reactions such as oxidation or dehydrogenation. However, in spite of numerous studies on these types of reactions, except for a work by Shibata *et al.*,[1] who determined the acidity of CuO calcined at 773 K to be 0.170 mmol g^{-1} or 0.113 mmol m^{-2} by liquid-phase adsorption of butylamine, no direct studies on the acidic or basic nature of the oxides have been conducted.

Cu_2O is known as an industrial catalyst for the oxidation of propene into acrolein.[2] The reaction is of the first order with respect to oxygen and independent of propene pressure.[3] Thus, the adsorption of oxygen appears to be a rate-determining step and an allylic species formed by the abstraction of a hydrogen atom from a propene molecule by adsorbed oxygen is an intermediate for the oxidation.

The adsorption of oxygen,[4,5] carbon monoxide,[3-6] and propene[7] on Cu_2O and the adsorption of carbon monoxide on CuO[8] have been reported.

Silver is known as a catalyst for partial oxidation to produce ethylene oxide from ethylene and formaldehyde from methanol. Under the reaction conditions, silver oxide is unstable and silver is in a metallic state.

Silver oxide catalyzes the hydration of ethylene with steam in a vapor phase.[9] Over a temperature range of 370−430 K, silver oxide on an alumina carrier gave conversions to glycol ranging from 20%−30%, with selectivity of 80%−90%. This yield is affected by catalyst age, increasing to an approximately constant value of 80% after 5 h of operation.

The oxides of gold are in quasi-stable phases, and metallic gold is the subject of investigation for catalysis. Gold adsorbs hydrogen,[10-12] oxygen,[13-15] carbon monoxide,[16] acetylene.[16] Gold catalyzes the oxidation of ethylene and methanol,[17] oxygen exchange between carbon monoxide and carbon dioxide,[18] and hydrogen exchange between benzene and cyclohexane.[19] Recently, gold supported on Co_3O_4, α-Fe_2O_3 or NiO was reported to be an excellent catalyst for catalytic oxidation of carbon monoxide.[20] The acid-base character of gold oxides has not been reported.

REFERENCES

1. K. Shibata, T. Kiyoura, K. Tanabe, *J. Res. Inst. Catal.*, Hokkaido Univ., **18**, 189 (1970).
2. C. N. Satterfield, in: *Heterogeneous Catalysis in Practice*, McGraw-Hill Book., New York, 1980 p.191.
3. V. M. Belousov, Ya. B. Grakhovskii, M. Ya. Rubanik, *Kinet. Katal.*, **3**, 221 (1962).
4. W. E. Garner, T. J. Gray, F. S. Stone, *Proc. Roy. Soc.*, **A197**, 294 (1974).
5. W. E. Garner, F. S. Stone, P. F. Tiley, *Proc. Roy. Soc.*, **A211**, 472 (1952).
6. D. O. Hayward, B. M. W. Trapnell, in: *Chemisorption*, Butterworth, London 1964, p.269.
7. V. G. Mikhal'chenko, V. D. Sokolovskii, A. A. Filippova, A. A. Dovydov, *Kinet. Katal.*, **14**, 1253 (1973).
8. J. W. London, A. T. Bell, *J. Catal.*, **31**, 32(1973).
9. R. R. Cartmell, J. R. Galloway, R. W. Olson, J. M. Smith, *Ind. Eng. Chem.*, **40**, 390 (1948).
10. R. J. Mikovsky, M. Boudart, H. S. Taylor, *J. Am. Chem. Soc.*, **76**, 3814 (1954).
11. B. J. Wood, H. Wise, *J. Phys. Chem.*, **65**, 1976 (1971).
12. H. Wise, K. M. Sancier, *J. Catal.*, **2**, 149 (1963).
13. W. R. Patterson, C. Kemball, *J. Catal.*, **2**, 465, (1963).
14. N. V. Kul'kova, L. L. Levchenko, *Kinet. Katal.*, **6**, 688 (1965).
15. W. R. MacDonald, K. E. Hays, *J. Catal.*, **18**, 115, (1970).
16. B. M. W. Trapnell, *Proc. Roy. Soc.*, **A218**, 566 (1953).
17. M. D. Thomas, *J. Am. Chem. Soc.*, **42**, 867 (1920).
18. D. Y. Cha, G. Parravano, *J. Catal.*, **18**, 200 (1970).
19. G. Parravano, *J. Catal.*, **18**, 320 (1970).
20. M. Haruta, T. Kobayashi, H. Sano, N. Yamada, *Chem. Lett.*, **1987**, 405.

3.1.10 ZnO, CdO

A. ZnO

ZnO is usually of wurtzite structure, Zn being coordinated with four oxide ions. At a very high pressure ($>10^8$Pa) it is transformed to a rocksalt structure. Under low partial pressure of oxygen, oxygen is evolved and Zn(I) ion goes into interstitial position (Zn_i, eq. 1). $Zn_{1+x}O$ thus formed become n-type semiconductor (band gap: 3.2 eV),[1] and photoconduction as well as photocatalysis is observed.

$$ZnO \longrightarrow Zn_i + 1/2O_2 \qquad (1)$$

The surface area varies from 1 to 15 m^2g^{-1} depending on the preparation method and starting materials.

Dihydrogen is adsorbed in two types besides molecular adsorption. One is dissociative and reversible adsorption (eq. 2). The adsorbed hydrogen has been directly observed in IR at H_2; 4050 cm^{-1}, Zn–H; 1705 cm^{-1} and O–H; 3490 cm^{-1}.[2]

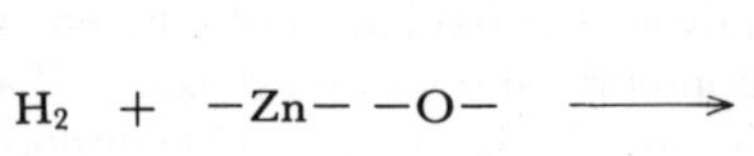
(2)

The other is slow and irreversible adsorption which is IR-inactive and also inactive for hydrogenation.[2] Adsorption of CO_2 to form carbonate has been indicated by IR.[3] On the other hand, it has been reported that CO is not adsorbed on ZnO at room temperature, but is at 77 K.[4] The heats of adsorption were calculated from isotherms to be 60–120 kJ mol^{-1} for CO_2 and 60–80 kJ mol^{-1} for NH_3.[5]

ZnO is considered to be amphoteric[6] and both basicity and acidity have been experimentally shown. The IR spectrum of NH_3 adsorbed indicated the presence of Lewis acidity.[7] The basicity on the basis of the IR study on the adsorption of Brønsted acids (hydrocarbons, alcohols, and ammonia) has been reported.[8] Those acids with pK_a's greater than 36 did not dissociate. pK_a of propene is 35 and that for ammonia 36.[9] Therefore, it was concluded that the surface of ZnO posseses a basicity comparble with the conjugate base anions of Brønsted acids with $pK_a = 36$. The dissociative adsorption of Brønsted acids with pK_a less than 19 was later reconfirmed.[10]

ZnO produces H_2 and CO_2 from formic acid. This selectivity indicates that ZnO is a solid base. It has been demonstrated by IR that the decomposition of formic acid proceeds via a formate species as shown below.[11]

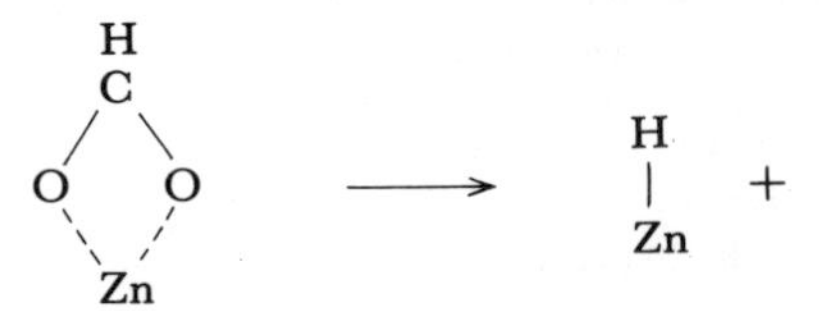
(3)

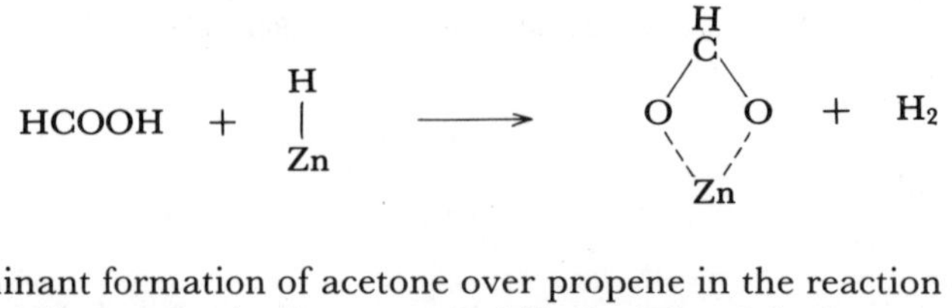
(4)

The dominant formation of acetone over propene in the reaction of isopropyl alcohol[12] also indicates the basic nature of ZnO. The dehydrogenation of isopropyl alcohol[13] and the decomposition of methanol[14] over ZnO have also been investigated by IR. According to this study, the former reaction proceeds as shown in eq. (5), the rate-determining step being the second step of the dehydrogenation of a surface alkoxide (dissociation of β-C−H) to form an enol-type adsorbate and a dihydrogen molecule.[13)]

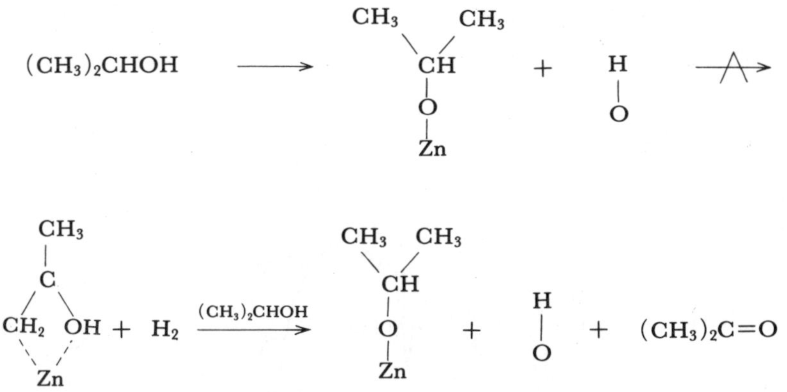

(5)

It has been reported that alkoxide species are formed from alcohols at the surface density of $1-2 \times 10^{14}$ molecules cm^{-2}.[15)] Upon thermal desorption those alkoxides from C_2-C_4 alcohols decomposed to aldehydes (or ketones) and olefins at 480 − 550 K.[15)] The selectivities to olefins were 0.2 to 0.4 except for ethanol for which the ratio was 0.9. Interactions and thermal desorption of several molecules for different surfaces of a ZnO single crystal have been studied in ultra-high vacuum.[16,17)] Fig. 3.30 shows the crystal planes examined. (Electronic properties and surfaces geometry of ZnO crystal have been briefly reviewed.[18)]) The strength of the interaction of oxygen-containing

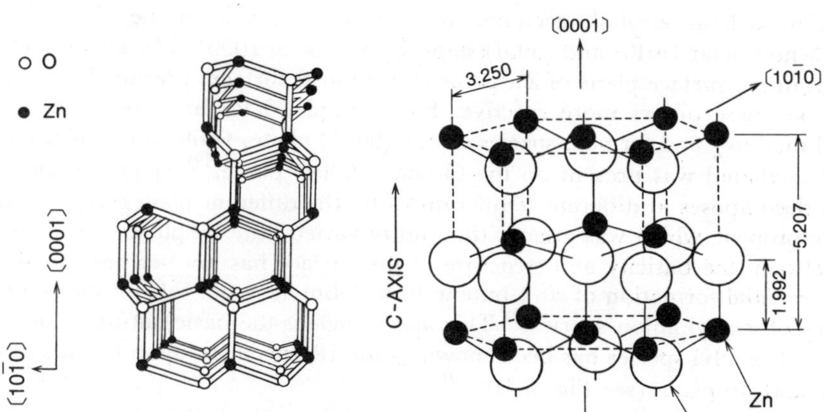

Fig. 3.30 Schematic representation of the different ZnO surfaces.
(A) Stepped ZnO (vertical left-side plane),
(B) Surfaces without steps, e.g. Zn-polar (0001) and O-polar (000$\bar{1}$) surfaces.

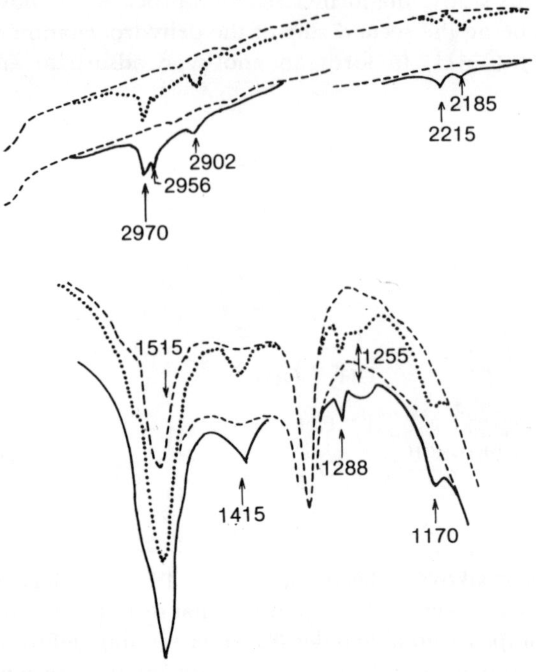

Fig. 3.31 Spectrum of chemisorbed propylene ($CD_3-CH=CH_2$ and $CH_3-CH=CD_2$) : *dottedline* ; chemisorbed $CD_3-CH=CH_2$ on zinc oxide, *solid line*; chemisorbed $CH_3-CH=CD_2$ on zinc oxide. A band at 1415 cm^{-1} is assigned to ν (C-C-C)of π-allyl.

molecules such as alcohol, acetone, and formic acid was in the order: Zn-polar (0001) > non-polar (10$\bar{1}$0) and (50$\bar{5}$1)(stepped) > O-polar (000$\bar{1}$). Zn ions directed outward from the surface plane of Zn-polar and stepped surfaces presumably act as acid and make these planes more reactive. For example, XPS indicated that methanol formed methoxy and formate species on the (0001) plane, while only molecularly adsorbed methanol was present on the O-polar (000$\bar{1}$) plane.[17] Isopropyl alcohol adsorbed decomposes at different temperatures by the different planes, but the ratio of acetone/propene which was greater than unity varied little by plane. So the relationship between the basicity and structure of the surface has not yet been made clear.

Preferential formation of *cis*-2-butene from 1-butene (*cis*/1 ≒ 10) implies the intermediacy of π-allyl anion species.[2] This also indicates the basic nature of the surface of ZnO. The allyl species has been shown in the IR spectrum upon the adsorption of butene and propene. (see Fig. 3.31)[2,19]

Acid strength distribution has been measured by Tanabe and coworkers by means of titration with Hammett indicators as shown in Table 3.15.[20] It can be seen that ZnO is very weak acid. As for the basicity, it was reported that the base amount was 0.05 mmole g^{-1} for $H_0 \geq 4.0$ in the scale of the acid strength of conjugate acid[2] (see Chapter 2). It was also reported that the acid-base properties are sensitive to pretreatment or environment.[22]

TABLE 3.15 Acid strength distribution of ZnO

Calcination temp/K	Acid amount/mmole g^{-1}					
	$H_0 \leq 6.0$	$H_0 \leq 4.8$	$H_0 \leq 4.0$	$H_0 \leq 3.3$	$H_0 \leq 1.5$	$H_0 \leq -3$
473	0.37	0.29	0.005	0		
573	2.07	1.71	—	0.15	0	
673	2.29	1.81	—	0.52	0.22	0
773	0.024	0.018	—	0.017	0	

(Reproduced with permission from *et al., Bull. Chem. Soc. Jpn.*, **45**, 49(1972)).

The acidity and basicity increase when ZnO is mixed with other oxides. $ZnO-TiO_2$ becomes more acidic and reveals high activity for hydration of ethylene.[20] Both acid and base amounts increase upon mixing with SiO_2 and isomerization of butene is catalyzed by the acid sites of $ZnO-SiO_2$[23] Addition of K to ZnO increased the base amount, and the catalytic activity for oxidation of formic acid sharply increased.[24] A mixed oxide catalyst consisting of ZnO and Fe_2O_3 is a good basic catalyst for the methylation by methanol of phenol to 2,6-xylenol.[25]

B. CdO

CdO has a hexagonal structure and it sublimes above about 1000 K. It is an n-type semiconductor with interstitial Cd as in the case of ZnO. Reaction of H_2O_2 with $Cd(NH_3)_4^{2+}$ forms CdO_2, which decomposes to CdO and O_2 at 400 – 500 K.

Isomerization between *cis*-2-butene and 1-butene preferentially takes place over isomerization between *cis* and *trans* and between *trans* and 1-butene (*cis/trans* = 20 from 1-butene and 1/*trans* = 10 from *cis*-2-butene).[26] High selectivity to *cis*-2-butene and 1-butene in the hydrogenation of butadiene has also been observed.[26] All of these reactions proceed via a π-allyl intermediate and these facts indicate that the surface of CdO is basic (see Section 4.1). However, no direct measurement of the basicity has been reported.

REFERENCES

1. W. D. Kingergy, H. K. Bowen, D. R. Uhlmann, in: *Introduction to Ceramics*, 2nd Ed., Chapter 17, John Wiley & Sons, New York, 1976.
2. R. J. Kokes, A. L. Dent, *Advan. Catal. Relat. Subj.*, **22**, 1 (1972).
3. D. G. Rethwish, J. A. Dumesic, *Langmuir*, **2**, 73 (1986).
4. K. Klier, *Advan. Catal.*, **31**, 243 (1982); A. A. Tsyganenko, L. A. Denisenko, S. M. Zverev, V. N. Filimonov, *J. Catal.*, **99**, 10 (1985).
5. I. Yasumoto, *J. Phys. Chem.*, **88**, 4041 (1984).
6. K. Tanabe, *Solid Acids and Bases*, Kodansha, Tokyo and Academic, New York, 1970.
7. M. C. Kung, H. H. Kung, *Catal. Rev. Sci. Eng.*, **27**, 425 (1985).
8. R. J. Kokes, *Intra-Science Chem. Rep.*, **6**, 77 (1972); cited in Ref. (7).
9. D. J. Cram, *Fundamentals of Carbanion Chemistry*, Academic Press, New York, 1965.
10. R. N. Spitz, J. E. Barton, M. A. Barteau, R. H. Staley, A. W. Sleight, *J. Phys. Chem.*, **90**, 4067 (1986).
11. Y. Noto, K. Fukuda, T. Onishi, K. Tamaru, *Trans. Faraday Soc.*, **63**, 3081 (1967).

12. O. Krylov, *Catalysis by Nonmetals*, Academic Press, New York, 1970.
13. O. Koga, T. Onishi, K. Tamaru, *J. Chem. Soc. Faraday I*, **76**, 19 (1980); E. Akiba, M. Soma, T. Onishi, K. Tamaru, *Z. Phys. Chem. N. F.*, **119**, 103 (1980).
14. A. Ueno, T. Onishi, K. Tamaru, *Trans. Faraday Soc.*, **67**, 3585 (1971).
15. M. Bowker, R. W. Petts, K. C. Waugh, *J. Catal.*, **99**, 53 (1986).
16. W. H. Cheng, H. H. Kung, *Surf. Sci.*, **122**, 21 (1982); S. Akhter, W. H. Cheng, K. Lui, H. H. Kung., *J. Catal.*, **85**, 437 (1984); K. Lui, S. Akhter, H. H. Kung, in: ACS Symp. Series 279, *Solid State Chemistry in Catalysis* (R. K. Grasselli, J. F. Brazdil, Eds.), p.205 (1985).
17. J. M. Vohs, M. A. Barteau, *Surf. Sci.*, **176**, 91 (1986); *J. Phys Chem.*, **91**, 4766 (1987); *Surf. Sci.*, **197**, 109 (1988).
18. K. Jacobi, *Surf. Sci.*, **132**, 1 (1983).
19. T. Nagashima, H. Miyata, Y. Kubokawa, *J. Chem. Soc. Faraday Trans I*, **81**, 2409 (1985).
20. K. Tanabe, C. Ishiya, I. Matsuzaki, I. Ichikawa, H. Hattori, *Bull. Chem. Soc. Jpn.*, **45**, 47 (1972).
21. K. Tanabe, in: *Catalysis, Science and Technology* (J. R. Anderson, M. Boudart, Eds.) Vol. 2, Springer-Verlag, New York, 1981, p.248.
22. H. Vinek, J. Lercher, H. Noller, *React. Kinet. Catal. Lett.*, **15**, 21 (1980).
23. T. Sumiyoshi, K. Tanabe, H. Hattori, *Bull. Jap. Petrol. Inst.*, **17**, 65 (1975).
24. M. Ai, *J. Catal.*, **50**, 291 (1977).
25. T. Kotanigawa, M. Yamamoto, K. Shimokawa, Y. Yoshida, *Bull Chem. Soc. Jpn.*, **44**, 1961 (1971).
26. T. Okuhara, K. Tanaka, *J. Catal.*, **61**, 135 (1980).

3.1.11 Oxides of B, Al, Ga

A. B_2O_3

B_2O_3 is usually used as a catalyst by dispersing it on support materials such as Al_2O_3, SiO_2 and TiO_2. It has been reported that $B_2O_3-Al_2O_3$ exhibited very strong acidity ($H_0 \leq -8.2$) at low levels of loading due to the specific reaction between surface OH groups and boric acid.[1] At high loading levels, moderately strong acid sites prevailed, strong acid sites being covered by B_2O_3.

The catalytic performance for Beckmann rearrangement of cyclohexanone oxime to ε-caprolactum was compared between $B_2O_3-SiO_2$ prepared by impregnation and by chemical vapor decompositon.[2,3] The latter resulted in uniform dispersion of B_2O_3 and showed much improved selectivity (95 mol%), as shown in Fig. 3.32. Over $B_2O_3-Al_2O_3$ prepared by impregnation, the selectivity was lower.

B. Aluminum Oxide

Aluminum oxide is widely used as adsorbents and catalysts. As industrial catalysts, it is mostly used as catalyst supports for metals (Pt, Pd, etc.) and metal oxides (Cr, Mo, etc.).

a. Structure, preparation

Aluminum oxide (alumina or Al_2O_3) occurs in various crystallographic modifications. Completely anhydrous aluminum oxide is α-Al_2O_3 (corrundum), which is the most stable. It has a structure of hexagonal closest packing of oxide ion, with Al ion occupying two-thirds of the octahedral sites. α-Alumina is prepared by heat treatment above *ca.* 1470 K of aluminum hydroxides, oxyhydroxides, or alumina hydrates, which are obtained by neutralization of solution of aluminum salts or hydrolysis of aluminum alkoxides. Heat treatments at lower temperatures form various kinds of transi-

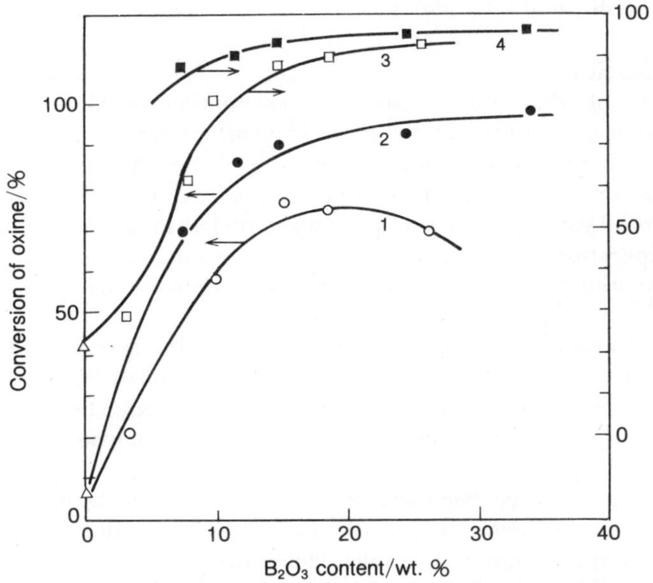

Fig. 3.32 Effect of B_2O_3 content. (1) Oxime conversion of impregnation B_2O_3-SiO_2, (2) oxime conversion of vapor decomposition B_2O_3-SiO_2, (3) lactam selectivity of impregnation B_2O_3-SiO_2, (4) lactam selectivity of vapor decomposition B_2O_3-SiO_2; reaction temperture, 423 K.
(Reproduced with permission by S. Sato *et al.*, *J. Catal.*, **102**, 99 (1986)).

tion alumina such as γ, η, χ, θ, δ, κ, etc., depending on the precursors and conditions of heat treatments (Fig. 3.33).[4] The additives and surface area influence the transformation processes.[5] Most of these aluminas contain water, proton and/or alkali in their structure. Among these transition aluminas, γ- and η-alumina are most important as catalysts. These two have defect spinel structures.[4-7] The differences between the two are degree of tetragonal distortion of crystal structure ($\gamma > \eta$), regularity of stacking of hexagonal layers ($\gamma > \eta$), and Al-O bond distance ($\eta > \gamma$; difference being 0.05 − 0.1 nm). The surface of the small particles of these spinel-type alumina was reported to

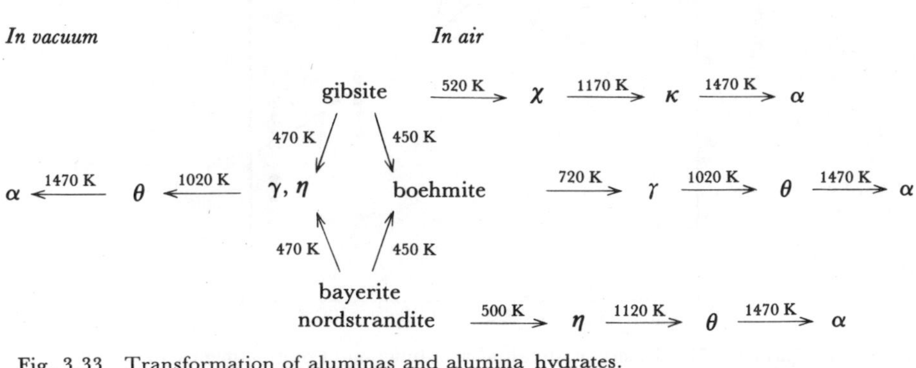

Fig. 3.33 Transformation of aluminas and alumina hydrates.
(Reproduced with permission by T. Foger, *Catalysis Science and Technology*, **6**, 231 (1984)).

consist of (100), (110), and (111) panes.[8)]

b. Surface properties

Surface areas of aluminas obtained by calcination of alumina hydrates at 550 − 1100 K are usually 100 − 300 m^2 g^{-1}. The surface area of γ- and η-alumina is 150 − 250 m^2 g^{-1} and that of α-alumina a few m^2 g^{-1}. For the former average micropore is commonly 1 − 10 nm and pore volume 0.4 − 0.7 cm^3 g^{-1}. Pore size distribution can be controlled by varying the size of primary particles, for example, by a pH swing method.[9)] Incorporation of combustible small particles such as carbon or cellulose and subsequent calcination in air produces alumina with two maxima in pore-size distribution (bimodal).[10)]

Activity for adsorption and catalytic function of alumina is revealed by partial dehydroxylation of its surface. The variation of the density of surface hydroxyl groups is shown inf Fig. 3.34.[6)] The isoelectric point has been reported to be about 7.[11)] Proton exchange was hardly observed in the reaction of alumina in aqueous solution of sodium acetate.[12)]

Adsorption properties have been widely studied. Water adsorbs physically (desorbing at 373 − 393 K), and chemically (*ca.* 10^{14} molecules cm^{-2}, desorbing at about 573 K) and by surface hydroxylation. Complete dehydroxylation requires heat treatment at about 1300 K. NH_3 molecules adsorb very strongly and extensively, the strength

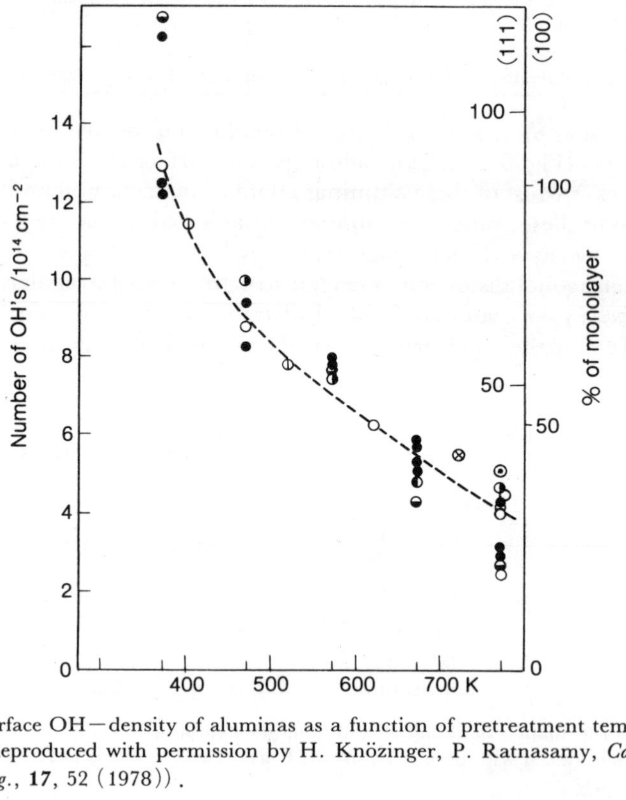

Fig. 3.34 Surface OH−density of aluminas as a function of pretreatment temperature. (Reproduced with permission by H. Knözinger, P. Ratnasamy, *Catal. Rev. Sci. Eng.*, **17**, 52 (1978)).

and the amount being comparable with silica-alumina.[13] Most of NH_3 molecules adsorb by coordination to Al ion on the surface; a small part of them dissociates and forms NH_2-Al and $H-O$.[14,15] Pyridine also absorbs by coordination to Al ion. These are confirmed by IR and demonstrate the Lewis acidity of the alumina surface as described later. Two different kinds of adsorption of olefins are shown by temperature programmed desorption (TPD).[16,17] Electron transfer to the surface occurs in the adsorption of polyacene molecules.[18] Two different kinds of adsorption[19a] as well as lateral interaction between adsorbed CO[19b] were also indicated for CO adsorption. One is weak and reversible, giving an IR band at 2203 − 2215 cm^{-1}. The other is is strong adsorption with a band at 2244 cm^{-1} and perturbs a basic OH band at 3786 cm^{-1}. The number of adsorption sites are estimated to be *ca.* 2×10^{13} cm^{-2}. TPD of H_2 detected five different states of adsorbed H_2.[20] Recently, H_2 and CH_4 molecules which adsorbed in a polarized form on η-Al_2O_3 were detected by IR at low temperature.[21] As for the adsorption of CO_2, the following six types have been indicated[14] (The wave numbers characteristic of the species are also given in parentheses).

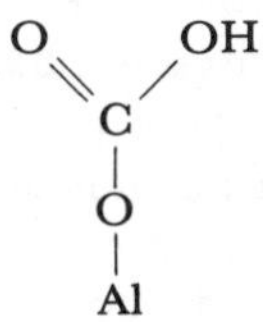

bicarbonate
(1290−1410 and 1620−1660 cm^{-1})

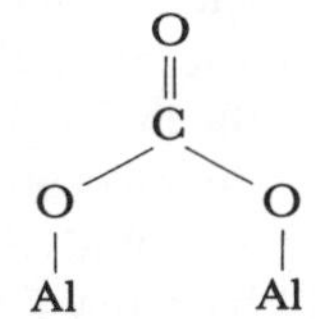

"organic" bridging type
(1750−1870 and 1150−1280)

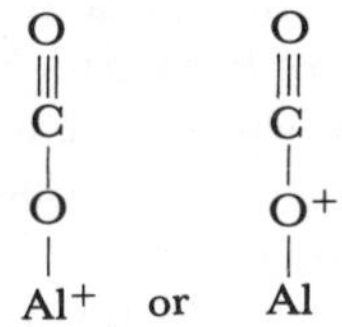

(1820 and 1780)

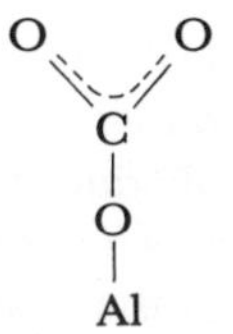

unidentate carbonate
(1300−1370 and 1470−1530)

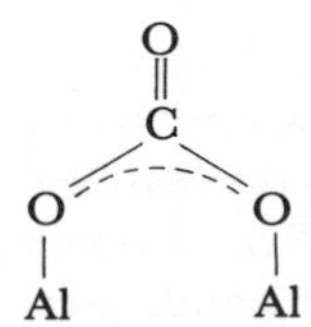

bidentate carbonate
(1590−1630 and 1260−1270)

CO_3^{2-}

carbonate
(1020−1090 and 1420−1470)

Adsorption of metal ions and oxo ions is an important process for the preparation of supported catalysts. Adsorption of Mo ion is reported to be as follows.[22]

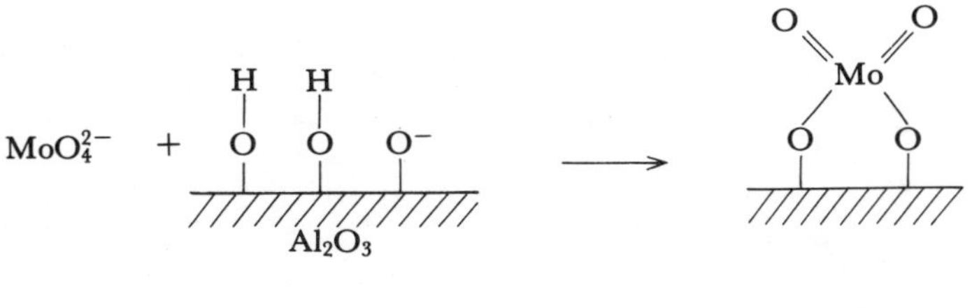

(pH>8) (1a)

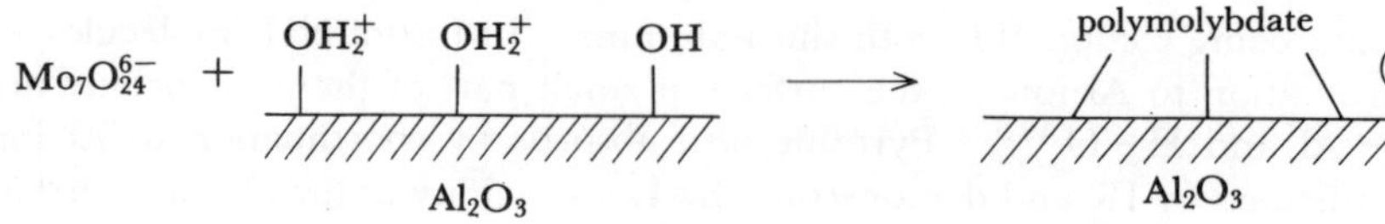

(pH<6) (1b)

Similarly, reactions of organometallic molecules proceed as shown for example below.[23)]

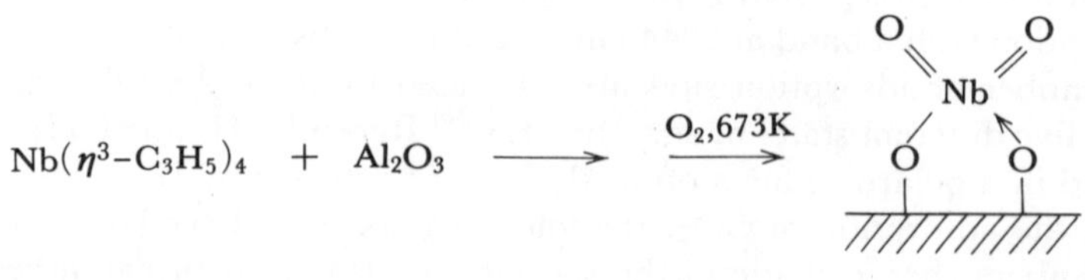

(2)

c. Acidic and basic properties

Surface of aluminas activated by heat treatment above 670 K, usually γ- and η-alumina, posseses both acid sites and basic sites. Their presence has been demonstrated by strong adsorption of basic and acidic molecules or by the poisoning effect of those molecules on various reactions.[14,24)] The coloration of an indicator also showed the presence of strong acid sites,[25)] but it was not confirmed in some cases whether the color change was really due to an acid-base reaction or not. For example, the adsorption of *p*-nitroaniline (PNA) did not show the UV band which is due to protonated species (245 nm), but a band at 450 nm assignable to PNA adsorbed on Lewis acid sites.[26)]

In the adsorption of pyridine, most investigations agree that IR did not detect any protonated pyridine (BPy-band), but detected pyridine coordinated to a Lewis acid site (LPy-band), and weakly hydrogen-bonded pyridine (HPy-band).[14)] Therefore, at least strong Brønsted sites are absent on the surface of alumina. Brønsted sites indicated by NMR and IR of adsorbed pyridines[27,28)] are probably due to pyridine adsorbed on very weak acid sites or hydrogen-bonded pyridine. The same conclusion has been obtained by the absence or IR bands ascribable to NH_4^+ in the adsorption of NH_3.[14)] The wave numbers of LPy-bands (19b: 1447 – 1464 and 8a: 1600 – 1634 cm^{-1}) tend to increase with the increase of the acid strength of Lewis sites.[14)] 4-Methyl and 2,6-dimethyl pyridine adsorb more weakly than pyridine, although they are stronger bases than pyridine. This indicates the presence of steric hindrance of methyl groups adjacent to N atom in the adsorption of methylpyridines.[14,28)] On γ- and η-Al_2O_3, the following reaction has also been reported (eq. (3)). Presence of Lewis acidity was demonstrated also for α-Al_2O_3 by the pyridine adsorption.[29)]

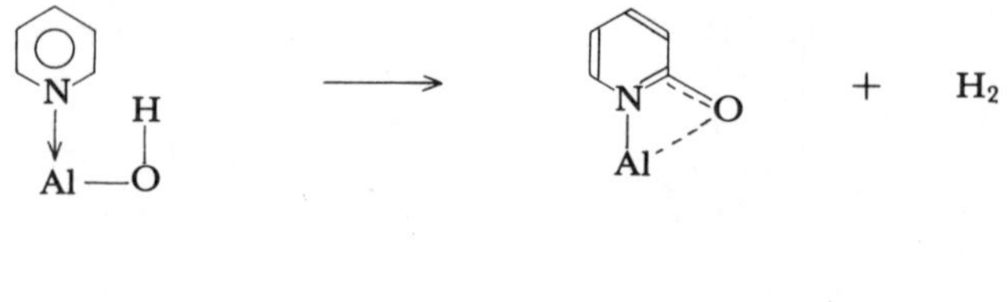

(3)

Table 3.16 shows the acid strength distribution of the Lewis sites on the surface of alumina as determined by stepwise TPD of pyridine combined with IR.[30] After pyridine was adsorbed at 383 – 423 K, the sample was evacuated by increasing the temperature stepwise from 383 to 637 K. The intensity of the LPy band was measured after each evacuation step. The results showed that the amount of strong acid sites are comparable or greater than the amount of Brønsted + Lewis sites of silica-alumina. The same conclusion was obtained by calorimetric titration using NH_3 adsorption. The amount of acid sites thus measured (heat of NH_3 adsorption > 70 kJ mol^{-1}) was 0.69 mmol g^{-1}.[13]

TABLE 3.16 Acid strength distribution of several aluminas

Alumina†1	Acid amount/μ mol g^{-1} †2			
	Acid strength (T_d/K)†3			
	>383	>473	>573	>673
ALO-1	215	116	47	26
ALO-2	276	156	60	34
ALO-3	119	42	9	2
ALO-4	188	106	42	24
ALO-5	289	125	45	21

†1 ALO-1~5 were aluminas of the reference catalysts of Catalysis Society of Japan.
†2 Amount of pyridine which remained after evacuation at the temperature indicated, T_d.
†3 Acid strength represented by evacuation temperature.

The presence of basic OH groups (3800 cm^{-1}) has been shown by the formation of bicarbonate ion upon the adsorption of CO_2 (eq. 4).[31] Adsorption of $Mn(CH_3)(CO)_5$ also indicates the presence of basic sites.[41] The IR band obtained for the above species was very similar to the IR spectrum observed for the molecule coordinated to $AlBr_3$ (compare A and B of eq. 5).

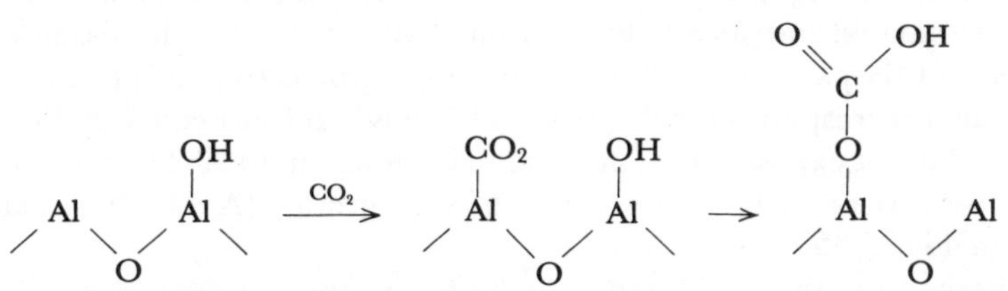

(4)

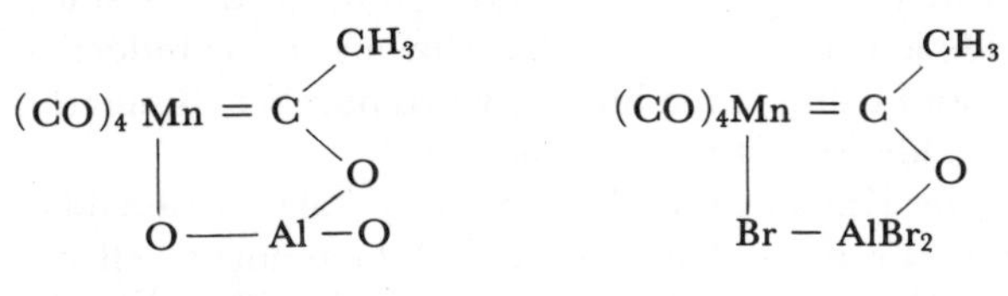

(5)

A B

The base strength distribution measured by the TPD of bicarbonate species by use of IR (1640 − 1645 and 1238 − 1241 cm^{-1}) is shown in Table 3.17.[32] Acid and base strength distributions have been measured by titration with indicators as well.[25]

Acid strength increases and Brønsted acidity is revealed when alumina is treated with halogen-containing molecules such as hydrogen halides. It was reported for HF − Al_2O_3 catalysts that the amount of strongly adsrobed NH_3 increased with the HF content.[24,33] Acidity modification by acid treatment has also been attempted.[34] According to this study, the amount of strong acid sites ($H_0 \leq -8.2$) was 0.46 mmol g^{-1} and basic sites ($H_- > 26.5$) 0.51 mmol g^{-1}.

TABLE 3.17 Strength distribution of basic OH sites

Strength region $T_i < T_d{}^\dagger < T_j/K$		Amounts of basic OH sites/μ mol g^{-1}				
		ALO-1	ALO-2	ALO-3	ALO-4	ALO-5
	≤293	93	50	12	79	66
293	353	58	29	45	44	42
353	413	34	31	30	36	40
413	473	25	12	11	22	9
473<		0	0	0	4	0
Total		210	122	98	185	157

† Temperature of desorption

d. Surface models of alumina

Surface activity of alumina is revealed by partial dehydroxylation of the surface. For the partially dehydroxylated surface of spinel-type alumina, two models have been proposed.

Peri[35] proposed a model for γ-alumina. This model has explained well the five IR bands for stretching modes of surface OH groups and the variation of their intensities with heat treatment. He assumed a fully hydroxylated (100) plane of γ-alumina with Al ions in octahedral sites being located below the hydroxylated surface plane. The dehydroxylation was simulated statistically by a Monte Carlo method, with assumption of random removal of OH pairs; initially without creating defective sites (up to 67% dehydroxylation which corresponds the degassing at 770 K) and subsequently forming defects comprising adjacent exposed Al atoms and oxide ions (up to 90.4% dehydroxylation at 940 K). By this procedure, five kinds of OH groups (A − E sites) can be formed as shown in Fig. 3.35.

These sites correspond to five OH bands which appear upon degassing of γ-Al_2O_3 at 873 − 973 K as indicated in Fig. 3.35. The OH group of the A site in which the number of adjacent oxide ions is four is most basic owing to the inductive effect from the adjacent oxide ions and exhibits the highest wave number. Similarly, the C site, which has no adjacent oxide ion, is most acidic.

The model proposed by Knözinger and Ratnasamy[6] also reasonably explains the IR bands of the surface OH groups. They considered the inductive effect from the adjacent oxide ions on the OH groups, which was assumed by Peri,[35] to have a minor influence, because the effect has to be transmitted through four bonds. It was also con-

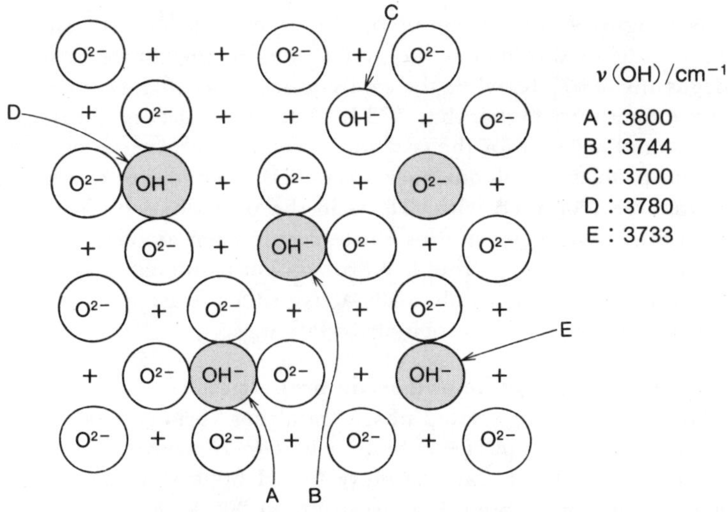

Fig. 3.35 Suggested scheme for acidic and basic sites on γ-alumina.
Letters A to E identify the different types of isolated hydroxyl ions, and "+" denotes an Al^{3+} ion on the layer below the surface
(Reproduced with permission by J. Peri, , *J. Phys. Chem.*, **69**, 223 (1965)).

sidered that the surface of alumina is not only the (100) plane. According to this model, the IR frequency of the surface OH groups is predominantly influenced by their net charge, which is determined by the configuration or coordination of the surface OH groups. The net charge can be calculated by the Pauling's electrostatic valence rule.[36)] The surface dehydroxylation was assumed to take place so as to minimize the net charge during the dehydroxylation; for example, water is formed from acidic proton and basic OH. Examples of proposed models as well as the net charges and assigned OH frequencies are shown below (Td: tetrahedral, Oh: octahedral).

The number of sites produced by the ''regular'' dehydroxylation described above is of the order of 10^{14} sites cm^{-2}, while the number of catalytically active sites estimat-

terminal OH coordinated to Td Al	OH bridging Td and Oh Al	OH coordinated to Al in Oh site	OH coordinated to 3 Oh Al	OH liking two Oh Al

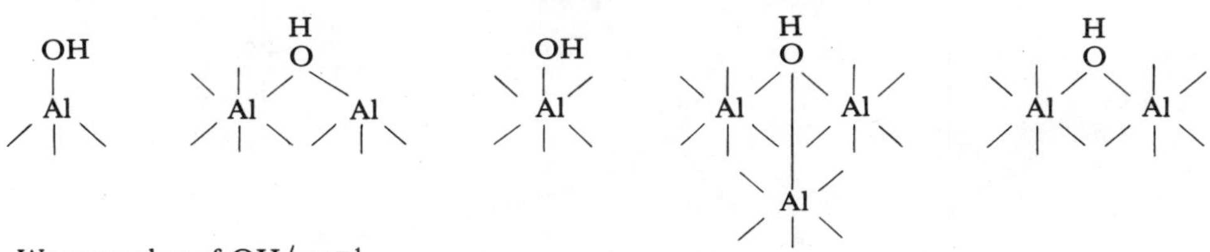

Wave number of OH/cm^{-1}

3760–3780	3730–3735	3785–3800	3700–3710	3740–3745
	Net charge at OH			
−0.25	+0.25	−0.5	+0.5	0
(Ia)	(IIa)	(Ib)	(III)	(IIb)

ed by the poisoning experiments is only on the order of $10^{12}-10^{13}$ sites cm^{-2}. Furthermore, according to this model the active sites of the above order are already formed by degassing at 573 K, while the catalytic activities often show maxima by heat treatment at a temperature higher than 573 K. Therefore, the regular dehydroxylation by this model cannot account for the creation of catalytically active sites. Peri's model also has a similar problem. So Knözinger ascribed the active sites to a special configuration which may be identified with defects in the partially dehydroxylated surface, that is, with multiple vacancies and/or clusters of oxygen atoms in a certain special environment. It is probable that the dehydroxylation assumed as above proceeds on the surface of alumina, but the catalytically active sites are ascribed to highly strained configurations produced by severe dehydroxylation.

e. Catalysis in relation to acidic and basic properties

Alumina, usually γ-, η- or related phase, catalyzes various reactions. In general, activation (bond dissociation) of H–H bond, e.g., *o*-/*p*-H_2 conversion and H_2-D_2 exchange, proceeds at 78–150 K, activation of C–H bond (for example, CH_4-CD_4 isotopic exchange and double bond isomerization) at 300 K, activation of C–C bond (skeletal isomerization of hydrocarbons) at 600 K and of C–O bond (dehydration of alcohol) at about 400 K.[6] It is remarkable that the surface of Al_2O_3 is capable of dissociating the C-H bond of alkanes at a low temperature.[37] The dissociation is likely to be $RH \rightarrow R^- + H^+$, since the rate decreases with the increase in the pK_a value of alkanes.[38]

Figure 3.36 shows the catalytic activity (normalized to surface area) of Al_2O_3 with

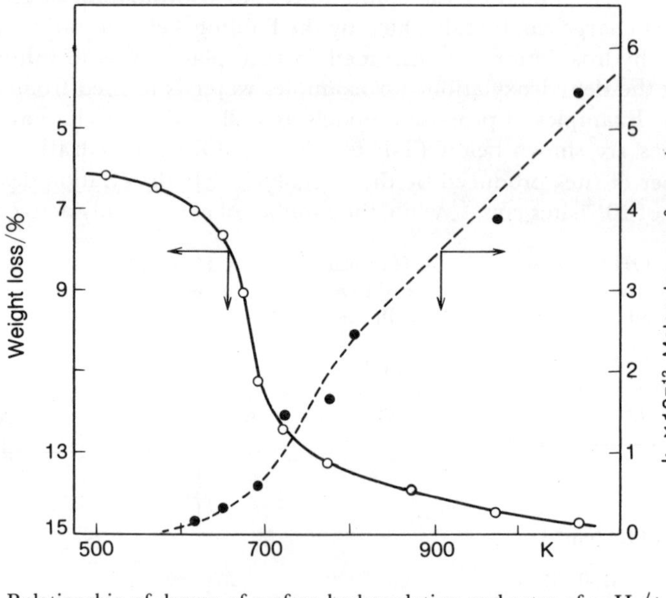

Fig. 3.36 Relationship of degree of surface hydroxylation and rates of *o*-H_2/*p*-H_2 conversion at 78K with pretreatment temperature of alumina.
(Reproduced with permission by F. H. van Cauwelaert, W, K, Hall, *Trans, Faraday Soc.*, **66**, 454 (1970)).

heat treatment in the case of *o*-/*p*-H_2 conversion.[39] Since the surface area decreases by treatment at high temperatures, the rate per unit weight shows a maximum about 750 K. Dehydrohalogenation, hydration, deamination, aldol condensation, etc. are among the reactions catalyzed by alumina. Various organic reactions which proceed selectively on the surface of alumina are known. Similarly, reagents doped on alumina exhibit high stereoselectivity.[40] For example,

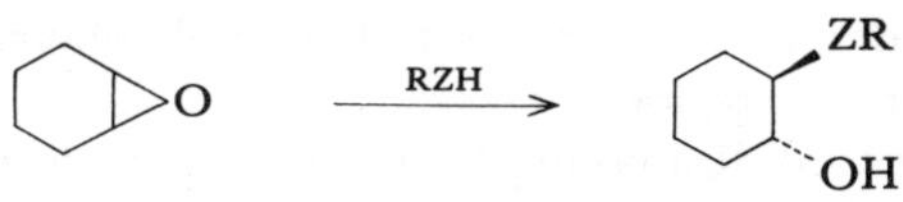

(6)

Catalytically active sites apparently differ considerably in structure and nature between the reactions catalyzed. For example, dissociation of H_2 and CH_4 seems to be catalyzed by highly strained sites with high gradient of electrostatic field; these are produced by severe dehydroxylation as described in the previous section. The carbanionic nature of the intermediates for the dissociation of C – H bond of alkanes[38] indicates that the active sites have basic character. On the other hand, strong Lewis acid plays an important role in skeletal isomerization of hydrocarbons, and impurity contained in alumina such as alkali strongly affects the activity.[41,42] In some reactions both acidic and basic sites take part concertedly. If one considers the high stability of C – H bonds of alkanes, it may be more reasonable to assume that the dissociation mentioned above is actually made possible by the concerted action of the basic and acidic sites of Al_2O_3.

Thus, various kinds of sites which are potentially catalytically active exist on the surface of alumina, and some reactions are catalyzed by typical acid and/or base and some by very special sites formed by severe dehydroxylation, though the latter may also be regarded in a sense to be acid and/or basic sites. The number of active sites has been estimated to be 10^{13} cm^{-2} for dehydration of 1-butanol, 8×10^{12} cm^{-2} for isomerization of cyclohexane[43] and 5×10^{12} cm^{-2} for $CH_4 - CD_4$ isotopic exchange.[37] The numbers are comparable to or slightly larger than the amount of chemisorption of H_2 or CO, but much smaller (less than one-tenth) than the number of surface Al ions or OH groups (*ca.* 10^{14} sites cm^{-2}).

Poisoning effects on catalytic reactions have been frequently attempted to reveal the nature of the active sites. The results differ depending on the reactions and reaction conditions, and sometimes between investigators.[14] However, the following may be stated in general. Both acidic and basic poisons exhibit retarding effects in dehydration of alcohols.[14] This poisoning experiment as well as the presence of deuterium isotope effect[44] and stereospecific anti elimination[41,45] led to the conclusion that dehydration of secondary alcohols proceeds concertedly on pair sites of acid and base. As for the acid sites in this mechanism, some consider Lewis acid sites and others assume surface OH sites.[14,46]

$CH_4 - D_2$ exchange reaction has been studied over several aluminas prepared by different methods,[47] and little correlation was found between the acid or base properties measured by titration and the catalytic activity, although NH_3 retarded the reaction. An $Al^{3+} - O^{2-}$ pair site was postulated to be the active site. Interesting correlations were found between the rates of isotopic exchange of $CH_4 - D_2$,

$C_2H_2-D_2$, cyclo-$C_3H_6-D_2$ and 1-butene isomerization over these alumina catalysts.[48] In the case of COS hydrolysis, a poisoning experiment showed that basic sites were essential.[49]

CO_2 does not retard the isomerization of olefins, but does in H_2/D_2 exchange,[50] so the active sites for these two reactions are different. The presence of four different active sites was postulated on a γ-alumina, namely A and B sites for 1-butene chemisorption, I sites for 1-butene isomerization and E sites for H_2-D_2 exchange.[17] A fifth site was further identified by the fact that double-bond shift and *cis/trans* isomerization can occur on independent sites.[6]

Therefore, as Knözinger and Ratnasamy[6] suggested, the site active for heterolytic cleavage of H−H and C−H may be strong acid-base pair sites in high gradient of electric field. Possible structures of these sites are very positive Al sites adjacent to basic OH groups which are produced by evacuation above 773 K. These sites strongly adsorb CO and the number is of the order of $10^{12}-10^{13}$ cm^{-2}. The difference in configuration or degree of strain of the site is probably the origin of the difference in the catalytic nature.

Dehydrohalogenation was also indicated to be catalyzed by acid-base pair sites.[51] Table 3.18 shows the product distribution from dehydrochlorination of 1,1,2-trichloroethane over alumina as well as over silica-alumina (a typical solid acid) and $KOH-SiO_2$ (a typical solid base). Silica-alumina and $KOH-SiO_2$ showed products typical of acid and base catalysts, respectively. On the other hand, the products from alumina are different from the others and well explained by a concerted mechanism catalyzed by both acid and base.

The relationship between acid strength and catalytic activity has been quantitatively investigated.[54] The acid strength distribution (only Lewis sites were present on the aluminas) measured by thermal desorption of pyridine is shown in Table 3.16, and the rate of dehydration of 2-butanol was measured over the aluminas of which the acid

TABLE 3.18 Product distribution of dehydrochlorination of 1,1,2-trichloroethane

	$SiO_2-Al_2O_3$	$KOH-SiO_2$	Al_2O_3
Product (A)	86 %	6 %	69 %
(B)	14	9	10
(C)	0	85	21

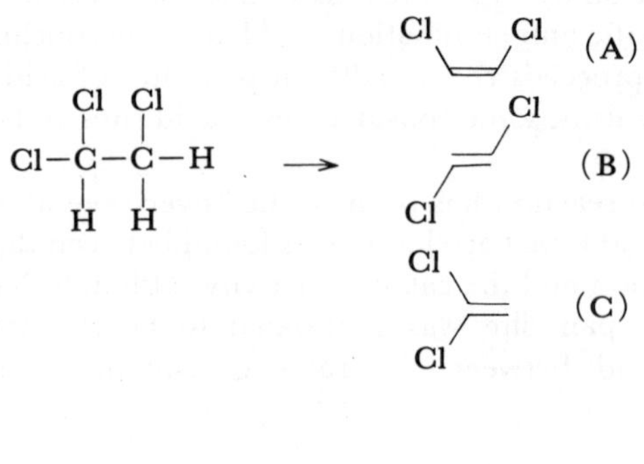

sites had been poisoned by known amounts. The acid strength was converted from the temperature of pyridine desorption, T_d, to heat of pyridine adsorption, Q_{py}, by quantitative comparison of T_δ with the data of differential calorimetric titration[30]. The rate constant (turnover frequency), $k(j)$, over one Lewis site in a certain range (j) of acid strength was calculated for each region of acid strength by the regional analysis proposed by Yoneda.[52] The rate constant of a given alumina, k, is given by the acid amount, $n(j)$, in the jth region of acid strength as

$$k=\sum_j k(j)\times n(j) \tag{7}$$

For each alumina a linear relationship was found between the logarithm of the rate constant $k(j)$, and the acid strength, Q_{py}, as shown in Fig. 3.37. That is, the Brønsted rule of catalysis holds for each alumina. The fact that three linear plots were obtained means that the acid site having the same acid strength had different catalytic activity from one alumina to another. They explained further this difference between the three catalysts by the difference in the basicity of the aluminas shown in Table 3.16. The rate constant for acid strength $k(j)$ becomes greater for alumina with higher basicity-to-acidity ratio. This implies that the acid-base pair sites are the active sites.

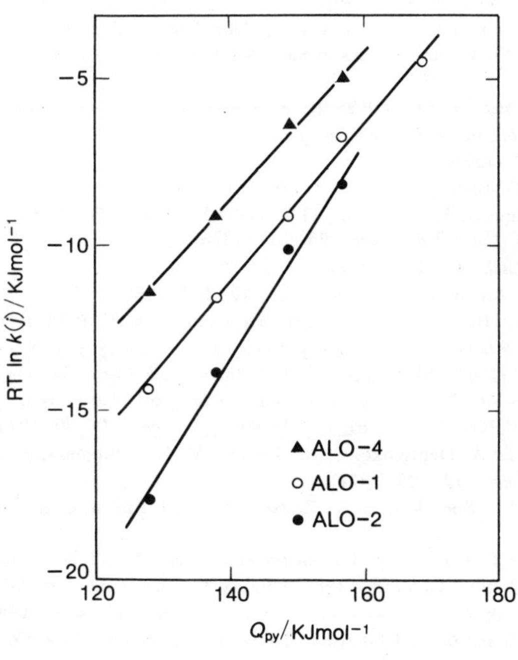

Fig. 3.37 Dependency of turnover frequency $k(j)$ for butanol dehydration upon Lewis acid strength. Q_{py} heat of adsorption of pyridine.

C. Ga_2O_3

Ga_2O_3 activated above about 800 K behaves like Al_2O_3 for olefin isomerization; a π-allyl intermediate is formed on basic sites.[53] However, when it is activated at 573 K, it shows a broad IR band at 2940 cm^{-1} assigned to OH stretching of α-GaO(OH) and a band at 3650 cm^{-1} due to surface OH, and exhibits peculiar catalytic activity for isotopic exchange between D_2 and hydrocarbon, which cannot be explained by mechanisms such as alkyl reversal and π-allyl intermediate mechanisms. For example, direct *cis-trans* isomerization of *n*-alkenes was totally selective below 433 K. Novel mechanisms involving σ-bonded alkyls and vinyls adsorbed via bonding to oxide ions on the surface have been proposed.[53]

REFERENCES

1. Y. Izumi, T. Shiba, *Bull. Chem. Soc. Jpn.*, **38**, 1797 (1964).
2. S. Sato, K. Urabe, Y. Izumi, *J. Catal.*, **102**, 99 (1986).
3. H. Sakurai, S. Sato, K. Urabe, Y. Izumi, *Chem. Lett.*, **1985**, 1783.
4. B.C. Lippens, J. J. Steggerda, in: *Physical and Chemical Aspects of Adsorbents and Catalysts*, Academic Press, 1970, p.171; K. Foger, in: *Catalysis*, Vol. 6, Springer, 1984, p.227; C. A. Spitler, S. S. Pallack, *J. Catal.*, **69**, 241 (1981); G. Marcelin, R. F. Vogel, *J. Catal.*, **81**, 252 (1983).
5. e.g., P. Burtin, J. P. Brunnelle, M. Pijolat, M. Soustelle, *Appl. Catal.*, **34**, 225 (1987).
6. H. Knozinger, P. Ratnasamy, *Catal. Rev.-Sci. Eng.*, **17**, 31 (1978).
7. A. J. Leonard, P. N. Semaille, J. J. Fripiat, Proc. Br. Ceram. Soc., **103** (1969), cited in ref.(6).
8. N. D. Parkins, Proc. 5th Intern. Congr. Catal., Palm Beach, 1972, **1**, 255 (1973).
9. T. Ono, Y. Ohguchi, O. Togari, Studies in Surf. Sci. Catal., Vol. 16, *in Preparation of Catalysis III*, Elsevier, 1983, p.631.
10. D. Masmadijan, *J. Catal.*, **1**, 547 (1962); R. E. Tisher, *J. Catal.*, **72**, 255 (1981).
11. G. A. Parks, *Chem. Rev.*, **65**, 177 (1965); K. Jiratova, *Appl. Catal.*, **1**, 165 (1981); J. P. Brunnelle, *Pure Appl. Chem.*, **50**, 1211 (1978).
12. V. C. F. Holm, G. C. Bailey, A. Clark, *J. Phys. Chem.*, **63**, 129 (1959).
13. T. Masuda, H. Taniguchi, K. Tsutsumi, H. Takahashi, *Jap. J. Petrol. Inst.*, **22**, 67, (1969).
14. H. Knozinger, *Advan. Catal. Relat. Subj.*, **25**, 184 (1976).
15. Y. Amenomiya, *J. Catal.*, **46**, 326 (1977).
16. R. J. Cvetanovic, Y. Amenomiya, *Advan. Catal.*, **17**, 103 (1967).
17. M. P. Rosynek, J. W. Hightower, Proc. 5th Intern. Congr. Catal., 1972, Palm Beach, **2**, 851 (1973).
18. F. E. Shephard, J. J. Rooney, C. Kemball, *J. Catal.*, **1**, 379 (1962); J. J. Rooney, R. C. Pink, *Trans, Faraday Soc.*, **58**, 1632 (1962); M. Okuda, T. Tachibana, *Bull. Chem. Soc. Jpn.*, **33**, 863 (1960); B. D. Flockhart, J. A. N. Scott, R. C. Pink, *Trans. Faraday Soc.*, **67**, 730 (1966).

19a) G. Della Gatta, B. Fubini, G. Ghiotti, C. Morterra, *J. Catal.*, **43**, 90 (1976).
b) A. A. Tsyganenko, L. A. Denisenko, S. M. Zverev, V. N. Filimonov, *J. Catal.*, **94**, 10 (1985).

20. Y. Amenomiya, *J. Catal.*, **22**, 109 (1971).
21. V. B. Kazansky, Vo Yu, Borovkov, A. V. Zaitsev, Proc. 9th Intern. Congr. Catal., Calgary, Vol. 3, 1988, p.1426.
22. W. K. Hall, Proc. 4th Intern. Congr. Chemistry and Usage of Molybdenum (P. C. H. Mitchell and H. F, Barry, eds.), Golden, Colorado, 1982, Climax Molybd. Co., Ann Arbor, 1982.
23. M. Nishimura, K. Asakura, K. Iwasawa, *J. Chem. Soc., Chem. Commun.*, **1986**, 1661.
24. K. Tanabe, *Solid Acids and Bases*, Kodansha, Tokyo and Academic, New York, 1970.
25. W. Kania, K. Jurczyk, *Appl. Catal.*, **34**, 1 (1987).
26. K. Mizuno, J. Take, Y. Yoneda, *Bull. Chem. Soc. Jpn.*, **49**, 634 (1976).
27. R. M. Pearson, *J. Catal.*, **46**, 279 (1977); *ibid.*, **53**, 173 (1978); H. Knozinger, *ibid*, **53**, 171 (1978).

28. A. Corma, C. Rodellas, V. Fornes, *J. Catal.*, **88**, 374 (1987).
29. C. Morterra, S. Coluccia, A. Chiorino, F. Boccuzzi, *J. Catal.*, **54**, 348 (1978).
30. J. Take, H. Matsumoto, S. Okada, H. Yamaguchi, K. Tsutsumi, H. Takahashi, Y. Yoneda, *Shokubai (Catalyst)*, **23**, 344, (1981) (in Japanese).
31. N. D. Parkins, *J. Chem. Soc.*, **1969**, 410.
32. K. Nakacho, J. Take, Y. Yoneda, National Meeting of Chem. Soc. Jpn., 3M28, April, 1983.
33. A. N. Webb, *IEC*, **49**, 261 (1965); V. C. F. Holm, A. Clark, *IEC, Prod. Res. Develop.*, **2**, 38 (1963); H. R. Gerberich, W. K. Hall, *J. Catal.*, **6**, 209 (1966); H. Hoffkes, E. Baumgarten, D. Hollenberg, *ibid.*, **77**, 257 (1982); G. B. McViker, G. M. Kramer, J. J. Ziemiak, *ibid.*, **83**, 286 (1983); A. Corma, V. Fornes, E. Ortega, *ibid.*, **92**, 284 (1985); J. R. Schlup, R. W. Vaughan, *ibid.*, **99**, 304 (1986).
34. H. Homs, P. Ramirez de la Piscina, J. E. Sueiras, *J. Catal.*, **89**, 531, (1984).
35. J. B. Peri, *J. Phys. Chem.*, **69**, 220 (1965).
36. L. Pauling, *The Nature of the Chemical Bond*, 3rd Ed., Cornell Univ. Press, Ithaca, 1960, p.548.
37. J. G. Larson, W. K. Hall, *J. Phys. Chem.*, **69**, 3050 (1965).
38. P. J. Robertson, M. S. Scurrel, C. Kemball, *J. Chem. Soc. Faraday I*, **1975**, 903.
39. F. H. Van Cauwelaert, W. Keith Hall, *Trans. Faraday Soc.*, **66**, 454 (1970).
40. R. Baird, S. Winstein, *J. Am. Chem. Soc.*, **79**, 4328 (1957); G. H. Posner, *Angew. Chem. Int. Ed. Engl.*, **17**, 487 (1978); M. E. Alonso, A. Morales, *J. Org. Chem.*, **45**, 4532 (1980); J. Yamawaki, T. Kawate, T. Ando, T. Hanabusa, *Bull. Chem. Soc. Jpn.*, **56**, 1885 (1983).
41. H. Pines, *J. Catal.*, **78**, 1 (1982).
42. Y. Amenomiya, G. Plsizer, *J. Catal.*, **76**, 345 (1982); S. Siddan, K. Narayanan, *J. Catal.*, **65**, 353 (1981).
43. H. Pines, W. O. Haag, *J. Am. Chem. Soc.*, **82**, 2471 (1960).
44. H. Knözinger, A. Schegliba, *J. Catal.*, **17**, 252 (1970).
45. H. Pines, J. Manassen, *Advan. Catal.*, **16**, 49 (1966); C. L. Kibby, S. S. Lande, W. Keith Hall, *J. Am. Chem. Soc.*, **94**, 214 (1972).
46. H. Arai, Y. Saito, Y. Yoneda, *J. Catal.*, **9**, 146 (1967); V. Moravek, H. Kraus, *ibid.*, **87**, 452 (1984).
47. M. Uchiyama, K. Tanabe, *Shokubai (Catalyst)*, **20**, 211 (1978).
48. M. Uchiyama, H. Hattori, K. Tanabe, *Bull. Chem. Soc., Jpn.*, **54**, 2521 (1981).
49. H. Itoh, C. V. Hidalgo, T. Hattori, M. Niwa, Y. Murakami, *J. Catal.*, **85**, 339 (1984).
50. J. W. Hightower, W. K. Hall, *Trans. Faraday Soc.*, **66**, 477 (1970); *J. Catal.* **13**, 161 (1969).
51. I. Mochida, Y. Anju, A. Kato, T. Seiyama, *J. Org. Chem.*, **39**, 3785 (1974); I. Mochida, A. Uchino, H. Fujitsu, K. Takeshita, *J. Catal.*, **43**, 264 (1976); I. Mochida *et al.*, *Bull Chem. Soc. Jpn.*, **44**, 3305 (1971).
52. Y. Yoneda, *J. Catal.*, **9**, 51 (1967).
53. F. B. Carlton, T. A. Gilmore, J. J. Rooney, Proc. 6th Intern. Congr. Catal., 1976, London, The Chemical Society, London, 1976, p.291.
54. K. Nakacho, J. Take, Y. Yoneda, National Meeting of Chem. Soc. Jan., 2K26, April, 1982.

3.1.12 SiO_2, GeO_2, SnO_2, PbO, PbO_2

A. SiO_2

a. Silanol groups on silica gel

The surface of silica gel consists of a layer of silanol groups (SiOH) and physically adsorbed water. Most of the water is removed upon drying in air at 400 – 500 K. Silanol groups are left on the surface and exist in three different configurations, i.e. isolated (a), geminal (b) and vicinal (c).

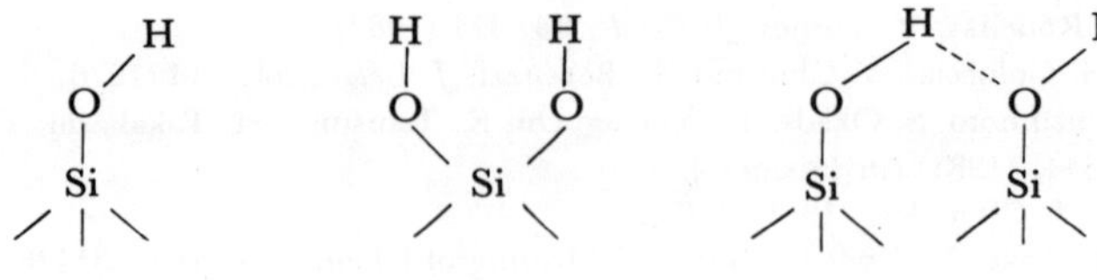

(a) (b) (c)

Silanol groups are progressively lost with increasing temperature to form siloxane groups on the surface.

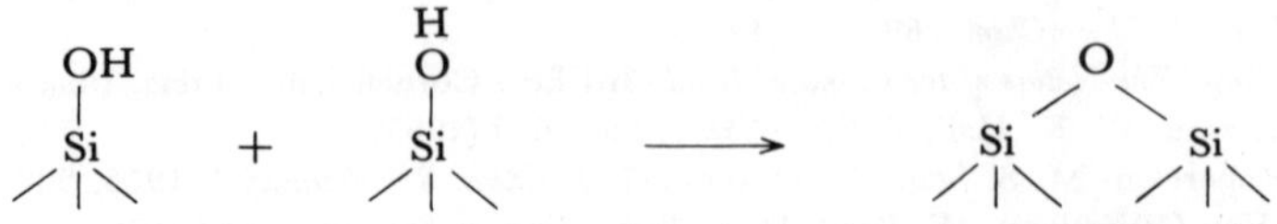

The newly formed siloxane bonds are very reactive since dehydration leaves the surface in a strained condition. The rehydration is completely reversible up to 673 K.[1] On heating to higher temperatures, reorientation of the SiO_4 tetrahedra occurs to relieve the strain. Thus the sites become nonsusceptible to rehydration.[1]

The number of silanol groups on silica gel has been estimated using a variety of techniques. Fig. 3.38 shows the surface concentration of silanol groups as a function of the temperature of heat-treatment under vacuum. The concentration of silanol groups on silica gel treated at 800 K is *ca.* 3 μmol m^{-2}.[2]

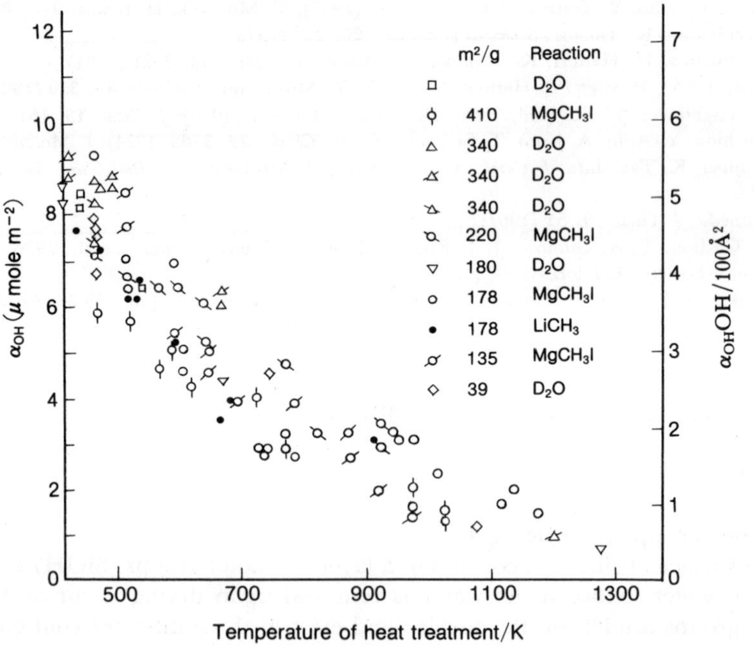

Fig. 3.38 **Change in the surface concentration of silanol groups on silica gel with temperarure of heat treatment**
(Reproduced with permission by S. Ogasawara, *Shokubai,* **18**, 124 (1976)).

b. Acidic and basic properties of silica gel.

Silanol groups of silica gel are very weakly acidic in nature.

Schindler and Kamber[3] determined the pK_a-value of the surface silanol group to be 6.8 by titration, while Strazhesko *et al.*[4] determined it to be 7.1 ± 0.5 also by titrating with various bases, the exact values depending on the titrating base.

Hair and Hertl[5] proposed a method for determining the acid strength of surface OH groups from the shift of OH band frequency of adsorbed phenol. The pK_a value of the silanol groups of silica gel was determined to be 7.1 by this method.[5]

The acidity and the distribution of silica gel was determined by titration with butylamine by using indicators. The cumulative acid amounts for varying acid strength are as follows (in mmol g^{-1}).[6]

0 ($H_0 = -5.6$), 0.041 ($H_0 = -3.0$), 0.052 ($H_0 = -1.5$)
0.091 ($H_0 = 3.3$), 0.101 ($H_0 = 4.8$), 0.204 ($H_0 = 6.8$)

Another set of results was also reported.[7]

0 ($H_0 = 1.5$), 0.066 ($H_0 = 3.3$), 0.109 ($H_0 = 4.0$),
0.264 ($H_0 = 4.8$).

The acidity of silica gel was determined by irreversible adsorption of basic molecules.[8] Pyridine (p$K_a = 5.25$) gave the value of 0.206 mmol g^{-1}, while aniline (p$K_a = 4.6$) gave the value of 0.01 mmol g^{-1}.

The basicity of silica gel was determined similarly by irreversible adsorption of acidic molecules. Acrylic acid (p$K_a = 4.25$) and phenol (p$K_a = 9.9$) gave the basicity values of 0.093, 0.012 mmol g^{-1}, respectively.[9]

It should be noted that the acidity or basicity of silica gels may be affected by the presence of impurities such as aluminum and sodium, which may be contained in the starting materials for preparing silica materials.

c. Acidity of crystalline silicates

Crystalline silicates are much stronger acid than silica gel. The results of titration by butylamine with Hammett indicators are given in Table 3.19.[10] $H_2Si_{14}O_{29}{\cdot}5H_2O$ is the strongest acid with acidic strength of $H_0 = -5$ — -3. The silicates, $H_2Si_8O_{17}{\cdot}0.5H_2O$, $H_2Si_8O_{17}{\cdot}1.1H_2O$ and $H_2Si_4O_9{\cdot}1.1H_2O$, have acid centers of $H_0 = 3$ — 1.5.

The acid strength of silicates depends on the dehydration temperature.[10] While $H_2Si_{14}O_{29}{\cdot}5H_2O$ has the acid strength of -5 — -3 in an air-dried sample, the acid strength is weakened to $H_0 = -3$ — 1.5 by treatment at 573 K, where the interlayer water molecules are completely lost.

d. Catalytic properties of silica gel

As described above, silica gel is very weakly acidic. The basicity is also small. Thus, the activity of silica gel for dehydration of 4-methyl−2-pentanol is about 20 times lower than alumina.[6] Though both the acidic and basic strength of silica gel is weak, this is sometimes profitable as catalysts for many organic reactions. The reactants can be activated (or polarized), and extensive side reactions avoided. The use

TABLE 3.19 Acidities of crystalline silicates

$H_0 \approx pK_I$	$H_2Si_2O_5$ mmol g^{-1}	mmol $mmol^{-1}$	$H_2Si_8O_{17}\cdot0.5H_2O$ mmol g^{-1}	mmol $mmol^{-1}$	$H_2Si_8O_{17}\cdot1,1H_2O$ mmol g^{-1}	mmol $mmol^{-1}$	$H_2Si_{11}O_{29}\cdot5,4H_2O$ mmol g^{-1}	mmol $mmol^{-1}$
<1.5	0	0	0.041	0.021	0.010	0.005	0.008	0.008
1.5–2.3	0	0	0.112	0.057	0.050	0.026	0.042	0.041
2.3–3.3	0.205	0.028	0.484	0.248	0.49	0.256	0.460	0.448
3.3–4.0	0.045	0.006	0	0	0	0	0	0
4.0–4.9	0	0	0	0	0	0	0	0
4.9–6.8	0.447	0.066	0.338	0.173	1.35	0.705	1.49	1.452
<1.5–6.8	0.727	0.100	0.975	0.499	1.90	0.992	2.00	1.949

(Reproduced with permission by Von H. J. Werner *et al.*, *Z. anorg allg. Chem.*, **470**, 121(1980)).

of silica gel as catalysts has been summarized.[11,12]

Kamimori *et al.* reported that *t*-butylation of phenol by *t*-butyl chloride occurred easily even at room temperature in the presence of silica gel or other inorganic supports.[13] No alkylation products were obtained in the absence of a catalyst even under reflux conditions. The products are mixtures of mono-, di- and tri-alkylated phenols.

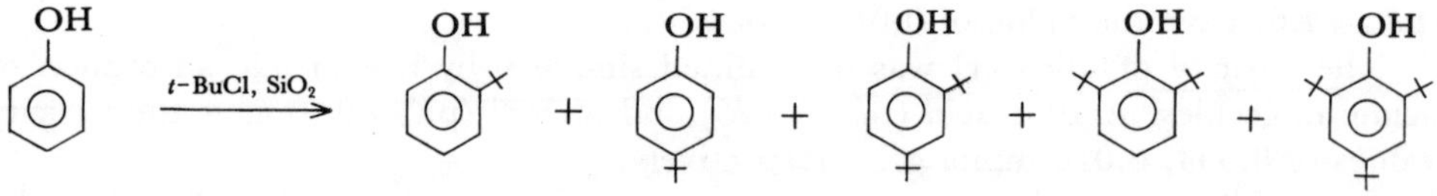

No *t*-butyl phenyl ether is formed in contrast to the case of *t*-butylation by *t*-butyl chloride (or isobutene) in the presence of $AlCl_3$ or HF. 2-*t*-Butyl, 2,6-di-*t*-butyl, and 2,4,6-tri-*t*-butyl-phenols, which are hard to obtain directly by the Friedel-Crafts process, are prepared easily by this one-step process.

Biphenyl-4,4′-diol and 2-naphtol can be alkylated by *t*-butyl chloride by the same method. Several other alkyl halides can also be used; for example, 2,6-di-*t*-butylphenol is alkylated by chloromethyl methyl ether afford I in a high yield.[13]

I

The alkylation of heterocyclic compounds such as furan and thiophene, which can not be alkylated by the conventional Friedel-Crafts, are also performed with this method.[13]

Aldoximes can be isomerized to corresponding amides in dry toluene with silica gel under reflux conditions.[14)] The list of aldoximes are given in Table 3.20. This method is much better than other methods for simplicity and higher conversion. The nitrile is not an intermediate in this isomerization unlike acid catalyzed reactions.[14)] Intramolecular cyclization has been proposed as a possible pathway of the isomerization.[14)]

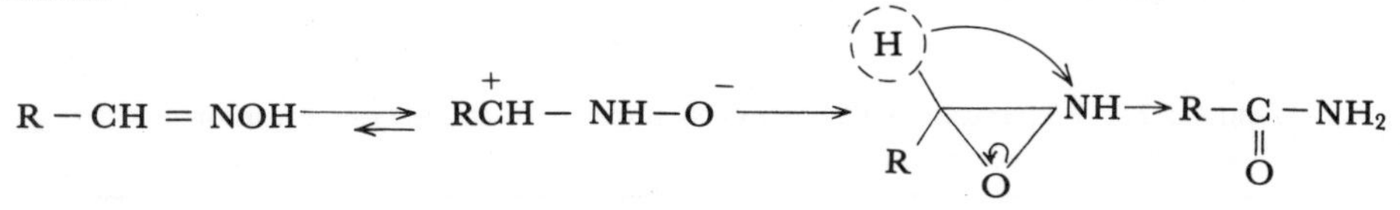

TABLE 3.20 Silica-gel induced isomerization of aldoximes to amides

Aldoxime from	Time (h)	Yield[†1] %
Benzaldehyde	69	92
p-Chlorobenzaldehyde	59	91
p-Hydroxybenzaldehyde	61	84
p-Methoxybenzaldehyde	52	82
Piperonal	64	93
Cinnamaldehyde	66	79
Salicylaldehyde	73	83
2,4-Dihydroxy-benzaldehyde	68	61[†2]
Acetaldehyd	57	89

[†1] Conversions are quantitative, the yield denoted here is the actual amount isolated.[†2] Unconverted oxime has been recovered.

(Reproduced with permission by J. B. Chattopadhyaya and A. V. Rama Rao, *Tetrahedron*, **30**, 2900 (1974)).

Beckmann rearrangement of cyclohexanone oxime *p*-toluenesulfonate is catalyzed by silica gel at room temperature.[15)] A small amount of water, 50 − 60 μl/g-catalyst, is beneficial for the transformation.

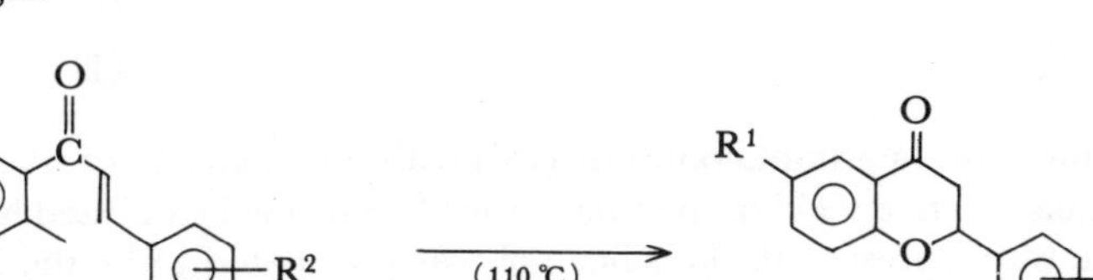

The isomerization of 2′-hydroxychalcone to the corresponding flavanones is catalyzed by silica gel.[16)]

The isomerization was carried out by adsorbing substituted 2′-hydroxychalcones on silica gel and heating at 383 K for 6–8 h in a hot-air oven. The amount of silica gel used was ten times the weight of the reactant. The product was extracted with hot ethyl acetate. The flavanones were purified by column chromatography. Various flavanones thus synthesized are indicated in Table 3.21.[16)]

TABLE 3.21 Preparation of flavanones by silica-gel catalyzed isomerization of 2′-hydroxychalcones

R^1	R^2	Chalcone	Product	Yield (%)
H	H	2′-Hydroxychalcone	Flavanone	40
Br	H	5′-Bromo-2′-hydroxychalcone	6-Bromoflavanone	75
Br	3-OMe	5′-Bromo-2′-hydroxy-3-methoxychalcone	6-Bromo-3′-methoxyflavanone	55
Br	4-OMe	5′-Bromo-2′-hydroxy-4-methoxychalcone	6-Bromo-4′-methoxyflavanone	52
Cl	H	5′-Chloro-2′-hydroxy-cychalcone	6-Chloroflavanone	58
COEt	4-OMe	2′-Hydroxy-4-methoxy-5′-propionylchalcone	4′-Methoxy-6-propionyl-flavanone	55
COPh	H	5′-Benzoyl-2′-hydroxy-chalcone	6-Benzoylflavanone	60
COPh	3-OMe	5′-Benzoyl-2′-hydroxy-3-methoxychalcone	6-Benzoyl-3′-methoxy-flavanone	60
COPh	4-OMe	5′-Benzoyl-2′-hydroxy-4-methoxychalcone	6-Benzoyl-4′-methoxy-flavanone	45

(Reproduced with permission by N. K. Sangwan, *et al.*, *Chem Ind.*, **1984,** 271)

Kato *et al.*[17)] have described a novel rearrangement in which II is converted to III in almost quantitative yield by treatment with silica gel.

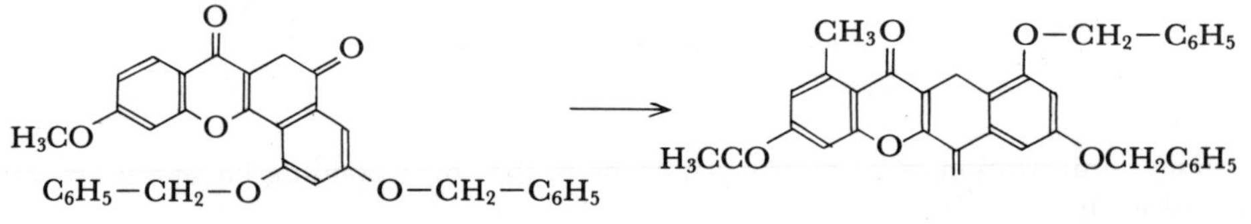

(II) (III)

The treatment of the thiomalonimide (IV) with a suspension of silica gel in chloroform gave a quantitative yield of the thiazoline.[18)] Silica gel catalyzed hydrolysis of the lactam by the water present in the silica gel was postulated to be the first step in the transformation.[16)]

(IV)

Some other reactions catalyzed by silica gel are as follows.

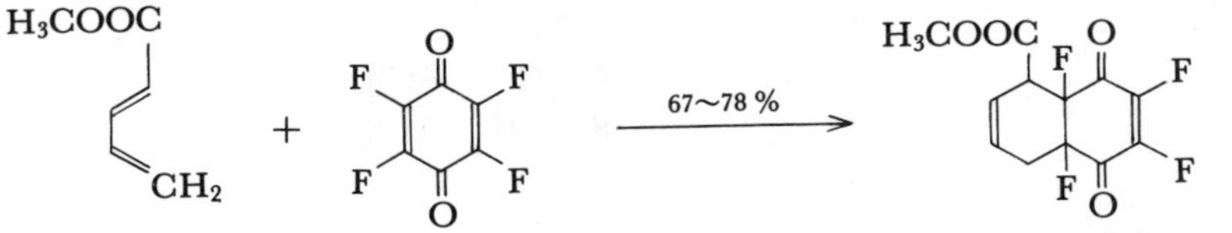

(ref. 19)

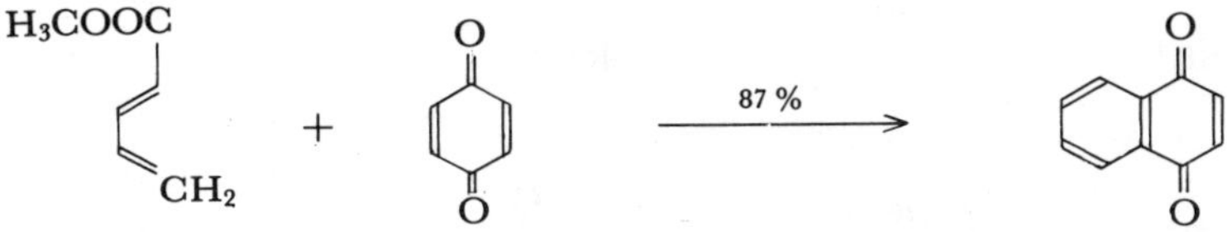

(ref. 19)

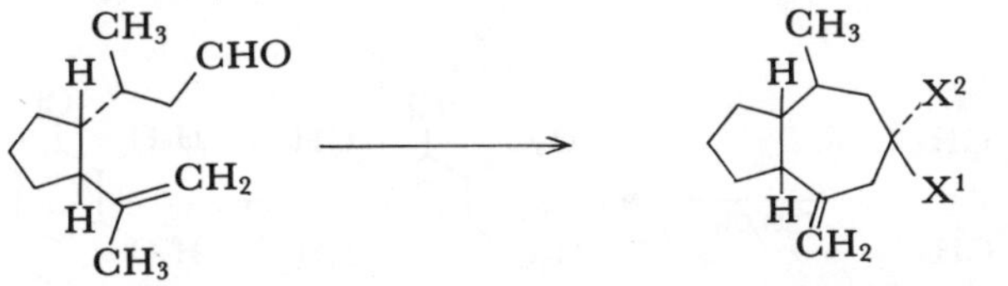

(ref. 20)

$X^1=H$, $X^2=OH$ 66 %
$X^1=OH$, $X^2=H$ 9 %

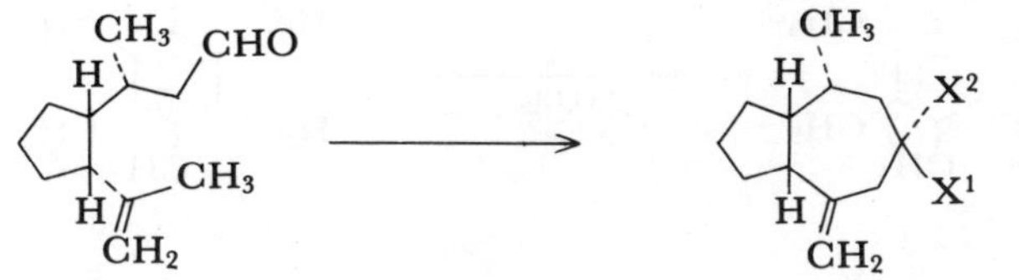

(ref. 20)

$X^1=H$, $X^2=OH$ 86 %
$X^1=OH$, $X^2=H$ 2 %

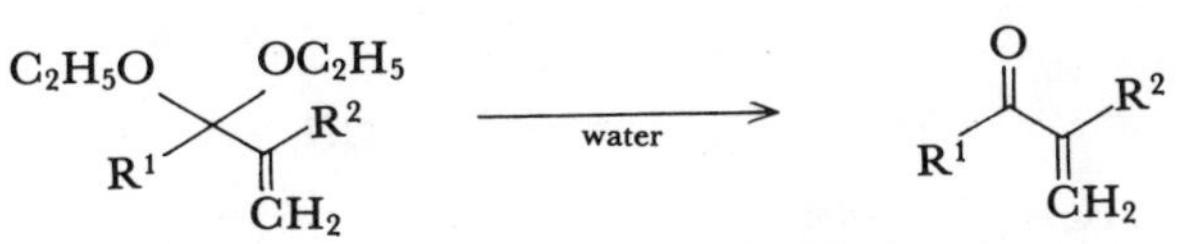

(ref. 21)

$R^1=H_3C$, $R^2=-CH(CH_3)_2$ 95 %
$R^1=H_3C$, $R^2=-(CH_2)_3-CH_3$ 95 %
$R^1=C_2H_5$, $R^2=-(CH_2)_3-CH_2$ 92 %

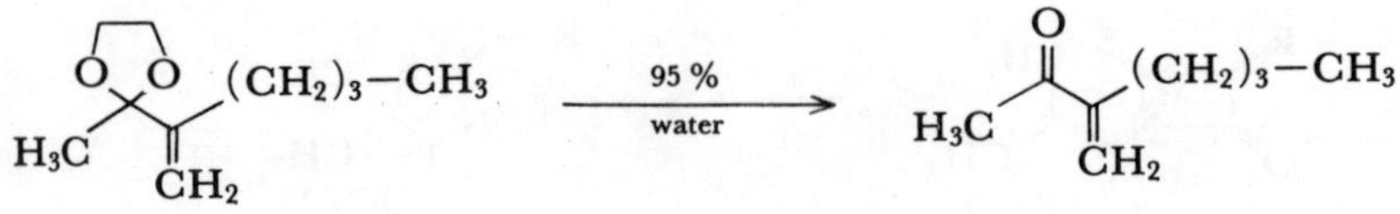

(ref. 21)

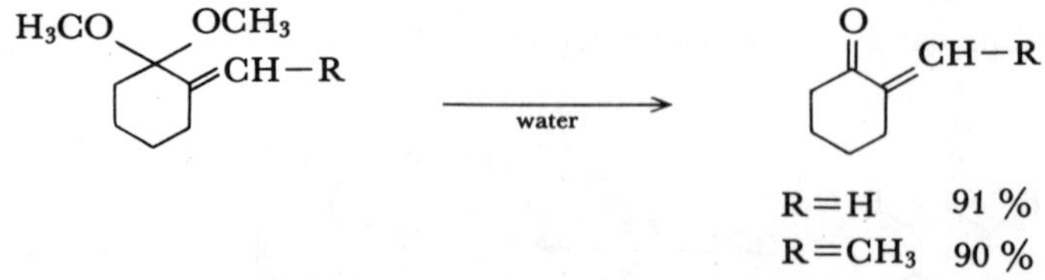

(ref. 21)

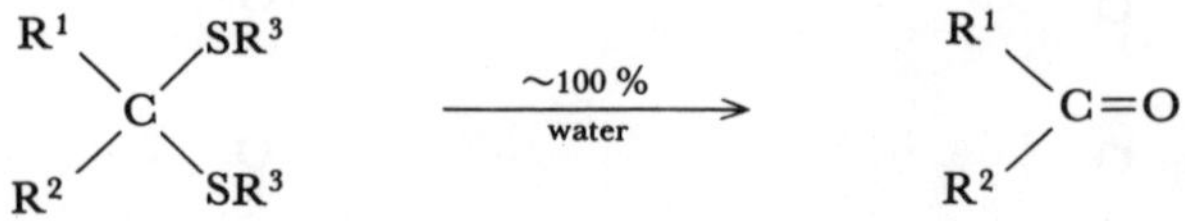

(ref. 22, 23)

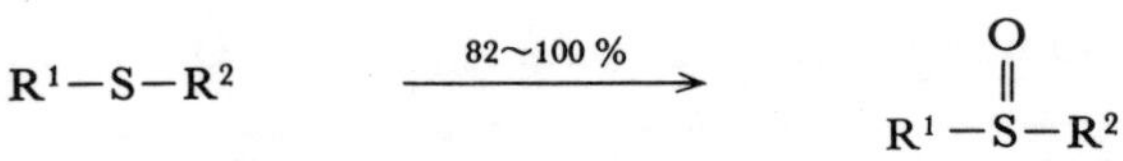

(ref. 24)

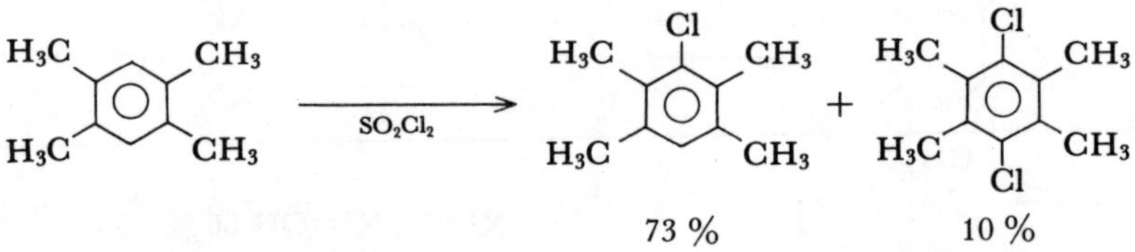

(ref. 25)

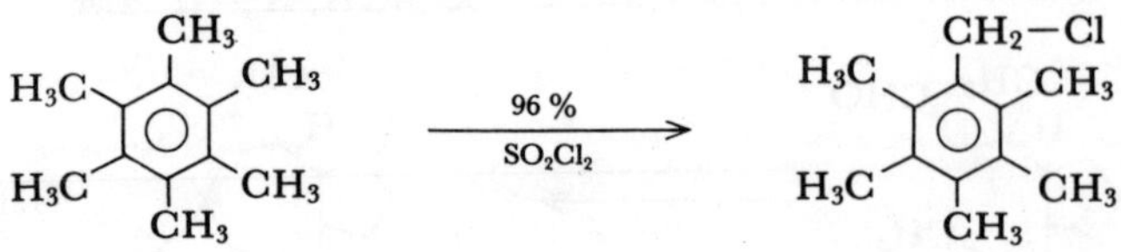

(ref. 26)

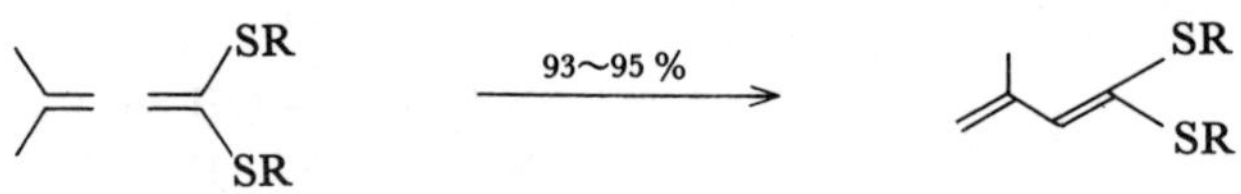

(ref. 27)

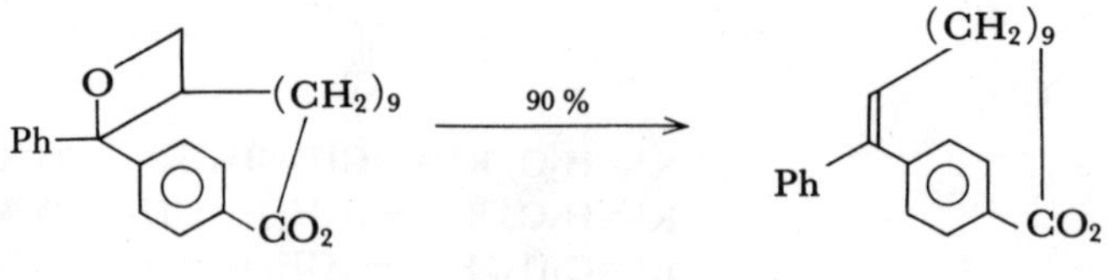

(ref. 28)

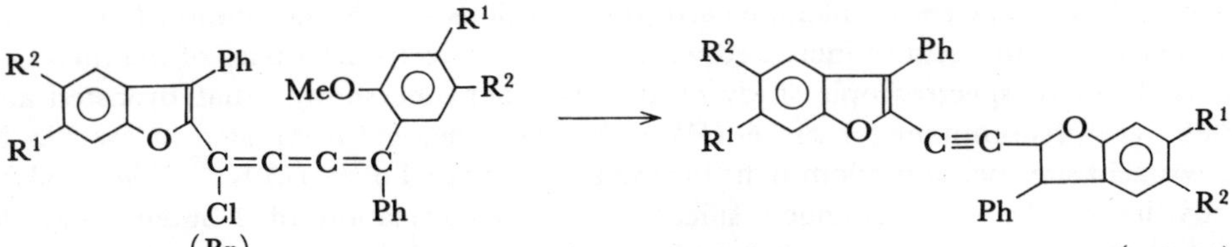

(ref. 29)

e. Modification of silica gel

Ion-exchange with metal cations Since the silanol groups are weakly acidic, the protons can be ion-exchanged with various metal cations. The ion-exchange capacities or the equilibria of the ion-exchange reactions have been determined for various cations.[30-32)]

$$n\text{SiO}-\text{H} + \text{M}^{n+} \longrightarrow (\text{SiO})_n - \text{M}^{n+} + n\text{H}^+$$

The introduction of metal cations onto the surface can be accomplished also with metal complexes.[33-35)] For example, the following equilibrium was postulated for the reaction with copper (II) ammine complex.[33-35)]

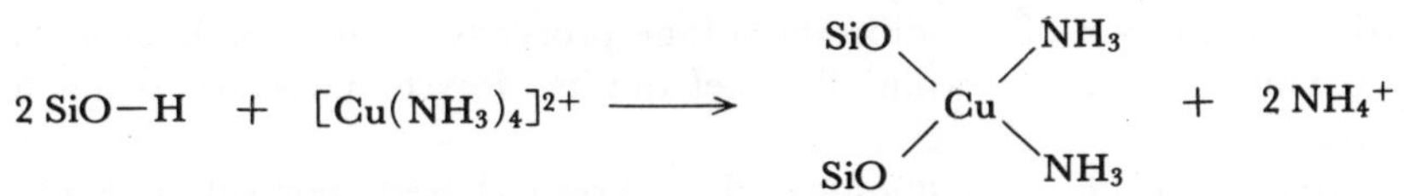

The states of the metal cations fixed on silica gel surfaces have been studied by a variety of spectroscopic techniques.[31,32,36)]

The acid and base properties of catalytic activities of ion-exchanged silica-gel were also studied. Table 3.22 shows the acidity of the cation-exchanged silica gel which was evacuated at 373 or 573 K.[37)] The acidities were determined by an indicator method.

TABLE 3.22 Acidity of cation-exchanged silica gel

Cation	Amount of cations/m mol g^{-1}	Acidity/m mol g^{-1} 373 K[†1], 573 K[†1]		Electronegativity of cation
Mn(II)	0.43	trace	0.03	7.5
Co(II)	0.47	0.08	0.06	9.0
Ni(II)	0.42	0.05	0.08	9.0
Cu(II)	0.44	0.08	0.11	9.5
Zn(II)	0.45	0.10	0.15	8.0
Cd(II)	0.52	0.06	0.04	8.5
Ca(II)	0.55	trace	0	5.0
Mg(II)	0.37	trace	0	6.0

[†1] evacuation temperature

(Reproduced with permission by K. Taniguchi *et al.*, *Nippon Kagaku Zasshi*, **91**, 613(1970)).

Silica gel exchanged with alkaline earth metal cations showed no acidity. For the rest of the cations, the acidity increased with increasing electronegativity of the metal cations. Infrared spectroscopic study of adsorbed pyridine showed that Brønsted acid sites were predominant for nickel (II) exchanged-silica gel pretreated below 523 K, Lewis acid sites being predominant for the gel pretreated over 573 K.[37)] The catalytic activities of the ion-exchanged silica gel for isomerization of 1-butene and dehydrobromination of isopropylbromide were studied.[37)] The isomerization of propylene oxide to propanal, acetone and alkyl alcohol was carried out over porous glass exchanged with a variety of metal cations and the selectivity of the isomerization was discussed in terms of the acid−base properties of the ion exchanged materials.[38)]

A remarkable effect of alkali cation exchange of silica gel on the acid-base properties and the selectivity of dehydrobromination of 2-bromobutane has been observed.[39,40)] The reaction proceeds with anti-elimination over Cs- and K-exchanged silica gel in which basic sites play a primary role. On the other hand, the reaction proceeds with syn elimination over silica gel and Na- and Li-exchanged silica gels, all of which are weakly acidic. The observed dependence of the steric courses of the reaction on the acid-base properties was explained by the variation of the reaction mechanism from a concerted elimination over basic surfaces to a carbenium-ion like mechanism over acidic surfaces (Section 4.8). In conformity with this conclusion, the dehydrobromination proceeds with syn elimination over silica gels exchanged with Ba(II), Ca(II), Mg(II) and Ni(II) ions.[41)]

Dehydrobromination of 2,3-dibromobutane proceeds by anti mode over Na-, K- and Cs-exchanged silica gels, while silica gel and its Li-exchanged form is inactive for the reaction.[39)]

Silica gel having sulfo groups Cationic exchange resins based upon sulfonated styrene-divinylbenzene copolymers are strong solid acids and used as catalysts for a variety of industrial reactions (Section 3.6.2). However, the thermal stability of the ion-exchange resins limits their application to about 400 K. In order to obtain catalysts of higher thermal stability, the introduction of sulfo groups to the surface of silica gel has been attempted, because the dissociation energy of Si−C bond is relatively high.

Saus and coworkers[42,43)] prepared silica gel having sulfobenzyl groups by the reaction of silanol groups with a silane coupling reagent (trimethoxybenzylsilane) and subsequent sulfonation.

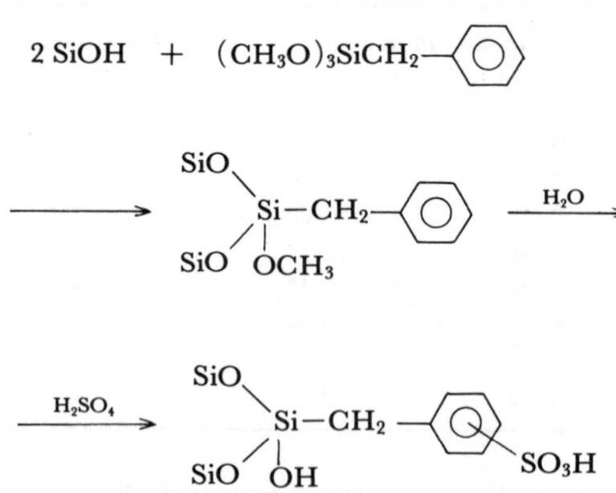

The material thus prepared had the acid amount of 0.68 mmol g^{-1} with Hammett acidity of $-0.62 > H_0 > -6.6$. The thermal stability on heating under nitrogen was high; 240 h at 573 K gave no loss in capacity; 100 h at 623 K gave 5% loss of sulfo groups; 70 h at 673 K completely destroyed the structure. They used the material as catalysts for alkylation of benzene with propene[42] and oligomerization of isobutene.[43]

Suzuki and Ono[44] prepared silica gel having sulfo groups with three different methods and compared the catalytic activities and the stability of the materials.

(I) reaction of silanol groups with 1,3-propanesultone

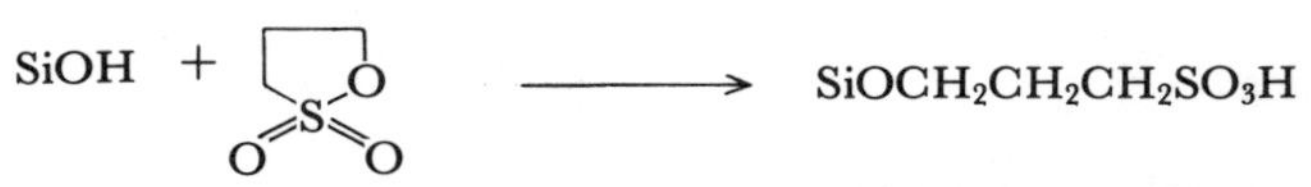

(II) reaction of silanol groups with sodium diethyl (3-sulfopropyl) phosphinate

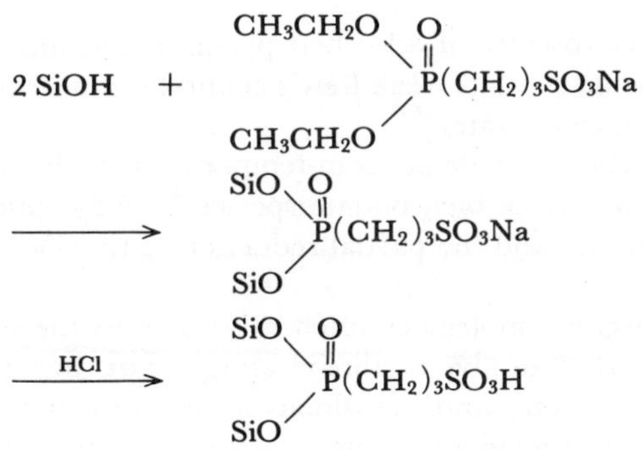

(III) reaction of silanol groups with trichloro phenetyl silane followed by sulfonation

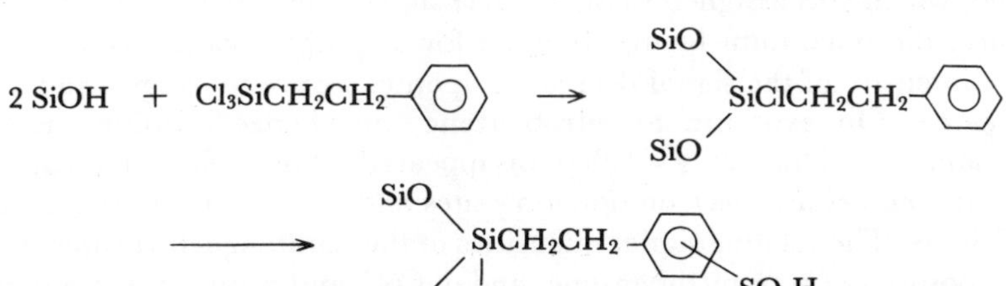

The numbers of acid sites as determined by a titration method were 0.3, 0.2 and 2.0 m mol g^{-1} for the silica gel modified by methods I, II, and III, respectively. The order of the catalytic activities of the three catalysts for both dehydration of isopropyl alcohol and the reaction of isobutene with methanol was in agreement with the order of the number of acid sites.

Silica-like materials having sulfo groups were prepared by co-condensation of tetra alkoxysilane and trialkoxyl- or trichloro-organosilane and subsequent sulfonation.[35,36] The characteristics of the sulfonated polyorganosiloxanes (SPOS) such as surface areas, pore size distribuion, and ion exchange capacities were determined. SPOS materials generally have more sulfo groups than silica gels modified by silane coupling reagents. Poly((sulfo phenyl) siloxane) and poly((sulfopropyl) siloxane) are thermally more stable than a cation-exchange resin. The catalytic activities of SPOS materials for vapor phase dehydration of alcohols and the liquid phase esterification were almost the same as those of Amberlyst-15. For the vapor phase nitration of benzene with nitrogen dioxide, SPOS materials showed excellent activities in contrast with negligible activity of the ion exchange resin.[37]

B. Tin Oxides

a. Surface properties of tin oxides

Thornton and Harrison studied the infrared spectra of tin (IV) oxide as a function of evacuation temperature.[48] Molecular water is largely removed by 320 K and fully removed at 473 K. Hydrogen-bonded and isolated hydroxyl groups exist after evacuation at 773 K.

The infrared spectroscopy of adsorbed pyridine and ammonia revealed that the surface of tin (IV) oxide exhibits weak Lewis acidity, but does not show Brønsted acidity even in the presence of water.[49]

Adsorption of carbon dioxide at room temperature leads to the formation of a surface carbonate and a surface bicarbonate species.[48] Adsorption of carbon monoxide gives a carbonate species and the partial reduction of tin (IV) oxide to a tin (II) oxide species.[48]

Adsorption of organic molecules often gives rise to the oxidation of adsorbates. Methanol is chemisorbed to give methoxyl groups, but they are readily oxidized to a surface formate.[50] Acetone and acetaldehyde are adsorbed predominantly as acetates.[50] Hexachloroacetone gives trichloroacetate.[33] Nitriles RCN [$R = CCl_3$, CH_3, $C(CH_3)_3$] are converted to surface acetoamido anions, $RCONH^-$, on adsorption.[51]

Itoh *et al.*[52] showed that tin (IV) oxide evacuated above 673 K gave an ESR signal at $g = 1,900$, which was assigned to Sn^{3+}. The signal intensity varied with evacuation temperature, the maximum being observed for a sample evacuated at 773 K (Fig. 3.39). The intensity of the signal decreased upon exposure to oxygen and a signal of O_2^- appeared. On exposure to nitrobenzene, nitrobenzene anion radicals were produced and the signal at $g = 1.900$ disappeared. These facts indicate that the paramagnetic centers are electron donating sites and most of them are located on the surface of SnO_2. The relative signal intensities of the paramagnetic centers before and after the exposure to O_2 or nitrobenzene, and of O_2^- and nitrobenzene anion radicals formed are shown as a function of evacuation temperature in Fig. 3.39.[52] The authors also measured the conductivity of Sn(IV) oxides at varying evacuation temperatures. The dependence of the conductivity on evacuation temperature was quite similar to that of the amount of surface paramagnetic centers. On exposure to oxygen or nitrobenzene, the conductivity became almost zero, indicating that the thermal reduction of tin (IV) oxide occurs only on the surface layer, but not in the bulk of the solid.

The acidity of tin (IV) oxide calcined at 773 K was reported to be 0.133 mmol as determined by butylamine titration using methyl red ($pK_a = 4.8$) as an indicator.[7,53]

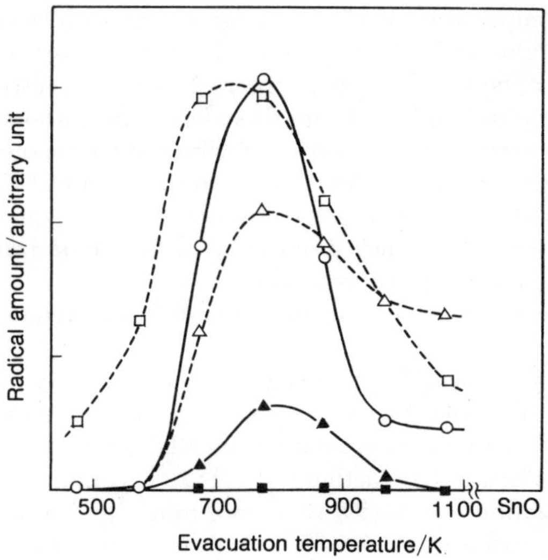

Fig. 3.39 Amounts of O_2^- (△— —), nitrobenzene anion radicals (□— —) and the radicals of g=1.900 (○—) of SnO_2(I)evacuated at various temperatures, the radicals of g=1.900 after the exposure to O_2 (▲—) and nitrobenzene (■—). (Reproduced with permission by M. Itoh *et al*, *J. Catal.*, **43**, 197 (1976)).

No acid sites stronger than $H_0 = 4.0$ were, however, detected.[53)]

The amounts of irreversible adsorption of ammonia and carbon dioxide on tin (IV) oxide (surface area 37.6 m^2g^{-1}) evacuated at 773 K were 4×10^{-6} and 3×10^{-6} mol m^{-2}, respectively.[53)]

b. Catalytic properties

Tin oxides are active catalysts for oxidation and are also used as a component of oxidation catalysts. However, reports on acidic or basic catalysis of the oxides are not abundant.

The reaction of 2-butanol over tin (IV) oxide at 573 K gives only methyl ethyl ketone, a dehydrogenation product; butenes, dehydration products, are not formed.[53)] Tin oxides have catalytic activities for the decomposition of diacetone alcohol at 303 K.[53)] These facts indicate that tin (IV) oxide is a basic rather than an acidic oxide.

The results for butene isomerization are controversial. Kemball *et al.*[54)] reported that the isomerization of 1-butene occurred readily at temperatures *ca.* 300 K, with concomitant formation of significant amounts of butadiene. The initial *cis/trans* product ratio was 1.2 – 1.5. They postulated the reaction mechanism involving a butadiene surface species formed by the simultaneous loss of two hydrogen atoms from adjacent carbon atoms on the adsorbed 1-butene molecule. They observed different characteristics on the *cis*-2-butene isomerization. Exclusive *cis-trans* isomerization was observed with no detectable double-bond migration or butadiene formation. An intramolecular mechanism involving a secondary carbonium ion as an intermediate was assumed for the isomerization of *cis*-2-butene.

Itoh *et al.*[52)] studied the isomerization of 1-butene at 573 K. The rate was maximum at evacuation temperature of 773 K in agreement with the temperature where the maximum intensity of an ESR signal at $g = 1.900$ was observed. The rate as well as the concentration of the paramagnetic centers greatly decreased upon exposure of the oxide to oxygen or nitrobenzene. From these facts, the authors concluded that the active sites for the isomerization are associated with the paramagnetic centers, which have an electron donating character. In contrast to the results of Kemball *et al.*,[54] the *cis/trans* product ratios of as high as 19 were observed. This is typical of base-catalyzed isomerization of 1-butene. Thus they proposed that the isomerization proceed via π-allyl carbanion at the paramagnetic centers.

Sn(II) oxide has been reported to be not active for isomerization or dehydrogenation of 1-butene.[52]

The acidic property of tin (IV) oxide is greatly enhanced by incorporating sulfate ions into the oxide.[55] The tin (IV) oxide containing sulfate ions are prepared by immersing tin hydroxide in an aqueous solution of $(NH_4)_2SO_4$, followed by drying and calination at 773 K. The catalytic activity of a SO_4^{2-}-containing tin (IV) oxide for cyclopropane isomerization was 4.25×10^{-5} mol min^{-1}g^{-1} at 373 K, while that of a SO_4^{-2}-free tin (IV) oxide was 3.7×10^{-7} mol min^{-1}g^{-1} even at 573 K.

Tatke and Rooney[56] noted that temperatures of 473 K were required for the isomerization of 1-butene over tin (IV) oxide, but that the reaction proceeded smoothly over tin (IV) oxide containing a small amount of sulfide.

C. Lead Oxides and Germanium Oxide

Lead oxides are used as oxidation catalysts or as a component of the catalysts. Studies on the acidic or basic properties of the surfaces of lead oxides are few. The surface acidity of lead (II) oxide was determined to be 0.7 mmol g^{-1} by titration with butylamine using methyl red ($pK_a = 4.8$) as an indicator.[7]

The catalytic activity per unit surface area of lead (II) oxide for dehydrogenation of 2-propanol is 5.3 times higher than that of zinc oxides.[57]

No acidic or basic character of germanium oxide has been reported.

REFERENCES

1. G. J. Young, *J. Colloid Sci.*, **13**, 67 (1958).
2. S. Ogasawara, *Shokubai*, **18**, 124 (1976) (in Japanese).
3. P. Schindler, H. R. Kamber, *Helv. Chim. Acta*, **51**, 1781 (1968).
4. D. N. Strazhesko, V. B. Strelko, V. N. Belyakov, S. C. Rubank, *J. Chromatography*, **102**, 191 (1974).
5. M. L. Hair, W. Hertl, *J. Phys. Chem.*, **74**, 91 (1970).
6. T. Yamaguchi, K. Tanabe, *Bull. Chem. Soc. Jpn.*, **47**, 424 (1974).
7. K. Shibata, T. Kiyoura, J. Kitagawa, T. Sumiyoshi, K. Tanabe, *Bull. Chem. Soc. Jpn.*, **46**, 2985 (1973).
8. J. M. Campelo, A. Garcia, J. M. Gutierrez, D. Luna, J. M. Marinas, *Can. J. Chem.*, **61**, 2567 (1983).
9. J. M. Campelo, A. Garcia, D. Luna, J. M. Marinas, *Can. J. Chem.*, **62**, 638 (1984).
10. Von H.-J. Werner, K. Beneke, G. Lagaly, *Z. anorg. allg. Chem.*, **470**, 118 (1980).
11. A. Mckillop, D. W. Young, *Synethsis*, **1979**, 401.
12. M. Hojo, R. Masuda, *J. Synth. Org. Chem. Jpn.*, **37**, 557, 689 (1979) (in Japanese).
13. Y. Kamimori, M. Hojo, R. Masuda, T. Izumi, S. Tsukamoto, *J. Org. Chem.*, **49**, 1416 (1984).
14. J. B. Chattopadhyaya, A. V. R. Rama Rao, *Tetrahedron*, **30**, 2899 (1974).
15. J. M. Riego, A. Costa, P. Deya, J. V. Sinsterra, J. M. Marinas, *React. Kinet. Catal. Lett.*, **19**, 61 (1982).

16. N. K. Sangwan, B. S. Varma, K. S. Dhindsa, *Chem. Ind.*, **1984**, 271.
17. T. Kato, N. Katagiri, J. Nakano, H. Kawamura, *J. Chem. Soc., Chem. Comm.*, **1977**, 645.
18. M. D. Bachi, J. Vaya, *Tetrahedron Lett.*, **1977**, 2209.
19. M. Hudlicky, *J. Org. Chem.*, **39**, 3461 (1974).
20. J. A. Marshall, N. H. Anderson, P. C. Johnson, *J. Org. Chem.*, **35**, 186 (1970).
21. F. Huet, M. Pellet, J. M. Conia, *Tetrahedron Lett.*, **1977**, 3505.
22. M. Hojo, R. Masuda, *Synthesis*, **1976**, 678.
23. M. Hojo, R. Masuda, T. Saeki, K. Fujimori, S. Tsutsui, *Tetrahedron Lett.*, **1977**, 3883.
24. M. Hojo, R. Masuda, *Tetrahedron Lett.*, **1976**, 613.
25. M. Hojo, R. Masuda, *Synth. Comm.*, **5**, 169 (1975).
26. H. Hart, J. L. Reilly, J. B. -C. Jiang, *J. Org. Chem.*, **42**, 2684 (1973).
27. J.-C. Client, S. Julia, *J. Chem. Research*, **1978**, (S) 125; (M) 1714.
28. D. Bichan, M. Winnik, *Tetrahedron Lett.*, **1974**, 3857.
29. S. Kobayashi, M. Shinya, H. Taniguchi, *Tetrahedron Lett.*, **1971**, 71.
30. D. L. Dugger, J. M. Stanton, B. N. Irby, B. L. McConnel, W. W. Cummings, R. W. Maatman, *J. Phys. Chem.*, **68**, 757 (1964).
31. K. Taniguchi, M. Nakajima, S. Yoshida, K. Tarama, *Nippon Kagaku Zasshi*, **91**, 524 (1970) (in Japanese).
32. H. Tominaga, M. Kaneko, Y. Ono, *J. Catal.*, **50**, 400 (1977).
33. A. Kozawa, *J. Inorg. Nucl. Chem.*, **21**, 315 (1961).
34. H. Tominaga, Y. Ono, T. Keii, *J. Catal.*, **40**, 197 (1975).
35. M. Shimokawabe, N. Takezawa, H. Kobayashi, *Appl. Catal.*, **2**, 379 (1982).
36. K. Taniguchi, M. Nakajima, S. Yoshida, K. Tarama, *Nippon Kagaku Zasshi*, **91**, 529 (1970) (in Japanese).
37. K. Taniguchi, M. Nakajima, S. Yoshida, K. Tarama, *Nippon Kagaku Zasshi*, **91**, 612 (1970) (in Japanese).
38. T. Imanaka, Y. Hayashi, S. Teranishi, *Nippon Kagaku Kaishi*, **1973**, 889 (in Japanese).
39. M. Misono, Y. Aoki, Y. Yoneda, *Bull. Chem. Soc. Jpn.*, **49**, 627 (1976).
40. M. Misono, T. Takizawa, Y. Yoneda, *J. Catal.*, **52**, 397 (1978).
41. Y. Aoki, M. Misono, Y. Yoneda, *Bull. Chem. Soc. Jpn.*, **49**, 3437 (1976).
42. A. Saus, B. Limbacker, R. Brulls, R. Kunkel, in: *Catalysis by Acids and Bases* (B. Imelik *et al.*, eds.) Elsevier, Amsterdam, 1985, p.383.
43. A. Saus, E. Schmidl, *J. Catal.*, **94**, 187 (1985).
44. S. Suzuki, Y. Ono, *Nihon Kagaku Kaishi*, **1985**, 1111 (in Japanese).
45. S. Suzuki, Y. Ono, *J. Mol. Catal.*, **43**, 41 (1987).
46. S. Suzuki, Y. Ono, S. Nakata, S. Asaoka, *J. Phys. Chem.*, **91**, 1659 (1987).
47. S. Suzuki, K. Tohmori, Y. Ono, *Chem. Lett.*, **1986**, 747.
48. E. W. Thornton, P. G. Harrison, *J. Chem. Soc., Faraday Trans.*, **1**, 71, 461 (1975).
49. P. G. Harrison, E. W. Thornton, *J. Chem. Soc., Faraday Trans.*, **1**, 71, 1013 (1975).
50. E. W. Thornton, P. G. Harrison, *J. Chem. Soc., Faraday Trans.*, **1**, 71, 2468 (1975).
51. P. G. Harrison, E. W. Thornton, *J. Chem. Soc., Faraday Trans.*, **1**, 72, 2484 (1976).
52. M. Itoh, H. Hattori, K. Tanabe, *J. Catal.*, **43**, 192 (1976).
53. G.-W. Wang, H. Hattori, K. Tanabe, *Bull. Chem. Soc. Jpn.*, **56**, 2407 (1983).
54. C. Kemball, H. F. Leach, I. R. Shannon, *J. Catal.*, **29**, 99, (1973).
55. G.-W. Wang, H. Hattori, K. Tanabe, *Chem. Lett.*, **1983**, 959.
56. D. G. Tatke, J. J. Rooney, *Chem. Commun.*, **1969**, 612.
57. O. V. Krylov, in: *Catalysis by Nonmetals*, Academic Press, New York and London, 1970, p.115.

3.1.13 Oxides of P, As, Sb, Bi

A. Solid Phosphoric Acid

Ipatieff[1)] developed the solid phosphoric acid catalyst which has been employed for the catalytic oligomerization of propene and butenes to liquid, gasoline boiling polymers.[2] The common catalyst is about 60% P_2O_5 − 40% SiO_2 (kieselguhr).[2] The typical procedure for catalyst preparation is as follows.[1] Orthophosphoric acid (75−100%) and a small amount of zinc oxide and zinc chloride are added to kieselguhr and heated at 450 − 570 K for 20 − 60 h. By this treatment, most of the orthophosphoric acid is converted into pyrophosphoric acid. The solid is then crushed and formed. It is known that pyrophosphoric acid is the most active catalyst, while orthophosphoric acid is fairly active and metaphosphoric acid almost inactive. The change in the acid composition on the catalyst surface with heat treatment was studies by Ohtuka and Aomura.[3] About 75% of orthophosphoric acid was converted into pyrophosphoric acid by heat treatment at 473 K for 4 h. When heated at 573 K, pyrophosphoric acid formation is always accompanied by metaphosphoric acid formation.

The catalyst can be deactivated by the modification of the acid components if it does not contain the proper amount of water.[2] Excessive water causes the catalyst to get soft and the pellet to collapse, increasing the pressure drop in the bed. Too little water dehydrates the phosphoric acid to inactive polyphosphoric acids; therefore small quantities of water are added to the reaction mixture to maintain the optimum degree of hydration. 350 − 400 ppm H_2O is the range for 473 − 593 K. The solid phosphoric acid catalysts are not easily regenerable.

Solid phosphoric acid catalysts are also used for producing isopropylbenzene from a mixture of benzene and a refinery stream containing propene and propane.[4] The carbonylation of alkenes[5] and the hydration of olefins[6,7] are also effected by supported phosphoric acid.

The acid strength of H_3PO_4/SiO_2 was determined to be −5.6 to −8.2 in H_0-value by a titration method by Benesi.[8]

Hashimoto and Mitsutani[9,10] found that phosphoric acid supported on silica, which were calcined at higher temperature (973 − 1473 K) had a high activity and selectivity for the preparation of isoprene by the decomposition of 4,4-dimethyl-meta-dioxane (MDO).

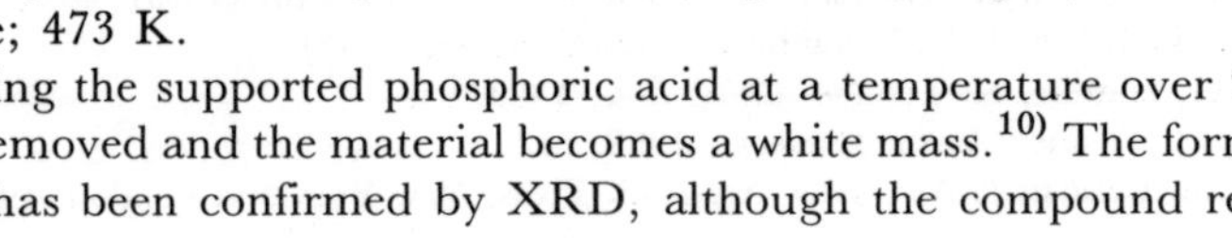

The conversion of MDO, yield of isoprene, selectivity for isoprene and the surface area of the catalyst are shown in Fig. 3.40.[10] The reaction conditions are as follows: catalyst; 6.9 g, MDO; 11.0 gh^{-1}, H_2O; 18.0 gh^{-1}, nitrogen; 1.0 dm^3h^{-1}, reaction temperature; 473 K.

By treating the supported phosphoric acid at a temperature over 973 K, most of the acid is removed and the material becomes a white mass.[10] The formation of a new compound has been confirmed by XRD, although the compound remains unidentified.[10]

B. Oxides of As, Sb, Bi

Oxides of arsenic, antimony and bismuth are very important as components of mixed metal oxides for oxidation of propene and butenes. However, the acid or base properties of the individual oxides have rarely been studied.

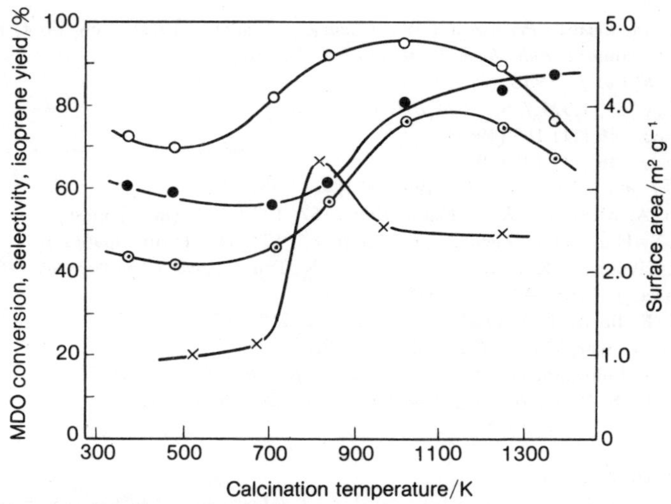

Fig. 3.40 Effect of calcination temperature on the catalytic activity for the decomposition of MDO and the surface area of the catalyst.
○; MDO conversion, ●; selectivity, ⊙; isoprene yield, ×; surface area.
(Reproduced with permission by A. Mitsutani and Y. Hamamoto, *Kogyo Kagaku Zashi*, **67**, 1232(1964)).

Shibata *et al.*[11] determined the acidities of Sb_2O_3 and Bi_2O_3 using an indicator method. The acidity of Sb_2O_3 was estimated to be 55 μmol g^{-1}, all the acid sites having acid strength of $3.3 < H_0 < 4.0$. The acidity of bismuth oxide was estimated to be 250 μmol g^{-1}, all the sites having strength of $4.8 < H_0 < 4.0$. Thus both oxides have very weak acid sites.

Ai and Ikawa[12] measured the acidity of Bi_2O_3 by irreversible adsorption of ammonia or pyridine and found that the number of acid sites is negligible. They also found the number of basic sites of Bi_2O_3 as determined by irreversible adsorption of acetic acid or carbon dioxide to be about 2.2 or 80 μmol g^{-1} respectively. These results show that Bi_2O_3 is a basic rather than an acidic oxide.

1,5-Hexadiene is formed when propene is passed over Bi_2O_3[13-15] or Sb_2O_4[15] at 750 − 823 K, indicating that allylic species are formed on the surface. The oxides are reduced through the reaction.

Isobutylaldehyde is oxidized with an oxide of arsenic (As_2O_5), antimony (Sb_2O_5, Sb_2O_4) or bismuth (Bi_2O_3) to give methacrylaldehyde.[16]

$$2CH_3CH(CH_3)CHO + As_2O_5 \longrightarrow 2CH_2{=}C(CH_3){-}CHO + As_2O_3 + 2H_2O$$

Oxygen may be fed simultaneously to obtain a catalytic type of oxidation.

REFERENCES

1. US Patent 1,993,512, 1,993,513, 2,018,065, 2,018,066, 2,020,649

2. C. L. Thomas, in: *Catalytic Processes and Proven Catalysts*, Academic Press, New York, 1970, p.67.
3. H. Ohtsuka, K. Aomura, *Bull. Japn. Petrol. Inst.*, **4**, 3 (1962).
4. H. W. Grote, *Oil Gas J.*, **1958**, (3) 73.
5. US Patent 1,924,763, 1,924,766, 1,924,767, 1,924,768
6. *Hydrocarbon Process*, **46** (11) 168 (1967).
7. *Hydrocarbon Process*, **46** (11) 195 (1967).
8. H..A. Benesi, *J. Am. Chem. Soc.*, **78**, 5490 (1957).
9. Y. Hamamoto, A. Mitsutani, *Kogyo Kagaku Zasshi*, **67**, 127 (1964) (in Japanese).
10. A. Mitsutani, Y. Hamamoto, *Kogyo Kagaku Zasshi*, **67**, 1231 (1964) (in Japanese).
11. K. Shibata, T. Kiyoura, J. Kitagawa, T. Sumiyoshi, K. Tanabe, *Bull. Chem. Soc. Jpn.*, **46**, 2985 (1973).
12. M. Ai, T. Ikawa, *J. Catal.*, **40**, 203 (1975).
13. H. E. Suift, J. E. Bozik, J. A. Ondrey, *J. Catal.*, **21**, 212 (1971).
14. F. Massoth, D. A. Scarpiello, *J. Catal.*, **21**, 225 (1971).
15. V. Fattore, Z. A. Fuhrman, G. Manara, B. Natori, *J. Catal.*, **37**, 215 (1975).
16. C. W. Hargis, H. S. Yong, *Ind. Eng. Chem. Prod. Res. Dev.*, **5**, 72, (1966).

3.1.14 Oxides of Se, Te

Like antimony or bismuth oxides, oxides of selenium or tellurium are often used as components of mixed oxides used for the catalysts of olefin oxidations. These oxides plausibly act as active sites for dehydrogenation of olefins. Thus, when propene is passed over SeO_2 at $573-603$ K, it is converted to acrolein and water. The oxide is reduced to elemental selenium.[1)]

Dehydrogenation activities of the oxides of selenium and tellurium were observed by the pulse technique.[2,3)] Thus, the zirconium oxides which are loaded with selenic acid or telluric acid and calcined in air can dehydrogenate 2-propanol to acetone[2)] and hexane to benzene.[3)] In a typical reaction of 2-propanol, the conversion into acetone decreases continuously after the third pulse, probably owing to a decrease in the amount of oxygen on the catalyst surface.[2)] Poisoning experiments with injection of CO_2, H_2O or butylamine at 523 K before reaction had no effect on the yield of acetone. Thus this dehydrogenation process appears to be an oxidative dehydrogenation.[2)] No studies on the acidic or basic character of oxides of selenium and tellurium oxides have been reported.

REFERENCES

1. N. Kominami, *Kogyo Kagaku Zasshi*, **65**, 1514 (1982) (in Japanese).
2. M. Hino, K. Arata, *J. Chem. Soc., Chem. Commun.*, **1984**, 1037.
3. M. Hino, K. Arata, *Chem. Lett.*, **1985**, 1483.

3.2 MIXED METAL OXIDES

3.2.1 Mechanism of Acidity Generation

A new hypothesis regarding the acidity generation of binary oxides has been proposed by Tanabe *et al.*; the hypothesis predicts what kinds of binary oxides will show acidic properties (Brønsted or Lewis acid) and provides insight regarding the structure of the acid sites.[1)] According to the hypothesis, acidity generation is caused by an excess of a negative or positive charge in the model structure of a binary oxide. The model structure is pictured according to the following two postulates: i) The coordination number of a positive element of a metal oxide, C_1, and that of a second metal oxide, C_2, are maintained even when mixed; ii) The coordination number of a negative element (oxygen) of a major component oxide is retained for all the oxygens in a binary oxide.

For example, the structure of TiO_2-SiO_2, where TiO_2 is the major component oxide, and that of SiO_2-TiO_2, where SiO_2 is the major component oxide, are shown in Fig. 3.41. In Fig. 3.41 the coordination numbers of the positive elements in the component single oxides remain 4 for Si and 6 for Ti when they are mixed, whereas those of the negative elements should be 3 and 2 respectively, according to postulates i) and ii) above. In the case of Fig. 3.41a, the four positive charges of the silicon atom are distributed to four bonds, i.e. a positive charge is distributed to each bond, while the two negative charges of the oxygen atom are distributed to three bonds, i.e. $-2/3$ of a valence unit is distributed to each bond. The difference in charge for one bond is $+1-2/3=+1/3$, and for all the bonds the valence unit of $+1/3\times 4=+4/3$ is excess. In this case, the Lewis acidity is assumed to appear upon the presence of an excess of positive charge. In Fig. 3.41b, four positive charges of the titanium atom are distributed to six bonds, i.e. $+4/6$ of a valence unit to each bond, while two negative charges of the oxygen atom are distributed to two bonds, i.e. a negative charge to each bond. The charge difference for each bond is $+4/6-1=-1/3$, and for all the bonds the valence unit of $-1/3\times 6=-2$ is excess. In this case, Brønsted acidity is assumed to appear, because two protons are considered to associate with six oxygens to keep

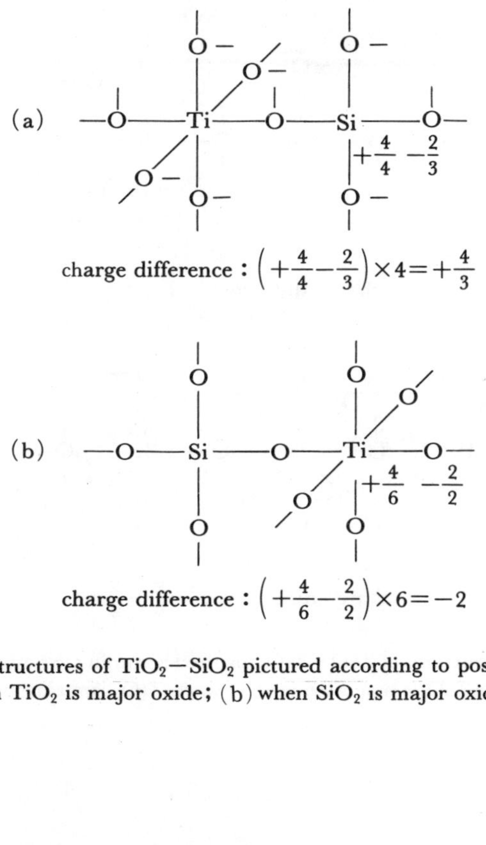

Fig. 3.41 Model structures of TiO_2-SiO_2 pictured according to postulates i) and ii)
(a) when TiO_2 is major oxide; (b) when SiO_2 is major oxide

electric neutrality. In any case, TiO_2-SiO_2 is expected to show acidic property because of the excess of a positive or negative charge. In fact, it exhibited very high acidity.[2)]

Let us examine another example. In $ZnO-ZrO_2$, there is no excess charge in any part of its composition according to our model structure written by postulates i) and ii), as illustrated in Fig. 3.42. Therefore, the binary oxide is not expected to show any acidic property. This prediction agrees with the experimental result that $ZnO-ZrO_2$ does not show acidity larger than the sum of the acidities of the component oxides.[3)]

The validity of the hypothesis was examined for 31 kinds of binary oxides. The case where the hypothesis predicts that a binary oxide should generate acidity is marked by an open circle in the fifth column of Table 3.23. On the other hand, the case where a binary oxide should not generate acidity is shown by an × in the same column. Experimental results cited from the literature are shown in the next column, where open circles mark when the acid amounts at a certain acid strength per unit surface area of any binary oxides are larger than the sum of the acid amounts at the same acid strength divided by the sum of the surface areas of the component single oxides, while ×'s mark when no acidity is generated. The results indicate that new acid sites which differ from those of single oxides are created on the surface of 26 species of binary oxides. Cases where the result predicted by the hypothesis agrees with the experimental result are marked by open circles in the last column of Table 3.23, and the cases of disagreement, by ×'s. As can be seen in the table, agreement between the prediction of Tanabe *et al.*'s hypothesis with the experimental results was found for 29 of the 32 kinds of binary oxides. Thus, the validity of their hopothesis is 91 percent. The validity

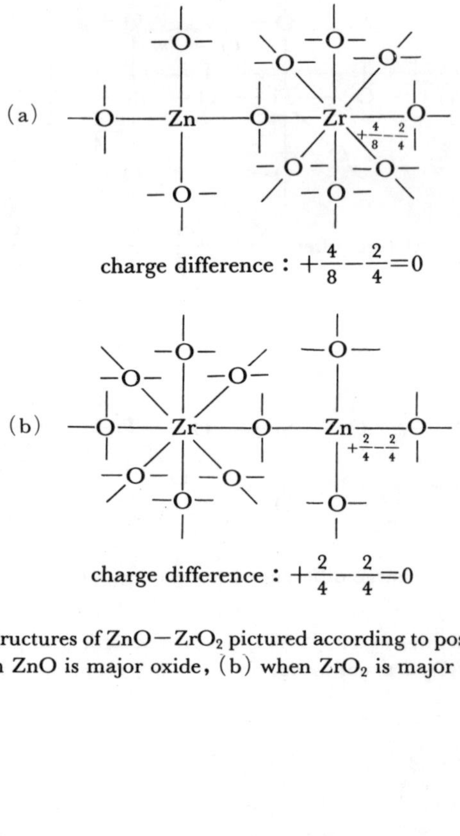

Fig. 3.42 Model structures of $ZnO-ZrO_2$ pictured according to postulates i) and ii), (a) when ZnO is major oxide, (b) when ZrO_2 is major oxide.

TABLE 3.23 Validity of hypotheses for acidity prediction
V: valence of positive element, C: coordination number of positive element

Mixed-Oxides	$\alpha=V/C$		Acidity increase			Validity of hypotheses	
1 2	α_1	α_2	Thomas	Tanabe *et al.*	Experi-mental results	Thomas	Tanabe *et al.*
TiO_2-CuO	4/6	2/4	×	○	○	×	○
TiO_2-MgO		2/6	○	○	○	○	○
TiO_2-ZnO		2/4	×	○	○	×	○
TiO_2-CdO		2/6	○	○	○	○	○
TiO_2-Al_2O_3		3/6	○	○	○	○	○
TiO_2-SiO_2		4/4	×	○	○	×	○
TiO_2-ZrO_2		4/8	×	○	○	×	○
TiO_2-PbO		2/8	×	○	○	×	○
TiO_2-Bi_2O_3		3/6	○	○	○	○	○
TiO_2-Fe_2O_3		3/6	○	○	○	○	○
ZnO-MgO	2/4	2/6	×	○	○	×	○
ZnO-Al_2O_3		3/6	×	×	○	×	×
ZnO-SiO_2		4/4	○	○	○	○	○
ZnO-ZrO_2		4/8	○	×	×	×	○
ZnO-PbO		2/8	○	○	×	×	×
ZnO-Sb_2O_3		3/6	×	×	×	○	○
ZnO-Bi_2O_3		3/6	×	×	×	○	○
Al_2O_3-MgO	3/6	2/6	○	○	○	○	○
Al_2O_3-B_2O_3		3/3	○	○	○	○	○
Al_2O_3-ZrO_2		4/8	×	×	○	×	×
Al_2O_3-Sb_2O_3		3/6	×	×	×	○	○
Al_2O_3-Bi_2O_3		3/6	×	×	×	○	○
SiO_2-BeO	4/4	2/4	○	○	○	○	○
SiO_2-MgO		2/6	×	○	○	×	○
SiO_2-CaO		2/6	×	○	○	×	○
SiO_2-SrO		2/6	×	○	?	?	?
SiO_2-BaO		2/6	×	○	?	?	?
SiO_2-Ga_2O_3		3/6	×	○	○	×	○
SiO_2-Al_2O_3		3/4, 3/6	○	○	○	○	○
SiO_2-Y_2O_3		3/6	×	○	○	×	○
SiO_2-La_2O_3		3/6	×	○	○	×	○
SiO_2-ZrO_2		4/8	○	○	○	○	○
SiO_2-Fe_2O_3		3/6	×	○	○	×	○
ZrO_2-CdO	4/8	2/6	×	○	○	×	○

Thomas' hypothesis 15/32=47 % correct.
Tanabe *et al.*'s hypothesis 29/32=91 % correct.

of the old but well known Thomas' hypothesis[4] is only 47 percent.

Although Thomas' hypothesis cannot be applied to Lewis acids[4], Tanabe *et al.*'s hypothesis also predicts the type of acid sites (Brønsted or Lewis), as has been mentioned above. According to the hypothesis, TiO_2-ZnO should show Brønsted acidity when TiO_2 is a major component oxide and Lewis acidity when ZnO is a major component oxide. An infrared study of pyridine adsorbed on the binary oxide revealed that TiO_2 including 5 percent ZnO exhibited Brønsted acidity alone, while ZnO including 5 percent TiO_2 exhibited Lewis acidity alone.[5]

As can be seen in the model structure (Figs. 3.41 and 3.42) pictured according to postulates i) and ii), the hypothesis is applicable to chemically mixed binary oxides, but not to mechanically mixed oxides. Since the binary oxides given in Table 3.23 were prepared by calcining mixtures of co-precipitated hydroxides at a high temperature (770 K), they are not mechanically mixed oxides. The X-ray diffraction diagrams of the binary oxides showed no or only weak diffraction lines, and almost all of them were amorphous. Thus, the structures are different from those of the single component oxides.

The Tanabe *et al.*'s hypothesis predicts which combinations of oxides in the periodic table will generate acidity and at what compositions the Brønsted or Lewis acidity will appear, but it does not predict the acid strength. The prediction of acid strength will be discussed further on.

Recently, Seiyama presented a different model for the acidity generation of binary metal oxides.[6] He assumes that acidity appears at the boundary where two oxides contact. In the case of $ZnO-ZrO_2$ (Fig. 3.43), the oxygen by which ZnO combines with ZrO_2 has a negative charge, since $+2/4$ charge of Zn and $+4/8$ charge of Zr are distributed to the boundary oxygen which has two negative charges and, hence, the charge difference, Δ, around the boundary oxygen becomes -1. Therefore, a Brønsted acid site should appear according to the same argument as mentioned in Tanabe *et al.*'s hypothesis. In fact, the acidity generation of binary oxides such as $ZnO-ZrO_2$ strongly depends on the preparation method. The acidity generation of $ZnO-ZrO_2$ was observed in Seiyama's expriment,[6] but not in Tanabe *et al.*'s experiments.[1] Seiyama's model may be applied to binary oxides in which chemical mixing is not adequate, while Tanabe *et al.*'s model can be applied to amorphous binary oxides.

Concerning the acid strength of binary oxides, the highest acid strengths are found to increase with the increase of the algebraically averaged electronegativities as shown in Fig. 3.44.[7] The correlation in the figure, which is useful for predicting the acid strengths of unknown binary oxides, indicates that electronegativity controls the acid

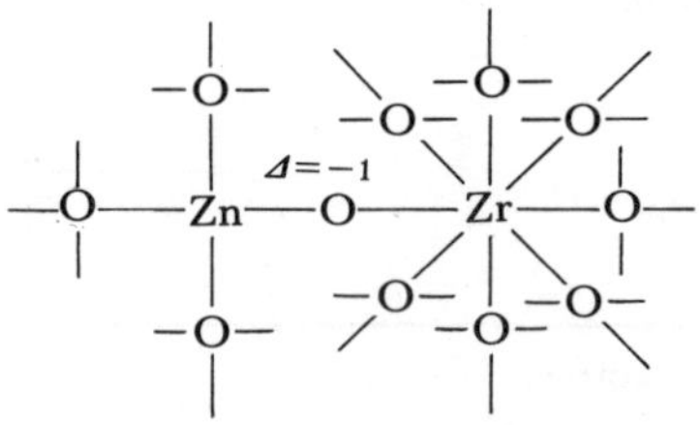

charge difference : $\left(+\frac{2}{4}+\frac{4}{8}\right)-2=-1(\Delta)$

Fig. 3.43 Model of acidity generation at the boundary between two oxides

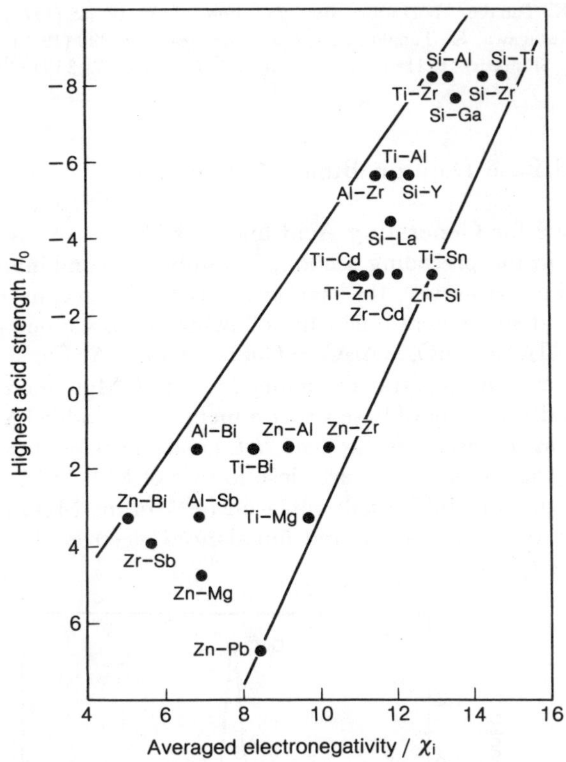

Fig. 3.44 **Highest acid strength *vs.* averaged electronegativity of metal ions of binary oxides (molar ratio=1)**
(Reproduced with permission from *J. Catalysis - Science and Technology* (**J. R. Anderson and M. Boudart eds.) Vol. 2. P. 269, Springer, 1981**)

strength.

The basicity increase caused by mixing two metal oxides has been reported for Al_2O_3-MgO,[8,9] TiO_2-MgO,[10] SiO_2-ZnO,[11] TiO_2-ZrO_2,[12] and Al_2O_3-ZnO.[13] no general rules about the basicity increase have yet been found.

REFERENCES

1. K. Tanabe, T. Sumiyoshi, K. Shibata, T. Kiyoura, J. Kitagawa, *Bull. Chem. Soc. Jpn.*, **47**, 1064 (1974).
2. M. Itoh, H. Hattori, K. Tanabe, *J. Catal.*, **35**, 225 (1974).
3. K. Shibata, T. Kiyoura, J. Kitagawa, T. Sumiyoshi, K. Tanabe, *Bull. Chem. Soc. Jpn.*, **46**, 2985 (1973).
4. C. L. Thomas, *Ind. Eng. Chem.*, **41**, 2564 (1949).
5. K. Tanabe, C. Ishiya, I. Matsuzaki, I. Ichikawa, H. Hattori, *Bull. Chem. Soc., Jpn.*, **45**, 47 (1972).
6. T. Seiyama, *Metal Oxides and their Catalytic Actions*, Kodansha, Tokyo, 1978.
7. K. Tanabe, *in: Catalysis-Science and Technology,* (J. R. Anderson and M. Boudart, eds.) Vol. 2, Chapt. 5, p.269, Springer-Verlag, 1981.
8. S. Miyata, T. Kumura, H. Hattori, K. Tanabe, *Nippon Kagaku Zasshi*, **92**, 514 (1971) (in Japanese).
9. N. Yamagata, Y. Owada, S. Okazaki, K. Tanabe, *J. Catal.*, **47**, 358 (1977).
10. K. Tanabe, H. Hattori, T. Sumiyoshi, K. Tamaru, T. Kondo, *J. Catal.*, **53**, 1, (1978).

11. T. Sumiyoshi, K. Tanabe, H. Hattori, *Bull. Jpn. Petrol. Inst.*, **17**, 65 (1975).
12. K. Arata, S. Akutagawa, K. Tanabe, *Bull. Chem. Soc. Jpn.*, **49**, 390 (1976).
13. K. Tanabe, Ke. Shimazu, H. Hattori, Ke. Shimazu, *J. Catal.*, **57**, 35 (1979).

3.2.2 Acid and Base Data on Binary Oxides

A. Combinations for Generating Acid and Base Sites

As described in the preceding section, a number of combination of metal oxides generates acid sites. However, the combinations for generating base sites are fewer than those for acid site generation. The following combinations generate base sites: TiO_2-MgO, Al_2O_3-ZnO, Al_2O_3-CaO, $TiO_2-Al_2O_3$, TiO_2-ZrO_2, and SiO_2-ZnO. Incorporation of transition metal ions into MgO increases base sites. The relation between the amount of base sites on metal cation-added MgO and ionic radii of the metal cations is shown in Fig. 3.45.[1)] Increase in base amount is prominent as the ionic radii of the metal cations are close to that of Mg^{2+}. Metal ions whose ionic radii are close to that of Mg^{2+} easily replace Mg^{2+} in the MgO lattice. The replacement results in a deformed lattice and unbalanced electron charge distribution, increasing the basicity.

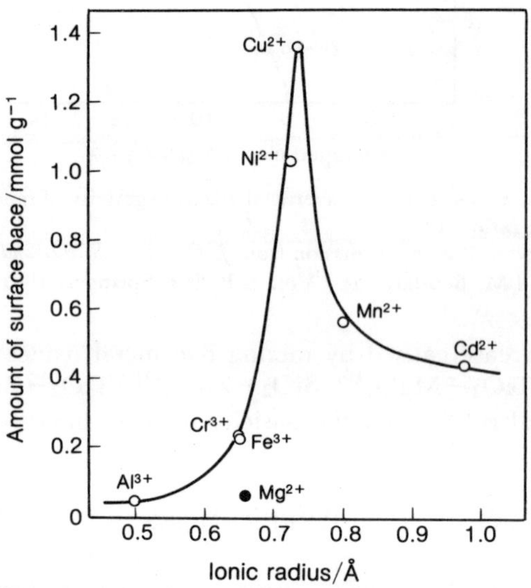

Fig. 3.45 **Basicity variation of MgO as a function of ionic radius of added metal cations.**

B. Factors Determining Acid and Base Sites Generation

a. Composition

For a certain combinations of metal oxide, the number of acid and base sites generated usually depends upon the composition of the binary oxide. Examples are shown in Fig. 3.46 for acid site generation of SiO_2-MoO_3,[2)] and Fig. 3.47 for base site

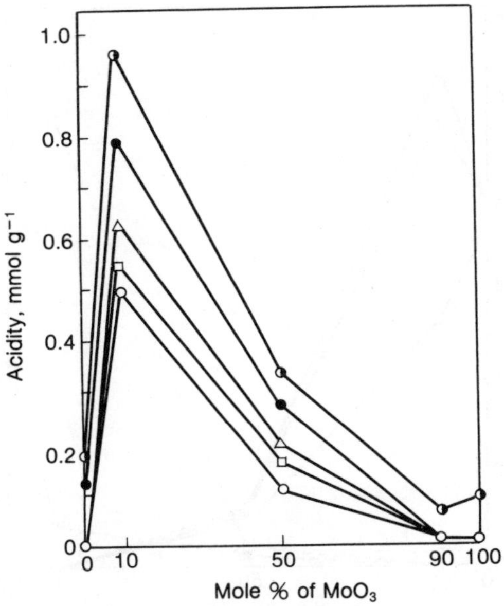

Fig. 3.46 Acidity of SiO_2-MoO_3 of different compositions.
$H_0=-3.0$ (○), $+1.5$ (□), $+3.3$ (△), $+4.8$ (●), $+6.8$ (◑).
(Reproduced with permission by K. Murayama *et al.*, *Bull. Chem., Soc., Jpn.*, **50**, 88, (1979)).

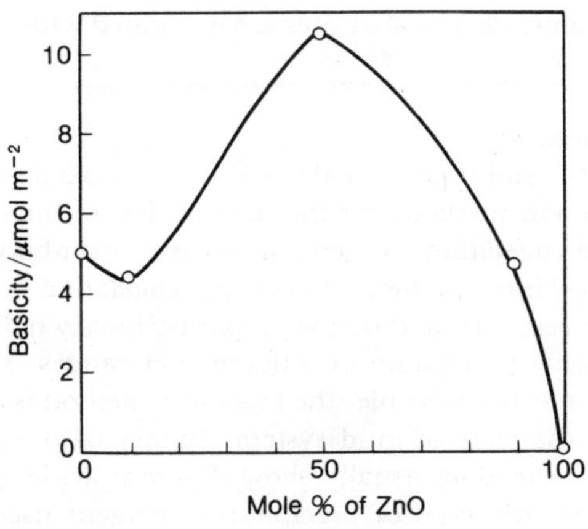

Fig. 3.47 Change in basicity at $pK_a=12.2$ of Al_2O_3-ZnO catalysts calcined at 773 K with change in ZnO content.
(Reproduced with permission by K. Tanabe *et al.*, *J. Catal.*, **57**, 37 (1979)).

generation of Al_2O_3-ZnO.[3] For SiO_2-MoO_3, the maximum number of acid sites is generated at a composition of 90% SiO_2. For Al_2O_3-ZnO, the maximum is obtained at a composition of 50% Al_2O_3. An example of the generation of both acid and base sites is shown in Fig. 3.48 for TiO_2-MgO.[4] At high MgO content, base sites

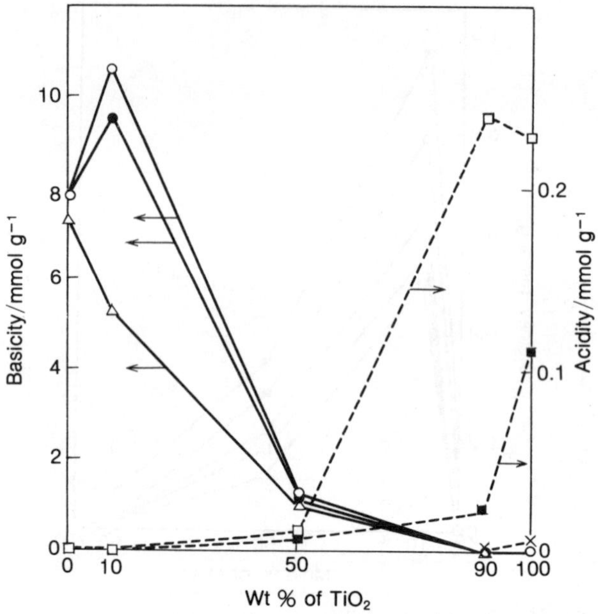

Fig. 3.48 Acidity and basicity of $MgO-TiO_2$ of different compositions: ○, $H_-=12.2;2.2$; ●, $H_-=15.0$; △, $H_-=17.2$; □, $H_0=6.8$; $H_0=4.8$; ×, $H_0=1.5$.

are generated, and becomes maximum at a composition of 90% MgO. At high content of TiO_2, base sites diminish and acid sites are generated with maximum at a TiO_2 content of 90%.

b. Preparation method

Amount, strength, and type of acid site on binary oxides are affected by preparation methods. Common methods for the preparation of binary oxides are kneading, coprecipitation, and cogelation. In general, good contact between two components is attained by the cogelation method. However, cogelation is not always applicable. Coprecipitation is a versatile method for preparing binary oxides except in some cases where two components precipitate in different pH ranges. For the binary oxides in which coprecipitation is not possible, the kneading method is applicable. Two components are mixed in the state of mud system. Binary oxides prepared by cogelation, coprecipitation, and kneading usually show different acidic properties.

In coprecipitation, the type of precipitation reagent used also affects the acidic properties of the resulting binary oxide. An example is shown in Fig. 3.49 for TiO_2-ZnO.[5,6] One TiO_2-ZnO is prepared using urea as the precipitation reagent while another is prepared using aqueous ammonia. As urea is heated, it decomposes into ammonia and carbon dioxide:

$$NH_2-CO-NH_2 \longrightarrow NH_3 + CO_2 \uparrow$$

Ammonia dissolves into the solution and carbon dioxide evolves. As a result, the pH of the solution increases. The pH values at different parts in solution at a certain time

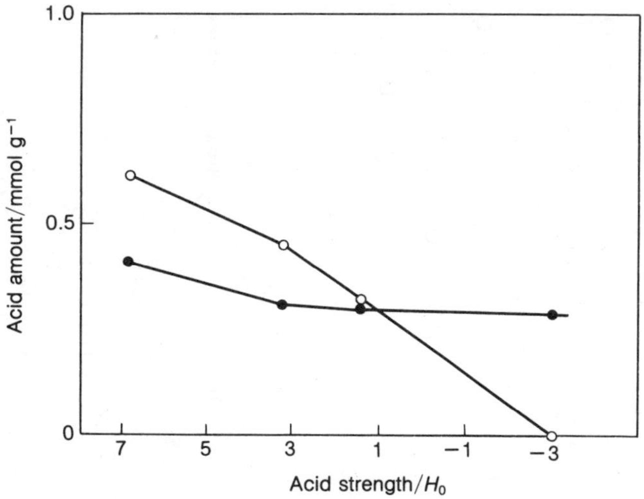

Fig. 3.49 Acidity distribution of TiO_2-ZnO prepared by homogeneous precipitation (○ : urea) and by heterogeneous precipitation (● : ammonia)

are the same, because urea decomposes at every point in the solution. This method is called "homogeneous precipitation." On the other hand, when aqueous ammonia is used as the precipitating reagent, the pH values differ from one point to another in the solution during precipitation. This method is called "heterogeneous precipitation."

The acid site distrubution of TiO_2-ZnO prepared by homegeneous and heterogeneous precipitation are different as shown in Fig. 3.49. Heterogeneous precipitation generates acid sites stronger than $H_0 = -3.0$ while homogeneous precipitation gives acid sites weaker than $H_0 = -3.0$. The same tendency is observed for the TiO_2-SiO_2 binary oxide.[6)]

c. Pretreatment temperature

The pretreatment temperature is crucial for acid site generation. Wet precipitate, a precursor of binary oxide, usually contains an NH_4^+ ion to balance the charge. For acid site generation, it is necessary to heat the precursor to convert NH_4^+ into H^+ as well as dehydrate to form oxide. An example of the change in structure and acid site generation during heat treatment is schematically illustrated in Fig. 3.50 for $SiO_2-Al_2O_3$ as an example. In the precipitating solution, condensation of Si−OH with Al−OH occurs to form Si−O−Al linkage. On heating, conversion of hexacoordinated Al into tetracoordinated Al occurs accompanied by liberation of water. Heating at higher temperatures causes decomposition of NH_4^+ into H^+ and NH_3. The NH_3 is desorbed from the surface and H^+ is retained on the surface. The H^+ acts as Brønsted acid. Further heat treatment causes dehydroxylation forming Lewis acid site. In the last stage of heat treatment, the binary oxide is in an amorphous form. Heat treatment at higher temperatures facilitates crystallization resulting in stabilization of surface state and decreases in surface area. Therefore, acid sites decrease in number and strength.

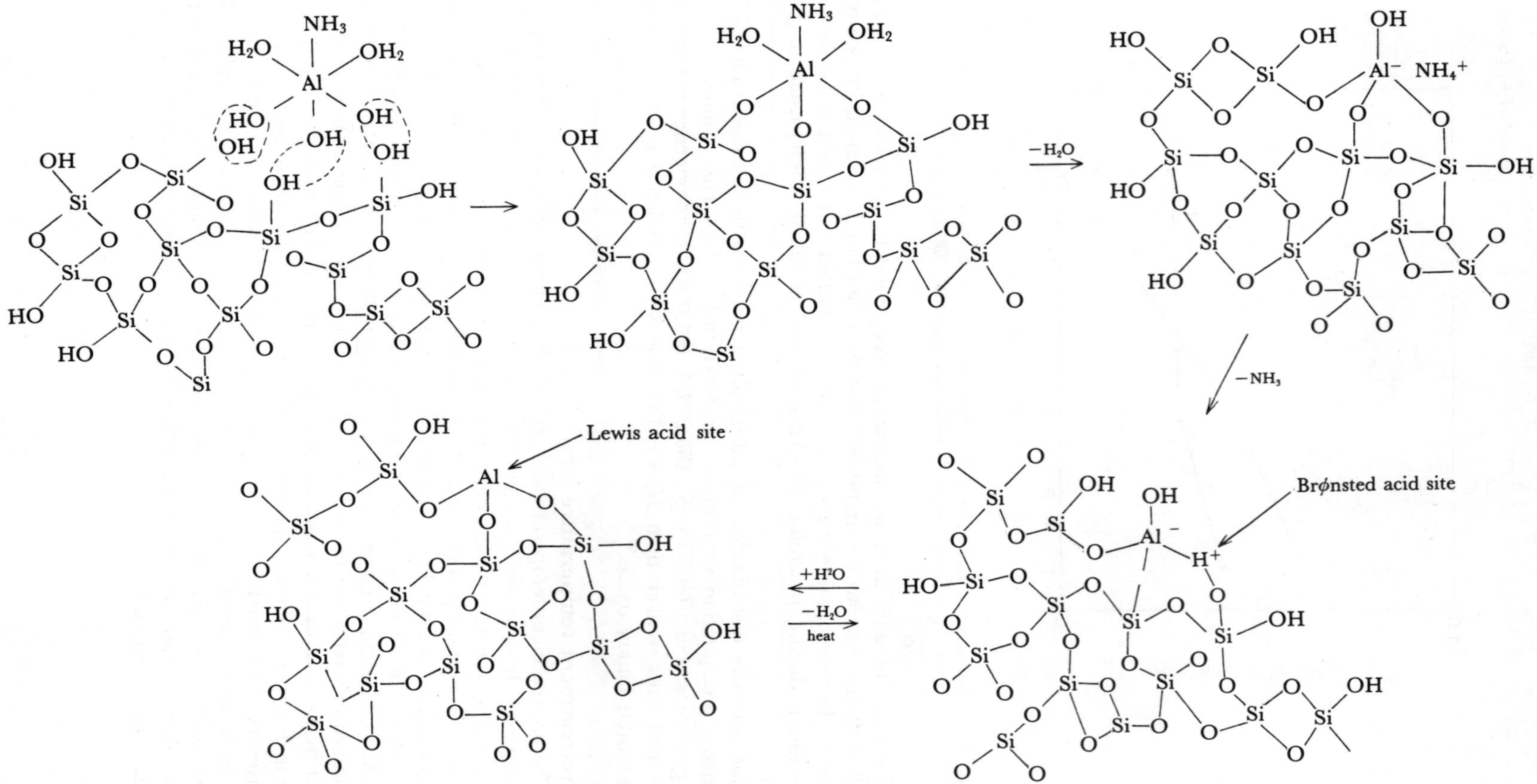

Fig. 3.50 Generation of acid sites on $SiO_2-Al_2O_3$

The above pattern of acid site generation and elimination is observed not only for $SiO_2-Al_2O_3$ but for most cases.

C. Acid-Base Properties of Some Binary Oxides

a. Binary oxides containing SiO_2

$SiO_2-Al_2O_3$ Silica-alumina is a representative acidic binary oxide which has been extensively studied. The concept of the solid acid catalyst was established through studies on silica-alumina. Cumulative studies have corroborated that the acidic centers on the solid surfaces act as catalytically active sites. Among the many reasons establishing the concept of solid acids, the following four are of primary importance.

i) The existence of acid sites has been confirmed by indicator color change or IR studies on adsorbed molecules, and their catalytic activities correlate with the number of acid sites.

ii) The catalytic activity of the solid is poisoned by basic molecules.

iii) The solid catalyzes the reaction well-known as acid-catalyzed reactions in homogeneous acidic media.

iv) Mechanistic studies of the reactions by e.g. tracer study and product distribution indicate that reaction intermediates are cations formed by the interaction of acid sites with the reactants.

All the above studies were conducted using silica-alumina.

The acid sites on $SiO_2-Al_2O_3$ are stronger than $H_0 = -8.2$,[7] and of both Brønsted and Lewis acid types.[8] The Lewis acid sites increase when the pretreatment temperature is raised. They convert into Brønsted acid sites on adsorption of water.[9,10]

Silica-alumina catalyzes a wide variety of reactions, and good correlations between the number of acid sites and catalytic activities are observed in many cases including propene polymerization,[11] cumene cracking,[11] and *o*-xylene isomerization.[12]

Mechanistic studies of butene isomerization on $SiO_2-Al_2O_3$ were extensively performed in tracer studies using deuterium and ^{14}C.[13-15] The isomerization invoves intermolecular hydrogen transfer, and the reaction intermediates are sec-butyl cations formed by the addition of an H^+ on the surface to the carbon atom in the butene molecule. The active sites are Brønsted acid sites.

SiO_2-TiO_2 Silica-titania shows strongly acidic properties.[16] Basic molecules such as ammonia, pyridine and butylamine are strongly adsorbed on the surfaces. The amounts of these molecules retained on the surfaces at the outgassing temperatures 273 − 823 K are larger than those on silica-alumina. The strongest acid sites exceed acid strength of $H_0 = -8.2$. Though the number of acid sites stronger than $H_0 = -3.0$ shows maximum at the composition of 50 mol % TiO_2, the catalytic activities for amination of phenol with ammonia, and 1-butene and *cis*-2-butene isomerization show maxima at the composition of TiO_2 90%.

Existence of base sites on SiO_2-TiO_2 pretreated at temperatures higher than 823 K is demonstrated by the catalytic features for butene isomerization.[17] The ratios of *cis* to *trans*-2-butene produced in 1-butene isomerization are high, and an intramolecular hydrogen transfer is involved in the reaction. Appearance of base sites is caused by the reduction of Ti^{4+} to Ti^{3+} on high temperature treatment.

SiO_2-MoO_3 As shown in Fig. 3.46, SiO_2-MoO_3 shows acidic properties. The SiO_2-MoO_3 containing 10% MoO_3 shows the maximum number of acid sites in the

H_0 range $-3.0 \sim +6.8$. The acidities do not change much with pretreatment temperature. Acid sites of strength H_0 $-3.0 \sim 6.8$ appeared on pretreatment at 573 K, and remained unchanged up to 773 K, and slightly decreased at 873 K. The presence of Brønsted acid sites is suggested by the incorporation of surface D atoms into the products as well as by the involvement of an intermolecular H transfer in butene isomerization. The acid sites are eliminated on reduction of Mo^{6+} in the binary oxide to Mo^{5+}.

Corresponding to the maximum acidity, the catalytic activities for paraldehyde depolymerization and *cis*-2-butene isomerization show maxima at the composition of 10% MoO_3. The active sites for butene isomerization are poisoned by ammonia but not by CO_2, indicating that the acid sites are the active sites for the reaction.[18] This is in contrast to the catalytic behavior of the single component oxide MoO_3 in which the active sites are poisoned by CO_2 but not by ammonia.

SiO_2-ZnO Both acid and base sites are present on the surface of SiO_2-ZnO.[19,20] The number of acid sites stronger than $H_0=1.5$ are the highest at the composition of ZnO 30%, while those of $H_0 < -3.0$ are highest at ZnO 70%. The acid sites are of the Lewis acid type as determined by IR measurement of adsorbed pyridine.

The catalytic properties for butene isomerization were examined. By running the coisomerization of d_0/d_8 butene, it was found that an intramolecular H transfer is involved in the reaction. The active sites are poisoned by basic molecules such as ammonia and pyridine, but not by an acidic molecule CO_2. Lewis acid sites are considered to be the active sites for butene isomerization. The reaction is initiated by the abstraction of an H^- from the reactant molecule by Lewis acid sites on the catalyst.[20]

SiO_2-MgO Silica-magnesia has a large number of acid as well as base sites. The number of acid sites exceeds that of $SiO_2-Al_2O_3$.[7] However, the strength of the acid sites is weaker than $H_0= -3.3$.[7] The maximum number of acid sites is obtained at the composition of MgO 50%, while the number of base sites increases with the content of MgO.[22] On adsorption of pyridine, the band at 1540 cm^{-1}, which is ascribed to pyridinium ion, is not found, indicating that the surface OH groups are not such strong acid sites as to convert pyridine into pyridinium ion.[21] The acidic strength of OH groups on the silica magnesia is determined to be between those of SiO_2 and MgO by means of IR band shift on adsorption of acetone.[21]

Typical acid-catalyzed reactions such as cracking of hydrocarbons and dehydration of alcohols are catalyzed by SiO_2-MgO. The catalytic activity for hydrogen transfer between ethanol and acetone correlates with the number of base sites. The formation of 1,3-butadiene from ethanol occurs on SiO_2-MgO.[22] Silica magnesia of 85% MgO composition shows the maximum activity. At this composition, both acid and base sites exist on the surface, and the reaction proceeds by acid–base bifunctional action. The formation of 1,3-butadiene from ethanol involves several successive steps. Each step proceeds on acid sites and base sites independently. Silica magnesia containing 85% MgO possesses well balanced acid and base sites and catalyzes the total reaction effectively.

Other binary oxides containing SiO_2 Silica-iron oxide possesses maximum acid sites at the composition of SiO_2 10%. At this composition, the maximum activity for 1-butene isomerization is observed.[23]

Silica-zirconia shows strongly acidic properties. The reported acid strengths measured by indicator method are $H_0 = -8.2$[24] and -5.6.[25] The maximum acidity is observed at a calcination temperature of 773 K. At a pretreatment temperature of 773 K, the catalytic activity for paraldehyde decompositon also shows maximum. The basic sites measured by calorimetric titration with trichloroacetic acid amount to 0.60 mmol g^{-1}. The catalytic activity for decomposition of hydrogen peroxide correlates with the basicity.[26)]

Appearance of acid sites is also reported for the binary oxides, SiO_2-ThO_2 and SiO_2-WO_3.[27)]

b. Binary oxides containing Al_2O_3

Al_2O_3-MgO Both acid and base sites are present on the binary oxide Al_2O_3-MgO.[28-31)] The acid sites measured by IR spectroscopy of absorbed pyridine are of Lewis acid type; no Brønsted acid sites are detected by pyridine adsorption method. Three different types of Lewis acid sites are ascribed to Al^{3+} cations in the alumina phase, cations in $MgAl_2O_4$ phase, and cations in MgO phase. The base strength is weakened by increasing the MgO content. The OH groups on the surface are not capable of donating H^+ to the pyridine molecule, but show acidic character. The acid strength of the OH groups determined by IR study of adsorbed acetone decreases with increasing MgO content.

$Al_2O_3-TiO_2$ The acidic properties of $Al_2O_3-TiO_2$ vary with preparation method.[32)] For the binary oxides prepared by heterogeneous coprecipitation, the composition of 90% Al_2O_3 and 10% TiO_2 gives the maximum acid sites. In contrast, the binary oxide prepared by homogeneous coprecipitation with urea possesses small numbers of acid sites, in particular, the binary oxide containing 90% Al_2O_3 has no acid sites. The acid sites are mostly of Lewis acid type which do not convert into Brønsted acid sites on adsoprtion of water. The basic sites appear only in the presence of the appropriate amount of water, addition of 40 μmol m^{-2} H_2O giving maximum basicity.

The binary oxides catalyze butene isomerization and 2-butanol dehydration. For butene isomerization, the alkyl cation mechanism is the main path. Contribution of the carbanion mechanism increases with increased TiO_2 content. A good correlation is observed between the acidity and the activity for dehydration.

Al_2O_3-ZnO Both acid and base sites are generated on the binary oxide Al_2O_3-ZnO. The numbers of acid sites on the binary oxides of different compositions are higher for the samples prepared from chlorides than for those prepared from nitrates. As the content of ZnO increases, the number of acid sites decreases monotonously. The number of basic sites stronger than $H_- = 12.2$ show maximum at the composition of ZnO 50% as shown in Fig. 3.47.

The oxides show catalytic activities for alkylation of phenol with methanol, butene isomerization, and carene isomerization. The alkylation activity correlates with acidity, while butene isomerization activity correlates with basicity. 3-Carene undergoes selective double bond migration to 2-carene on the binary oxide containing 90% ZnO, while menthadiene and cymene are produced by a three-membered ring opening over the catalyst containing small amounts of ZnO, acidic binary oxides.

$Al_2O_3-MoO_3$ MoO_3/Al_2O_3 is widely used a hydrodesulfurization and hydrodenitrogenation catalyst by modification with Co or Ni. Alumina and both Co- and Ni-impregnated alumina contain only Lewis acid sites,[33)] whereas alumina impregnated with Mo, either in the presence or absence of Co or Ni, contains both Lewis

and Brønsted acid sites.[33,34]

The number and strength of the acid sites vary with MoO_3 content. Most of acid sites on $MoO_3-Al_2O_3$ containing less than 12.5 wt% MoO_3 are stronger than $H_0=-3.0$, while acid site strength widely distribute in $H_0=6.8\sim-3.0$ for $MoO_3-Al_2O_3$ containing more than 12.5 wt% MoO_3.[35] Reduction of $MoO_3-Al_2O_3$ increases in the sites irreversibly adsorbing ammonia. The structural change on reduction and ammonia adsorption model are proposed as shown in Fig. 3.51.[35] Structure *A* represents Brønsted acid site. On reduction, anion vacancies are generated (structures *B* and *C*) and act as Lewis acid to adsorb ammonia (structures *D* and *E*).

c. Binary oxides contianing TiO_2

TiO_2-MgO The binary oxide TiO_2-MgO shows both acidic and basic properties. As shown in Fig. 3.48, acid sites prevail on the binary oxides rich in TiO_2, whereas

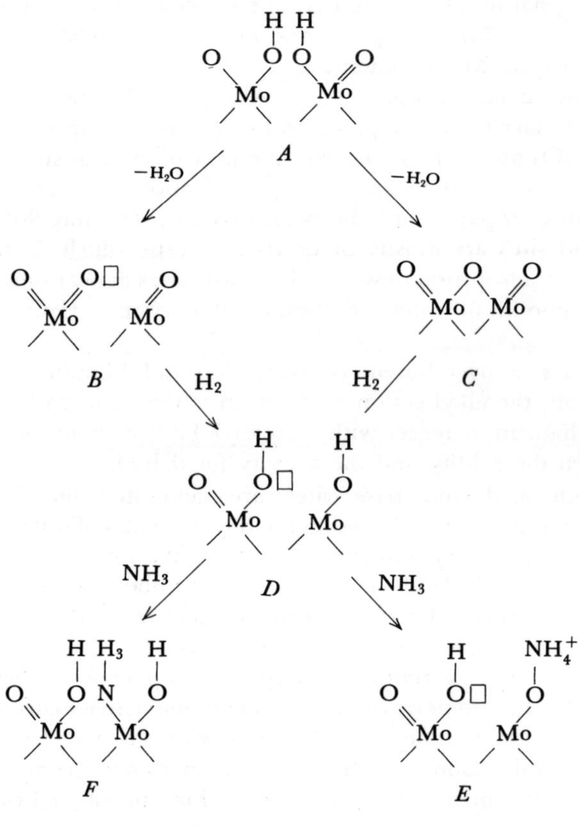

Fig. 3.51 Generation of acid sites on MoO_3/Al_2O_3 by reduction with hydrogen, and adsorption of ammonia.
(Reproduced with permission by M. Yamada, *et al.*, *Nippon Kagaku Kaishi* **1976** 230)

basic sites are present on the oxides rich in MgO. At the composition TiO_2:MgO 1:1, acid sites and base sites coexist. The acid strength is at most H_0=3.3, and the base strength H_-=17.2.

The variations of the activity for decomposition of diacetone alcohol, dehydration of 4-methyl-2-pentanol, and alkylation of phenol are shown as a function of the composition of the binary oxide in Fig. 3.52.[4] The catalytic activity for decomposition of diacetone alcohol correlates with the number of base sites, while that for the dehydration of 4-methyl-2-pentanol correlates with the number of acid sites. The alkylation of phenol with methanol is most effectively catalyzed by a binary oxide possessing both acid and base sites, the oxide containing TiO_2 and MgO in a 1:1 ratio being the most active.

The OH groups on the binary oxides of different compositions show different acidic strength. The acidic strength, measured by O−H frequency shift in IR absorption on adsorption of acetone, decreases with increasing MgO content.[31]

TiO_2-ZrO_2 Strong acid sites, H_0= −5.6, appear on the oxide together with base sites.[36,37] In particular, the oxide containing equal amounts of TiO_2 and ZrO_2 shows the maximum amount of strong acid and base sites. The oxide shows activity for methylcyclohexene oxide,[37] and nonoxidative dehydrogenation of ethylbenzene.[38,39] For both reactions, acid-base pair sites operate as the active sites.

TiO_2-ZnO The acidic properties of the binary oxide TiO_2-ZnO vary with the preparation method, as shown in Fig. 3.49.[5] Heterogeneous coprecipitation gives stronger acid sites than homogeneous coprecipitation. The former method produces

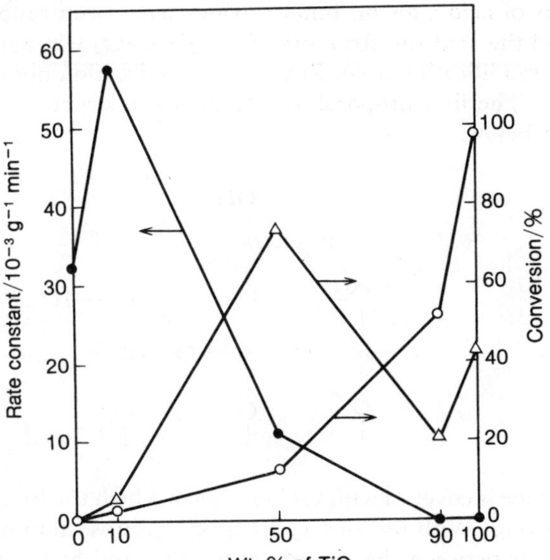

Fig. 3.52 Activity variations of $MgO-TiO_2$ of different compositions.
●; Decomposition of diacetone alcohol, ○; Dehydration of 4-methyl-2-pentanol, △; Alkylation of phenol with methanol.

acid sites stronger than $H_0 = -3.0$ while the latter produces acid sites weaker than $H_0 = -3.0$. The acid sites are of both Brønsted and Lewis types.

A characteristic feature of the oxide is high activity for ethylene hydration.[5] The high activity as well as high selectively for ethanol are due to the moderate strength of the acid sites.

TiO_2-SnO_2 Acid sites of $H_0 < -3.0$ are generated on the surface of the binary oxide TiO_2-SnO_2.[40] The number of acid sites become maximum at the composition TiO_2 50%. At this composition, the oxide shows the maximum catalytic activity for butene isomerization, which proceeds by the carbenium ion mechanism. The basic sites on component oxides, which are associated with Ti^{3+} and Sn^{3+}, are eliminated in the binary oxides.

Others The binary oxide ZrO_2-SnO_2 has both acid and base sites.[41] The maximum strength of the acid sites is observed to be $H_0 = -3.0$ at the composition ZrO_2 90%. The acid sites on oxides of other compositions are weaker than $H_0=1.5$. The number of base sites on the binary oxide is less than on the component oxide ZrO_2. The catalytic activities for cyclopropane ring opening and 2-butanol dehydration correlate with the acidity, while those for 1-butene isomerization and diacetone alcohol decomposition correlate with basicity.

Tungsten oxide mounted on ZrO_2 has acid sites as strong as $H_0 = -14.52$.[42] The oxide shows activity for acylation of toluene with benzoic anhydride at 303 K, and butane skeletal isomerization to isobutane at 373 K. The maximum activity is obtained when the oxide is calcined at 1073 – 1273 K.[42]

D. Structural and Quantum Chemical Studies for Acid Sites on Binary Oxides

The structures of acid sites on binary oxides have been proposed exclusively for $SiO_2-Al_2O_3$, and the acid site structure of $SiO_2-Al_2O_3$ was applied to other binary oxides with some modification according to the coordination numbers and valences of the metal cations. The first proposal for the acid site structure was made by Hansford,[43] as shown below.

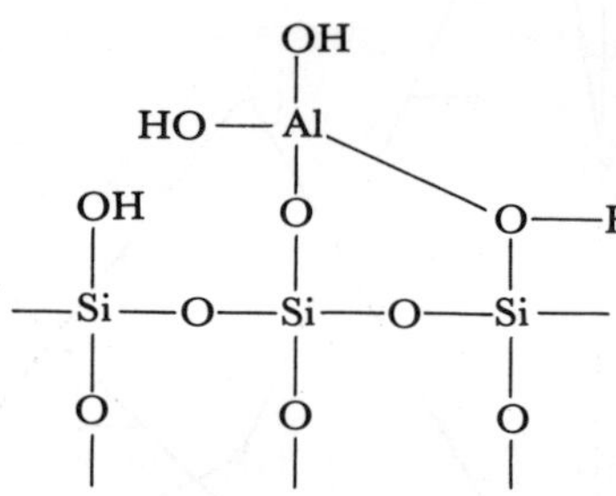

The silica gel surface is covered with OH groups in which the binding power of oxygen to hydrogen is strong. With the surface hydroxyl groups, aluminum hydroxide react to split out of water between the aluminum hydrate and hydroxyl group of the silica gel surface. The binding power of oxygen is weakened by coordination with aluminum of the hydroxyl oxygen. As a result, the hydrogen tends to act as an acid.

Thomas[44] proposed the acid site structure shown below three years after the proposal by Hansford.

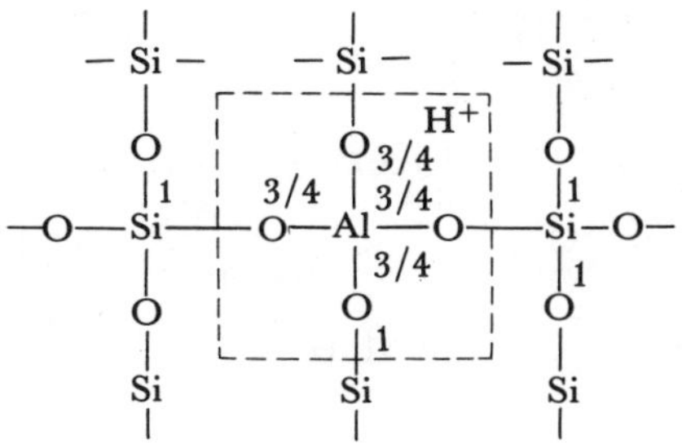

In the structure of silica, a silicone atom is removed and replaced with a tetrahedral aluminum atom. The AlO_4 part is unsatisfied by a whole valence unit. The positively charged hydrogen ion must be associated with four oxygen atoms to balance the electrostatic neutrality.

Tamele proposed the structure shown below.[11)]

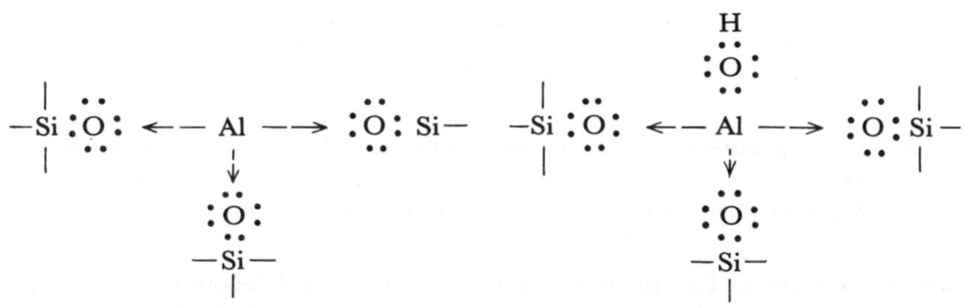

Lewis acid **Brønsted acid**

Coordination of Al atoms carrying three positive charges with oxygens attached to Si carrying four positive charges results in the displacement of electrons by the proximity of Si ions, as indicated by the arrows. In aqueous medium, the electron pair is denoted by water, the hydroxyl group becomes a part of the solid structure, with hydrogen held not too strongly by electrostatic attraction. In the thoroughly dehydrated state the surface becomes Lewis acid.

Definite identification of the acid site of $SiO_2-Al_2O_3$ was difficult because of its amorphous structure. However, the appearance of well-defined crystalline structures of zeolites enabled further detailed studies of the acid site structures. The acid site structures revealed on zeolites, such as a dislodged AlO_2^- unit, may be feedbacked to the acid site structure of amorphous binary oxides. The acid site structures of zeolites are described in Section 3.4.

Molecular orbital calculations (CNDO/2) on cluster models were performed to elucidate the acid-base properties of several solid acids and bases such as silica, metal ion-exchanged silicas, and silica–alumina.[45)] Charge density, LUMO, and HOMO were regarded as measures of the strengths of protonic acidity, basicity, and Lewis acidity, respectively. In the models of silica–alumina, Al atom was substituted for Si atom. To make the model structure neutral, one hydrogen atom was added. The stablest structure was the one in which hydrogen is located just below the bridging oxygen (O_b). The cluster models of silica–alumina are shown in Fig. 3.53. Charge densities

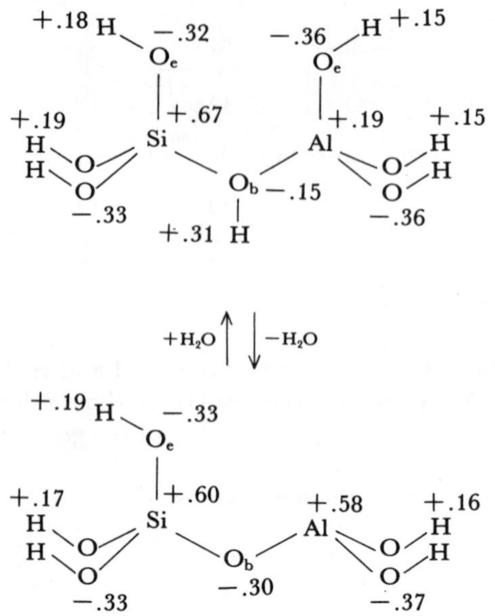

Fig. 3.53 Charge densities calculated for models of silica-alumina and its dehydrated form.
(Reproduced with permission by W. Grabowski *et al.*, **61**, 160 (1981)).

are shown in the figure. The positive charge on H located below the O_b is large and comparable to that in hydronium ion (H_3O; +0.32 by CNDO/2).

Quantum mechanical calculation was also done by Yoshida *et al.* for the structure of acid sites of $SiO_2-Al_2O_3$, in particular for the definite position of the H^+.[46] Model structures are shown in Fig. 3.54. Structure (I) is the structure of SiO_2 on which the calculation is based. In the model structure, H's other than H_s are placed instead of the SiO_2 unit of the actual structure to simplify calculations.

The strength of acid site is expressed by the energy required to dissociate H_s^+, ΔE. Calculation on structure (II) reveals that the ΔE is lowered by the replacement of Si atom by Al atom, and that the bond of Al to the O_s atom bridging the Si atom is very weak as compared with the other Al−O and Si−O bonds. The calculation implies that the H_s becomes protonic by coordination of Al to the O_s. The generation of Brønsted acid site is explained by the coordination of the Al atom to the O_s. The coordination of the Al atom to the O_s atom is of Lewis acid base interaction, and the stronger the interaction, the stronger the Brønsted acid of H_s. Lewis acid strength is expressed by the level of LUMO. The LUMO level is higher for $Mg(OH)_2$ than for $Al(OH)_3$ by 1.7 eV, indicating that Lewis acidity is weaker for $Mg(OH)_2$ than for $Al(OH)_3$. It is expected that the Brønsted acid generated by the coordination of $Mg(OH)_2$ to the O_s is weaker than that generated by the coordination of $Al(OH)_3$. Calculation based on structure (III) suggests the generation of a weaker Brønsted acid compared with the Brønsted acid of structure (II). Experimental results indicating that the acid sites on SiO_2-Al_2O are stronger than those on SiO_2-MgO support this view.

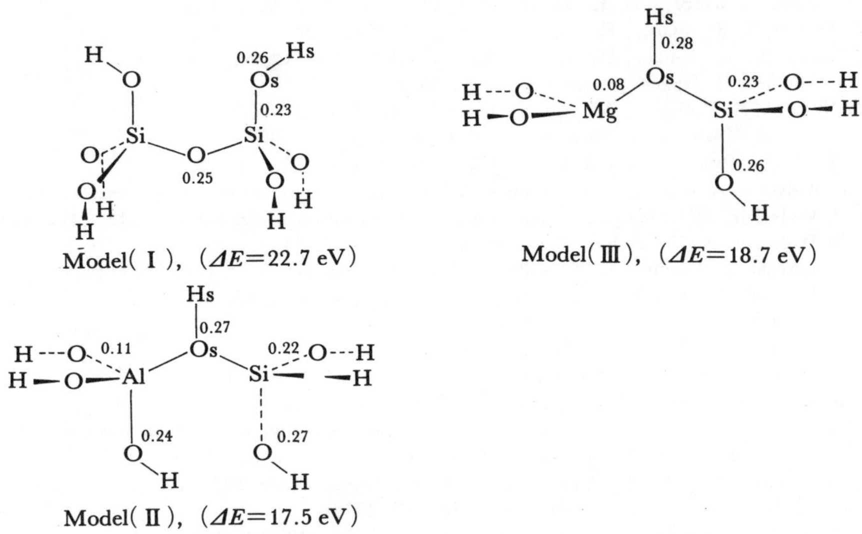

Fig. 3.54 **Models for calculation of acid strength.**

The quantum-chemical cluster models of acid base sites of binary oxides have recently been reviewed by Zhidomirov and Kazansky.[47)] Concerning the Brønsted acid sites of the binary oxide silica-alumina, they summarize as follows. The acid sites acting in typical acid-catalyzed reactions are bridged hydroxyl groups and water molecules coordinated on a trigonal aluminum atom. These centers are characterized by approximately equal surface densities and atomic catalytic activities.

REFERENCES

1. W. Ueda, Y. Moro-oka, T. Ikawa, *Shokubai* (Catalyst) **28**, 208 (1986) (in Japanese).
2. K. Maruyama, H. Hattori, K. Tanabe, *Bull. Chem. Soc. Jpn.*, **50**, 86 (1977).
3. K. Tanabe, K. Shimazu, H. Hattori, K. Shimazu, *J. Catal.*, **57**, 35 (1979).
4. K. Tanabe, H. Hattori, T. Sumiyoshi, K. Tamaru, T. Kondo, *J. Catal.*, **53**, 1 (1978).
5. K. Tanabe, C. Ishiya, I. Matsuzaki, I. Ichikawa, H. Hattori, *Bull. Chem. Soc. Jpn.*, **45**, 47 (1972).
6. K. Tanabe, M. Itoh, Morishige, H. Hattori, in: *Preparation of Catalysts* (B. Delmon, P. A. Jacob, G. Poncelet, eds.) Elsevier, Amsterdam (1976) p.65.
7. H. A. Benesi, *J. Am. Chem. Soc.*, **78**, 5490 (1956); *J. Phys. Chem.*, **61**, 970 (1957).
8. E. P. Parry, *J. Catal.*, **2**, 371 (1963).
9. M. R. Basila, T. R. Kantner, K. H. Rhee, *J. Phys. Chem.*, **68**, 3197 (1964).
10. M. R. Basila, T. R. Kantner, *J. Phys. Chem.*, **70**, 467 (1967).
11. K. Tamele, *Disc. Faraday Soc.*, **8**, 270 (1950).
12. J. Ward, R. C. Hansford, *J. Catal.*, **13**, 354 (1969).
13. J. W. Hightower, W. K. Hall, *J. Am. Chem. Soc.*, **89**, 7778 (1967).
14. J. W. Hightower, H. R. Gerberich, W. K. Hall, *J. Catal.*, **7**, 57 (1967).
15. J. W. Hightower, W. K. Hall, *J. Phys. Chem.*, **71**, 1015 (1967).
16. M. Itoh, H. Hattori, K. Tanabe, *J. Catal.*, **35**, 225 (1974).
17. H. Hattori, M. Itoh, K. Tanabe, *J. Catal.*, **43**, 192 (1976).

18. H. Hattori, K. Maruyama, K. Tanabe, *Bull. Chem. Soc. Jpn.*, **50**, 2181 (1977).
19. K. Tanabe, T. Sumiyoshi, H. Hattori, *Chem. Lett.*, **1972**, 723.
20. T. Sumiyoshi, K. Tanabe, H. Hattori, *Bull. Jpn. Petrol. Inst.*, **17**, 65, (1975).
21. J. A. Lercher, H. Noller, *J. Catal.*, **77**, 152 (1982).
22. H. Niiyama, E. Echigoya, *Bull. Jpn. Petrol. Inst.*, **14**, 83 (1972).
23. T. Iizuka, H. Tatsumi, K. Tanabe, *Aust. J. Chem.*, **35**, 919 (1982).
24. V. A. Dzisko, Proc. 3rd Intern. Congr. Catal., Amsterdam, 1964, p.422.
25. S. P. Walvekar, A. B. Halgeri, S. Ramanna, T. N. Srinivasan, *Revue Roumanaine Chim.*, **21**, 237 (1976).
26. S. P. Walvekar, A. B. Halgeri, S. Ramanna, T. N. Srinivasan, *Fertilizer Tech.*, **13**, 241 (1976).
27. S. P. Walveker, A. B. Halgeri, *Technology* (India), **11**, 73 (1974).
28. J. A. Lercher, C. Colombier, H. Noller, *J. Chem. Soc., Faraday Trans. 1*, **80**, 949 (1984).
29. J. A. Lercher, C. Colombier, H. Noller, *React, Kinet. Catal. Lett.*, **23**, 365 (1983).
30. J. A. Lercher, *React. Kinet. Catal. Lett.*, **20**, 409 (1982).
31. J. A. Lercher, *Zeit. Phys. Chem. Neu. Folg.*, **129**, 209 (1982).
32. E. Rodenas, H. Hattori, T. Yamaguchi, K. Tanabe, *J. Catal.*, **69**, (1981).
33. F. E. Kiviat, L. Petrakis, *J. Phys. Chem.*, **77**, 1232 (1973).
34. J. Laine, J. Brito, S. Yunes, Proc. 3rd Intern. Conf. Chemistry and Uses of Molybdenum, (H. F. Barry, P. C. H. Mitchel, ed.), Ann. Arbor, 1979, p.111.
35. M. Yamadaya, T. Kabe, M. Oba, Y. Miki, *Nippon Kagaku Kaishi*, **1976**, 227 (in Japanese).
36. J. L. Lercher, *Zeit. Phys. Chem. Neue Folg.*, **118**, 209 (1979).
37. K. Arata, K. Akutagawa, K. Tanabe, *Bull. Chem. Soc. Jpn.*, **49**, 390 (1976).
38. I. Wang, W.-F. Chang, R.-J. Shiau, J.-C. Wu, C.-S. Chung, *J. Catal.*, **83**, 42 (1983).
39. J.-C. Wu, C.-S. Chung, C.-L. Ay, I. Wang, *J. Catal.*, **87**, 98 (1984).
40. M. Itoh, H. Hattori, K. Tanabe, *J. Catal.*, **43**, 192 (1976).
41. G.-W. Wang, H. Hattori, K. Tanabe, *Bull. Chem. Soc. Jpn.*, **56**, 2407 (1983).
42. K. Arata, M. Hino, Proc. 9th Intern. Congr. Catal., 1988 Calgary, p.1727.
43. R. C. Hansford, *Ind. Eng. Chem.*, **39**, 39 (1947).
44. C. L. Thomas, *Ind. Eng. Chem.*, **41**, 2564 (1949).
45. W. Grabowski, M. Misono, Y. Yoneda, *J. Catal.*, **61**, 103 (1980).
46. H. Hawakami, S. Yoshida, T. Yonezawa, *J. Chm. Soc., Faraday Trans.*, **2**, 80, 205 (1984).
47. G. M. Zhidomirov, V. B. Kazansky, *Adv. Catal.*, **34**, 131 (1986).

3.3 CLAY MINERALS

3.3.1 Sheet Silicates

A. Structure of Sheet Silicate[1)]

Sheet silicates can be classified into two main groups, two-layered silicates and three-layered silicates. Two-layered silicates such as kaolinite have idealized formulas $Al_2Si_2O_5(OH)_4$ and may be considered to be the condensation products of $Al(OH)_6$ octohedral sheets with tetrahedral sheets of $Si_2O_3(OH)_2$.

In three-layered silicates, an octahedral layer is sandwiched between two tetrahedral layers. They are further divided into those having a dioctahedral structure and those having a trihedral structure. The former have the idealized formula $Al_2(Si_4O_{10})(OH)_2$ in electrically neutral structure, where only two-thirds of all possible octahedral sites are occupied by Al^{3+}. The latter have the idealized formula $Mg_3(Si_4O_{10})(OH)_2$, and the Mg^{2+} ions occupy all three such sites in a unit cell. The diversity of clays arises from deviations with respect to the ideal formulas. Aluminum can be substituted for silicon in the tetrahedral layer to a maximum of 15%. In the smectites, a number of metallic cations such as Li^+, Mg^{2+}, and Fe^{2+} can replace Al^{3+} in octahedral layers and Al^{3+} in octahedral layers and Al^{3+} can replace Si^{4+} in tetrahedral layers. By the substitution, three-layered sheets take on a surplus negative charge. In the synthetic smectites it is relatively easy to arrange for the OH^- ions to be replaced by F^-. Formulas of some important sheet silicates are listed in Table 3.24. The structure of montmorillonite is given in Fig. 3.55.

TABLE 3.24 Idealized formulas of some important smectites

minerals	idealized formula
montmorillonite	$(Al_{2-y}Mg_y)Si_4O_{10}(OH)_2$ nH_2O
beidellite	$Al_2(Si_{4-x}Al_x)O_{10}(OH)_2$ nH_2O
nontronite	$Fe(III)_2(Si_{4-x}Al_x)O_{10}(OH)_2$ nH_2O
saponite	$Mg_3(Si_{4-x}Al_x)O_{10}(OH)_2$ nH_2O
hectrite	$(Mg_{3-y}Li_y)Si_4O_{10}(OH)_2$ nH_2O
sauconite	$Zn_3(Si_{4-x}Al_x)O_{10}(OH)_2$ nH_2O

B. Pillared Interlayered Clays

The intercalation of smectites with polar organic molecules is long established and well documented. Recently, versatile methods for preparing novel types of molecular sieves, designated as cross-linked smectites or pillard interlayered clays have been developed. The method involves cross-linking of smectites unit layers with well-defined oligomeric species derived from metal hydroxides or other inorganic compounds. The interlayer distance in the resulting structure is determined by the molecular dimensions of the cross-linking agent, while the lateral distance can be regulated by the charge density of the smectite and/or by the extent of cross linking. The formation of the pillared smectites is shown schematically in Fig. 3.56.[2,3)] The polymeric aluminum cation, $(Al_{13}O_4(OH)_{24+x}(H_2O)_{12-x})^{+(7-x)}$ is intercalated in the interlayers by ion-exchange technique. The material is then calcined to convert the polymeric cations into the oxide. The chemical process during calcination has been followed by ^{27}Al and ^{29}Si MMR technique.[4-6)] In this case, the basal spacing of 1.7 to 1.8 nm or interlayer spacing of 0.7 – 0.8 nm can be attained as expected from the size of the aluminum polymer. Besides aluminum polymers, polymeric cations such as $[Zr_4(OH)_{6-n}(H_2O)]_{n+8}$, [7)] $[Fe_3(OCOC_3H_7)_7OH]^+$, [8)] $[Nb_6Cl_{12}]^{+}_{2}{}^{+}_{-3}$, [9)] $[Ta_6Cl_{12}]^{2+}$ [9)] are also known to be useful for expanding the interlayer spacing.

3.3.2 Acidity of Sheet Silicate and Pillared Clays

Acidity of clays stems from different sources. The water molecules belonging to the hydration shell of exchangeable cations are subjected to a strong electrical polarizing field; therefore they have a degree of dissociation several orders of magnitude larger than liquid water.[10)]

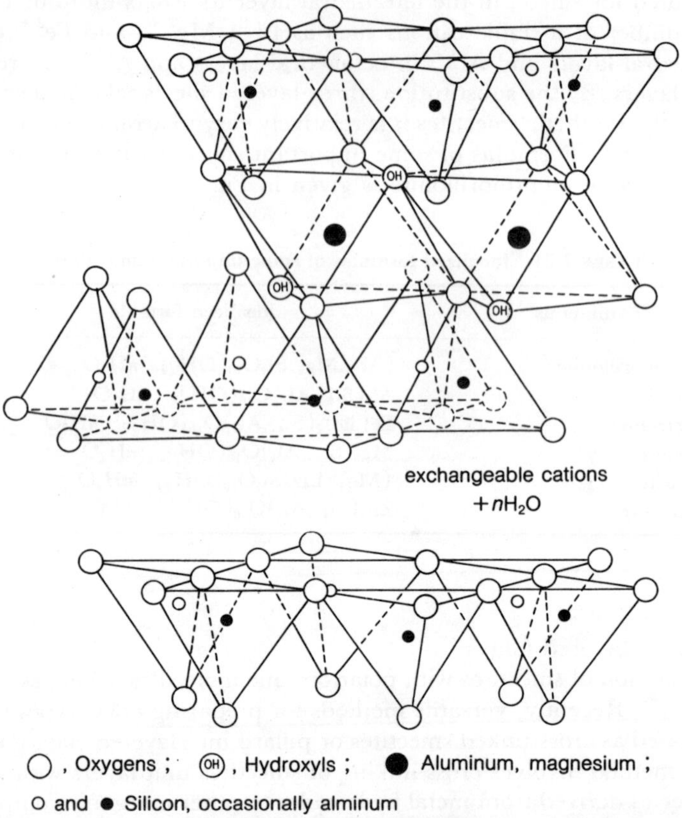

Fig. 3.55 Structure of montmorillonite.

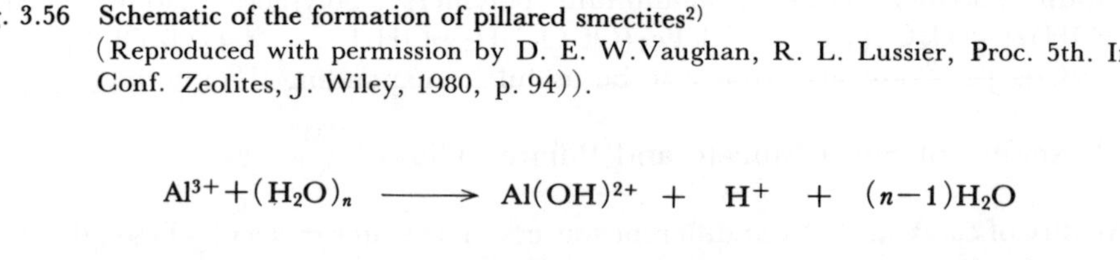

Fig. 3.56 Schematic of the formation of pillared smectites[2]
(Reproduced with permission by D. E. W. Vaughan, R. L. Lussier, Proc. 5th. Int. Conf. Zeolites, J. Wiley, 1980, p. 94)).

$$Al^{3+} + (H_2O)_n \longrightarrow Al(OH)^{2+} + H^+ + (n-1)H_2O$$

Brønsted acid sites are also generated by the dehydroxylation of pillars;[2)]

$$[Al_{13}O_4(OH)_{24}(H_2O)_{12}]^{7+} \longrightarrow 7H^+ + 6.5Al_2O_3 + 20.5H_2O$$

The acidity of clays and pillared clays has been studied by infrared spectroscopy using pyridine as a probe molecule. Fig. 3.57 shows the infrared spectra of pyridine adsorbed on beidelite pillared with aluminum hydroxide oligomers after thermal treatment under vacuum at increasing temperatures.[11)] The intercalated beidellite contains Lewis acid sites and Brønsted acid sites characterized by 1454 cm^{-1} and 1540 cm^{-1}, respectively. The intense band at 1454 cm^{-1} is associated to Lewis acid sites on the pillars, since the band does not show up in the spectra of proton exchange clays.

In Fig. 3.58 is shown the integrated intensities of the band at 1540 cm^{-1} for pillared beidellite and montmorillonite against the outgassing temperature.[11)] Increasing the calcination temperature prior to pyridine adsorption results in a steep drop in the proton content in the case of pillared montmorillonite, while pillared beidellite keeps its acidity. The steep drop of the Brønsted acid sites observed for pillared montmorillonite was attributed to the fact that, upon thermal activation, the protons migrate into the octahedral layer of the clay, where they induce a premature dehydroxylation.[12)] Thus, the acidity is mainly of the Lewis type for samples treated at higher temperatures. A similar result was reported also for bentonites pillared with alumina clusters.[13)]

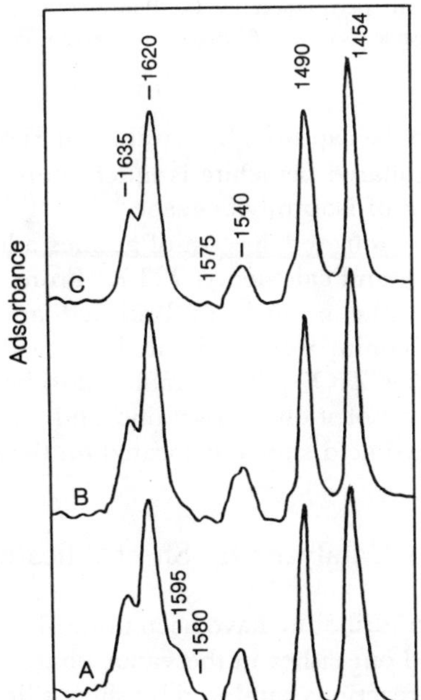

Fig. 3.57 IR spectra of pyridine adsorbed on pillared beidellites previously outgassed at 573 K) after heating under vacuum at 423 K (A), 523 K (B) and 583 K (C).[11)] (Reproduced with permission by G. Poncelet A. Schutz, *Chemical Reactions in Organic and Inorganic Constrained Systems* (R. Setton, ed.)D. Reidel Pub., 1986, p. 172.)

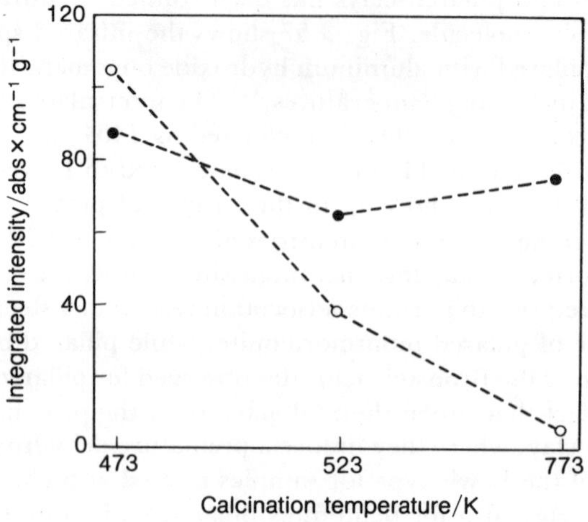

Fig. 3.58 Integrated intensities of the band at 1540 cm^{-1}-for pillared beidellite (●) and pillared montmorillonite (○) at different temperatures of precalcination[11] (Reproduced with permission by G. Poncelet A. Schutz, *Chemical Reactions in Organic and Inorganic Constrained Systems* (R. Setton, ed.) D. Reidel Pub., 1986, p. 173.)

In beidellite, protons can be captured by tetrahedral Si – O – Al linkages, as occurs in Y zeolites.[4,12] In fact, pillared beidellite is much more active than pillared montmorillonite for the cracking of isopropylbenzene.[12]

Take *et al.*[14] studied the infrared spectra of amines adsorbed on montmorillonite pillared with Al_2O_3 or ZrO_2 and calcined at 723 K. From the intensities of the bands due to adsorbed pyridine, the number of Brønsted and Lewis acid sites on the Al_2O_3 – pillared montmorillonite were estimated to be 0.36 and 0.93 mmol g^{-1}, respectively, and those on the ZrO_2-pillared material to be 0.32 and 0.92 mmol g^{-1}, respectively. Competitive adsorption of pyridine and isoamyl amine revealed that about 30 percent of Brønsted acid sites was located on the external surface of the pillared montmorillonites.

3.3.3 Organic Reactions Catalyzed by Sheet Silicates

Sheet silicates like montmorillonite have been utilized as catalysts for many organic reactions, which are carried out either in the vapor phase or in the liquid phase. The following are some typical reactions catalyzed by sheet silicates. For the complete list, readers should refer to recent reviews and references cited therein.[15–17]

A. Cracking

Montmorillonite was used as the initial catalyst for the catalytic cracking of gas oil as early as 1930. Various efforts were made to modify the clays in order to improve the yield of gasoline fractions and the life of the catalysts. The catalytic activity and the thermal stability of natural montmorillonite was improved by hot acid treatment. This treatment was considered to leach out almost half of the octahedral aluminum in the lattice and to deposit them as Al_2O_3 on the catalyst surface.[18,19] The major drawback of the sheet silicates was their hydrothermal instability. Clays were completely superseded by synthetic silica-alumina and later by zeolites.

B. Elimination Reactions

Dehydration from primary alcohols over Al(III)-exchanged montmorillonite at 473 K gives mainly di-(alk-1-yl) ethers with little alkene production, whereas secondary and tertiary alcohols, other than propan-2-ol, undergo facile intramolecular dehydration exclusively to the corresponding alkenes.[20]

$$CH_3CH_2CH_2OH \longrightarrow \underset{67.2\ \%}{(CH_3CH_2CH_2)_2O} + \underset{7.9\ \%}{CH_3CH_2CH_2O\overset{CH_3}{\overset{|}{C}}HCH_3} + \underset{2.5\ \%}{C_3H_6}$$

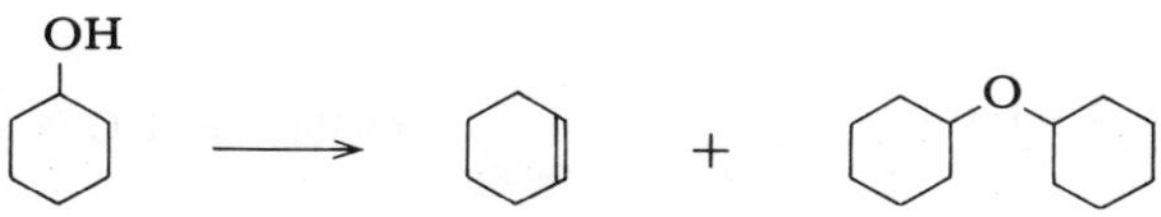

89 % 4 %

Diethyleneglycol is cyclised to yield 1,4-dioxane in addition to oligomerization.[20]

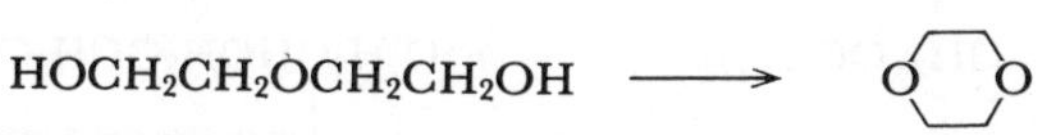

60 %

Primary amines and thiols undergo similar elimination with Al(III)-exchanged montmorillonites. Primary amines are converted at 473 K to di-alk-1-ylamine by loss of ammonia.[21] Likewise, hydrogen sulfide is eliminated from thiols to form di-alk-1-yl thioethers.[22]

Pyrrolidine is transformed at 473 K to yield 4-(1-pyrrolidyl)-butanamine **I** and 1,4 di-(1-pyrrolidyl) butane **II**.[21]

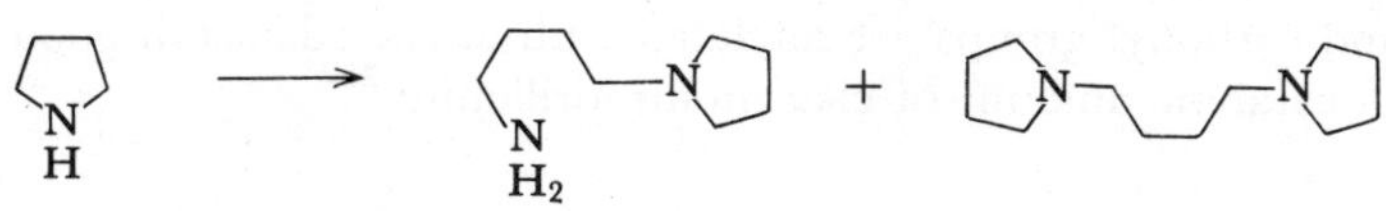

I 41 %

II 10 %

C. Condensation Reactions

Acetals are easily prepared by the use of acid-treated or cation-exchanged montmorillonite. Cyclic acetals are obtained from carbonyl compounds[23] and 1,2-diols, or from enol ethers and 1,2-diols.[24]

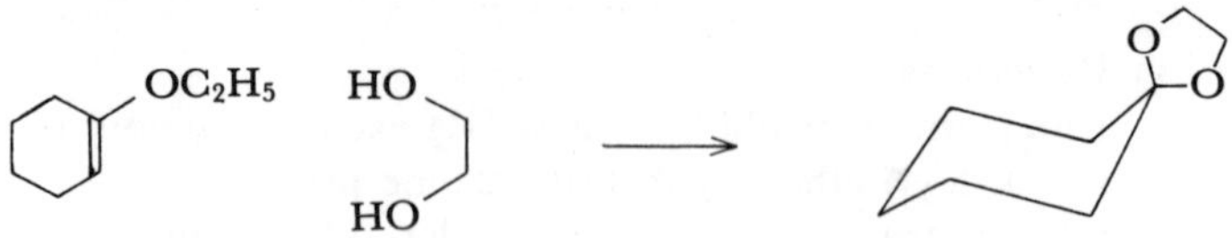

70 %

Trimethyl orthoformate absorbed on montmorillonite is an effective reagent for the rapid, high yield conversion of a variety of carbonyl compounds into their respective acetals.[25]

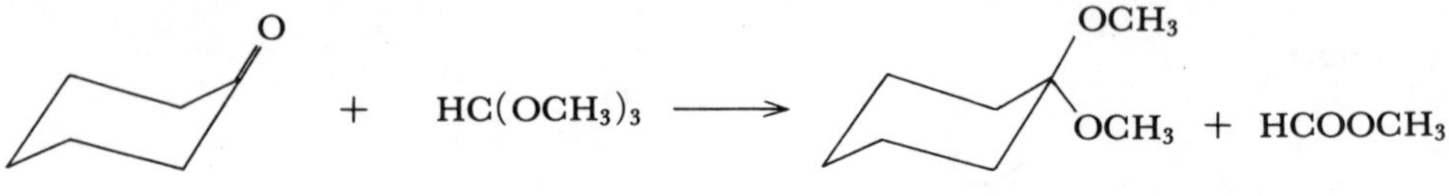

100 %

Acetal derivatives of alcohol groups can be prepared by alcohol exchange in the presence of montmorillonite.[26]

$$n\text{-}C_4H_9\underset{\displaystyle C_2H_5}{\underset{|}{C}}HCH_2OH + CH_2(OC_2H_5)_2 \longrightarrow n\text{-}C_4H_9\underset{\displaystyle C_2H_5}{\underset{|}{C}}HCH_2OCH_2OC_2H_5 + C_2H_5OH$$

Acetals of formaldehyde can be prepared from the alcohol, dichloromethane and aqueous sodium hydroxide in the presence of a quaternary-ammonium exchanged montmorillonite as a phase transfer catalyst.[27]

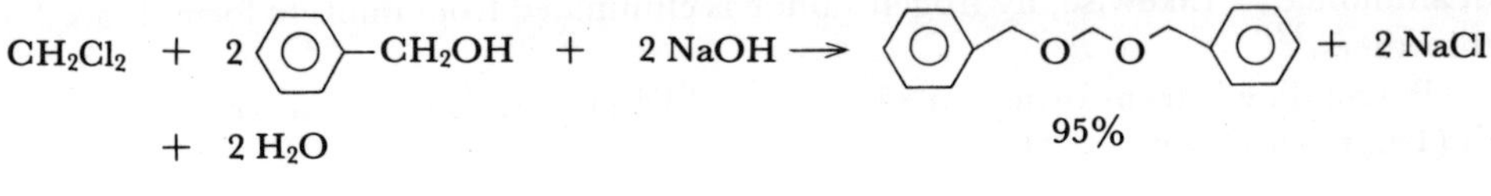

Acetals and carbonyl groups are allylated with allylic silanes in good yield in the presence of a catalytic amount of clay montmorillonite.[28]

$$n\text{-}C_7H_{15}CH(OCH_3)_2 + CH_2{=}CHCH_2Si(CH_3)_3 \longrightarrow n\text{-}C_7H_{15}CH(OCH_3)CH_2CH{=}CH_2$$

$$n\text{-}C_7H_{15}CHO + CH_2{=}CHCH_2Si(CH_3)_3 \longrightarrow n\text{-}C_7H_{15}CH(OH)CH_2CH{=}CH_2$$

Enamines can be prepared by the condensation of carbonyl compounds with secondary amines in the presence of acid treated montmorillonite in refluxing benzene.[29]

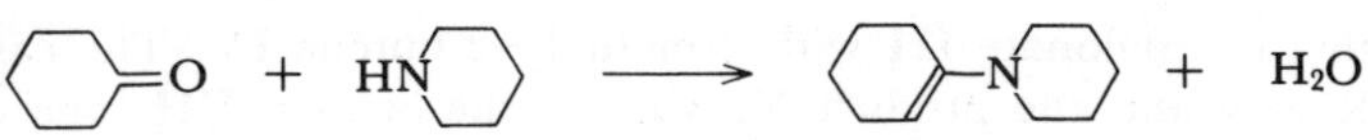

80 %

D. Addition Reactions

Water can be added to alkenes to give secondary alcohols and di-alk-2-yl ethers over Cu(II)-exchanged montmorillonite.[30)]

Alcohols can be added to alkenes in the presence of montmorillonite. For example, isobutene and methanol react at 363 K to give a good yield of methyl-*t*-butyl ether.[31)]

Ethene and acetic acid react in the interlamellar regions of Al^{3+}-exchanged montmorillonite to yield ethyl acetate and a variety of carboxylic acid can be added to C_2-C_8 alkenes at temperatures above 373 K to yield the corresponding esters in high and selective yields.[32)] The K-10 bentonite clay (an industrial catalyst for cracking) doped with Fe(III) catalyzes the Diels-Alder reaction between furan or dimethyl-2,5-furan and either arolein or methylvinyl ketone.[33)]

The catalytic activities of two types of sheet silicates, kaolinite and montmorillonite were examined for the ene reactions.[34)] Table 3.25 summarizes the results of the reac-

TABLE 3.25 Reaction of diethyl oxomalonate with 2-methyl 2-butene

Reactions conditions†1			Yield†2	Product distribution, % †3				
Catalyst	temp.K	time, h	%	V	VI	VII	VIII	X
Kaolin†4	345	19	82	37	1	3		59
Kaolin†4	398	19	82	63	2	5		29
Kaolin†4	363	120	78	5	19	49		26
Mont. Camp Berteau	351	24	70			18	36	45
Mont. Camp Berteau	371	48	76		5	64	18	13
Mont. K10†5	353	72	80			49	44	7
Al^{3+}-mont.	353	8	70			38	32	30
Cu^{2+}-mont.	353	8	60			40	23	37
Cr^{3+}-mont.	353	8	70			68	36	5
Fe^{3+}-mont.†6	353	8	65			60	30	9
H^+-mont.†7	293	8	94			60	30	10
H^+-mont.†7	293	8	98			90	6	4

†1 Amixture of 2 g of catalyst, 15 mmol of IV, 10 mmol of III was heated without solvent, for the time indicated.

†2 Sum of isolated yields of all components based on III.

†3 Determined by GLC and ^1H NMR.

†4 From Prolabo.

†5 From Fluka A.G.

†6 $FeCl_3$ doped K10 montmorillonite (0.6 mmol $FeCl_3$/1 g K10).

†7 H_2SO_4 doped K10 montmorillonite (3 mmol H_2SO_4/1 g K10).

(Reproduced with permission by J. F. Roudier, A. Foucard, *Chemical Reactions in Organic and Inorganic Constrained Systems* (R. Setton, ed.), Q, Reidel Pub., 1986, p. 232)

tion of diethyl oxomalonate **III** with 2-methyl−2-butene **IV**. The case of kaolin at 343−363 K gave the ene product **V**, **VI** and the lactone **VII** (one diastereomer). Compound **VI** was formed by a secondary ene reaction with the ene product **V** (Scheme I).

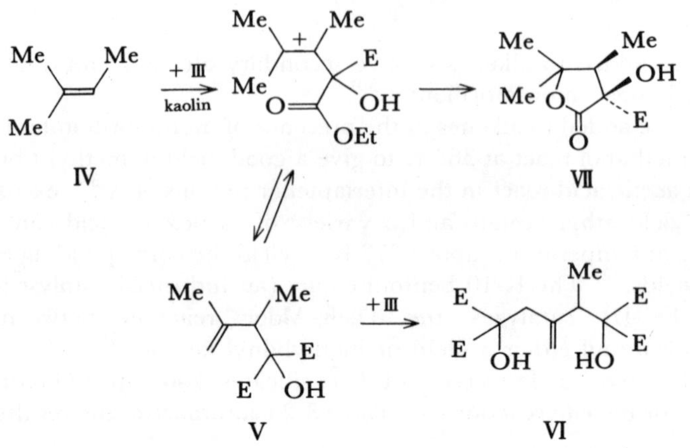

Scheme I Reaction of 2-methyl-2-butene **IV** with diethylmalonate **III** over kaolin ($E=CO_2C_2H_5$)

The ene reaction catalyzed by acidic montmorillonite gave solely the lactone **VII** and **VIII** (two diastesomers in a ratio of 50:50). The lactone **VIII** is plausibly formed by the ene reaction of **III** with 2-methyl-1-butene **IX**, which arises from isomerization of **IV**. The reaction scheme is as follows (Scheme II).

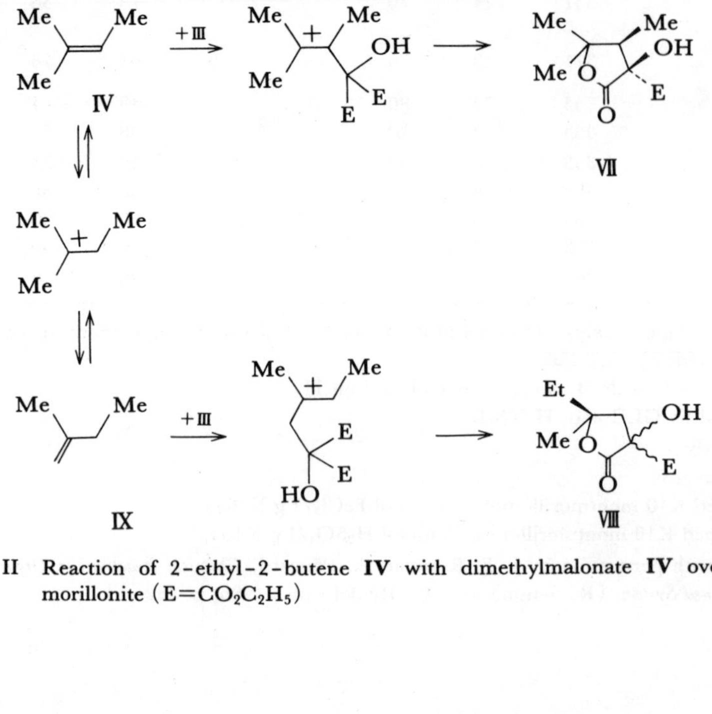

Scheme II Reaction of 2-ethyl-2-butene **IV** with dimethylmalonate **IV** over montmorillonite ($E=CO_2C_2H_5$)

Besides the ene products, diethyloxomalonate **X** hydrate is also found in the products.

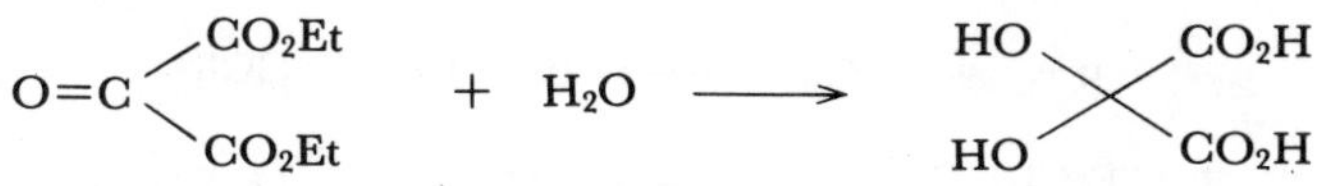

X

Acetals are added to enol ethers to form precursors of α,β-unsaturated aldehyde in the presence of montmorillonite.[35)]

$$C_2H_5CH(OC_2H_5)_2 + H_2C = CHOC_2H_5 \xrightarrow{\text{clay}} C_2H_5CH(OC_2H_5)CH_2CH(OC_2H_5)_2$$

$$\xrightarrow{H^+} C_2H_5CH{=}CHCHO$$

Cross aldol adducts are obtained from silyl ene ethers and aldehyde or acetals.[36)]

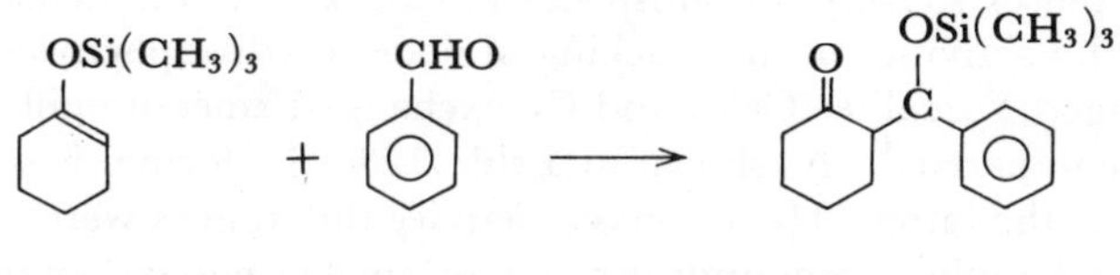

The diastereoselectivity in the reaction is significantly sensitive to the solvent used.

Al^{3+} ion-exchanged montmorillonite efficiently catalyzes the aldol reactions of silyl ketene acetals with carbonyl compounds and acetals.[37)]

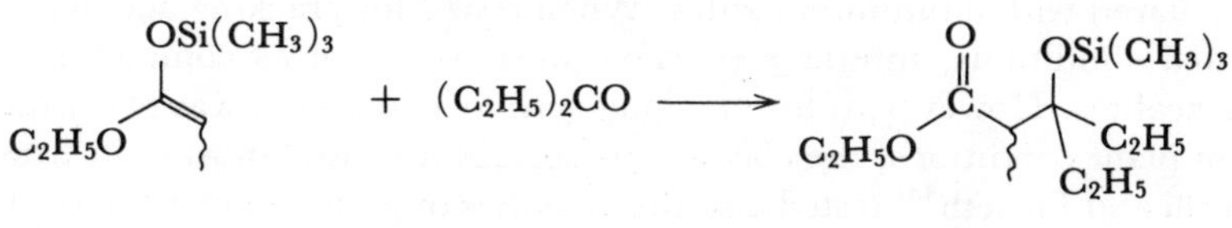

Silyl ketene acetals and silyl enol ether react with α,β-unsaturated esters and α,β-unsaturated ketones to afford the Michel adducts in the form of silyl ketene acetals and silyl enol ethers in good yields.[38)]

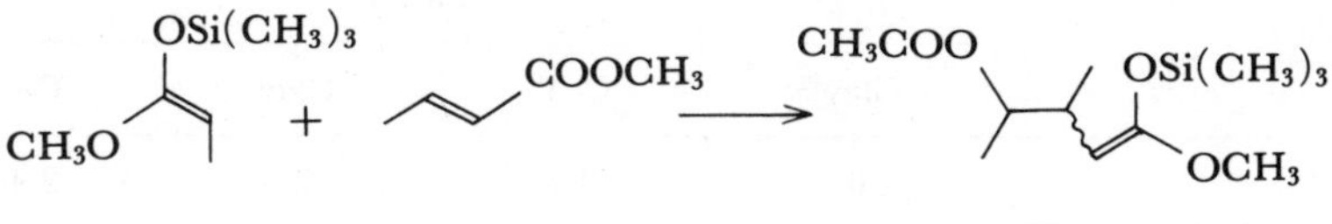

84 %

(syn : anti=27 : 73)

Chlorination of adamantane by carbon tetrachloride is catalyzed by the Fe(III)-doped K-10 clays.[39)] Good yields of 1-adamantyl chloride and of 1,3-adamantyl dichloride were obtained. The intermediacy of a tertiary 1-adamantyl cation is postulated as in the following scheme.

$$(FeCl_{3-i})^{i+} + R{-}H \longrightarrow (FeCl_{3-i}H)^{(i-1)+} + R^+$$

$$R^+ + CCl_4 \longrightarrow R{-}Cl + CCl_3^+$$

$$CCl_3^+ + (FeCl_{3-i}H)^{(i-1)+} \longrightarrow CHCl_3 + (FeCl_{3-i})^{i+}$$

The arylation of adamantane can be achieved by changing from carbon tetrachloride to an aromatic solvent. With benzene as solvent, good yields of 1-phenyl- and of 1,3-diphenyladamantane were obtained.[39)]

3.3.4 Catalysis by Pillared Clays

Catalytic activities of pillared clays have been tested mainly for vapor-phase reactions at high temperatures. Shabtai and coworkers compared the catalytic activities of pillared smectites with those of rate-earth exchange Y-zeolites (REY) for many reactions.[3,40)] The activity of pillared clays are consistently higher than REY, and the relative activity increases sharply for substrates having kinetic diameters σ greater than 0.9 nm. Thus, the activities of the cracking of dodecahydrotiphenylene ($\sigma \sim 1.15$ nm) with Ce-exchanged Y zeolite (CeY) and Ce-exchanged smectite pillared with aluminum oxide are compared.[40)] As shown in Table 3.26, the former is some hundreds of times as active as the latter. The observed activity differences were attributed to free penetration of such bulky compounds in the interlameller network of the pillared clays, as opposed to their exclusion from the channel system of Y-zeolites, suggesting a pronounced advantage of pillared smectites over the conventional zeolites for cracking of large polycyclic molecules, such as those found in heavy oil fractions and synfuels.

Occelli[41)] studied the cracking of a gas oil having a 533 – 699 K boiling range over bentonite pillared with aluminum oxides. When tested for cracking activity after mild hydrothermal conditions, interlayered clays were as active as commercial catalysts containing zeolites. However, if the cracking activity evaluation was performed under typical pilot plant conditions, the clay lost its surface area and most of its catalytic activity. Occelli and Finseth[42)] tested also the activities of pillared hectrites for the crack-

TABLE 3.26 Comparison of cracking conversions of dodecahydrotriphenylene (DHT) with pillared clay and Ce-exchanged Y-zeolite catalysts[†1]

	Catalyst			
	Clay[†2]	CeY	Clay	CeY
LHSV/h^{-1}	0.2	0.2	2.4	2.4
Conversion of DHT at 673 K/%	58.6	1.3	29.1	0.04
Relative conversion	45.1		728	

†1 In each experiment 5 g of a 3.5 % solution of DHT in benzene and 0.2 g of catalyst in admixture with 2.0 g of carborundom were used.

†2 Ce-exchanged form of montmorillonite pillared with aluminum oxide.

(Reproduced with permission by DECHEMA, 8th Inter. Cong. Catal. process., **4**, 744 (1984).

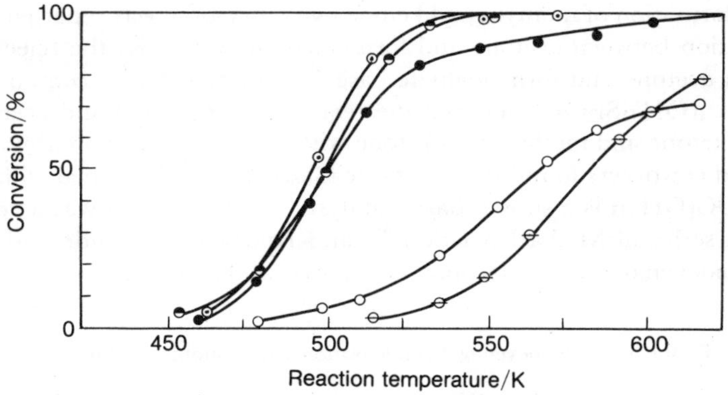

Fig. 3.59 Temperature-programmed reaction of hexane for ⊙; ultrastable Y-zeolite, ◓; H–ZSM-5, ●; pillared beidellite, ○; pillared montmorillonite, ⊖; silica-alumina.
(Reproduced with permission by G. Poncelet, A. Schutz, *Chemical Reactions in Organic and Inorganic Constrained Systems* (*R.Setton, ed.*), *D. Reidel pub.*, 1986. p. 175.)

ing of the gas oil and found that they were not as active as pillared bentonites.

Hydroconversion of heptane (H_2/n-$C_7 = 16$) was carried out under temperature-programmed conditions over pillared beidellite, pillared montmorillonite, a ultrastable Y-zeolite, H–ZSM-5 and silica-alumina, all containing 1 wt% of Pt.[11] The conversions of heptane over five catalysts are shown as a function of the reaction temperature. As shown in Fig. 3.59, pillared beidellite is more active than pillared montmorillonite, and the former is almost as active as the zeolites. Silica–alumina is the least active.

Kikuchi *et al.*[43] studied the disproportionation of 1,2,4-trimethylbenzene over montmorillonite pillared by aluminum oxide at 473 K. The catalyst was selective for the production of 1,2,4,5-tetramethylbenzene and xylenes. The high selectivity to 1,2,4,5-tetramethylbenzene was attributed to be the result of transition-state selectivity induced by space restriction. Urabe *et al.* reported that a saponite pillared with aluminum oxide was as active as pillared montmorillonite for the alkylation of toluene with methanol.[44]

3.3.5 Catalysis by Other Clays

Kitayama reported that Mn^{2+}-exchanged sepiolite can be a selective catalyst for the synthesis of butadiene from ethanol.[45] Butadiene can be produced also from alkali cation containig hectrite.[46]

Chrysotile $Mg_3(OH)_4Si_2O_5$ is a layered silicate without any intercalating character. Suzuki and Ono[47] found that the catalytic activity of chrysotile is very dependent on the Na_2O/SiO_2 ratio of the starting gel for the synthesis. When Na_2O/SiO_2 ratio is less than 1.5, the physical form of synthesized material is flakes and shows a high activity for the dehydration of 2-propanol at 513 K. When Na_2O/SiO_2 ratio is more than 2, the material is thick-wall tubes and much less active than the flake-type material. Besides dehydration, dehydrogenation proceeds over the tube-like material to yield acetone at higher temperature.

Selective formation of methyl vinyl ketone over chrysotile was reported through aldol condensation between acetone and formaldehyde at 673 K, the selectivity being 98% on both acetone and formaldehyde base.[48] Synthetic Co^{2+}-containing chrysotile, $Co_xMg_{3-x}(OH)_4Si_2O_5$, catalyzed the reaction of methanol and acetone to give methyl vinyl ketone and methyl ethyl ketone with 75% selectivity on acetone basis.[48]

Laszlo and coworkers found that xonotlite $Ca_6Si_6O_{17}(OH)_2$ doped with *t*-butoxide ions from $KOC(CH_3)_3$ is a superb basic catalyst.[17] The material was a very efficient catalyst for a series of Michael reactions[49] and Knoevenagel condensations.[50] The results for Knoevenagel condensations are shown in Table 3.27.

TABLE 3.27 Knoevenagel condensation with xonotlite-*t*-butoxide

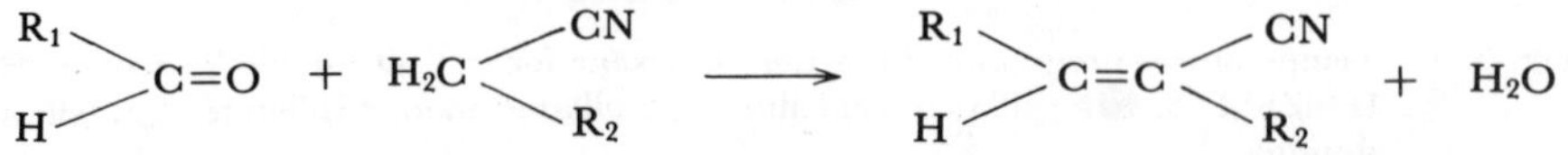

R_1	R_2=CN	$R_2=COOC_2H_5$	$R_3=COOCH_3$
R-C_6H_4			
p-H	43 %	82 %	84 %
p-CH_3	57	80	80
p-OCH_3	77	86	81
p-$N(CH_3)_2$	35	78	73
p-Cl	34	82	86
p-CN	—	92	88
m-CN	60	—	—
p-$NO_2C_6H_5$	40	83	79
2-naphtaldehyde	60	89	95

reaction time=1 h, aldehyde 10 mmol, methylene compound 10 mmol, 100 mg of catalyst (xonotlite /*t*-butoxite =15 g/12 g)
(Reproduced with permission by S. Chalais *et al.*, *Tetrahedron Lett.*, **26**, 4454(1985)).

REFERENCES

1. J. M. Thomas, in: *Intercalation Chemistry* (M. S. Whittingham, A. J. Jacobson, eds.), Academic Press, New York, 1982, p.55.
2. D. E. W. Vaughan, R. J. Lussier, Proc. 5th Int. Conf. Zeolites, Naples, 1980, (L. V. C Rees, ed.), J. Willy, Chichester, 1980, p.94.
3. J. Shabtai, R. Lazar, A. G. Oblad, Proc. 7th Int. Congr. Catalysis, Tokyo, 1980, (T. Seiyama, K. Tanabe, eds.), Kodansha, Tokyo and Elsevier, Amsterdam, 1980, p.826.
4. D. Plee, F. Borg, L. Gatineau, J. J. Fripiat, *J. Am. Chem. Soc.*, **107**, 2362 (1985).
5. D. Tilak, B. Tennakoon, W. Jones, J. M. Thomas, *J. Chem. Soc., Faraday Trans. 1*, **82**, 3081 (1986).
6. A. Schutz, W. E. E. Stone, G. Poncelet, J. J. Fripiat, *Clays Clay Miner.*, **35**, 251 (1987).
7. S. Yamanaka, G. W. Brindley, *Clays Clay Miner.*, **27**, 119 (1979).
8. S. Yamanaka, T. Doi, S. Sako, M. Hattori, *Mat. Res. Bull.*, **19**, 161 (1984).

9. S. P. Christiano, J. Wang, T. J. Pinnavaia, *Inorg. Chem.*, **24**, 1222 (1985); *J. Solid State Chem.*, **64**, 232 (1986).
10. J. J. Fripiat, A. N. Jelli, G. Poncelet, J. Andre, *J. Phys. Chem.*, **69**, 2185 (1965).
11. G. Poncelet, A. Schutz, in: *Chemical Reactions in Organic and Inorganic Constrained Systems*, (R. Setton, ed.), D. Reidel Publishing, Comp. Dortrecht, 1986, p.165.
12. B. Chourabi, J. J. Fripiat, *Clays Clay Miner.*, **29**, 260 (1981).
13. M. L. Occelli, J. E. Lester, *Ind. Eng. Chem., Prod. Res. Dev.*, **24**, 27 (1985).
14. J. Take, T. Yamaguchi, K. Miyamoto, N. Ohyama, M. Misono, Proc. 7th Intern. Zeolite Conf., 1986, Tokyo, (Y. Murakami, A. Iijima, N. W. Ward, eds.), Kodansha, Tokyo and Elsevier, Amsterdam, 1986, p.495.
15. J. M. Thomas, in: *Intercalation Chemistry* (M. S. Whittingham, A. J. Jacobson, eds.) Academic Press, New York, 1982, p.55.
16. J. A. Ballantine, in: *Chemical Reactions in Organic and Inorganic Constrained Systems* (R. Setton, ed.) D. Reidel Publishing, Comp. Dortrecht, 1986, p.197.
17. P. Laszlo, *Acc. Chem. Res.*, **19**, 121 (1986).
18. C. L. Thomas, J. Hickey, G. Strecker, *Ind. Eng. Chem.*, **42**, 866 (1950).
19. G. A. Mills, J. Holms, E. B. Cornelius, *J. Phys. Colloid. Chem.*, **54**, 1170 (1950).
20. J. A. Ballantine, M. Davies, I. Patel, J. H. Purnell, M. Rayanakorn, K. J. Williams, J. M. Thomas, *J. Mol. Catal.*, **30**, 373 (1985).
21. J. A. Ballantine, J. H. Purnell, M. Rayanakorn, K. J. Williams, J. M. Thomas, *J. Mol. Catal.*, **30**, 373 (1985).
22. J. A. Ballantine, R. P. Galvin, R. M. O'Neil, J. H. Purnell, M. Rayanakorn, J. M. Thomas, *J. Chem. Soc., Chem. Commun.*, **1981**, 695.
23. J. Y. Conan, A. Natat, D. Privolet, *Bull. Soc. Chim. Fr.*, **1976**, 1935.
24. T. Vu. Moc, H. Petit, P. Maitte, *Bull. Chem. Soc. Fr.*, **1979**, 264.
25. E. C. Taylor, C. Chaing, *Synthesis*, **1977**, 467.
26. U. Schafer, *Synthesis,* **1981**, 794.
27. A. Cornelis, P. Laszlo, *Synthesis*, **1982**, 162.
28. M. Kawai, M. Onaka, Y. Izumi, *Chem. Lett.*, **1986**, 381.
29. S. Hunig, K. Hubner, E. Benzing, *Chem. Ber.*, **95**, 931 (1962).
30. J. M. Adams, J. A. Ballantine, S. H. Graham, R. J. Laub, J. H. Purnell, P. I. Reid, W. Y. M. Shaman, J. M. Thomas, *J. Catal.*, **58**, 238 (1979).
31. A. Bylina, J. M. Adams, S. H. Graham, J. M. Thomas, *J. Chem. Soc., Chem. Commun.*, **1980**, 1003.
32. J. A. Ballantine, M. Davies, H. Purnell, M. Rayanakorn, J. M. Thomas, K. J. Williams, *J. Mol. Catal.*, **26**, 57 (1984).
33. P. Laszlo, J. Lucchetti, *Tetrahedron Lett.*, **39**, 4387 (1984).
34. J. F. Roudier, A. Foucaud, in: *Chemical Reactions in Organic and Inorganic Constrained Systems*, (R. Setton ed.) D. Reidel Publishing, Comp. Dortrecht, 1976, p.239.
35. D. Fishman, J. T. Klug, A. Shani, *Synthesis,* **1981**, 137.
36. M. Kawai, M. Onaka, Y. Izumi, *Chem. Lett.*, **1986**, 1581.
37. M. Onaka, R. Ohno, M. Kawai, Y. Izumi, *Bull. Chem. Soc. Jpn.*, **60**, 2689 (1987).
38. M. Kawai, M. Onaka, Y. Izumi, *J. Chem. Soc., Chem. Commun.*, **1987**, 203.
39. S. Chalais, A. Cornelis, A. Gertmans, W. Koldziejski, P. Laszlo, A. Mathy, P. Metra, *Helv. Chem. Acta*, **68**, 1196 (1985).
40. J. Shabtai, F. E. Massoth, M. Tokarz, G. M. Tsai, J. McCauley, Proc. 8th Intern. Congr. Catalysis, Berlin, 1984, Verlag Chemie, Weinheim, Vol. 4, p.735.
41. M. L. Occelli, *Ind. Eng. Chem., Prod. Res. Dev.*, **22**, 553 (1983).
42. M. L. Occelli, D. H. Finseth, *J. Catal.*, **99**, 316 (1986).
43. E. Kikuchi, T. Matsuda, H. Fujiki, Y. Morita, *Appl. Catal.*, **11**, 331 (1984).
44. K. Urabe, H. Sakurai, Y. Izumi, *Chem. Soc., Chem. Commun.*, **1986**, 1074.
45. Y. Kitayama, A. Michishita, *J. Chem. Soc., Chem. Commun.*, **1981**, 401.
46. E. Suzuki, S. Idemura, Y. Ono, *Appl. Clay Sci.*, **3**, 123 (1988).
47. S. Suzuki, Y. Ono, *Appl. Catal.*, **10**, 361 (1984).
48. E. Suzuki, S. Idemura, Y. Ono, *Chem. Lett.*, **1987**, 1843.
49. P. Laszlo, P. Pennetreau, *Tetrahedron Lett.*, **26**, 2645 (1985).
50. S. Chalais, P. Laszlo, A. Mathy, *Tetrahedron Lett.*, **26**, 4453 (1985).

3.4 ZEOLITES

3.4.1 Structure of Zeolites

Zeolites are crystalline aluminosilicates that develop uniform pore structure having minimum channel diameter of 0.3 to 1.0 nm. The size depends primarily upon the type of zeolite. Zeolites provide high activity and unusual selectivity in a variety of acid-catalyzed reactions. Most of the reactions are caused by the acidic nature of zeolites. This section will discuss the acidic properties of zeolites.

The structure of zeolites consists of a three-dimensional framework of SiO_4 or AlO_4 tetrahedra, each of which contains a silicon or aluminum atom in the center. The oxygen atoms are shared between adjoining tetrahedra, which can be present in various ratios and arranged in a variety of ways. The framework thus obtained contains pores, channels, and cages, or interconnected voids.

Zeolites may be represented by the general formula,

$$M_{x/n}[(AlO_2)_x(SiO_2)_y]wH_2O$$

where the term in brackets is the crystallographic unit cell. The metal cation of valence n is present to produce electrical neutrality since for each aluminum tetrahedron in the lattice there is an overall charge of -1.

The frameworks of zeolites used most frequently as adsorbent or catalyst are shown in Figs. 3.60 – 3.63. The Al or Si atoms are located at the intersection of lines that represent oxygen bridges. The X and Y zeolites are structually and topologically related to the mineral faujasite and frequently referred to as faujasite-type zeolites. The two materials differ chemically by their Si/Al ratios, which are $1-1.5$ and $1.5-3.0$ for X and Y zeolite, respectively. In faujasites, large cavities of 1.3 nm in diameter (supercages) are connected to each other through apertures of 1.0 nm.

In type A zeolite (Fig. 3.61), large cavities are connected through apertures of

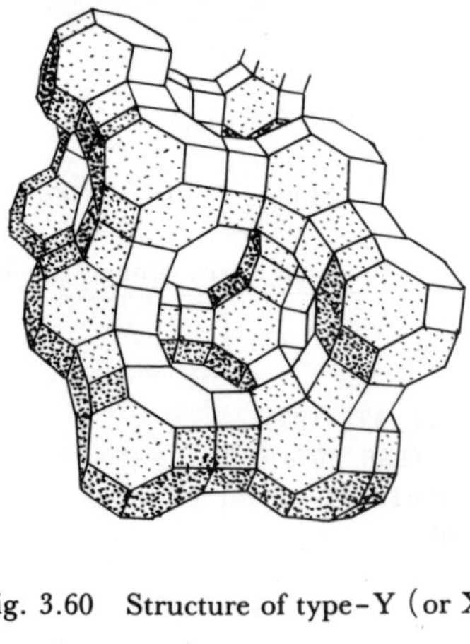

Fig. 3.60 Structure of type-Y (or X) zeolite

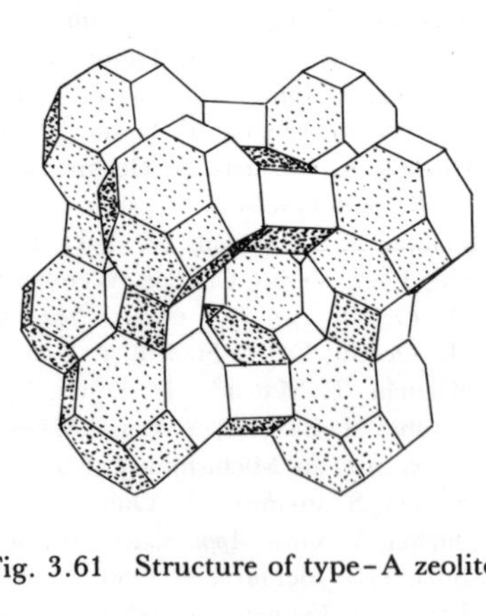

Fig. 3.61 Structure of type-A zeolite

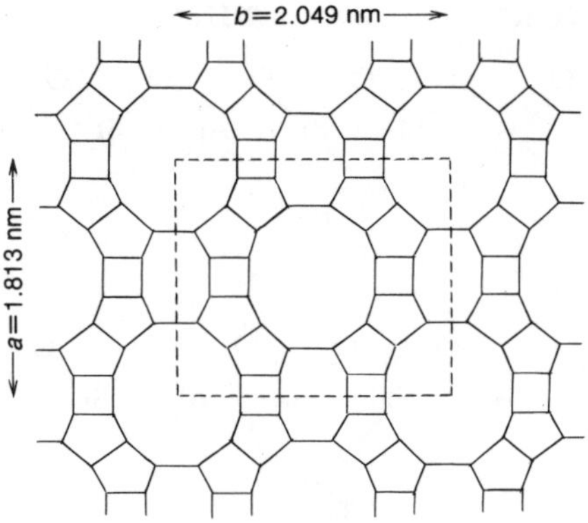

Fig. 3.62 Skeletal diagram of the (001) face of mordenite

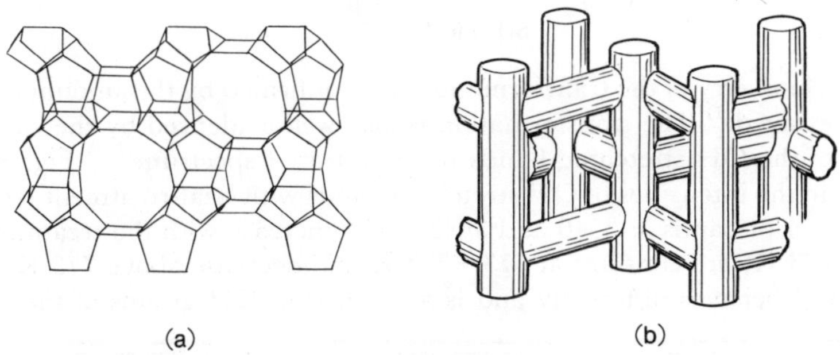

Fig. 3.63 Structure of ZSM-5 (a) Skeletal diagram of the (010) face of ZSM-5 (b) channel network

0.5 nm, determined by eight-membered rings.

The mordenite pore structure (Fig. 3.62) consists of elliptical and noninterconnected channels parallel to the *c*-axis of the orthorhombic structure. Their openings are limited by twelve-membered rings (0.6 – 0.7 nm). ZSM-5 zeolite (Fig. 3.63) shows a unique pore structure that consists of two intersecting channel systems: one straight and the other sinusoidal and perpendicular to the former (Fig. 3.63). Both channel systems have ten-membered-ring elliptical openings (*ca.* 0.55 Å in diameter)

3.4.2 Acidity of Zeolites

The investigation of the acidic properties of zeolites started with ammonium ion-exchanged Y-zeolites (NH_4Y). Ammonium ion-exchanged Y-zeolite evolves ammonia and water by heat-treatment at 650 – 600 K, and 770 – 820 K, respectively. The transformation of NH_4Y can be schematically expressed as follows.

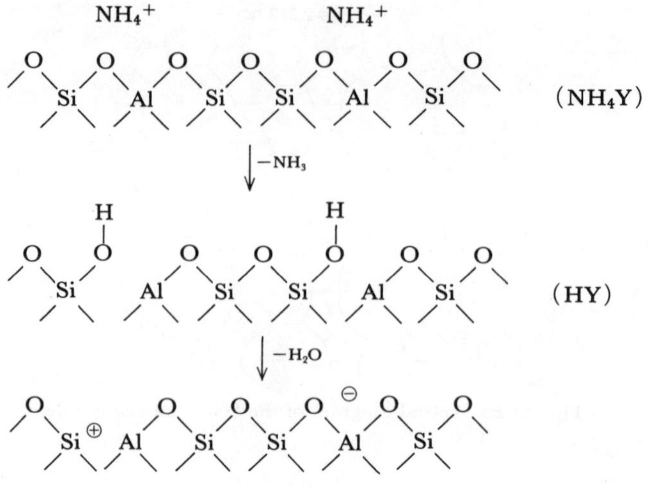

(I) (II)

Scheme A

The stoichiometry of the transformation was confirmed by the amount of ammonia and water evolved.[1)] The transformation is also well evidenced by the change in the intensities of the OH stretching bands in the infrared spectrum.[2,3)] Fig. 3.64 shows the change in the intensities of OH stretching bands with heat-treatment temperature. The intensities of bands at 3540 and 3643 cm^{-1} increase with the treatmet temperature up to 673 K, are constant at 673 − 773 K and decrease above 773 K. The band at 3740 cm^{-1} behaves differently and is attributed to OH groups of the amorphous

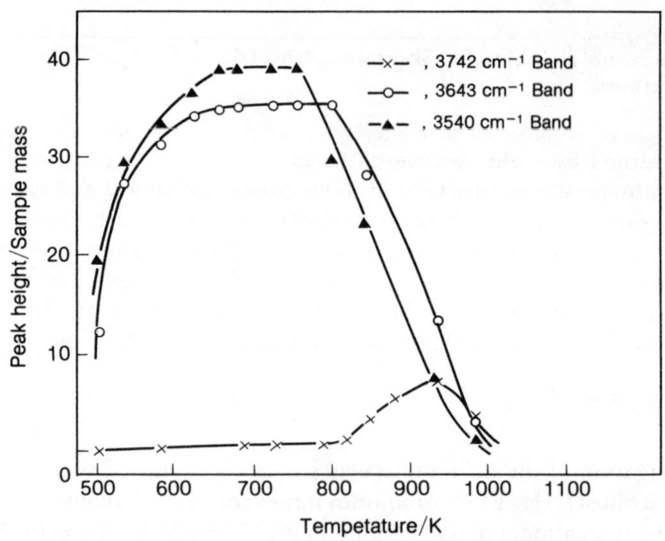

Fig. 3.64 Intensity of hydroxyl bands on NH_4 Y zeolite as a function of calcination temperature.[3)]
(Reproduced with permission by J. Ward, *J. Catal.*, **9**, 230 (1967)).

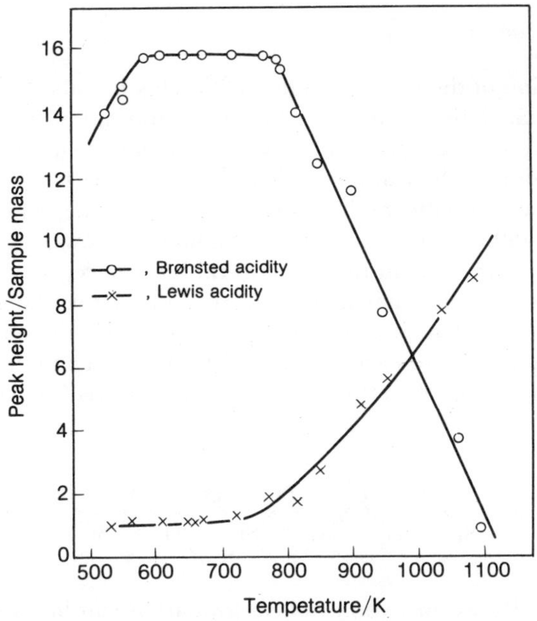

Fig. 3.65 Acid site population on NH_4Y as a function of calcination temperature. (Reproduced with permission by J. Ward, *J. Catal.*, **9**, 231 (1967)).

part of the material. The character of OH groups can be more explicity demonstrated by examining their interaction with pyridine, a base molecule.

Adsorption of pyridine on Brønsted sites and Lewis sites gives characteristic infrared bands at 1540 and 1420 cm^{-1}, respectively. Fig. 3.65 shows the change of acid site population (the intensities of 1540 and 1420 bands) with temperature of heat treatment of NH_4Y. It is clear that the change in the intensity of 1540 cm^{-1} band with the treatment temperature corresponds very well to the change in the OH band intensities. In fact, the OH bands at 3540 and 3643 cm^{-1} disappear on adsorption of pyridine. These facts give definite evidence that OH groups of HY zeolites (see Scheme A) are acidic. The acidic character of HY can be expressed by the following equilibrium.

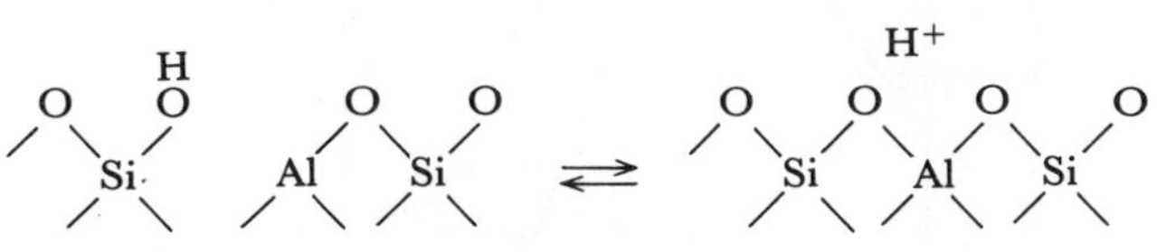

(1)

In fact, the intensities of OH bands decrease with temperature, indicating that equilibrium (1) shifts to the right side to reduce the number of OH groups at higher temperatures.[4)]

The OH band at 3740 cm^{-1} does not interact with pyridine, indicating that OH groups giving this band are not acidic. As is shown in Fig. 3.65, Lewis acid sites develop above 770 K. This was first correlated to tricoordinated aluminum in the local structure (I) formed by dehydroxylation of HY (Scheme A).

Adsorption of polynuclear aromatics such as perylene on dehydroxylated zeolite leads to the formation of their cation radicals.[5,6] This has also been regarded as evidence for tricoordinated aluminum. On the other hand, adsorption of molecules with high electron affinity such as tetracyanoethylene or trinitrobenzene on dehydroxylated HY leads to formation of their anion radicals.[6-8] The electron-transfer to adsorbed molecules is rationalized with the local structure (II) in the dehydroxylated zeolites (Scheme A). The number of radical cations or anions is, however, much smaller than the number of aluminum ions in the framework. Therefore, only a small part of the aluminum ions in the zeolites participates in the charge-transfer reactions.

It is now established that the local structure (I) in dehydroxylated zeolites is not stable and aluminum ions are easily dislodged from the zeolite framework and exist in the pores in the form of cationic species such as $(AlO)^+$ or their polymeric form.[9-11]

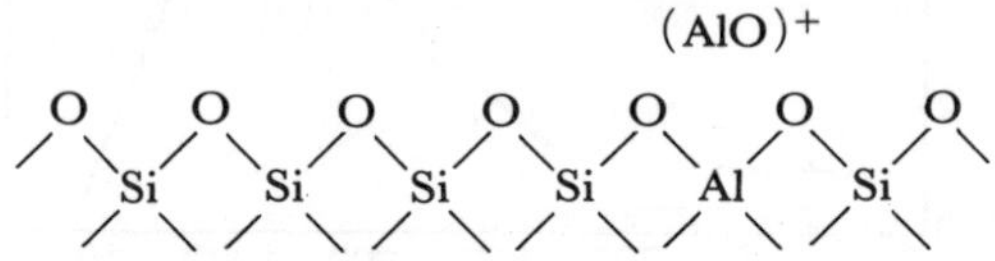

Thus, Lewis acid sites for cation-radical formation can be ascribed to aluminum cations formed by aluminum dislodgement. Recently, the importance of dislodged aluminum ions in the acidity of the zeolites has been stressed as will be described below (Section 3.4.5).

In most cases, the catalytic activity is related to the number of Brønsted sites rather than Lewis acid sites. Fig. 3.66 shows the catalytic activity of NH_4L zeolite for cracking of isopropylbenzene as a function of the pretreatment temperature.[12] It is clear that the activity is most pronounced in the temperature range in which acidic OH groups are present. A similar dependence on the pretreatment temperature is observed also in isomerization of *o*-xylene with NH_4Y.[13]

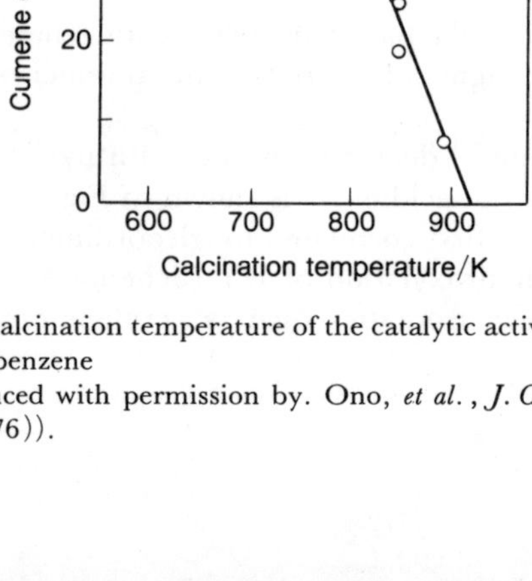

Fig. 3.66 Effect of calcination temperature of the catalytic activity of NH_4-L for cracking of isopropylbenzene
(Reproduced with permission by. Ono, *et al.*, *J. Chem. Soc. Faraday Transl.*, **72**, 2156 (1976)).

Acidic OH groups can be produced also by ion-exchange with polyvalent cations such as Ca^{2+}, Mg^{2+}, or La^{3+}.[14-17] The development of acidity can be expressed as

$$[Ca(OH_2)]^{2+} \longrightarrow [Ca(OH)]^{+} + H^{+}.$$

Thus, water molecules coordinated to polyvalent cations are dissociated by heat-treatment to give the following local structure.

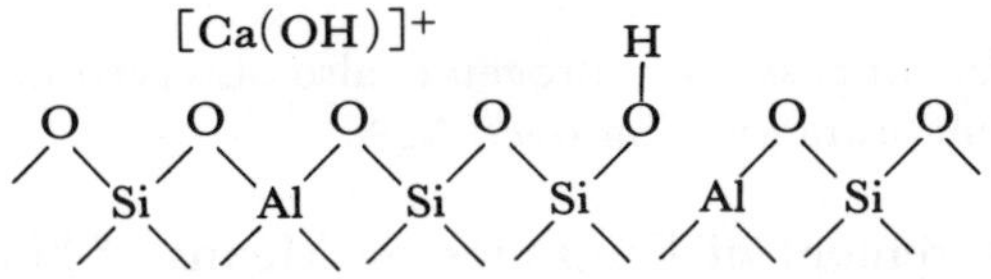

The absorption band due to OH group ion $[Ca(OH)]^{+}$ is observed in the infrared spectrum besides the bands of acidic OH groups, which are also present in HY.

Acidic sites are also formed by the reduction of transition metal cations.[17,18]

$$Cu^{2+} + H_2 \longrightarrow Cu^{0} + 2H^{+}$$

$$Ag^{+} + \frac{1}{2}H_2 \longrightarrow Ag^{0} + H^{+}$$

In the case of AgY zeolite, the catalytic activity is greatly enhanced by the presence of gaseous hydrogen.[19-21] The activity of AgY in the presence of hydrogen is much higher than that of HY, whose activity is not affected by hydrogen. Fig. 3.67 shows

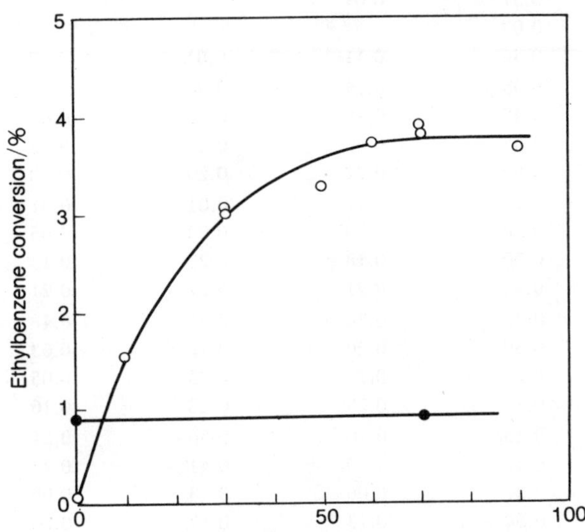

Fig. 3.67 Dependence of the catalytic of (○) AgY and (●) HY on the pressure of hydrogen. Reaction conditions: 473 K, $W/F = 7.62$ g h mol^{-1}, ethylbenzene $= 10.1$ kP_a
(Reproduced with permission by T. Baba, Y. Ono, *Zeolites*, **7**, 293 (1987)).

the dependence of the catalytic activity for ethylbenzene disproportionation on the pressure of hydrogen. The effect of gaseous hydrogen is reversible, i.e. elimination of hydrogen reduces the activity, which can be regained by reintroduction of hydrogen to the system. These facts show that there is a mechanism of interconversion of molecular hydrogen and proton. It is assumed that silver cluster ions are involved in the hydrogen-proton interconvention.

$$Ag_n^+ + H_2 \rightleftarrows Ag_n^0 + 2H^+$$

The enhancing effect of gaseous hydrogen is also observed in alcohol dehydration over AgY[20] and butene isomerization over AgA.[21]

3.4.3 Acidity Measurement of Faujasites by Means of Hammett Indicator

Klading[22] measured the surface acidity of Y zeolites containing Na^+, K^+, Ca^{2+},

TABLE 3.28 Accumulated acidities in ion-exchanged NaY after calcination at 773 K

	Butylamine titer, mmol/g, in H_0 range				Total in the range
Y zeolite†	+6.8 to +4.0	+4.0 to +1.5	+1.5 to −5.6	<−5.6	+6.8 to −5.6
Na	0.35	0.12			0.47
K(13)	0.40	0.05			0.45
K(78.6)	0.31	0.01			0.32
K(100)	0.03				0.03
Sr(21.1)	0.36	0.11	0.01		0.48
Sr(50.9)	0.43	0.14	0.06		0.63
Sr(56.2)	0.42	0.18	0.12	0.05	0.77
Sr(70.1)	0.63	0.20	0.24	0.08	1.15
Sr(86.2)	0.60	0.22	0.20	0.38	1.40
Ca(19.5)	0.35	0.11	0.01	0.01	0.48
Ca(52.2)	0.38	0.19	0.08	0.05	0.60
Ca(64.8)	0.50	0.18	0.22	0.13	1.03
Ca(70.1)	0.48	0.21	0.27	0.21	1.17
Ca(86.2)	0.45	0.30	0.35	0.48	1.58
La(17.9)	0.30	0.20	0.17	0.03	0.70
La(31.7)	0.27	0.17	0.23	0.05	0.72
La(54.5)	0.49	0.23	0.28	0.10	1.10
La(68.7)	0.43	0.10	0.36	0.34	1.23
La(76.9)	0.32	0.09	0.41	0.75	1.57
Gd(23.1)	0.37	0.06	0.13	0.09	0.65
Gd(42.3)	0.50	0.13	0.17	0.20	1.00
Gd(57.2)	0.41	0.21	0.27	0.41	1.30
Gd(61.5)	0.38	0.15	0.37	0.48	1.38
Gd(75.0)	0.45	0.22	0.43	0.80	1.90

† Numbers in parentheses indicate % ion exchange.
(Reproduced with permission by W. Klading. *J. Phys. Chem.*, **80**(3), 265(1976)).

Sr^{2+}, La^{3+} and Gd^{3+} by means of amine titrations using Hammett indicators the H_0 range of 6.8 to −8.2 in benzene solution. Table 3.28 gives the results of the acidity measurements of the zeolites after calcination at 773 K. As seen in the table, a completely K-exchanged Y zeolite has no acidity. In exchanging Na^+ against Sr^{2+} or Ca^{2+}, no change in acidity occurs at low exchange levels. By raising degree of exchange, the acidity becomes considerably higher than that of NaY. Fig. 3.68 gives the dependence of the acidity on the percentage of ion exchange of Na^+ ions by Ca^{2+} ions. A sharp rise in acidity at exchange greater than 50% indicates that the Ca^{2+} or Sr^{2+} first fills up the S_I sites of the zeolites. The observation that alkaline earth-exchanged zeolites at low exchange levels do not show much acidity is in accord with infrared investigation and the catalytic activities for cracking of *n*-hexane[23] or isopropylbenzene.[24]

The dependence of the acidity of LaY zeolites on the degree of the cation exchange is shown in Fig. 3.69. A strong enhancement of acidity occurs at exchanges greater than 70%, when strong acidic centers are formed.

The distribution of acid strength of HY, CaX, CaY, LaX and LaY by means of Hammett indicators were also reported by Otouma *et al.*[25]

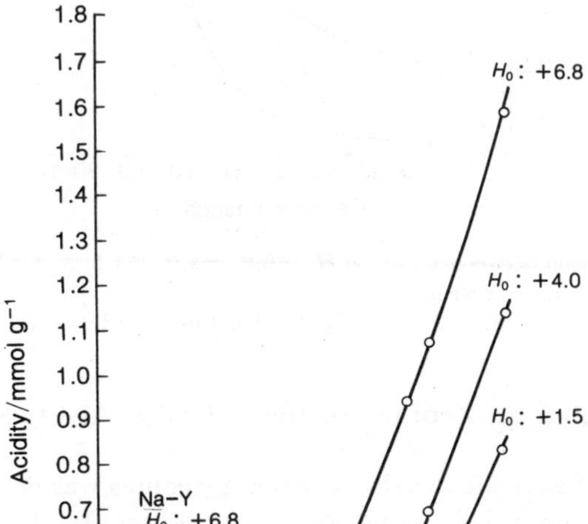

Fig. 3.68 Comparison between acidity at H_0 +6.8, +4.0, +1.5, and −5.6 with degree of exchange in CaNaY zeolite.
(Reproduced with permission by W. Klading, *J. Phys. Chem.*, **80** (3), 265 (1976)).

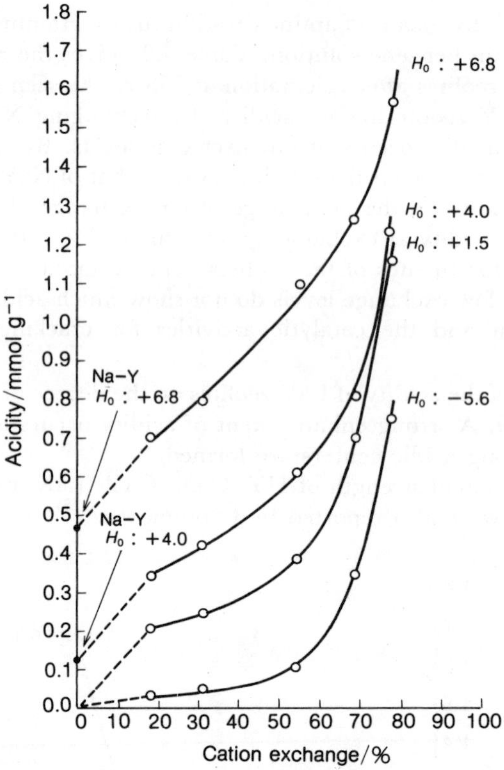

Fig. 3.69 Comparison between acidity at H_0 +6.8, +4.0, +1.5, and −5.6 with degree of exchange in LaNaY zeolite.
(Reproduced with permission by W. Klading, *J. Phys. Chem.*, **80** (3), 267 (1976)).

3.4.4 Acidity of Different Zeolites—Effect of SiO_2/Al_2O_3 Ratio

The mechanism of acid-site development in Y-zeolites was described in the previous section, but it is essentially same for other zeolites. In the case of more acid-resistant zeolites such as ZSM-5 or mordenite, proton can be introduced by direct exchange with dilute hydrochloric acid, though this often causes dealumination of the zeolite framework. The reason why NaY must be converted into NH_4Y to obtain HY is the instability of Y-type zeolites in acidic solutions.

The specific catalytic activity of an OH group changes from zeolite to zeolite. Mordenite is 17 times more active than Y-zeolite[26)] and the activity per OH group of HL is three times higher than that of HY.[12)] In general, the activity per OH group is higher with higher Al/Si ratio of zeolites.[27)] The fact that the nature of OH groups depends on the Si/Al ratio of zeolites was confirmed by the dependence of the shift of OH-stretching band on Al/Si ratio, as shown in Fig. 3.70.[27)] The wavenumber of OH-stretching bands decreases with increasing Si/Al ratio up to 5; further increase of the ratio does not affect the band position of OH-stretching.

Figure 3.70b shows the dependence of the turn-over frequency (activity per OH

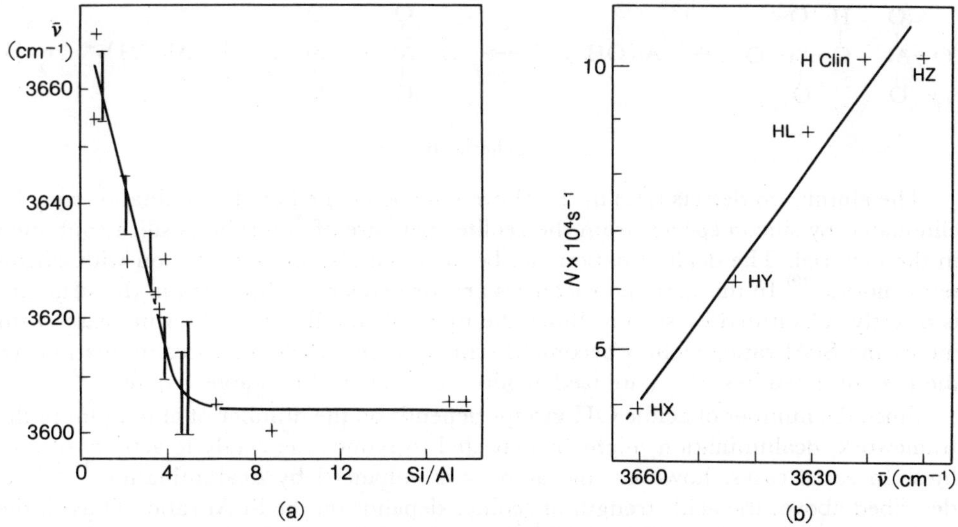

Fig. 3.70 (a) Change in wavenumber of the hydroxyl vibrating in the large cavities as a function of Si/Al ratio. From left to right : A, GeX, X, Y, chabazite, L, Ω, dealuminated Y, dealuminated Y, offretite, mordenite, clinoptilolite, dealuminated Y, dealuminated mordenite, ZSM−5.
(b) Turnover number (N) at 373 K for 2-propanol dehydration as a function of hydroxyl wavenumber
(Reproduced with permission by D. Barthomeuf, *Catalysis by Zeolites* (B. Imelik *et al.*, ed.), Elsevier, Amsterdam, 1980, p. 56.)

group) on the wavenumber of the OH-groups of zeolites. The turn-over number becomes larger with the decrease of the wavenumber or the increase of Al/Si ratio.[27]

Olson *et al.*[28] examined the catalytic activity for hexane cracking of ZSM-5 zeolites with varying Si/Al ratio and found that the activity per aluminum is independent of the ratio. This is expected from the relation as shown in Fig. 3.70(a), where the shift on OH-stretching band is independent of Si/Al ratio when it is over 5.

3.4.5 Effect of Dealumination on Acidic Properties

The Si/Al ratio of zeolites can be modified by removing aluminum from the framework. The dealumination can be achieved in various ways: treatment with acids, steam or EDTA.[29] The dealumination process can be expressed by the following.[29]

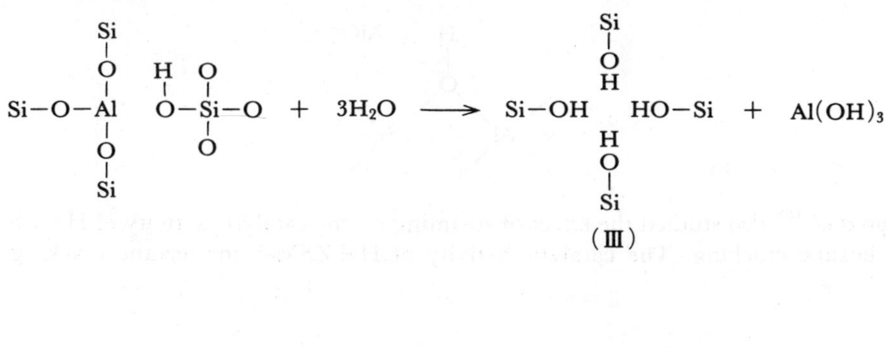

(III)

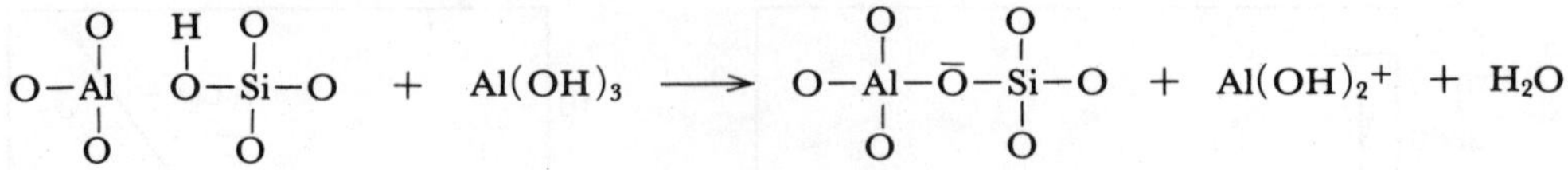

Scheme B

The aluminum defects (III) in the above scheme created by dealumination may be eliminated by silicon species from the zeolite structure of amorphous silica contained in the material. The dealumination can be achieved also by the reaction with silicon tetrachloride.[30] In this case, no vacancies are formed since aluminum in the structure is directly substituted by silicon. Since the thermal stability of zeolites increases with increasing Si/Al ratio, zeolites become thermally more stable after dealumination. In the case of Y-zeolites, the stabilized zeolites are called ultra-stable zeolites.

Since the number of acidic OH groups depends on the number of aluminum in the framework, dealumination might be expected to reduce the catalytic activity of zeolites. In some cases, however, the activity is enhanced by dealumination.[31,32] As described above, the acid strength of zeolites depends on the Si/Al ratio. Thus, if the effect of increase in the acid strength surpasses the effect of the decrease in the acidic centers, dealumination results in enhancement of catalytic activity.

The enhancement of the acid strength of OH groups is caused by their interaction with aluminum species dislodged from the framework and left in the cavities. Miradatos and Barthomeuf[33] revealed the presence of strong acid sites in mordenite dealuminated by steaming by means of a temperature programmed desorption (TPD) spectrum of ammonia and suggested that the strong acids are developed by the interaction of remaining OH groups with the dislodged aluminum species. Ashton *et al.* examined the catalytic activities for hexane cracking of ZSM-5 zeolites, which had been steamed at various steam pressures at 873 K for 2.5 h.[34] The results are given in Fig. 3.71, showing the ratio of the amount of dislodged aluminum and that of aluminum remaining in the framework. The extent of dealumination monotonically increases with the partial pressure of steam. On the other hand, the catalytic activity goes through a maximum at steam pressure of 60 Torr (= 60.8 kP_a). Thus, the partial dislodgement of aluminum gives a catalyst of the highest activity. The development of strong acid sites is confirmed by tpd spectrum of ammonia. Thus, the activity-steam pressure curve is very similar to the amount of strong acid sites-steam pressure curve, where the amount of strong acid sites are defined by the amount of ammonia desorbed above 713 K. Ashton *et al.*[34] inferred a mechanism for the enhancement of the acid strength of remaining OH groups similar to the one proposed by Miradatos and Barthomeuf.[33] Thus, the strong acid sites are tentatively expressed as

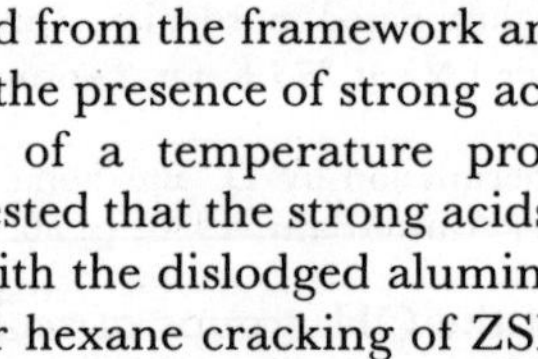

Lago *et al.*[35] also studied the effect of steaming on the catalytic activity of H−ZSM-5 for hexane cracking. The catalytic activity of H−ZSM-5 for hexane cracking was

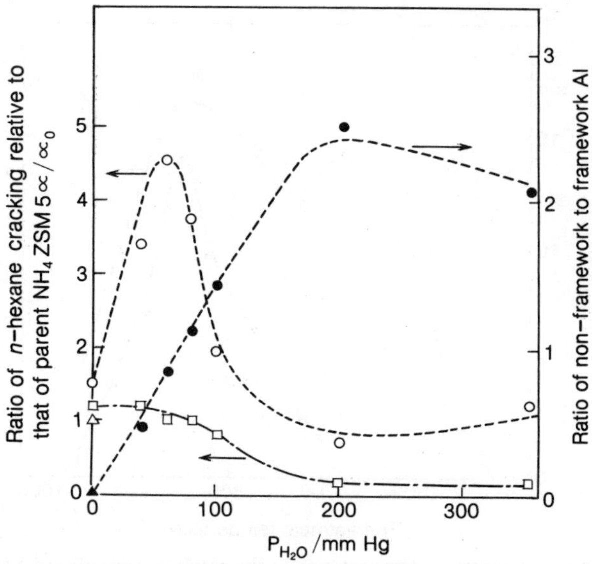

Fig. 3.71 *n*-Hexane cracking over HZSM-5 steamed at 873 K for 2.5 hours.
o-----o ; 1.0 min on stream, □-·—·-□ ; 20.0 min on stream, △; parent zeolite,
●-----● ; ratio of dislodged Al to framework Al, ▲; parent zeolite.
(Reproduced with permission by A. G. Ashton, *et al.*, *Catalisis by Acids and Bases*(B. Imelik *et al.*, eds.) Elsevier, Amsterdam, 1985, p. 106.)

strictly proportional to its aluminum content, when carefully prepared in the absence of water vapor.[24] While severe steaming reduces the catalytic activity, mild steaming produces catalysts of higher activity. They inferred that paired aluminum atoms in the framework are required for enhanced acidic centers, which have 45 – 75 times higher specific activity than normal acid sites.

The change in catalytic activity of NH_4Y with pretreatment temperature has been explained by the transformation of the zeolite as given in Scheme A (Section 3.4.2). However, some of the experimental facts cannot be explained by the transformation expressed in Scheme A. The optimum pretreatment temperature for hexane cracking was found to be 823 K, which is higher than the temperature giving the highest concentration of acidic OH groups.[36] This was ascribed to the formation of stronger acid sites by the partial dehydroxylation of the zeolite.[36] A similar effect of the pretreatment temperature was observed also for the isomerization and disproportionation of 1-methyl – 2-ethylbenzene.[37]

Sendoda and Ono studied the dependence of catalytic activity of ZSM-5 for alkane conversion on its pretreatment temperature.[38] The activity maximum for propane conversion was observed at the pretreatment temperature of 853 K (Fig. 3.72).[38,39] This is much higher than the temperature where NH_4 – ZSM-5 is transformed into H-ZSM-5, suggesting that ordinary OH groups are not active centers for propane conversion. In the case of hexane cracking, double maxima were observed at 673 K and 853 K (Fig. 3.72). The double maxima pattern was observed also for cracking of pentane and heptane, and for isomerization of *o*-xylene. These results show definitively the existence of two types of active centers. The one giving a maximum activity at 673

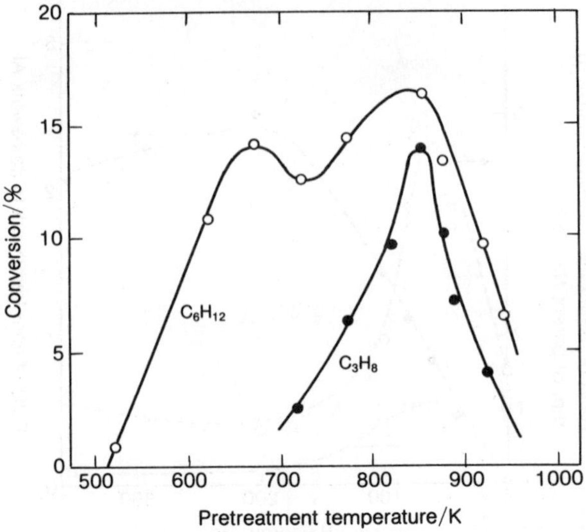

Fig. 3.72 **Effect of pretreatment temperature on the catalytic activities of $NH_4-ZSM-5$ for the conversion of hexane and propane.**
Hexane conversion : 513 K 36 kP_a, W/F=12.7 g h mol^{-1}
Propane conversion : 723 K 101.3 kP_a, W/F=7.0 g h mol^{-1}
(Reproduced with permission by Y. Sendoda, Y. Ono, *Zeolites,* **8**, 102 (1988)).

K is the ordinary OH group, and the other giving maximum activity at 853 K is similar to that postulated for the steamed zeolites. Only the latter is active for propane conversion. It was confirmed that 13% of the framework aluminum was dislodged after pretreatment at 853 K. It was shown that the activity for propane conversion was greatly enhanced by steaming.

It is important to note that partial dealumination can occur under ordinary pretreatment conditions. It also occurs during the preparation of H−ZSM-5 by the direct exchange of ZSM-5 (or mordenite) with acidic solution. Anderson found that H−ZSM-5 pepared by direct exchange is much more active than the one prepared from $NH_4-ZSM-5$.[40] Vedrine *et al.* found that Lewis acid sites are present even after pretreatment at temperature as low as 573K in H-ZMS-5.[41] These facts appear to be related to the dealumination during the preparation of H-ZSM-5 by the direct exchange with dilute acid.

3.4.6 Acidity of Metallosilicate

The synthesis of zeolites containing various elements such as B, P, or Ge has been carried out for a long time. Since the discovery of ZSM-5 (aluminosilicate) and silicalite, many attempts have been made to synthesize the metalloscilicate with the ZSM-5 structure. The isomorphous substitution of aluminum with other elements greatly modifies the acidic properties of the silicate. The elements introduced include, Be, B, Ti, Cr, Fe, Zn, Ga, Ge, and V. These elements were usually introduced by adding metal salts as one of the starting materials for the synthesis of the metallosilicate. It is also known that boron can be directly introduced by reacting ZSM-5 with boron

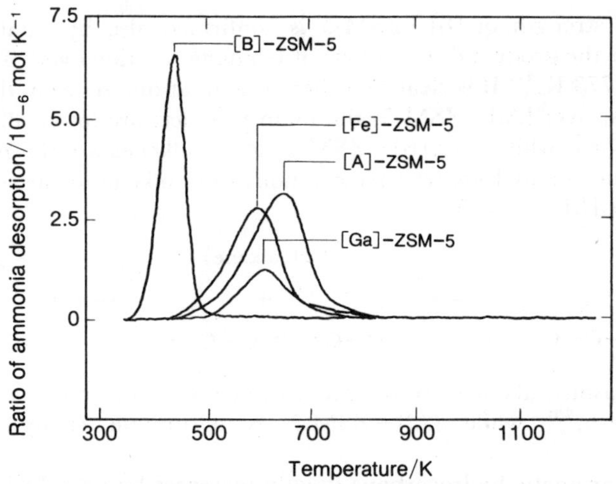

Fig. 3.73 **Temperature programmed desorption of ammonia from metallosilicates.** (Reproduced with permission by C. T-W. Chu, C. D. Chang, *J. Phys, Chem.*, **89**, 1571(1985)).

trichloride.[42] Metallosilicate with a ZSM-5 structure having metal M as a component will be denoted [M]−ZSM-5, hereafter. Silicate II (the framework topology of which is structually identical to that of ZSM-11) can be transformed into gallosilicate with its reaction with $NaGaO_2$ in an aqueous solution.[43]

Figure 3.73 shows the TPD spectrum of ammonia adsorbed on various metallosilicate.[44] The acid strength of metallosilicate changes in decreasing order as follows:

$$[Al]-ZSM\text{-}5 > [Ga]-ZSM\text{-}5 > [Fe]-ZSM\text{-}5 > [B]-ZSM\text{-}5$$

The band position of OH groups changes in conformity with TPD spectra. Thus, the OH band appears at 3610, 3620, 3630, and 3725 cm^{-1} for [Al]−, [Ga]−, [Fe]−, and [B]−ZSM-5, respectively.[44] The fact that the acid strength of [B]−ZSM-5 is much weaker than that of [Al]−ZSM-5 has been reported by several authors.[45,46]

TABLE 3.29 **Product distribution of the conversion of 1-butene over H-ZSM-5, H-[B]-ZSM-5 and Zn-[B]-ZSM-5**

Catalyst	H-[Al]-ZSM-5	H-[B]-ZSM-5	Zn-[B]-ZSM-5
conversion/%	77.3	71.7	81.2
Products/%†1			
$C_1 \sim C_4$ alkanes	41.3	5.1	6.3
$C_2H_4+C_3H_6$	14.6	38.3	21.1
C_4H_8†2	6.2	28.3	27.7
C_5+	2.4	25.3	7.0
aromatics	37.0	3.0	38.0

Reaction conditions, 773 K, W/F=5.3 g h mol^{-1}
1-butene=23.0 kPa
†1 carbon-number basis, †2 including 1-butene

Weaker acid strength of [B]−ZSM-5 is confirmed also by catalytic reactions. Table 3.29 shows the product distributions of 1-butene reaction over [B]−ZSM-5 and [Al]−ZSM-5 at 773 K.[47)] It is clear that there is a great difference in the product distirubution. Thus, over [Al]−ZSM-5, the main products are lower alkanes and aromatic hydrocarbons, while over [B]−ZSM-5, lower alkenes are the main products. This indicates that the hydride transfer reactions from alkene to carbenium ion does not proceed over [B]−ZSM-5.

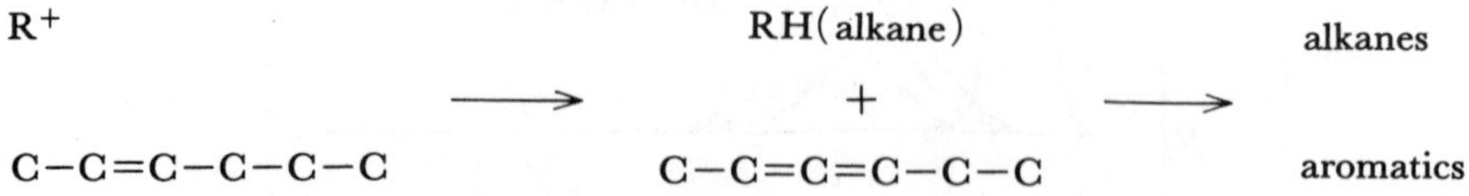

For the same reasons, alkenes are the main products in the conversion of methanol over [B]−ZSM-5,[48)] while [Al]−ZSM-5 is a unique catalyst for gasoline production.[49)]

The yield of aromatic hydrocarbons greatly increases by introducing zinc cations into [B]−ZSM-5 (Table 3.29).[47)] In this case, however, the yield of alkanes remains low. This is because the aromatics are formed by the direct dehydrogenation of olefins by the action of zinc species.[47,50)] As exemplified by this case, it is possible to achieve catalysis by metal cations at the same time suppressing catalysis by acid.

The acid strength of [Fe]−ZSM-5 can be inferred to be weak from the very low yield of alkanes and aromatics in the conversion of methanol or olefins.[51] Hölderich reported that ketone can be isomerized to aldehyde in a high yield over [B]−ZSM-5.[52)] ZSM-5 gave only low selectivity.[52)]

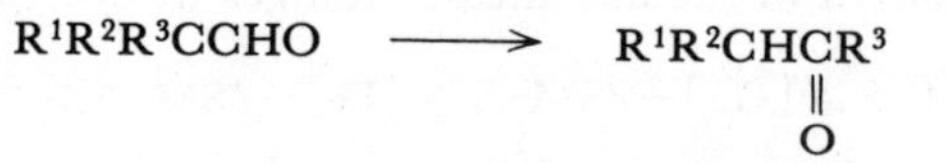

Since the acidic strength of [B]−ZSM-5 is weak, the role of the trace amount of aluminum impurity may not be negligible in their catalysts. Chu *et al.*[53)] examined the catalytic activities of [B]−ZSM-5 containing varying amounts of framework B for a number of acid-catalyzed reactions and concluded that the catalytic activity was due, if not entirely, to trace amounts (80−580 ppm) of framework aluminum.

3.4.7 $AlPO_4$-*n*, SAPO-*n* and Related Materials

A novel class of crystalline, micro-porous aluminophosphate phases was synthesized by Wilson *et al.*[54)] and named aluminophosphate molecular sieve, $AlPO_4$-*n*. (The suffix *n* denotes a specific structure type.) Their product composition expressed as oxide formula is xR Al_2O_3 $(1.0 \pm 0.2)P_2O_5$ $y5H_2O$, where R is an amine or a quarternary ammonium template. Calcination at a typical temperature of 773−873 K removes R and H_2O, and yields the microporous molecular sieve framework expressed as $AlPO_4$ or $(Al_{0.5}P_{0.5})O_2$. Though some of the materials are structually related to the zeolite family, most are novel. Typical structures of $AlPO_4$-*n* are listed in Table 3.30, where the pore sizes and pore volumes determined by oxygen and water adsorption are also shown. The frameworks of $AlPO_4$-*n* are neutral and thus have no ion exchange capacity. They exhibit only weakly acidic catalytic properties.

TABLE 3.30 Properties of selected $AlPO_4$ molecular sieves

Structure	Adsorption properties[†1]			
	Pore size, nm	Ring[†2] size	Intracrystalline pore vol, cm^3/g	
			O_2	H_2O
$AlPO_4$-5	0.8	12	0.18	0.3
$AlPO_4$-11	0.61	10	0.11	0.16
$AlPO_4$-14	0.41	8	0.19	0.28
$AlPO_4$-16	0.3	6	0	0.3
$AlPO_4$-17	0.46	8	0.20	0.28
$AlPO_4$-18	0.46	8	0.27	0.35
$AlPO_4$-20	0.3	6	0	0.24
$AlPO_4$-31	0.8	12	0.09	0.17
$AlPO_4$-33	0.41	8	0.23	0.23

†1 Determined by standard McBain-Baker gravimetric techniques after calcination (773-873 K in air); pore size determined from measurements on molecules of varing size; pore volumes near saturation, O_2 at 80 K, H_2O at room temperature.

†2 Number of tetrahedral atoms (Al or P) in ring that controls pore size

(Reproduced with permission by S. T. Wilson *et al.*, *J. Am. Chem. Soc.*, **104**, 1147(1982)).

Later, the synthesis of the family of crystalline silicoaluminophosphate (SAPO-*n*) was reported.[55)] Some of them have structures topologically related to zeolites or $AlPO_4$-*n*, some having novel structures. SAPO-*n* composition can be considered in terms of silicon substitution into a hypothetical aluminophosphate framework, the predominant substitution mechanism being silicon substitution for phosphorus.

The incorporation of various elements into aluminophosphate and silicoaluminophosphate frameworks has been accomplished recently, and they are denoted MeAPO-*n* and MeSAPO-*n*, where Me is metal cations such as Fe, Mg, Mn, Co or Zn.[56)] SAPO, MeAPO, and MeSAPO have anionic frameworks with a net negative charge with concomitant cation exchange properties and potential for Brønsted acid sites. As a probe of Brønsted acidity, the catalytic properties of these molecular sieve materials have been assessed with butane cracking. The pseudo-first order rate constants for a number of aluminophosphate-based materials are shown in Table 3.31. As expected, $AlPO_4$-*n* molecular sieves have only low activity. The activities of some of MeAPO-*n* and MeAPSO-*n* are higher than that of Y-type zeolite, but still much lower than ZSM-5. The catalytic activity for butane cracking may be considered as a measure of stronger Brønsted acid sites.

Pellet *et al.* reported on the reaction of propene over SAPO-*n* materials.[57)] Though the main products over ZSM-5 type aluminosilicate are aromatics and lower alkanes, oligomerization proceeds selectively over SAPO-11 and SAPO-31, indicating that these materials do not possess strong acid sites which are capable of promoting hydride transfer reactions.

Tapp *et al.* studied the TPD spectra of ammonia of CoAPO-5, SAPO-5 and $AlPO_4$-5 and found that the number of acid sites is highest with CoAPO-5, which ex-

TABLE 3.31 Catalytic activities of $AlPO_4$-n and related materials for butane cracking

	k		*k*		*k*		*k*
$AlPO_4$-5	~0.05	SAPO-5	0.2-16	BeAPO-34	3.7	SAPO-34	0.1-7.6
BeAPO-5	3.4	MAPSO-5	1.2	CoAPO-34	5-15	BeAPO-34	7.6
CoAPO-5	0.4	ZAPSO-5	1.5	FAPO-34	0.1-0.6	GAPSO-34	10.0
MAPO-5	0.5			MAPO-34	7-29		
MnAPO-5	1.2			MnAPO-34	2.5-5.2		
$AlPO_4$-11	<0.05	SAPO-11	0.5-3.5	ZAPO-34	13		
		SAPO-31	0.1-0.9			SAPO-44	1.2-2.4
		SAPO-41	1.3			ZAPSO-44	5.0
MAPO-36	11-24	MAPSO-36	11	chabasite	-7	NH_4Y	-2
CoAPO-36	11			erionite	4-5	ZSM-5	>40
MnAPO-36	68						

k: pseudo-first order rate constant at 773 K (cm^3g^{-1} min^{-1})
(Reproduced with permission by E. M. Flanigen *et al.*, *New Developmemt in Zeolite Science and Technology*, Kodansha-Elsevier, 1986, p. 110).

hibits the highest acid-strength.[58] They also showed that the acidic nature of CoAPO-11 exceeds that of CoAPO-5 in both number and strength.[58]

3.4.8 Zeolites as Base Catalysts

Table 3.32 shows the activities of faujasite-type zeolites for the reaction of γ-butyrolactone and hydrogen sulfide.[59]

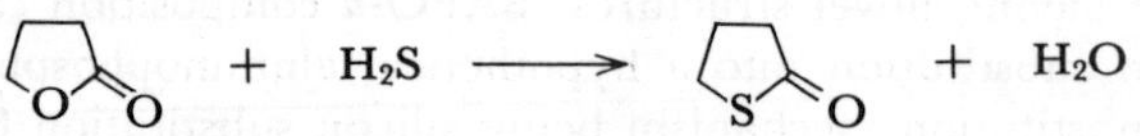

The following characteristics are clear from Table 3.32.

TABLE 3.32 Catalytic activities of Y-type zeolites for ring transformation of γ-butyrolactone into γ-butyrothiolactone

Catalyst	Exchanged (%)	Conversion (%)	Yield (%)
LiY	58	27	26
NaY	—	52	51
KY	97	45	45
RbY	64	51	51
CsY	64	79	78
NaX	—	99	86
KL	—	23	22
HY	66	4	1
MgY	56	2	2

Reaction conditions; 603 K, H_2S/lactone=6. W/F=6.26 g h mol^{-1}
(Reproduced with permission by K.Hatada *et al.*, *Bull. Chem. Soc. Jpn.*, **51** 448(1977)).

i) The activity changes with alkali cation in the decreasing order:
CsY > RbY > KY > NaY > LiY.
ii) NaX is more active than NaY.
iii) Acidic zeolites like HY or MgY have no activity.

These features are totally opposite to those found in acid-catalyzed reactions. The addition of pyridine to the system does not inhibit the reaction, but enhances the activity. These facts strongly indicate that the catalytic activity is not caused by protons, but by centers of basic character.

Alkali-exchanged faujasites are also active for the Claus reaction ($SO_2 + H_2S \rightarrow 3S + H_2O$)[60] and for the reaction of tetrahydrofuran and hydrogen sulfide,[61] indicating that the basic centers play decisive role in the activation of hydrogen sulfide.

The reaction of toluene with methanol over acidic zeolites gives xylenes. Sidorenko *et al.* found that the reaction over alkali-exchanged zeolites gave the products of the side chain alkylation: styrene and ethylbenzene.[62] Yashima *et al.* studied the reaction in detail and discovered the following. (1) Xylenes are the only products over LiX and LiY, while stylene and ethylbenzene are found over K-, Rb-, and Cs-exchanged faujasites. (2) X-type zeolites are more active than Y-type zeolites for the side chain alkylation. (3) Addition of aniline to the system enhances the formation of styrene and ethylbenzene, while addition of hydrogen chloride inhibits completely their production. (4) The basic character of KX or RbX is confirmed by color change of the indicators, cresol red and thymolphthalein.[63] Later, the cooperative nature of the acid and base centers was pointed out by Murakami and coworkers.[64,65]

The reactions, in which basic sites play a principal role, include aldol condensation of butanal[66] and dimerization of cyclopropene[67] and dehydrogenation of 2-propanol.[68]

Barthomeuf estimated the strength of basic sites of alkali-exchanged zeolite from the shift of the IR band at $\sim 3{,}200\ \mathrm{cm}^{-1}$ of adsorbed pyrole.[69] The strength determined by the shift increased with Al/(Al + Si) ratio and also with the size of the alkali cations. These trends are in good agreement with the change in oxygen charges determined using the Sanderson equalization principle.[70]

3.4.9 Shape Selective Reactions over Zeolites

Since zeolites have small and uniform pores and most of the active sites are located inside this pore system, the selectivities of the catalytic reactions often greatly depend upon the relative dimensions of the molecules and the pore openings. Actually, an infrared spectroscopic study of ZSM-5 zeolite revealed that only 5 – 10 percent of Brønsted sites are located on the external surface of the zeolite and the rest inside the pore systems.[71] The first report on shape-selective catalysis by Weisz and Frilette appeared in 1960.[72] Many applications are found in the petroleum and chemical industries for catalytic cracking and hydrocracking and aromatic alkylation.

In Table 3.33 the activities of CaX and CaA for dehydration of butyl and isobutyl alcohol are compared.[73] Over CaX, both alcohols are dehydrated rapidly in the temperature range of 503 – 533 K, with the isobutyl alcohol showing somewhat greater activity. This behavior is compatible with the fact that both are primary alcohols and should resemble each other. Both CaX and CaA show little difference in activity with butyl alcohol which can penetrate both crystals. However, the isobutyl alcohol, which

TABLE 3.33 Dehydration of primary butyl alcohols over CaA and CaX

Temp./K	Wt % dehydration			
	CaX		CaA	
	n-Butyl	Isobutyl	*n*-Butyl	Isobutyl
493	—	22	10	<2
503	9	46	18	<2
513	22	63	28	<2
533	64	85	60	<2
563	—	—	—	5

contact time=6s, 101.3 kPa
(Reproduced with permission by P. B. Weisz. *et al.*, *J. Catal.*, **1**, 310(1962)).

is excluded from the crystal interior of the CaA crystal, shows a dramatic lack of conversion, unless one goes to extremely high temperatures. This type of shape-selectivity is often called reactant selectivity, since the catalytic activities are determined by the size of the reactants.

Table 3.34 gives the ratios of isobutane to butane and isopentane to pentane in the products for hexane cracking over CaA and, for comparison, results with silica-alumina and CaX.[73)] Isoparaffins are essentially absent with CaA, while they are prevalent with silica-alumina and CaX. This "product selectivity" is caused by the fact that isoparaffins, cracking products, cannot pass out of the crystalline cavities.

TABLE 3.34 Isoparaffin/*n*-paraffin product ratios from *n*-hexane cracking

	5A	$SiO_2 \cdot Al_2O_3$	10X
iso-C_4/*n*-C_4	<.05	1.4	0.7
iso-C_5/*n*-C_5	<.05	10.0	1.0

(Reproduced with permission by P. B. Weisz. *et al.*, *J. Catal.*, **1**, 309 (1962)).

Csicsery proposed "restricted transition state-type selectivity," in which certain reactions are prevented because the transition state is too large for the cavities of the zeolite.[74)] However, neither reactants nor potential products are prevented from diffusing through the pores; only the formation of the transition state is hindered. A typical example can be found in acid-catalyzed transalkylation of dialkylbenzenes. This is a bimolecular reaction involving a diphenylmethane transition state. Csicsery found that, in the reaction of 1-methyl-2-ethylbenzene over mordenite, 1,3,5-substituted products were absent in the products.[75,76)] This fact was explained as follows: The formation of diarylmethane-type transition state, which might lead to their formation, is hindered by the spacial restriction of the mordenite pores. It should be noted that the 1,3,5-isomers undergo no steric hindrance for their diffusion through pores of mordenite.

Shape-selective reactions are widely applied in the chemical industry. Selectforming using Ni-containing erionite is a process by which normal alkanes in gasoline reformate fractions are hydrocracked essentially to propane.[77] The liquid-product shows an increased octane number because of the preferential conversion of its low-octane components. In the Mobil distillate dewaxing (MDDW), the pour point of a gas-oil distillate containing long-chain linear paraffins, isoparaffins, naphthenes, aromatic compounds, and highly branched paraffins is lowered by selectively cracking the linear paraffins and isoparaffins.[78] These are admitted into the intracrystalline volume of the zeolite, in which the reaction essentially takes place.

In the MTG (methanol-to-gasoline) process, methanol is converted effectively into hydrocarbons with gasoline-range boiling points over ZSM-5.[49] The reaction of methanol over ZSM-5 yields hydrocarbons with six to ten carbon atoms and $C_{10}{}^{+}$ aromatic compounds only in small or trace amounts. This can be explained by assuming that the upper size of the aromatic products is imposed by the ZSM-5 pore structure.

REFERENCES

1. J. Turkevich, S. Cibrorowski, *J. Phys. Chem.*, **71**, 3208 (1967).
2. J. B. Uytterhoeven, L. G. Christner, W. K. Hall, *J. Phys. Chem.*, **69**, 2117 (1965).
3. J. W. Ward, *J. Catal.*, **9**, 225 (1967).
4. J. W. Ward, *J. Catal.*, **9**, 396 (1967); **16**, 386 (1970).
5. O. N. Stamires, J. Turkevich, *J. Am. Chem. Soc.*, **86**, 749 (1963).
6. J. Turkevish, Y. Ono, *Adv. Catal.*, **20**, 135 (1969).
7. Y. Ono, H. Tokunaga, T. Keii, *J. Phys. Chem.*, **78**, 218 (1974).
8. B. D. Flockhart, L. Machaughlin, *J. Catal.*, **64**, 1093 (1968).
9. C. W. Breck, G. W. Skeels, Proc. 6th Intern. Congr. Catalysis, 1976, The Chemical Society, London, 1977, p.645.
10. W. H. Frank, G. W. Skeels, Proc. 5th Intern. Conf. Zeolites (L.V.C. Rees, ed.) 1980, Heydon, 1980, p.334.
11. P. A. Jacobs, H. K. Beyer, *J. Phys. Chem.*, **83**, 1174 (1979).
12. Y. Ono, M. Kaneko, K. Kogo, H. Takayanagi, T. Keii, *J. Chem. Soc., Faraday Trans. 1*, **72**, 2150 (1976).
13. J. Turkevich, Y. Ono, *Adv. Chem. Ser.*, **102**, 815 (1979).
14. L. G. Christner, B. Liengme, W. K. Hall, *Trans. Faraday Soc.*, **64**, 1093 (1968).
15. J. W. Ward, *J. Catal.*, **10**, 34 (1968).
16. J. W. Ward, *J. Phys. Chem.*, **72**, 4211 (1968).
17. C. Naccache, Y. Ben Taarit, *J. Catal.*, **22**, 171 (1971).
18. H. Beyer, P. A. Jacobs, J. B. Uytterhoeven, *J. Chem. Soc., Faraday Trans. 1*, **72**, 674 (1976).
19. T. Baba, Y. Ono, *Zeolites* **7**, 292 (1987).
20. T. Baba, S. G. Seo, Y. Ono, in: *Innovation in Zeolite Materials Science*, (P. J. Grobet *et al.*, eds.) Elsevier, Amsterdam, 1988, p.443.
21. T. Baba, Y. Ono, to be published.
22. W. Klading, *J. Phys. Chem.*, **80** (3), 262 (1976).
23. S. E. Tung, E. Mclininch, *J. Catal.*, **10**, 166 (1968).
24. K. Tsutsumi, H. Takahashi, *J. Catal.*, **24**, 1 (1972).
25. H. Otouma, Y. Arai, H. Ukihashi, *Bull. Chem. Soc. Jpn.*, **42**, 2449 (1969).
26. J. W. Ward, *J. Catal.*, **13**, 316 (1969).
27. D. Barthomeuf, *Catalysis by Zeolites* (B. Imelik *et al.*, ed.) Elsevier, Amsterdam, 1980, p.55.
28. D. H. Olson, W. O. Haag, R. M. Lago, *J. Catal.*, **61**, 390 (1980).
29. G. T. Kerr, *Adv. Chem. Ser.*, **121**, 219 (1979).

30. H. K. Beyer, I. Belenykaya, in: *Catalysis by Zeolites* (B. Imelik *et al.*, eds.) Elsevier, Amsterdam, 1980, p.203.
31. P. E. Ekerly, C. N. Kimberlin, *Ind. Eng. Chem. Prod. Res. Dev.*, **9**, 335 (1970).
32. P. B. Koradia, J. R. Kiovsky, M. Y. Asim, *J. Catal.*, **66**, 290 (1980).
33. C. Miradatos, D. Barthomeuf, *J. Chem. Soc., Chem. Commun.*, **1981**, 39.
34. A. G. Ashton, S. Batmanian, D. M. Clark, J. Dwyer, F. R. Fitch, A. Hinchcliffe, F. J. Michado, in: *Catalysis by Acids and Bases* (B. Imelik *et al.*, eds.) Elsevier, Amsterdam, 1985, p.101.
35. R. M. Lago, W. O. Haag, R. Mikovsky, D. H. Olson, K. D. Schmitt, G. T. Kerr, Proc. 7th Intern. Zeolite Conf. (Y. Murakami *et al.*, eds.), Kodansha, Tokyo and Elsevier, Amsterdam, 1986, p.677.
36. H. D. Hopkins, *J. Catal.*, **12**, 325 (1968).
37. D. A. Hickson, S. M. Csicsery, *J. Catal.*, **10**, 27 (1968).
38. Y. Sendoda, Y. Ono, *Zeolites*, **8**, 101 (1988).
39. H. Kitagawa, Y. Sendoda, Y. Ono, *J. Catal.*, **101**, 12 (1986).
40. R. A. Rajadhyaksha, J. R. Anderson, *J. Catal.*, **63** 510 (1980).
41. J. C. Vedrine, A. Auroux, U. Borus, P. Bejaifve, C. Naccache, P. Wiezchowski, E. Derouane, J. B. Nagy, J-P. Gilson J. H. C. van Hooff, J.P. van den Berg, J. Wolthuizen, *J. Catal.*, **59**, 248 (1979).
42. E. G. Derouane, L. Baltnis, R. M. Dessau, K. D. Schmit, *Catalysis by Acids and Bases*, (B. Imelik *et al.*, eds.) Elsevier, Amsterdam, 1985, p.135.
43. J. M. Thomas, X-S. Lin *J. Phys. Chem.*, **90**, 4843 (1986).
44. C. T-W. Chu, C. D. Chang, *J. Phys. Chem.*, **89**, 1569 (1985).
45. K.F.M.G. Scholle, A.P.M. Kentgens, W. S. Veeman, P. Frenken, G.P.M. van der Verden, *J. Phys. Chem.*, **88**, 5, (1984).
46. M. G. Howden, *Zeolites*, **5**, 334 (1985).
47. Y. Ono, H. Kitagawa, Y. Sendoda, *J. Chem. Soc., Faraday Trans. 1*, **83**, 2913 (1987).
48. W. Hölderich, H. Eichhorn, R. Lehnert, L. Marosi, W. Moss, H. Schlimper, Proc. 6th Intern. Zeolite Conf. (D. H. Olson, A. Bisio, eds.) Butterworth, Guildford, 1984, p.545.
49. C. D. Chang, A. J. Silverstri, *J. Catal.*, **47**, 549 (1977).
50. Y. Ono, H. Adachi, Y. Sendoda, *J. Chem. Soc., Faraday Trans. I*, **84**, 1091 (1988).
51. T. Inui, A.Miyamoto, H. Matsuda, H. Nagata, Y. Makino, K. Fukuda, F. Okazumi, Proc. 7th Zeolite Intern. Conf. (Y. Murakami *et al.*, eds.), Kodansha, Tokyo and Elsevier, Amsterdam, 1986, p.859.
52. W. Höldrich, Proc. 7th Zeolite Intern. Conf. (Y. Murakami *et al.*, eds.) Kodansha, Tokyo and Elsevier, Amsterdam, 1986, p.827.
53. C. T-W. Chu, G. H. Kuehl, R. Lago, C. C. Chang, *J. Catal.*, **93**, 451 (1985).
54. S. T. Wilson, B. M. Lok, C. A. Messina, T. R. Cannan, E. M. Franigen, *J. Am. Chem. Soc.*, **104**, 1146 (1982).
55. B. M. Lock, C. A. Messina, R. L. Patton, R. T. Gajek, T. R. Cannan, E. M. Flanigen, *J. Am. Chem. Soc.*, **106**, 6092 (1984).
56. E. M. Flanigen, B. M. Lok, R. L. Patton, S. T. Wilson, Proc. 7th Intern. Zeolite Conf. (Y. Murakami *et al.*, eds.), Kodansha, Tokyo and Elsevier, Amsterdam, 1986, p.103.
57. R. J. Pellet, G. N. Long, J. A. Rabo, Proc. 7th Intern. Zeolite Conf. (Y. Murakami *et al.*, eds.) Kodansha, Tokyo and Elsevier, Amsterdam, 1986, p.843.
58. N. J. Tapp, N. B. Milestene, J. L. Wright, *J.Chem. Soc., Chem. Commun.*, **1985**, 1801.
59. K. Hatada, Y. Takeyama, Y. Ono, *Bull. Chem. Soc. Jpn.*, **51**, 448 (1977).
60. Z. M. George, R. Tower, Proc. 5th Intl. Conf. Zeolites, (L.V.C. Rees, ed.) 1980, Naples, Heydon, London, 1980, p.850.
61. Y. Ono, T. Mori, K. Hatada, *Acta Phys. Chem.*, **24**, 233 (1984).
62. Y. N. Sidorenko, P. N. Galich, V. S. Gutynya, V. G. Ilins, I. E. Neimark, *Dokl. Akad. Hauk SSSR*, **173**, 132 (1967).
63. T. Yashima, H. Suzuki, N. Hara, *J. Catal.*, **33**, 486 (1974).
64. H. Itoh, T. Hattori, K. Suzuki, A. Miyamoto, Y. Murakami, *J. Catal.*, **72**, 170 (1981).
65. H. Itoh, T. Hattori, K. Suzuki, Y. Murakami, *J. Catal.*, **79**, 21 (1983).
66. Kh. Minachev, V. Garanin, T. Isakova, V. Kharlarmov, V. Bogomolov, *Adv. Chem. Ser.*, **102**, 441 (1971).
67. A. J. Schipperijin, J. Lukasa, *Rece. Trav. Chim. Pays-Bas.*, **92**, 572 (1973).

68. T. Yashima, H. Suzuki, N. Hara, *J. Catal.*, **33**, 486 (1974).
69. D. Barthomeuf, *J. Phys Chem.*, **88**, 43 (1984).
70. R. T. Sanderson, *Chemical Bonds and Bond Energy*, Academic Press, New York, 1976.
71. J. Take, T. Yamaguchi, K. Miyamoto, H. Ohyama, M. Misono, Proc. 7th Intern. Zeolite Conf. 1986, (Y. Murakami *at al.*, eds.), Kodansha, Tokyo and Elsevier, Amsterdam, 1986, p.495.
72. P. B. Weisz, V. J. Frilette, *J. Phys. Chem.*, **64**, 382 (1960).
73. P. B. Weisz, V. J. Frilette, R. W. Maatman, E. B. Mower, *J. Catal.*, **1**, 307, (1962).
74. S. M. Csicsery, in *Zeolite Chemistry and Catalysis* (J. A. Rabo, ed.) American Chemical Society, 1976, p.680.
75. S. M. Csicsery, *J. Catal.*, **19**, 394 (1970).
76. S. M. Csicsery, *J. Catal.*, **23**, 124 (1971).
77. N. Y. Chen, J. Maziuk, A. B. Schwartz, P. B. Weisz, *Oil Gas J.*, **66**, (47) 154 (1968).
78. N. Y. Chen, R. L. Gorring, H. R. Ireland, T. R. Stein, *Oil Gas J.*, **75**, (23) 165 (1977).

3.5 HETEROPOLY COMPOUNDS

3.5.1 General Remarks

Heteropolyanions are polymeric oxoanions which are formed by the condensation of more than two different oxoanions (e.g., eq.(1)). Polyanions consisting of one kind of oxoanions are called isopolyanions. The term "heteropoly compounds" is used to indicate heteropoly acids (free acid forms) and their salts.

$$12WO_4^{2-} + HPO_4^{2-} + 23H^+ \longrightarrow (PW_{12}O_{40})^{3-} + 12H_2O \qquad (1)$$

A variety of polyanion structures are known.[1-4] For example, the structure of the polyanion of so-called "Keggin structure" is shown in Fig. 3.74a. The heteropoly compounds having the Keggin structure are thermally more stable and rather easily obtained, so that investigations have been devoted mostly to this group. They are used as acid catalysts as well as oxidation catalysts in both hereto- and homogenous systems (cf. comprehensive review on heterogenous catalysis[5)]). The principal merits of heteropoly compounds when they are used as solid acid catalysts are as follows.[5-9]

i) Acidic properties as well as redox properties can be controlled by varying the constituent elements.

ii) Structure can be defined at the molecular level of heteropolyanions, so they are cluster models of mixed-oxide catalysts.

iii) Some heteropoly compounds exhibit pseudo-liquid behavior which endows those compounds with unique catalytic properties.

Acidic and redox properties of heteropoly compounds in the solid state are closely related to those in solution except for the effect of counter cation. In solution, 12-heteropoly acids (XM_{12}) are much stronger than $HClO_4$, H_2SO_4 and the oxoacids of the constituent elements (XM_{12} represents $H_xXM_{12}O_{40}$ or $XM_{12}O_{40}^{n-}$). The acid strengths estimated from the formation constants for complexes of polyanions with chloral hydrate,[10,11] are in the order,

$$PW_{12} > PMo_{12} > SiW_{12} \sim GeW_{12} > SiMo_{12} \sim GeMo_{12} \qquad (2)$$

The acid strength measured in organic media follows a similar order.[4] The general trends of W > Mo (for polyatom) and P^{5+} > Si^{4+}, Ge^{4+} (for heteroatom) are noted. The acid strength tends to decrease upon reduction or substitution of W, Mo, to V owing to an increased negative charge. The acidity function of Hammett, H_0, of concentrated solution of PW_{12} is higher by 1−1.5 units than that of $HClO_4$ and H_2SO_4.[4]

The softness of polyanions has been determined from the equilibrium of $AgI + XM_{12}O_{40}^{n-} \rightleftarrows AgXM_{12}O_{40}^{(n-1)-} + I^-$ as in eq. (3)[10,11]:

$$SiW_{12} > GeW_{12} > PW_{12} > PMo_{12} > SiMo_{12} > NO_3^- > TsO^- > SO_4^{2-} \qquad (3)$$

Heteropoly compounds are efficient catalysts for various reactions in solution, e.g. hydration, etheration and esterification. They usually exhibit much higher catalytic activities than mineral acids.[11,12] The high activities of heteropoly compounds are principally due to the strong acidity and the stabilization of reaction intermediates by complex formation.[11−14)]

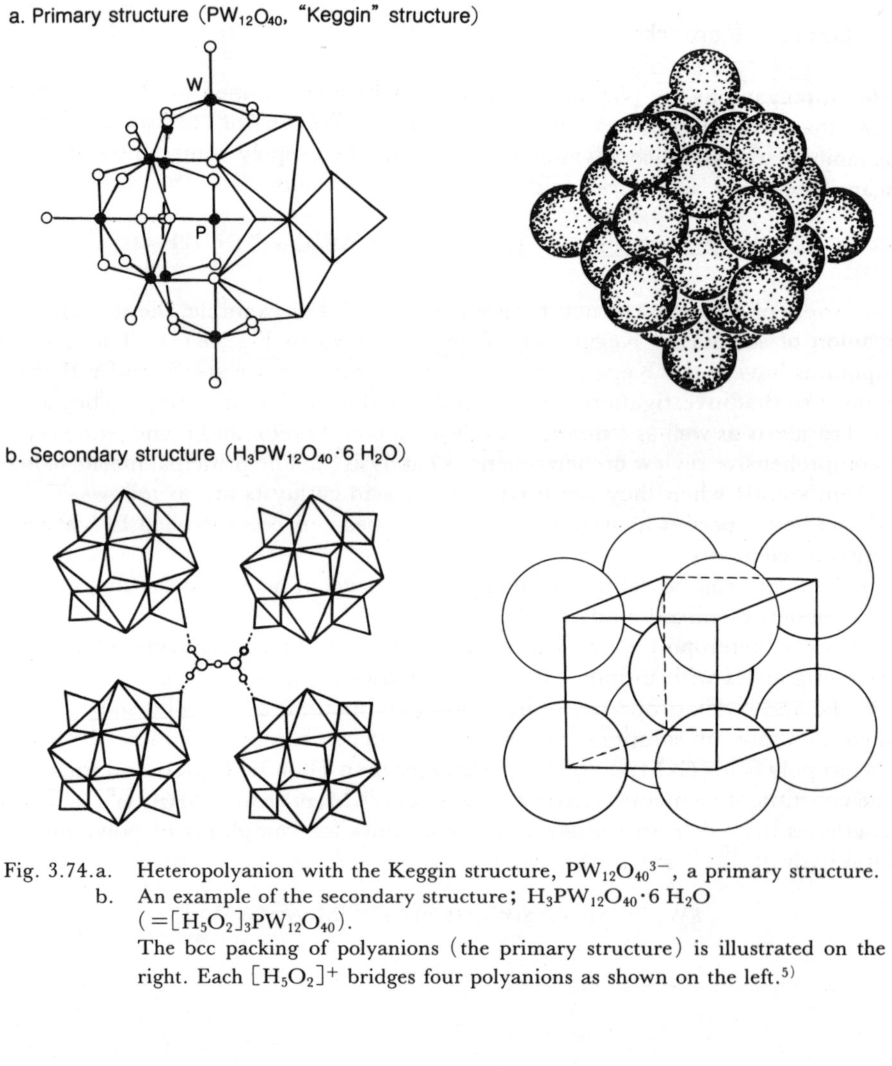

Fig. 3.74.a. Heteropolyanion with the Keggin structure, $PW_{12}O_{40}^{3-}$, a primary structure.
b. An example of the secondary structure; $H_3PW_{12}O_{40}\cdot 6\,H_2O$ ($=[H_5O_2]_3PW_{12}O_{40}$).
The bcc packing of polyanions (the primary structure) is illustrated on the right. Each $[H_5O_2]^+$ bridges four polyanions as shown on the left.[5]

3.5.2 Preparation and Physical Properties

a. Preparation

Heteropoly compounds are prepared in several ways.[1] Solids are obtained by either precipitation, recrystallization or drying depending on the structure and composition. Caution must be used during preparation processes for hydrolytic decomposition of polyanions and nonhomogeneity of the metal cation to polyanion ratio in the precipitates. More elaborate preparation and characterization are necessary for the preparation of polyanions with mixed addenda atoms.[15]

b. Primary and secondary structure

Heteropoly compounds in the solid state consist of heteropolyanions, cations, (protons, and metal or onium ions) and water of crystallization and/or other molecules. This three-dimensional arrangement of polyanions, etc. may be called the secondary structures and the heteropolyanions are denoted the primary structures.[16,17] It is important for the understanding of heteropoly compounds in the solid state to make a clear distinction between the primary and the secondary structure.

The primary structure having the Keggin structure is shown in Fig. 3.74a for the case of $PW_{12}O_{40}^{3-}$. Twelve WO_6 octahedra surround a central PO_4 tetrahedron. The central atom or heteroatom can be P, As, Si, Ge, B, etc. and most of the peripheral atoms, which are called poly or addenda atoms, are W or Mo. A few of the addenda atoms can be substituted by V, Co, Mn, etc. The secondary structure of $H_3PW_{12}O_{40}{\cdot}6H_2O \equiv [H_5O_2]_3PW_{12}O_{40}$ is shown in Fig. 3.74b,[18] where the polyanions are connected by $H^+(H_2O)_2$ bridges. This is the densest secondary structure of a bcc type (lattice constant 12 Å, $Z = 2$). The secondary structure of $Cs_3PW_{12}O_{40}$ has been presumed to be the same as $H_3PW_{12}O_{40}{\cdot}6H_2O$ in which each $H^+(H_2O_2)_2$ is substituted by Cs^+.[18] But the salts of Na, Cu, etc. have quite different secondary structures.

If one looks at the IR spectra which reflect the primary structure and the XRD powder patterns which depend on the secondary structure of 12-molybdophosphoric acid (PMo_{12}) having different amounts of water, as well as its salts,[19] the following important conclusion can be drawn[17,18]: In the solid state of a heteropoly compound, the primary structure is rather stable, but the secondary structure is very variable.

c. Thermal stability, water content and surface area

Changes in heteropoly compounds upon a heat treatment have been extensively studied by the use of TG, DTA, XRD, etc.[1,16,20–26] Acid forms are usually obtained with a large amount of the water of crystallization. Most of them are released below 370 K. Decomposition, which takes place at 620 – 870 K, is believed to be, e.g.

$$H_3PMo_{12}O_{40} \longrightarrow 1/2\,P_2O_5 + 12\,MoO_3 + 3/2\,H_2O.$$

$H_3PW_{12}O_{40}$ is much more thermally stable and more resistant to reduction than $H_3PMo_{12}O_{40}$.

The metal salts can be divided into two groups by their physical properties (groups A and B[24]). Group A consists of the salts of small cations such as Na^+ and Cu^{2+}.

The salts of larger ions such as Cs^+, Ag^+, NH_4^+, etc. are included in group B. The salts of group A resemble the acid forms in several respects, and the surface areas are usually $1-10\ m^2g^{-1}$. On the other hand, Cs salt has a very large surface area and is thermally much more stable.[20,22] The high surface area is due to its very fine primary particles.[27] The states of the protons as well as the water of crystallization have been investigated by NMR,[28-30] IR,[22,31] and electric conductivity measurements.[32]

d. Pseudo-liquid phase

Owing to the flexible nature of the secondary structure of the acid forms and group A salts, polar molecules such as alcohols and amines are readily absorbed into the solid bulk by substituting water molecules and/or by expanding the interdistance between polyanions.[16,17,19] Heteropoly acids which have absorbed a significant amount of polar molecules resemble in a sense a concentrated solution and are in a state between solid and solution. Therefore, this state is called the "pseudo-liquid phase".[16,17] Some reactions proceed mainly in this bulk phase. The tendency to form a pseudo-liquid phase depends on the kind of heteropoly compound and the molecules to be absorbed and on the reaction conditions.

3.5.3 Acidic Properties in the Solid State

As for the acidic properties (amount, strength and type of acid sites) of solid heteropoly compounds, "bulk acidity" as well as "surface acidity" must be considered, since acid catalysis often occurs in the solid bulk. These acidic properties are sensitive to the counter cations as well as to the constituent elements of polyanion. The acid forms are protonic acids and the acid strength reflects that in solution. In the case of salts, there are several possible mechanisms for the origins of the acidity.

a. Acid forms

The color changes of indicators showed that PW_{12} had an acidity stronger than -8.2 in H_0.[9] This observation was confirmed by other investigators.[25,33] The acid amount ($H_0 \leq -3.0$) measured by amine titration for acid form agrees with the stoichiometric number of protons,[9] but the distributions of acid strength for salts are reported to be broad and the acid amount ($H_0 \leq -5.6$) varies by pretreatment temperature.[34] The acidity measurements by thermal deserption (TD) of pyridine in combination with IR have also been reported.[16,22,23] When $H_3PMo_{12}O_{40}$ or $H_3PW_{12}O_{40}$ containing $5-6$ molecules of water of crystallization is placed in contact with pyridine vapor, several molecules of pyridine per polyanion are absorbed within 1 h. Upon evacuation at 298 K, the number of pyridine molecules becomes about six (twice the number of protons) and after evacuation at 303 K the number agrees with the number of protons (three). The IR spectra at the last stage show that water molecules are replaced by pyridine, forming a uniform pyridinium salt.

Typical TD results[23] for the samples prepared as described above are shown in Fig. 3.75. Those data demonstrate that heteropoly compounds in the acid form are strong protonic acids and that all the protons contribute to the acidity. For quantitative discussion it is necessary to confirm the establishment of equilibrium and the stability of the polyanion structure. The acid strength estimated by the TD of NH_3 for SiO_2-supported heteropoly acids paralleled that in solution.[35]

b. Metal salts

The following five mechanisms are present for the acidity of the salts.

i) Protons in the acidic salts (also deviation from the stoichiometry of neutral salts).

ii) Partial hydrolysis during the preparation process; e.g.,

$$PW_{12}O_{40}^{3-} + 3H_2O \longrightarrow PW_{11}O_{39}^{7-} + WO_4^{2-} + 6H^+ \quad (4)$$

iii) Acidic dissociation of water coordinated to metal ions; e.g.,

$$Ni(H_2O)_m^{2+} \longrightarrow Ni(H_2O)_{m-1}(OH)^+ + H^+ \quad (5)$$

iv) Lewis acidity of metal ions.

v) Protons formed by the reduction of metal ions; e.g.,

$$Ag^+ + 1/2H_2 \longrightarrow Ag^\circ + H^+ \quad (6)$$

In Fig. 3.75, TD results of pyridine for the acidic salts of Na ($Na_xH_{3-x}PW_{12}O_{40}$) and some other metals are also shown. TD of NH_3 gave similar results.[26] As the Na content increases, both the acid strength and amount decrease, showing that the acidic properties can be controlled by acidic salt formation. Greater acid amounts than nominal compositions, e.g., for $Na_3PW_{12}O_{40}$, are probably due to partial hydrolysis during preparation (eq. 4). Protonic acidity has been reported for Na,[23] Al,[36] and Ni salts.[37] In the case of $Cs_3PW_{12}O_{40}$, the hydrolysis proceeded only slightly and the acid amount was closer to zero. It was also reported that the acid strength and amounts decreased in the order H > Zr > Al > Zn > Mg > Ca > Na.[34]

The third mechanism was assumed for the salts of divalent metals.[38] Lewis acidity has been reported for Al salt from the IR spectra of absorbed NH_3,[36] but it was not detected in the case of Cu salt.[39] The last mechanism (eq. (6)) was proposed for the salts of Ag and Cu.[40,41]

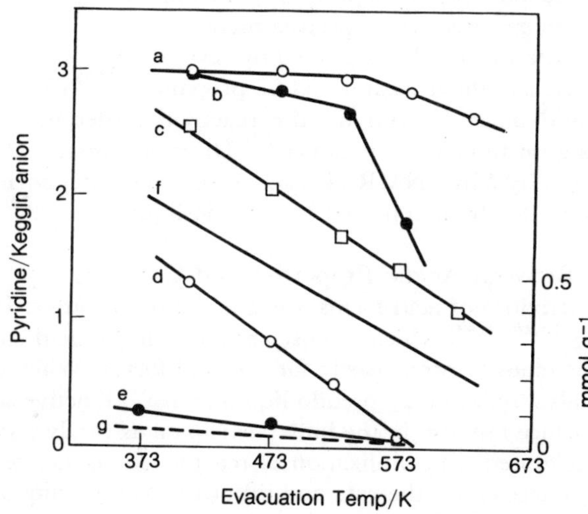

Fig. 3.75 Thermal desorption of absorbed pyridine from several 12-heteropoly tungstates and molybdate. Evacuated at each temperature for 1 h.
a; $H_3PW_{12}O_{40}$, b; $H_3PMo_{12}O_{40}$, c; $NaH_2PW_{12}O_{40}$, d; $Na_3PW_{12}O_{40}$, e; $Cs_3PW_{12}O_{40}$, f; $Cu_{3/2}PW_{12}O_{40}$, g; $SiO_2-Al_2O_3$.[5]

3.5.4 Acid Catalysis

It has been demonstrated for well characterized heteropoly acids that they are much more active catalysts for dehydration than ordinary solid acids such as zeolite and silica-alumina.[9,26] Catalytic tests reported so far indicate that heteropoly compounds are efficient for reactions of (or reactions in the presence of) oxygen-containing molecules (water, ether, and alcohol) such as hydration, etheration, esterification, and related reactions at relatively low temperatures. Superior performance of heteropoly compounds is often observed under conditions which favor pseudo-liquid behavior or the like to occur. They are also active for alkylation and transalkylation, but deactivation during reaction is usually significant, probably due to too high acid strength. The presence of oxygen bases seems to moderate the acid strength.

Typical examples of acid-catalysis of heteropoly compounds are as follows: Dehydration of methanol,[33,42] ethanol,[33,43,46] propanol[17,23,26,33,39,42,44 – 47] and butanol,[33,40] conversion of metanol or dimethyl ether to hydrocarbons,[25,40,49 – 52] etheration to form methyl *t*-butyl ether,[37,41,46,53] esterifications of acetic acid by ethanol[35,54] and pentanol,[55] decomposition of carboxylic acid[56] and formic acid,[23] alkylation of benzene by ethylene[35] and isomerization of butene,[9,22,23,37] *o*-xylene[40] and hexane.[52]

A. Bulk-type *vs.* Surface-type Catalysis

The acid catalysis of heteropoly compounds in the solid state is classified into "bulk-type" and "surface-type" reations.[23] The former type reactions proceed in the catalyst bulk and the latter only on the surface. Dehydration reactions of alcohols belong to the former and isomerization of butene to the latter. So the classification is closely related to the adsorption property of reactants. The activities for the surface-type reactions are more sensitive to pretreatment.

The bulk-type catalysis has been proved by several experiments such as i) a transient response analysis of the dehydration of 2-propanol,[44,57] ii) a "phase transition" of the pseudo-liquid phase,[45] and iii) the reactivity order of alcohols which was reversed depending on the partial pressure.[33] Unusual pressure dependence as well as direct observation by MAS-NMR of reaction intermediates such as protonated alcohol and alkoxide have been reported for pseudo-liquid phase.[29]

B. Relationship Between Acidic Properties and Catalysis

The catalytic activities of acid forms are usually in the order: $PW_{12} > SiW_{12} > PMo_{12} > SiMo_{12}$,[16,26,35,49] which almost agrees with the acid stength in solution. Bulk-type catalysis tends to occur easily for the acid forms. When catalytic reactions proceed in the catalyst bulk, i.e., "pseudo-liquid phase", i) active sites (protons, etc.) not only on the surface but also in the bulk participate as catalyst so that the reaction rate is greatly accelerated, ii) stabilization of reacting molecules or intermediates by complex formation accelerates the rate and iii) owing to the unique reaction environment a unique selectivity often results. Some examples in which very high activity was observed are shown in Table 3.35.

The acidic properties and, therefore, the acid-catalysis of metal salts sometimes vary in a complex manner, depending on several factors. The absorptivity and homogeneity, as well as the reduction and hydrolysis of polyanions, are particularly influential. When the salts are water-soluble (group A), the catalytic activity for bulk-type reactions parallels the bulk acidity measured by TD of pyridine (Fig. 3.76). In these cases, fair correlations between the activity of heteropoly compounds and the electronegativity of constituting metal ions have been reported.[24,38] A correlation was also found between the activity and the acid strength measured by indicators[33] as

TABLE 3.35 Comparison of catalytic activity of heteropoly acids with silica-alumina.

Reaction	Catalyst	Temperature/K	Ratio†1	Ref.
2-Propanol → Propene+H_2O	PW_{12}	398-423	30-100	9, 26
Ethanol → Ethylene+H_2O	PW_{12}	423-493	>300	†2
Isobutene+CH_3OH→MTBE	PM_{12}, PW_{12}	363	300	53
$CH_3COOH+C_2H_5OH \rightarrow CH_3COOC_2H_5$	PW_{12}/Carbon	423	4	35
Isobutyric Acid → Propene+CO+H_2O	PW_2, SiW_{12}	513	4	56
Benzene+CH_3OH → Toluene	PW_{12}	523	∞	23
2Toluene → Benzene+Xylene	PW_{12}	523	∞	23
Benzene+Ethylene → Ethylbenzene	PW_{12}/SiO_2	473	>6	35
Cyclohexylacetate → Acetic acid +Cyclohexene	$Cs_{2.5}H_{0.5}PW_{12}$	373	∞	†2

†1 The ratio of the catalytic activity of heteropoly compounds to that of silica-alumina.
†2 Unpublished work.
(Reproduced with permission from Proc. Intern. Symp. on Acid-Base Catalysis, Sapporo, 1989, Kodansha, Tokyo and VCH, Weinheim, 1989, p. 425).

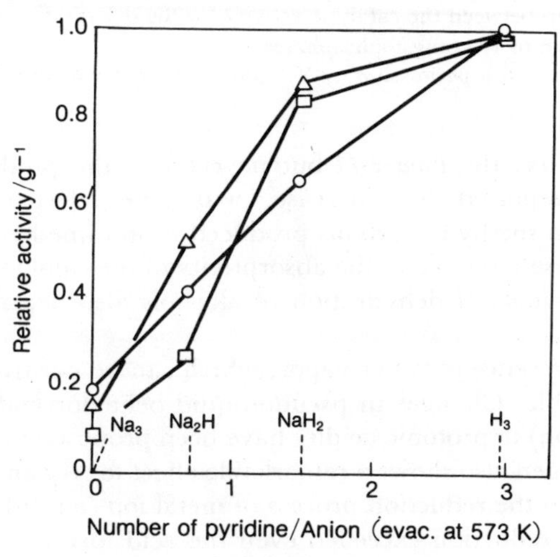

Fig. 3.76 Relationships between strong acidity and catalytic activity for acid-catalyzed reactions of $Na_xH_{3-x}PW_{12}O_{40}$.
○; dehydration of 2-propanol (373 K), □; conversion of methanol to hydrocarbons (558 K), △; decomposition of formic acid (423 K).
(Reproduced with permission from *J. Catal.*, **83**, 126 (1983)).

shown in Fig. 3.77. Results reported for the catalytic properties of group-B salts, which usually exhibit surface-type catalysis, are often inconsistent and somehow confusing. This is likely due to the deviation from the stoichiometry and the non-homogeneity of the salts. For example, the high activity of $Cs_{2.5}H_{0.5}PW_{12}O_{40}$ is mainly due to its very high surface area and proton concentration on the surface.[48] In a sense, this is a $H_3PW_{12}O_{40}$ thin film epitaxially formed on high-surface area $Cs_3PW_{12}O_{40}$.

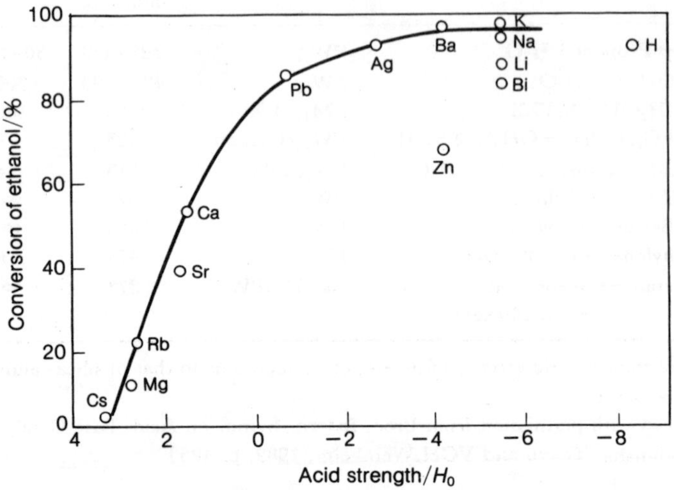

Fig. 3.77 Relationship between the catalytic activity for the dehydration of ethanol and the acid strength of 12-tungstophosphates.
(Reproduced with permission by Y.Saito, *et al.*, *J. Catal.*, **95**, 52 (1985)).

As for the selectivity, the *trans*-2-/1-butene ratio in the product of *cis*-2-butene isomerization was reported to increase with the electronegativity.[37] The olefin/paraffin ratios in the hydrocarbons produced from dimethyl ether over various salts of PW_{12} are inversely related to the absorptivity of the catalysts.[50] The olefin-to-ether ratio in the products of dehydration of alcohols also depends on the absorptivity.[50]

The effect of the presence of water vapor, which can be positive or negative,[24,39] is often very remarkable. Changes in pseudo-liquid behavior and transformation of Lewis acidity (metal ion) to protonic acidity have been proposed to explain the effects. The presence of hydrogen also shows a remarkable effect for Ag and Cu salts.[40,41] An induction period due to the reduction process of metal ion (eq. (6)) was observed and catalytic activity after reduction exceeded even the acid form.

C. Supported Heteropoly Compounds

Heteropoly compounds can be used dispersed on supports such as silica gel, kieselguhr, ion-exchange resin and active carbon. Some examples are shown in Table 3.36. The particles sizes of heteropoly acids are small and not detectable by XRD up to 20

TABLE 3.36 Reaction catalyzed by supported heteropoly acids

1) Esterification of acetic acid with ethanol at 423 K [35]

Heteropoly acid	Support	Conversion of AcOH/%	Selectivity/%		
			AcOH	Et_2O	Olefins
$H_3PW_{12}O_{40}$	SiO_2	90.1	91	9	0
$H_4SiW_{12}O_{40}$	SiO_2	96.2	88	12	0
$H_3PMo_{12}O_{40}$	SiO_2	55.4	91	9	0
$H_3PW_{12}O_{40}$	Carbon	48.0	100	0	0
$H_3PW_{12}O_{40}$	Al_2O_3	9.0	89	3	8
$H_3PW_{12}O_{40}$	TiO_2	97.0	74	26	trace
$SiO_2-Al_2O_3$	—	24.3	99	trace	1

2) Etheration of *t*-butanol and methanol at 363 K [53]

		Conversion/%	Selectivity/%
$H_3PMo_{12}O_{40}$	SiO_2	28.9	93.8
$H_3SiMo_{12}O_{40}$	SiO_2	29.7	94.7
$H_3PW_{12}O_{40}$	SiO_2	18.4	63.0
$H_4SiW_{12}O_{40}$	SiO_2	20.1	42.9
$SiO_2-Al_2O_3$	—	0.1	100(130 ℃)
H_3PO_4	SiO_2	14.0	94(110 ℃)

wt% on silica.[35] An increase in the surface area has a greater effect for the surface-type reactions than for the bulk-type reaction.[26] Heteropoly acids entrapped in micropores of active carbons can be used as isoluble solid acids.[54] These are also very selective catalysts for gas-phase esterification.[35] Supports such as alumina which show surface basicity give rise to decomposition of polyanions, so the use of non-aqueous solution for preparation is recommended in this case to minimize decomposition.[53]

REFERENCES

1. G. A. Tsigdinos, *Topics Curr. Chem.* **76**, 1 (1978).
2. M. T. Pope, *Heteropoly and Isopoly Oxometallates*, Springer, Berlin, 1983.
3. Y. Sasaki, K. Matsumoto, *Kagaku no Ryoiki*, **29**, 853 (1975) (in Japanese).
4. I. V. Kozhevnikov, K. I. Matveev, *Appl. Catal.*, **5**, 235 (1983) and references therein.
5. M. Misono, *Catal. Rev. Sci. Eng.*, **29**, 269 (1987).
6. M. Misono, *Kagaku no Ryoiki*, **35**, 43 (1981) (in Japanese); M. Misono in: Proc. Climax 4th Intern. Conf. Chemistry and Uses of Molybdenum, (H. F. Barry, P.C.H. Michell, eds.), Climax Molybdenum Co, Ann Arbor, 1982, p.289; *Materials Chem. Phys.*, **17**, 103, (1987).
7. Y. Izumi, M. Otake, *Kagaku Sosetsu*, (Chem. Soc. Jpn., ed.) No. 34, p.116, 1982 (in Japanese).

8a) M. Misono, in: *Catalysis by Acids and Bases*, (B. Imelik *et al.*, eds.), Elsevier, Amsterdam, 1985, p.147.
 b) M. Misono, T. Okuhara, N. Mizuno, *Hyomen*, **23**, 69 (1985) (in Japanese).

9. M. Otake, T. Onoda, *Shokubai (Catalyst)*, **18**, 169 (1976); **17**, 13P (1975) (in Japanese).

10. L. Barcza, M. T. Pope, *J. Phys. Chem.*, **79**, 92 (1975).
11. Y. Izumi, K. Matsuo, K. Urabe, *J. Mol. Catal.*, **18**, 299 (1983).
12. A. Aoshima, S. Tonomura, *U.S.-Japan Seminar on the Catalytic Activity of Polyoxoanions*, Shimoda, Japan, May 1985.
13. A. Aoshima, T. Yamaguchi, *Nippon Kagaku Kaishi*, **1986**, 514 (in Japanese).
14. H. Knoth, R. L. Harlow, *J. Am. Chem. Soc.*, **103**, 4265 (1981).
15. P. J. Domaille, *J. Am. Chem. Soc.*, **106**, 7677 (1984); D. E. Katsoulsi, M. T. Pope, *J. Am. Chem. Soc.*, **106**, 2737 (1984).
16. M. Misono, K. Sakata, Y. Yoneda, W. Y. Lee, in: Proc. 7th Intern. Congr. Catal.,Tokyo (T. Seiyama and K. Tanabe, eds.), 1980, Kodansha, Tokyo and Elsevier, Amsterdam, 1981, p.1047.
17. K. Sakata, M. Furuta, M. Misono, Y. Yoneda, *ACS/CSJ Chemical Congr.*, Honolulu, April, 1979.
18. G. M. Brown, M. R. Neo-Spiret, W. R. Busing, H. A. Levy, *Acta Cryst.*, **B33**, 1038 (1977).
19. M. Misono, N. Mizuno, K. Katamura, A. Kasai, Y. Konishi, K. Sakata, T. Okuhara, Y. Yoneda, *Bull. Chem. Soc. Jpn.*, **55**, 400 (1982); *Chem. Lett.*, **1981**, 391.
20. K. Eguchi, N. Yamazoe, T. Seiyama, *Nippon Kagaku Kaishi*, **1981**, 336 (in Japanese).
21. S. F. West, L. F. Andrieth, *J. Phys. Chem.*, **59**, 1069 (1955).
22. M. Furuta, K. Sakata, M. Misono, Y. Yoneda, *Chem. Lett.*, **1979**, 31; M. Misono, Y. Konishi, M.Furuta, Y. Yoneda, *Chem. Lett.*, **1978**, 709.
23. T. Okuhara, A. Kasai, N. Hayakawa, Y. Yoneda, M. Misono, *J. Catal.*, **83**, 121 (1983); *Shokubai (Catalyst)*, **22**, 226 (1980) (in Japanese).
24. H. Niiyama, Y. Saito, S. Yoshida, E. Echigoya, *Nippon Kagaku Kaishi*, **1982**, 569 (in Japanese).
25. H. Hayashi, J. B. Moffat, *J. Catal.*, **77**, 473 (1982); **83**, 192 (1983); B. K. Hodnett, J. B. Moffat, *J. Catal.*, **88**, 253 (1984).
26. N. Hayakawa, T. Okuhara, M. Misono, Y. Yoneda, *Nippon Kagaku Kaishi*, **1982**, 356 (in Japanese).
27. N. Mizuno, M. Misono, *Chem. Lett.*, **1987**, 967.
28. T. Wada, *C. R. Acad. Sci.*, **259**, 553 (1964).
29. M. Misono, T. Okuhara, T. Ichiki, T. Arai, Y. Kanda, *J. Am. Chem. Soc.*, **109**, 5535 (1987); K. Y. Lee, Y. Kanda, N. Mizuno, T. Okuhara, M. Misono, S. Nakata, S. Asaoka, *Chem. Lett.*, **1988**, 1175.
30. Y. Kanda, K. Y. Lee, S. Nakata, S. Asaoka, M. Misono, *Chem. Lett.*, **1988**, 139.
31. N. Mizuno, K. Katamura, Y. Yoneda, M. Misono, *J. Catal.*, **83**, 384 (1983).
32. O. Nakamura, I. Ogino, *Mat. Res. Bull.*, **17**, 231 (1882); *Chem. Lett.* **1979**, 17.
33. Y. Saito, P. N. Cook, H. Niiyama, E. Echigoya, *J. Catal.*, **95**, 49, (1985).
34. A. K. Ghosh, J. B. Moffat, *J. Catal.*, **101**, 238 (1986).
35. Y. Izumi, R. Hasebe, K. Urabe, *J. Catal.*, **84**, 402 (1983).
36. J. G. Highfield, J. B. Moffat, *J. Catal.*, **88**, 177 (1984).
37. K. Sugiyama, K. Kato, H. Miura, T. Matsuda, *J. Japan Petr. Inst.*, **26**, 24 (1983); T. Matsuda *et al.*, *J. Chem. Soc., Trans. Farad., I*, **77**, 3101 (1981).
38. H. Niiyama, Y. Saito, E. Echigoya, Proc. 7th Intern. Congr. Catal., **1980**, Kodansha ,Tokyo and Elsevier, Amsterdam, 1981, p.1416.
39. T. Okuhara, T. Hashimoto, T. Hibi, M. Misono, *J. Catal.*, **93**, 224 (1985).
40. T. Baba, H. Watanabe, Y. Ono, *J. Phys. Chem.*, **87**, 2406 (1983).
41. Y. Ono and T. Baba, Proc. 8th Intern. Congr. Catal., 1984, Vol. V, Verlag Chemie, Berlin, 1984, p.405.
42. J. G. Highfield, J. B. Moffat, *J. Catal.*, **98**, 245 (1986).
43. J. G. Highfield, J. B. Moffat, *J. Catal.*, **95**, 108 (1985).
44. T. Okuhara, T. Hashimoto, N. Mizuno, M. Misono, Y. Yoneda, H. Niiyama, Y. Saito, E. Echigoya, *Chem. Lett.*, **1983**, 573.
45. K. Takahashi, T. Okuhara, M. Misono, *Chem. Lett.*, **1985**, 841.
46. Y. Saito, H. Niiyama, E. Echigoya, *Nippon Kagaku Kaishi,* **1984**, 391 (in Japanese).
47. M. Ai. *J. Catal.* **71**, 88 (1981); *Appl. Catal.*, **4**, 245 (1982).
48. S. Tatematsu, T. Hibi, T. Okuhara, M. Misono, *Chem. Lett.*, **1984**, 865.
49. T. Baba, J. Sakai, H. Watanabe, Y. Ono, *Bull. Chem. Soc. Jpn.*, **55**, 2555 (1982).
50. T. Okuhara, T. Hibi, K. Takahashi, S. Tatematsu, M. Misono, *J. Chem. Soc., Chem. Comm.*, **1984**, 697; T. Hibi, K. Takahashi, T. Okuhara, M. Misono, Y. Yoneda, *Appl. Catal.*, **24**, 69 (1986).
51. Y. Ono, T. Baba, J. Sakai, T. Keii, *J. Chem. Soc., Chem. Commun.*, **1981**, 400.

52. Y. Ono, M. Taguchi, S. Suzuki, T. Baba, in: *Catalysis by Acids and Bases* (B. Imelik *et al.*, eds.), Elsevier, Amsterdam, 1985, p.167.
53. S. Igarashi, T. Matsuda, Y. Ogino, *J. Japan Petrol. Inst.*, **22**, 331 (1979); **23**, 30 (1980).
54. Y. Izumi, K. Urabe, *Chem. Lett.*, **1981**, 663.
55. T. Baba, Y. Ono, T. Ishimoto, S. Moritake, S. Tanooka, *Bull. Chem. Soc. Jpn.*, **58**, 2155 (1985).
56. M. Otake, T. Onoda, *J. Catal.*, **38**, 494 (1975).
57. T. Okuhara, T. Hashimoto, N. Mizuno, M. Misono, Y. Yoneda, *Shokubai (Catalyst)*, **24**, 318 (1982) (in Japanese).

3.6 ION-EXCHANGE RESINS

3.6.1 Structure of Ion-exchange Resins

A. Syrene-Divinylbenzene Copolymers

The most common formulation of ion-exchange resins is polystyrene cross-linked with divinylbenzene. The conventional styrene-divinylbenzene copolymer forms colorless transparent particles and consists of a homogeneous polymer phase. By changing the divinylbenzene content, one can modify the three dimesional networks of the copolymers. These resins are called gel-type copolymers.

The macroreticular resins are prepared by copolymerizing styrene and divinylbenzene in the presence of an organic compound that is a good solvent for the monomer but a poor swelling agent for the polymer.[1,2] They form opaque round particles and have large surface areas.

Various functional groups are introduced to the copolymers to form the cation or anion exchange resins. For example, the sulfonation of benzene nuclei with sulfuric acid yields cation-exchange resins of strong acidity.

Resins of weak acidity are obtained by introducing carboxy groups. Resins of strong basicity are obtained by introducing quaternary ammonium groups to the copolymer. The characteristics of some styrene-divinylbenzene ion exchange resins are listed in Table 3.37.

The cation exchange resins can be used up to 390 and 420 K for the gel types and the macroreticular types, respectively. The anion exchange resins can be used up to 343 – 370 K.

TABLE 3.37 Physical poroperties of styrene-divinylbenzene resins

	Type	Functional group	Specific surface area m^2/g-resin	Porosity ml/ml-resin	Exchange capacity meg/g-resin
Amberlyst 15	MR	$-SO_3^-M^+$	43	0.32	4.3
Amberlite IR-120	Gel	$-SO_3^-M^+$	<0.1	0.018	4.3
Amberlite IRA-900	MR	$-N^+(CH_3)_3X^-$	27	0.27	4.4
Amberlite IRA-400	Gel	$-N^+(CH_3)_3X^-$	<0.1	0.004	3.7
Amberlite IRA-93	MR	$-N(CH_3)_2$	25	0.48	4.6

B. Perfluorinated Resin Sulfonic Acid (Nafion-H®)

Nafion resins were first prepared by the Du Pont Company. The resins are perfluorinated polymers having sulfo groups in the amount of 0.01 to 5 meq/g-resin.

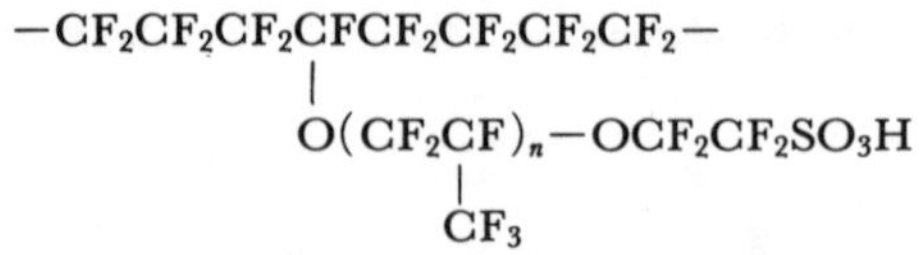

One method for preparing polymers of the above structure comprises polymerizing the corresponding perfluorinated vinyl compounds.

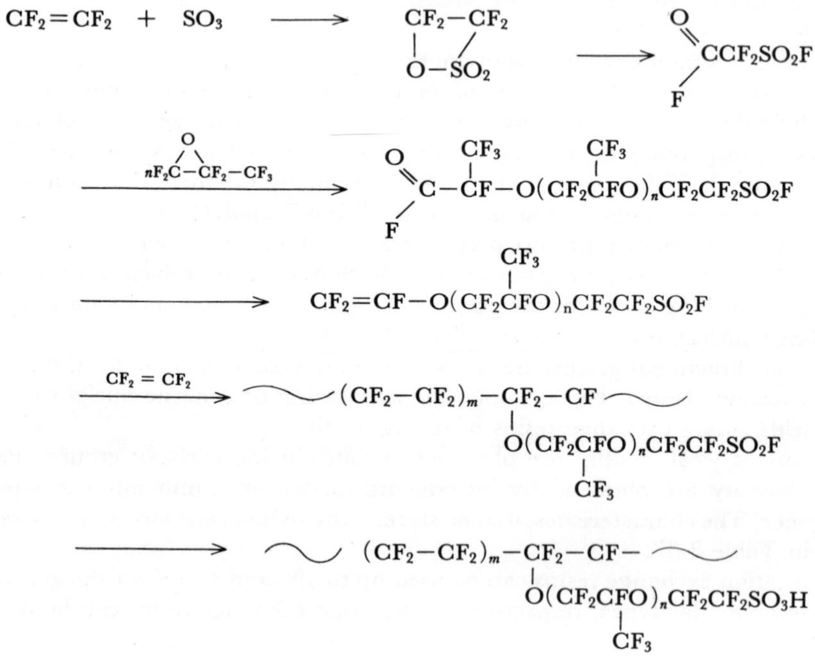

Nafion resins with equivalent molecular weight of 900-1200 contain the tetrafluoroethylene and the perfluorovinyl ether units in a ratio of 7:1. They are effectively insoluble in most solvent, but may be swollen by them.

Nafion resins have high chemical and thermal stability. The maximum continuous operation temperature of Nafion-H® is about 450 K in anhydrous systems. The maximum temperature in aqueous systems is 420 – 510 K. Besides its thermal and chemical stability, the feature that makes Nafion type polymers catalytically useful is their high acid strength of the acid form of polymers (Nafion-H®). Since the sulfo group is attached to a highly electron-withdrawn perfluoroalkyl backbone, a relatively high polarization of O-H bond results. Thus, Hammett H_0 acidity function value of Nafion-H® was estimated to be $-10 \sim -12$, which is comparable to or stronger than that of 96-100% sulfuric acid.[3)]

3.6.2 Characteristics of Styrene-Divinylbenzene Ion Exchange Resins as Catalyst

The ion exchange resins have various advantages over conventional acid or base catalysts. The use of aqueous acids has a number of drawbacks such as corrosion, side reactions, and environmental problems. These problems can be avoided by using ion-exchangers. Further advantages of ion-exchangers are that separation problems are manifestly simpler and the same catalyst can be used repeatedly. In some cases, it is even possible to distill the product directly in the presence of the cation-exchanger catalyst.

The resins also have disadavantages. They are less resistant to temperature and abrasion and are more expensive.

With the gel-type resins, catalytic groups such as sulfo groups in a resin-particle are almost entirely inaccessible to reactant molecules in the absence of strongly polar compounds such as water, which swell the resin network and allow the access of reactant molecules between polymer strands to the particle interior. Therefore, in non-aqueous or nonpolar systems, these resins have but a negligible pore structure and show little catalytic activity. For reactions involving weakly polar reactants nearly incapable of swelling, the macroreticular resins are much more useful. Kunin *et al.*[2] compared the catalytic activities of the macroreticular resin Amberlyst 15 and the acid forms of the gel-type resin Amberlite IR-120 for the decomposition of *t*-butyl acetate at 298 K. After 1 h, the activity of the former was such that 80% of equilibrium conversion was obtained and less than 1% was observed for the run with the latter. They obtained a similar result also for the synthesis of *t*-butyl methacrylate from isobutene and methacrylic acid at 273 K. Wesley and Gates[4] found that macroreticular resins had high catalytic acitvity for benzene propylation, but that gel-type resins had negligible catalytic activity.

Extent of crosslinking is also an important factor for the catalytic activity of the resins. Setinek[5] studied the effect of the crosslinking of both gel-type and macroporous resins on the catalytic activity for ester exchange reactions of aliphatic esters with alcohols in the gas and liquid phases. For gas-phase reactions with the macroporous exchangers, the reaction rate increased with increasing crosslinking. This trend was attributed to the change in the specific surface area of the resins. In the case of reactions in both the gas and liquid phases over gel-type resins, the reaction rates decreases with increasing crosslinking. The effect of crosslinking is more pronounced in the liquid-phase reactions. Here, the surface of the catalysts plays essentially no role and the rate of the catalytic reactions is determined only by the number of accessible functional groups within the mass of the polymer substance. As the degree of crosslinking increases, the permeability of the ion-exchanger particles diminishes and the rate slows down. In the case of the liquid-phase reactions, the reaction rate is also affected by the swelling of ion exchangers in the solvent. The ion exchangers with low degree of crosslinking swell much more, increasing the accessibility for reactants of their functional groups within the particles. In the case of the macroporous ion-exchangers, the rate of the liquid-phase reactions depended very little upon the degree of crosslinking. This was attributed to the mutual effects of varying surface area and of swelling of the polymer mass.

Anderianova[6] studied the decomposition of formic acid and esterification of acetic acid with ethyl alcohol in the vapor phase over gel-type resins of divinylbenzene content of 1% and 20%. At lower temperatures, the resin with lower degree of crosslinking was more active for both reactions. With increasing temperature, the difference in the rates decreased. This was attributed to the change in resin sorption capacity with temperature. On the other hand, it was reported that, for the liquid-phase dehydration of *t*-butyl alcohol with gel-type resins, the 8%-crosslinked resin had about twice the catalytic activity of the 2% resin.

Water present in the catalytic systems shows versatile effects on reaction rates. Heath and Gates[7] found the induction period in the dehydration of *t*-butyl alcohol and also noted that water addition reduced the induction time. This effect of water was attributed to the swelling of the resin network. The swelling reduces intraparticle resistance to mass transport, and makes an increasing fraction of the catalytic sites accessible to the reactant. Though water accelerated the reaction initially, it also inhibited the reaction.[7,8] The retardation with water was observed also in estrification of salicic acid with methanol[9] and benzene propylation.[4] The retarding effect of water was explained by a kinetic expression based on the Langmuir-Hinshelwood model, in which the competitive chemisorption of water and a reactant (alcohol or acid) is assumed.

Gates and Rodriguez[8] reported a transition from a catalysis by the bound sulfo groups (general acid catalysis) to a catalysis by the hydrated protons in the matrix (specific acid catalysis) in the dehydration of *t*-butyl alcohol. The rate is best expressed by the following equation.

$$r=\frac{kK_AC_A}{1+K_A C_A+(K_WC_W)^2}+k'C_A \qquad (1)$$

Here, C_A and C_W are the concentrations of the alcohol and water, respectively. At the values of $C_W<0.7$ M, the second term on the right hand side of eq. (1) is less than about 10% of the rate, while at the values of $C_W>11$ M, the first term is less than 10% of the rate. They inferred that, at high water concentrations, catalysis is predominantly due to hydrated protons, and at low water concentrations, catalysis occurs predominatly by undissociated sulfo groups. It follows that the catalysis is much more efficient by undissociated sulfo groups than by hydrated protons.

A similar retardation effect was observed also in the reactions of olefins and alcohols, as shown in Fig. 3.78.[10,11] At low methanol concentrations, the rate is proportional to the methanol concentration until a maximum rate is reached. Further increase of the methanol concentration gives a negative order in the reaction rates with respect to methanol and finally reaches a zero order. It was again explained that at low methanol concentrations, the reaction proceeds with a concerted mechanism on the associated network of sulfo groups. On the other hand, at higher methanol concentrations, it was assumed that the rate determining step is the protonation of the olefin by the solvated proton (ROH_2^+) formed by the following reaction,

$$-SO_3H + ROH \rightleftharpoons ROH_2^+ + -SO_3^-$$

which is completely shifted to the right at molar ratios of alcohol/olefin higher than 1.7.

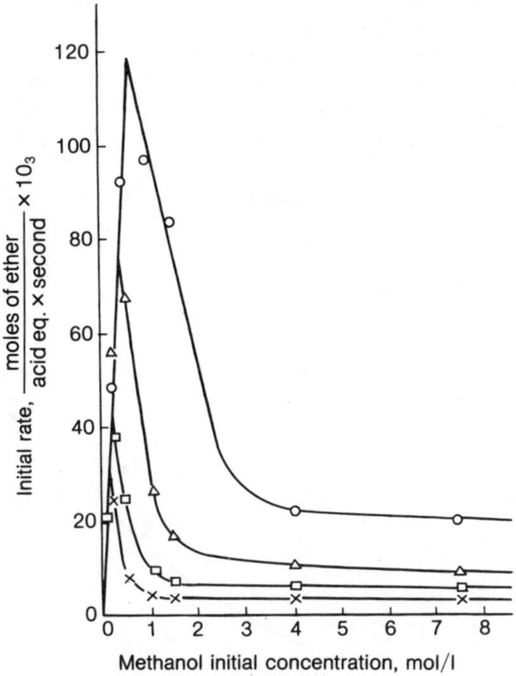

Fig. 3.78 Methyl-*t*-butyl ether (MTBE) synthesis. Initial rate dependence on initial methanol concentration at various constant isobutene contents. Temperature : 333 K catalyst : Amberlyst 15; isobutene (mol/*l*): ○; 6, ▲; 4, □; 2, ×; 1. (Reproduced with permission by F. Anclilloti *et al.*, *J. Mol Catal.*, **4**, 39(1985)).

The retardation of the rate by alcohols was observed also in the dehydration of isobutyl alcohol and isopopyl alcobol.[12] This was also attributed to the solvation of the network of acid groups by hydrogen-bonded alcohol.

The catalytic activity of sulfonic acid resin changes greatly with concentration of sulfo groups. Fig. 3.79 shows the dependence of the rate of *t*-butyl alcohol dehydration on the resin composition at three temperatures.[13] Here, the acid groups were successively replaced by the salt of one of the alkali metals. The rate at each temperature is represented by the following equation.

$$r = k_1 C_{SO_3H} + k_2 C^4_{SO_4H}$$

It was inferred that the first-order term indicates a mechanism involving a single $-SO_3H$ group as the catalytic site, while the fourth-order term indicates a mechanism involving approximately four $-SO_3H$ groups. Even seventh order dependence of the rate on the $-SO_3H$ group concentration was observed in benzene propylation reaction.[4] A large dependence of the rate on the sulfo group concentration was reported also for butene insomerization.[14]

According to Gates and coworkers, the intact network of hydrogen-bonded SO_3H groups is envisioned as sharing a proton deficiency and enveloping a carbenium ion.[4,12,13] In other words, a carbenium ion is stabilized by solvation by acid-groups.

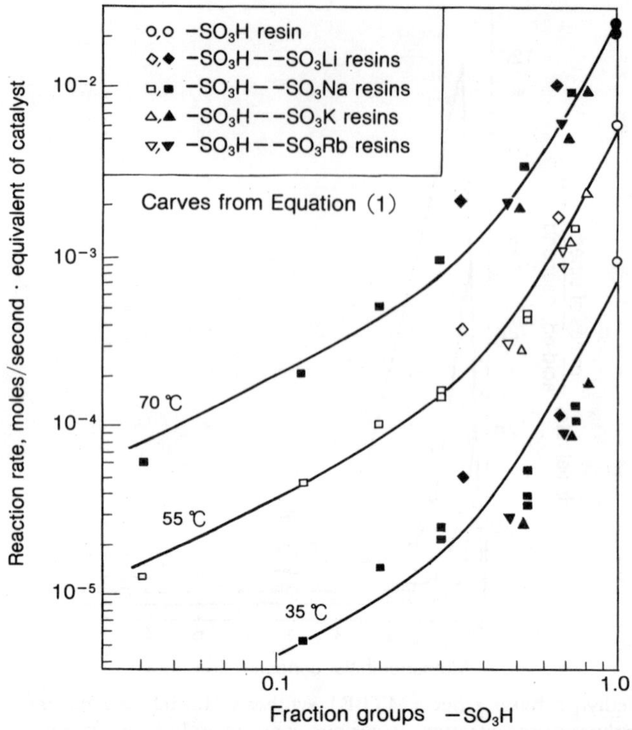

Fig. 3.79 Dependence of initial rate on temperature and resin composition in the dehydration of *t*-butyl alcohol.
(Reproduced with permission by B. C. Gates, *et al.*, *J. Catal.*, **24**, 323 (1972)).

Insertion of $-SO_3Na$ groups would cause break-up of the network, prevent the sharing of the proton deficiency and strongly reduce the proton donating tendency. Sorption of water or alcohol molecules may give a similar effect.

The sulfonic acid resins are much more active than *p*-toluenesulfonic acid, a typical homogeneous catalyst, for *t*-butyl dehydration[13] and the reaction of methanol and isobutene[10] at the same ratio of reactant concentrations to the number of acid equivalent. The high activity of the resin is also ascribed to the network structure of the sulfo groups in the resins.

3.6.3 Catalysis by Anion Exchange Resins

Anion-exchange resins are excellent catalysts for base-catalyzed reactions such as Knoevenagel condensations and aldol condensations.

Astle and Zaslowsky utilized a variety of anion-exchange resins to effect Knoevenagel condensations, aldol condensations, and cyanoethylation of alcohols.[15]

Knoevenagel condensation

$$RCHO + CH_3COCH_2COOC_2H_5 \longrightarrow \underset{\overset{\|}{CHR}}{CH_3COCOOC_2H_5} + H_2O$$

$$RCHO + NCCH_2COOC_2H_5 \longrightarrow \underset{\underset{CHR}{\|}}{NCCOOC_2H_5} + H_2O$$

$$RCHO + H_2C\begin{matrix} \diagup COOC_2H_5 \\ \diagdown COOC_2H_5 \end{matrix} \longrightarrow RCH = C\begin{matrix} \diagup COOC_2H_5 \\ \diagdown COOC_2H_5 \end{matrix} + H_2O$$

Aldol condensation

$$2\,RCH_2CHO \longrightarrow RCH_2\overset{\overset{OH}{|}}{C}H\,CHRCHO$$

$$RCH_2\underset{\underset{O}{\|}}{C}R^1 \longrightarrow RCH_2\overset{\overset{OH}{|}}{C}R^1CHR\overset{\overset{O}{\|}}{C}R^1$$

Cyanoethylation

$$CH_2{=}CHCN + ROH \longrightarrow ROCH_2CH_2CN$$

For Knoevenagel condensations and aldol condensations, weakly basic resins whose basicity is attributed to amino groups, were more effective than strongly basic resins bearing quanternary ammonium ions; strongly basic resins were more effective for cyanoethylation of alcohols.

Schmidle and Mansfield[16)] reported on cyanohydrin formation, benzoin condensation, nitroalcohol formation by the condensation of aldehydes with nitroparaffins, and cyanoethylation. Application to Michael reaction has also been investigated.[17)]

Anion-exchange resins are very useful catalysts for the production of silane by the disproportionation of dihydrodichlorosilane.[18)]

$$SiH_2Cl_2 \longrightarrow SiH_4 + HSiCl_3 + SiCl_4$$

Recently, Hodge *et al.*[19)] prepared polymer-bound chiral quaternary ammonium chlorides by the reaction of chlomethylated crosslinked polystyrene with cinchonidine, cinchonine or (−)-*N*-methylepledrine and carried out several Michael reactions. Catalysts prepared from cinchonidine(I) achieved optical yields of < 27% of the S-adduct in the addition of methyl-1-oxoindan-2-carboxylate(II) to methyl vinyl ketone.

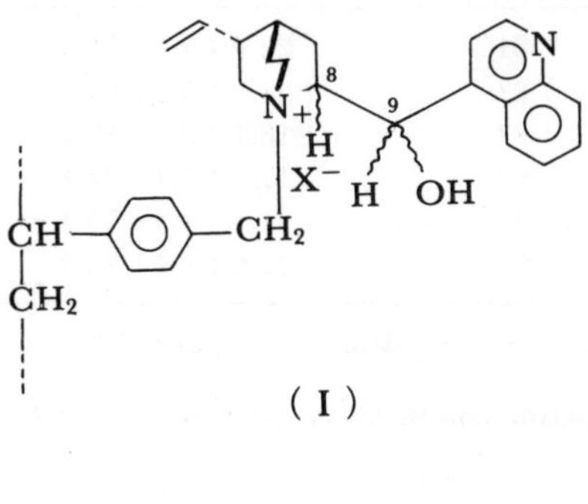

(I)

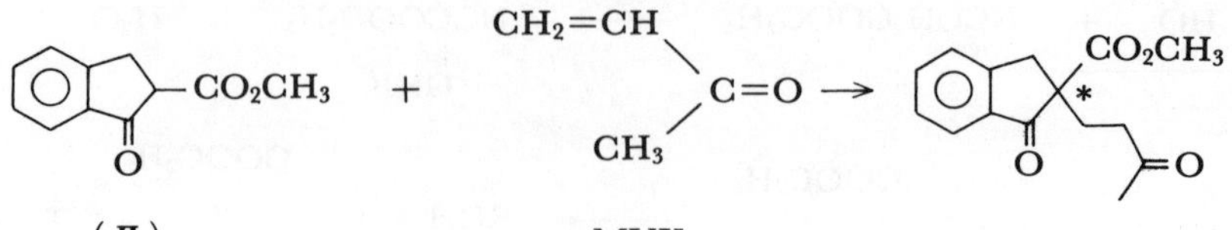

(II) MVK

3.6.4 Nafion-H® as a Catalyst for Organic Reactions

Nafion-H® resins can be used as a strong acid catalyst for a variety of organic reactions. A comprehensive review of the topic has been published.[3] Some examples are described below.

A. Acylation

Nafion-H® gives a convenient and efficient method of acylatión of benzene and substituted benzenes with aroyl chloride.[20] The acylation reactions were carried out by heating under reflux a stirred mixture of the corresponding benzoyl chloride, arene, and the Nafion-H® catalyst.

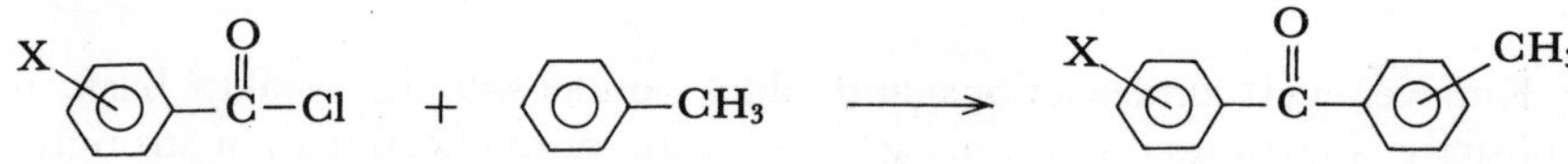

The reaction is general for aroyl halide (Table 3.38) and aromatics generally undergoing Friedel-Crafts acylations. Optimum yields were obtained when 10% – 30% of Nafion-H® was employed relative to aroyl halide. Friedel-Crafts acylation in solution chemistry generally requires molar amounts of catalyst which forms complexes with both the acylating agent and the carbonyl product. Work up, conseqeuently, is needed to decompose these complexes and the catalyst is usually unrecoverable. The reactions with Nafion-H® are very clean, with HCl, the only by-product, escaping during the reaction, and allow for easy work-up procedures.
Acylation reactions using Nafion-H® was reported also by Krepan.[21]

TABLE 3.38 Acylation of toluene with aroyl chlorides to methylbenzophenones

X	Yield† [%]	Isomer distribution $o:m:p$	$o:p$ Ratio
H	81	22.4 : 3.1 : 74.5	3.33
4-H_3C	83	28.7 : 3.1 : 68.2	2.34
2-F	87	16.7 : 2.9 : 80.4	4.81
3-F	82	19.5 : 3.4 : 77.1	3.95
4-F	87	21.5 : 3.4 : 75.1	3.49
3-Cl	82	22.6 : 1.1 : 76.3	3.38

† Yield based on aroyl chloride; identity and purity of products confirmed by G.L.C.
(Reproduced with permission by G. A. Olah *et al.*, *Synthesis*, **1987**, 673).

B. Diels-Alder Reactions

Nafion-H® is an efficent catalyst for Diels-Alder reactions (Table 3.39).[22] The reactions of anthracene with maleic anhydride, dimethyl maleate, and dimethyl fumalate were carried out at 333 – 353 K in the presence of Nafion-H® catalyst in either chloroform or benzene solvent. It should be noted that the reaction of dienophiles with very reactive dienes such as isoprene and 2,3-dimethylbutadiene can be carried out at room temperature to give the adduct in high yields. In usual systems, highly reactive dienophiles undergo polymerization during the desired reactions. In Diels-Alder reactions catalyzed by Friedel-Crafts Lewis acid catalysts, excess amounts of Lewis acid halides are required because of the formation of the complex between the halide and carbonyl oxygen atoms. Here again, Nafion-H® catalysts allow easy and clean separation of products and the catalysts are not destroyed upon work up.

TABLE 3.39 Nafion-H® catalyzed cycloaddition of dienophiles to dienes

Dienophile	Diene	Reaction time [h]	Yield [%] of adduct
maleic anhydride[†1]	anthracene	5	87
maleic anhydride[†2]	anthracene	5	91
p-benzoquinone[†1]	anthracene	2	92
dimethyl maleate[†1]	anthracene	15	95
dimethyl fumarate[†1]	anthracene	16	94
benzoquinone	isoprene[†3]	25	80
naphthoquinone	2,3-dimethylbutadiene	36	93
acrolein	1,3-cyclohexadiene	40	88

[†1] In refluxing benzene. [†2] In refluxing chloroform.
[†3] At room temperature, terachloromethane as solvent.
(Reproduced with permission by G. A. Olah *et al.*, *Synthesis*, **1978**, 271).

C. Other Reactions

Many other reactions have been proved to be catalyzed by Nafion-H®. The following are typical examples.

$$C_6H_5CH{=}CH_2 + C_6H_5CH_3 \longrightarrow C_6H_5\overset{\displaystyle CH_3}{\overset{|}{C}}HC_6H_4CH_3 \qquad \text{(ref. 23)}$$

$$C_6H_6 + (CH_3)_2CHCl \xrightarrow[\text{gas-phase}]{} C_6H_5CH(CH_3)_2 + HCl \qquad \text{(ref. 24)}$$

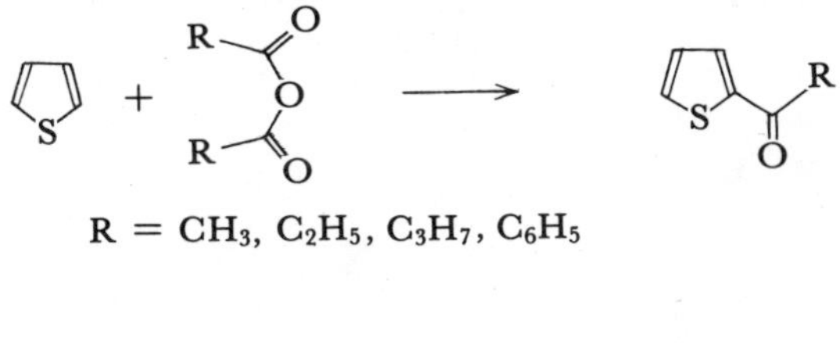

(ref. 25)

R = CH_3, C_2H_5, C_3H_7, C_6H_5

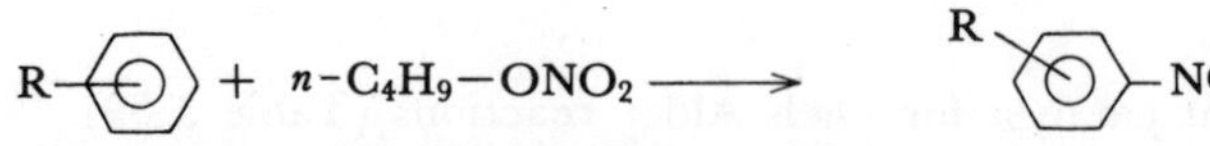

(ref. 26)

R = H, CH_3, OCH_3, Cl

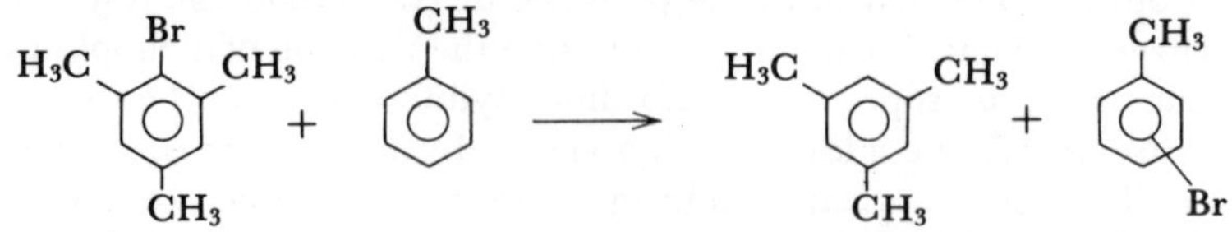

(ref. 27)

(ref. 28)

R = alkyl, arylalkyl, cycloalkyl

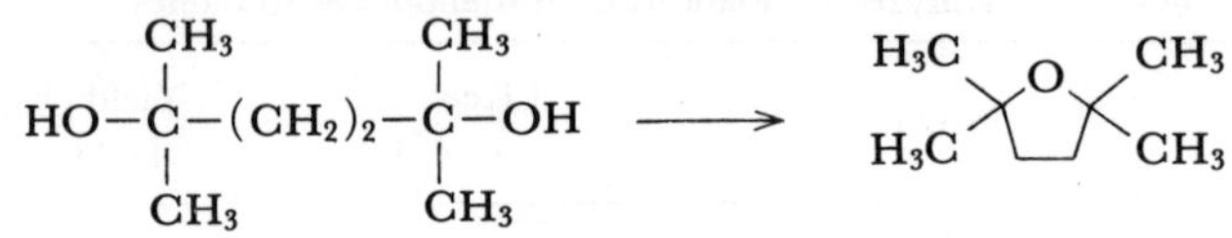

(ref. 29)

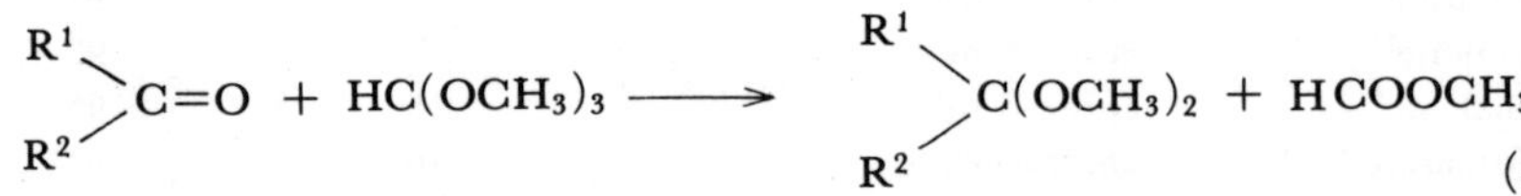

(ref. 30)

R^1, R^2 = H, alkyl, aryl

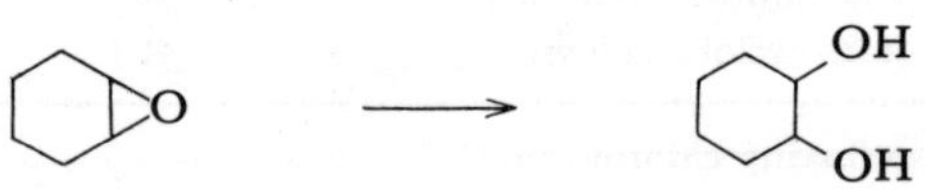

(ref. 31)

(*trans*)

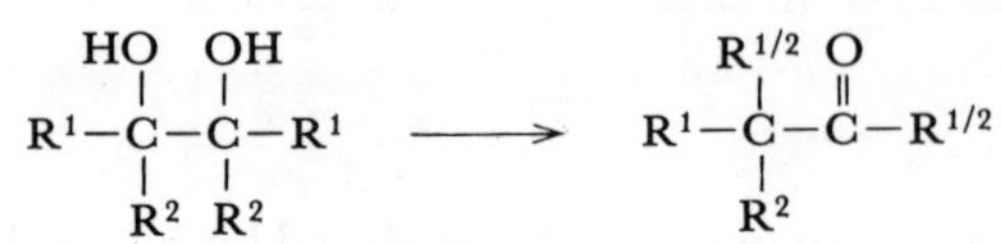

(ref. 32)

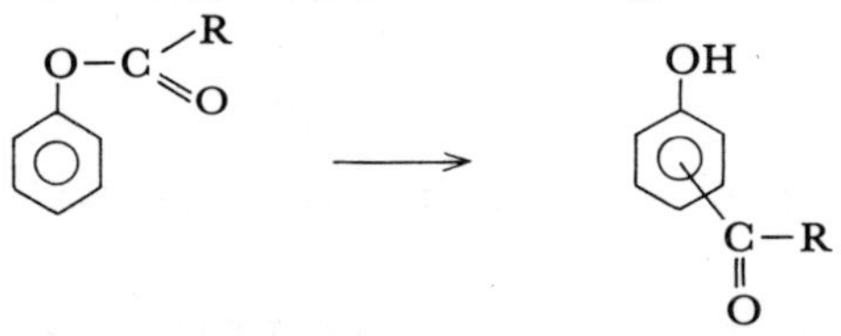

(ref. 33)

(ref. 34)

REFERENCES

1. R. Kunin, E. Meitzner, N. Burtnick, *J. Am. Checm. Soc.*, **84**, 305 (1962).
2. R. Kunin. E. Meitzner, J. A Oline, S. A. Fisher, N. Frisch, *Ind. Eng. Chem. Prod. Res. Dev.*, **1**, 140 (1962).
3. G.A. Olah, P.S. Iyer, G.K.S. Prakash, *Synthesis,* **1986**, 514.
4. R.B. Wesley, B.C. Gates, *J. Catal.*, **34**, 288 (1974).
5. K. Setinek, *Collect. Czeh. Chem. Comm.*, **42**, 979 (1977).
6. T.I. Andrianova, *Kinet. Katal.*, **5**, 927 (1963).
7. H. W. Heath, Jr., B.C. Gates, *AIChE J.*, **18**, 321 (1972).
8. B.C. Gates, W. Rodriguez, *J. Catal.*, **31**, 27 (1973).
9. M.B. Bochner, S.M. Gerber, W. Vieth, A.J. Rodger, *Ind. Eng. Chem., Fund.*, **4**, 314 (1965).
10. F. Ancillotti, M.M Mauri, E. Pescarollo, *J. Catal.*, **46**, 49 (1977).
11. F. Ancillotti, M.M Mauri, E. Pescarollo, L. Romagnoni, *J. Mol. Catal.*, **4**, 37 (1978).
12. R. Thornton, B.C. Gates, *J. Catal.*, **34**, 275 (1974).
13. B. C. Gates, J.S. Wisnoukas, H.W.Heath, Jr., *J. Catal.*, **24**, 320 (1972).
14. T. Uematsu, *Bull. Chem. Soc. Jpn.*, **45**, 3329 (1972).
15. M.J. Astle, J.A. Zaslowskt, *Ind. Eng. Chem.*, **44**, 2867 (1952).
16. C.J. Schmidle, R.C. Mansfield, *Ind. Eng. Chem.*, **44**, 1388 (1952).
17. E.D. Bermann, R. Corett, *J. Org. Chem.*, **21**, 107 (1956).
18. Japan Tokkyo Koho (Japan Patent) 1977-18678; Jpn. Kokai Tokkyo Koho, 1975-119798.
19. D. Hodge, E. Khoshdel, J. Waterhouse, *J. Chem. Soc., Perkin Trans. 1,* **1983**, 2205.
20. G.A. Olah, R. Malhotra, S.C.Narang, J.A. Olah, *Synthesis,* **1978**, 672.
21. C.G. Krespan, *J. Org. Chem.*, **44**, 4924 (1979).
22. G.A. Olah, D. Meidar, A.P. Fung, *Synthesis,* **1979**, 270.
23. H. Hasegawa, T. Higashimura, *Polymer J.*, **12**, 407 (1980).
24. G.A. Olah, D. Meidar, *Nouv. J. Chim.*, **3**, 269 (1979).
25. H. Konishi, K. Suetsugu, T. Okano, J. Kiji, *Bull. Chem. Soc. Jpn.*, **55**, 957 (1987).
26. G.A. Olah, S.C. Narang, *Synthesis*, **1978**, 690.
27. G.A. Olah, D. Meidar, J.A. Olah, *Nouv. J. Chim.*, **3**, 275 (1979).
28. G.A. Olah, A. Husain, B.P. Singh, *Synthesis,* **1983**, 892.
29. G.A. Olah, S.C. Narang, R. Mehrota, *Synthesis,*; **1981**, 474.
30. G.A. Olah, S.C. Narang, D. Meidar, G. Salem, *Synthesis,* **1981**, 282.
31. G.A. Olah, A.P. Fung, D. Meidar, *Synthesis,* **1981**, 280.
32. G.A. Olah, D. Maidar, *Synthesis*, **1978**, 358.
33. G.A. Olah, M. Arvanaghi, V.V. Krishnamurthy, *J. Org. Chem.*, **48**, 3359 (1983).
34. G.A. Olah, A.P. Fung, *Synthesis*, **1981**, 473.

3.7 METAL SULFIDES

Mo or W sulfides which are promoted by Co, Ni supported on alumina or silica-alumina are practically used in large amounts for hydrodesulfurization of heavy oil. Structure, properties and catalytic function of these catalysts have been studied extensively.[1-4] These reactions are not simple acid-base reactions. It has been suggested that there are two different active sites on Mo sulfide;[5] coordinatively unsaturated Mo (Lewis acid site) and Brønsted acid site induced by the interaction of H_2S with Mo sulfide under the reaction conditions. The latter appears to be the active site for hydrocracking.[5b] Ruthenium sulfide supported on Y zeolite and γ-alumina has been reported to be highly active for hydrodenitrogenation.[6] Mo sulfide is also applied in the liquefaction of coal.

Acidity and basicity of ZnS have been studied by means of titrations using n-butyl amine and benzoic acid.[8] According to the study, ZnS exhibited no basicity, but it contained a considerable amount of very weak acid sites. Upon heating at 570 − 770K, the acid amount increased to 1.1 or 2.5 mmole g^{-1} for $H_0 \leq 4.8$, but acid sites stronger than $H_0 \leq 1.5$ were not detected. Measurement of heat of adsorption of n-butylamine indicated the presence of 1.3 mmole g^{-1} of weak acidity on untreated ZnS.[8].

As for MoS_2 with a layered structure, the relationships between the degree of coordinative unsaturation of Mo and catalytic activities for isomerization and hydrogenation of olefins have been investigated in detail for the Mo ion located at the edge plane.[9] The study showed that those reactions proceeded in the presence of hydrogen on Mo at edge having certain coordinative unsaturation, while isomerization of 2-methyl-2-butene took place on the basal plane probably via an alkyl cation regardless of the presence of hydrogen. The latter fact indicates the presence or induction of weak Brønsted acid sites on the basal plane of MoS_2.

Adsorption of pyridine had retardation effect on the dehydrosulfurization over Mo sulfide and Mo − Co sulfide-alumina catalysts which were formed by sufidation of corresponding oxide precursors.[1,7] Adsortion of NH_3 is small for oxidized catalysts, but increases as they are reduced or sulfided.[11]

Pyridine and NH_3 block the vacant coordination sites of Mo. Those sites may be regarded to be Lewis acid sites in a broad sense. Based on studies of pyridine and NH_3 adsorption, two kinds of active sites were postulated:[10] (i) a strongly electrophilic site or strong acid site, responsible for hydrogenations, and (ii) a weak acid site which is poisoned only by NH_3 and primarily active for desulfurization. Adsortion of NH_3 was reported to be proportional to vacancies (coordinatively unsturated Mo site).[11] Isomerization of cyclopropane was studied,[12] and it was found that the catalytic activity increased with the concentration of vacancies. Hence, active sites consisting of a vacancy and neighboring Al − OH group were postulated. Acid strength distributions of Mo − Co oxide − Al_2O_3 catalysts measured by amine titration have been reported (see Table 3.13).[13]

Treatment with hydrogen sulfide of metal-ion-exchanged solid acids enhances the acidity and catalytic activity.[14]

REFERENCES

1. F.E. Massoth, *Advan. Catal. Relat. Subj.*, **27**, 265 (1978).
2. T.F. Hayden, J.M. Dumesic, *J. Catal.*, **103**, 366 (1987).
3. P. Grange, *Catal. Rev. Sci. Eng.*, **21**, 149 (1980).
4. G.C.A. Schuit, B.C., Gates, *AIChE J.*, **19**, 417 (1973); B. Delmon, Proc. 3rd Climax 3rd Intern. Conf. Chemistry and Uses of Molybdenum, 1979, p73.

5a) S. H. Yang, C. N. Satterfield, *J. Catal.*, **81**, 168 (1983);
 b) H. Shimada, T. Sato, Y. Yoshimura, J. Hiraishi, A. Nishijima, *J. Catal.*, **110**, 275 (1988).

6. T.G. Harvey, T.W. Matheson, *J. Catal.*, **101**, 253 (1986).
7. S.W. Cowley, F.E. Massoth, *J. Catal.*, **51**, 291 (1978).

8. K. Tanabe, Y. Yamaguchi, *J. Res. Inst. Catal., Hokkaido Unit.*, **9**, 179 (1964); *ibid.*, **14**, 93 (1966).
9. K. Tanaka, T. Okuhara, *Catal. Rev. Sci. Eng.*, **15**, 246 (1977).
10. P. Desikan, C.H. Amberg, *Can J. Chem.*, **42**, 843 (1964).
11. F.E. Massoth, *J. Catal.*, **36**, 164 (1975).
12. E.A. Lombardo, M. Lo Jacono, W.K. Hall, *J. Catal.*, **51**, 243 (1978).
13. P. Ratnasamy, D.K Sharma, L.D. Sharma., *J. Phys. Chem.*, **78**, 2069 (1974).
14. T. Hosotsubo, M. Sugioka, Y. Sanada, K.Aomura, *Nippon Kagaku Kaishi*, **1979**, 28; **1980**, 797 (in Japanese).

3.8 METAL SULFATES AND PHOSPHATES

3.8.1 Metal Sulfates

Many metal sulfates generate fairly large amounts of acid sites of moderate

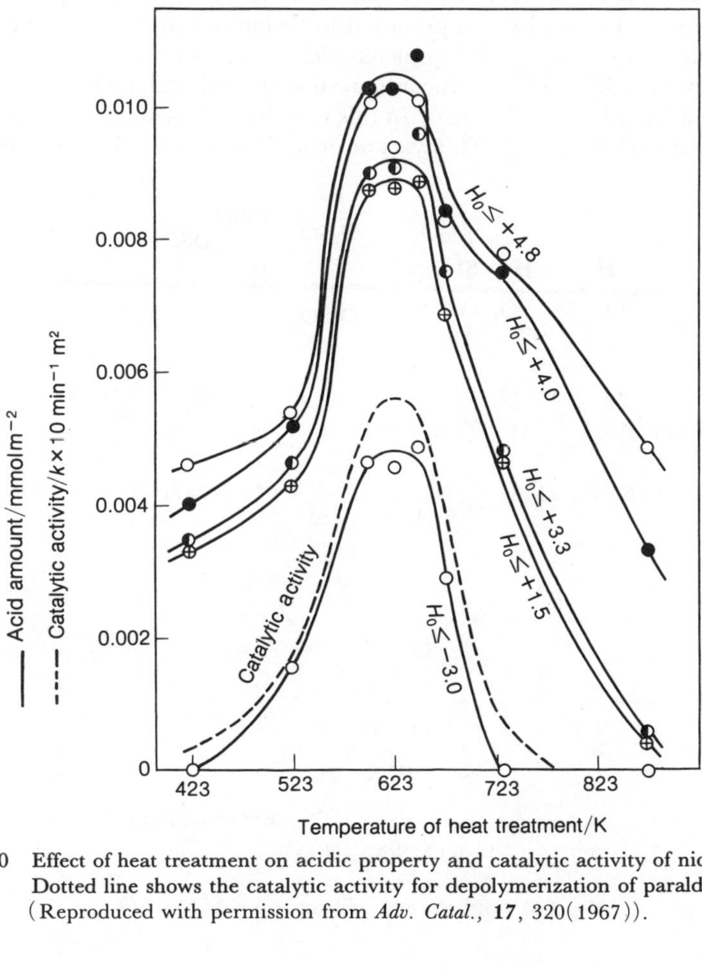

Fig. 3.80 Effect of heat treatment on acidic property and catalytic activity of nickel sulfate. Dotted line shows the catalytic activity for depolymerization of paraldehyde. (Reproduced with permission from *Adv. Catal.*, **17**, 320(1967)).

strength on their surfaces when they are calcined at around 623 K.[1] The moderate acid property of metal sulfates often gives high selectivities for diversified reactions[1,2] such as hydration, dehydration, polymerization, alkylation, esterification, isomerization, and condensation. Change in the acidic property with change of calcination temperature is shown in Fig. 3.80 for $NiSO_4 \cdot 7H_2O$, a typical sulfate.[1,3] The maximum acidities at any acid strength measured by the butylamine titration method using Hammett indicators are observed at 623 K, the strongest acid strength being $H_0 = -3$ corresponding to 50% H_2SO_4. As seen in Fig. 3.80, the acidity at $H_0 \leq 3$ is well correlated with the catalytic activity for depolymerization of paraldehyde. On the other hand, the maxima of Brønsted and Lewis acidities measured by the infrared method using pyridine appear at 523 and 673 K, respectively, as shown in Fig. 2.6. The catalytic activity for isomerization of α-pinene to camphene is correlated with the Brønsted acidity,[4] while the activity for hydrolysis of methylene chloride correlates with the Lewis acidity,[5] indicating that the active sites of nickel sulfate for the former and latter reactions are Brønsted and Lewis acid sites, respectively.

The structure of acid sites of nickel sulfate has been extensively studied using infrared, electron spin resonance, X-ray, nuclear magnetic resonance and Mossbauer effect techniques.[6] The acid site is proposed to be formed by an empty orbital of the nickel ion which appears in an incompletely dehydrated metastable transition structure (see (II) in Fig. 3.81).[1,2,5] This configuration is intermediate between the monohydrate (I) and anhydrous (III) forms. In this transitional form, nickel is pentavalent, and has a vacant sp^3d^2 orbital. This vacant orbital, and the resultant affinity for an

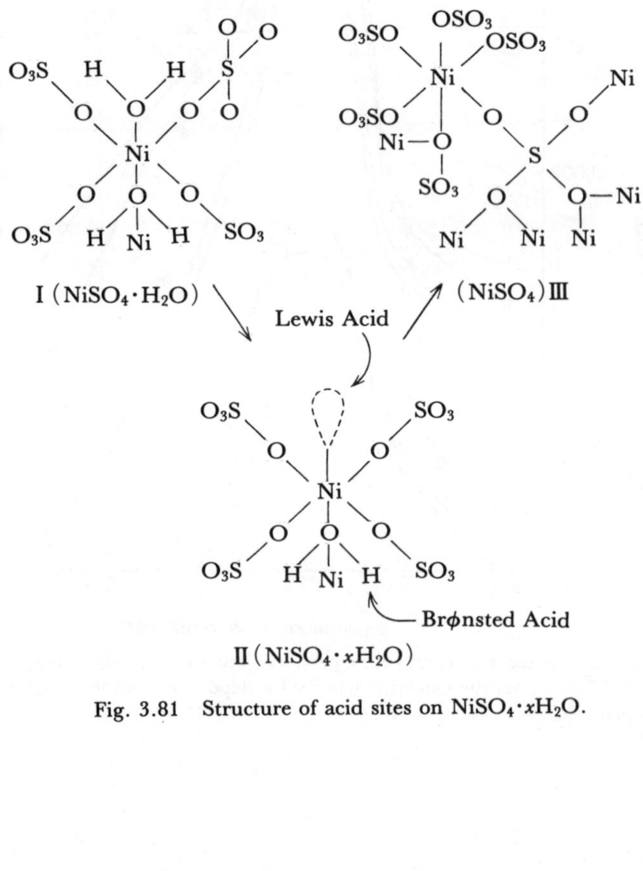

Fig. 3.81 Structure of acid sites on $NiSO_4 \cdot xH_2O$.

electron pair, accounts for the Lewis acid properties of nickel sulfate and its catalytic activity. The suggested structure is both strained and unstable, although the crystal network and the retained water do have some stabilizing influence. The Brønsted acidity arises from two sources. One is the water which is coordinated directly with a nickel ion in the above transitional form. The nickel tends to attract the oxygen atom, thus freeing a hydrogen ion. The other source is the surface water, acidified by the inductive effect of the neighboring cationic Lewis acid sites. The temperature dependence of the Brønsted and Lewis acidity, as derived from the infrared spectra of adsorbed pyridine, is shown in Fig. 2.6. Brønsted acid first appears when the vacant orbital of the nickel ion is formed by dehydration, and the amount increases progressively with increase in dehydration temperature. Since, however, the amount of water of hydration decreases as the temperature of heat treatment is increased, a temperature is eventually reached at which the Brønsted acidity begins to decline. Lewis acidity also increases with increasing heat treatment temperature, but only begins to decline when the metastable structure with the vacant orbital collapses and changes to the stable anhydrous structure at higher temperatures. The sum of the two acidity curves gives the total acid amount which can be measured by the amine titration method. $MgSO_4$, $MnSO_4$, $FeSO_4$, $CoSO_4$, $CuSO_4$, and $ZnSO_4$, which have monohydrate structures similar to that of nickel sulfate, are also considered to have acid centers which are comparable in structure to those of nickel sulfate .[7)]

Catalytic activities of the sulfates of Fe^{3+}, Al^{3+}, Sc^{3+}, Cu^{2+}, Ni^{2+}, Mg^{2+}, etc. on SiO_2 for isomerization of *n*-butene were reported to correlate well with the acid strengths of the sulfates expressed by the electronegativities of the metal ions or measured by the Hammett indicator method.[8)]

Compression of sulfate hydrates increases their surface acidities in the cases of $CdSO_4{\cdot}8H_2O$, $Ce_2(SO_4)_3{\cdot}8H_2O$, $Fe_2(SO_4)_3{\cdot}xH_2O$, $Al_2(SO_4)_3{\cdot}18H_2O$, and $KHSO_4$, while $BaSO_4$, $PbSO_4$, K_2SO_4, and $MgSO_4{\cdot}7H_2O$ exhibit no acidity at all,[9)] even at pressures 3000 kg cm^{-2}. Aluminum sulfate increases in acidity upon compression, but does not produce any acid sites at an acid strength of $H_0 = -3$ by application of pressures as high as 4,200 kg cm^{-2}. Thus compression does not generate as strong acid sites as does heat treatment. The weak acid sites on aluminum sulfate hydrate formed on compression are mainly Brønsted acid sites due to removal of anion water. The octahedron surrounding the Al^{3+} ion is deformed so that the distances between the Al^{3+} ion and the coordinated water become unequal, giving rise to two types of coordinated water, H_2O (α) and H_2O (β). O (β) is closer to Al^{3+} than before compression, thus increasing its covalent character, so that H_2O (α) tends to develop a protonic positive charge.[9)]

Gamma irradiation (^{60}Co, total dose: 5.7×10^7r) of a nickel sulfate previously heated at 423 K enhances its acidic property to values nearly equivalent to those of a nickel sulfate heat-treated at 623 K without irradiation.[10)] The acid sites formed by such irradiation do not decay.

REFERENCES

1. K. Tanabe, T. Takeshita, *Adv. Catal.*, **17**, 315 (1967).
2. T. Takeshita, R. Ohnishi, K. Tanabe, *Catal. Rev.*, **8**, 29 (1973).
3. K Tanabe, R. Ohnishi, *J. Res. Inst. Catal., Hokkaido Unit.*, **10**, 229 (1962).
4. R. Ohnishi, T, Takeshita, K. Tanabe, *Shokubai* (Tokyo), **7**, 306 (1965) (in Japanese).
5. T. Yamaguchi, K. Tanabe, Proc. 4th Intern. Congr. Catal., Moscow, 410, (1969).
6. T. Takeshita, R Ohnishi, T. Matsui, K. Tanabe, *J. Phys. Chem.*, **69**, 4077 (1965).
7. J. Coing-Boyat, G. Bassi, *C. R. Acad. Sci.*, Paris, **256**, 1482 (1963).
8. M. Misono, Y Saito, Y. Yoneda, *J. Catal.*, **9**, 135 (1967).
9. T. Kawakami, A. Konno, Y. Ogino, *Bull. Chem. Soc. Jpn.*, **44**, 1772 (1971).
10. K. Tanabe, T. Iizuka, M. Ogasawara, *J. Res. Inst. Catal., Hokkaido Univ.*, **16**, 532 (1968).

3.8.2 Metal Phosphate (Phosphorous Metal Oxides)

Various types of metal phosphates in amorphous or crystalline form are used as acid and base catalysts. The acid type catalysts are phosphorous oxides containing the following metals: Al, Ti, Zr, B, Si, Zn, Ca. Phosphorous oxides containing the following metals act as basic catalysts: Li, K, Ca, Cd. The phosphorous oxides containing V, Bi, Fe are oxidation catalysts.

TABLE 3.40 Reactions catalyzed by phosphorous oxides

Reaction types	Catalysts
Alcohol dehydration	ortho-phosphates [47], alkaline earth phospates [48], $AlPO_4$ [8, 49], BPO_4 [20, 26, 50, 51], $Ti(HPO_4)$ [52], $Zr(HPO_4)_2$ [35, 37, 53]
Alcohol dehydrogenation and dehydration	$CaPO_4$ [54], $Ca_{10}(PO_4)_6(OH)_2$ [41–43] $Zn_3(PO_4)_2$ [55], $Cd_3(PO_4)_2$ [56], $Al(H_2PO_4)_3$ [57], $Ca(PO_4)_2$ [58]
Alcohol dehydrogenation	$Ca_{10}(PO_4)_6(OH)_2$ [43], $Cd_3(PO_4)_2$ [71]
Hydrogen transfer	$Ca_{10}(PO_4)_6(OH)$ [43]
Alkenes isomerization	$AlPO_4$ [13–15], Metal ortho phosphates [59], $Zr(HPO_4)_2$ [32, 35]
Ethylene oxide and propylene oxide isomerization	Metal ortho phosphates [40]
Phenols alkylation	$AlPO_4$ [60], BPO_4 [61], $Ca_3(PO_4)_2$ [27], $Zn_3(PO_4)_2$ [72]
Toluene side chain alkylation	K_3PO_4 [46], $Ca_3(PO_4)_2$ [46]
Ethylene hydration	BPO_4 [62, 63]
Chlorobenzene hydrolysis	$Ca_{10}(OH)_2(PO_4)_6-Cu^{2+}$ [64–66], $LaPO_4$ [67]
Aldol condensation	Li_3PO_4 [68], $Ca_3(PO_4)_2$ [68]
Acetic acid vinylation	$Cd_3(PO_4)_2$ [69]
Diacetone alcohol decomposition	Alkali and alkaline earth phosphates [70]

Each metal phosphate has different crystal structures depending on the preparation and activation conditions. Acid−base properties and catalytic activities usually vary with the crystal structure. By changing the metals and preparetion conditions, it is possible to obtain wide varieties of catalysts of different acid−base properties, surface areas and, crystalline structure. This flexibility enables metal phosphates to be used in many types of reactions. Selected reactions catalyzed by phosphorous metal oxides are listed in Table 3.40. In this section, the acid−base properties and catalytic activities of aluminum phosphorous oxide, boron phosphorous oxide, zirconium phosphorous oxide, and calcium phosphorous oxide are described.

A. Aluminum Phosphorous Oxide

Aluminum phosphorous oxide is one of the representative acid catalysts and the most extensively studied phosphorous oxide. Aluminum phosphate has seven crystalline forms and is isostructural with SiO_2, the seven crystalline modifications of each being directly parallel. The ions, Al^{3+} and P^{5+}, replace Si^{4+}. Amorphous aluminum phosphate exists in addition to these seven crystalline forms.

Aluminum phosphrous oxide is usually prepared from an aqueous mixture of aluminum salts and phosphoric acid by precipitation with aqueous ammonia followed by calcination at high temperatures.[1,2] Control of the preparation conditions results in oxides of different crystalline structures. For the stoichiometric aluminum phosphorous oxide (aluminum phosphate), it is suggested that the pH of the precippitating solution be kept below 4.[3] Kearby prepared a transparent oxide, with high surface area and highly resistant to thermal treatment using ethylene oxide as a precipitation medium.[4] The resulting oxide has a surface area of 400 $m^2 \cdot g^{-1}$, and acid sites larger than those of $SiO_2-Al_2O_3$.

The acidic properties vary with the A1/ P ratio and content of OH groups. On the surface of stoichiometric aluminum phosphate (A1/P = 1), the existence of both Brønsted and Lewis acid sites is evidenced by IR study of adsorbed ammonia and pyridine of the sample pretreated below 873 K.[5] The hydroxyl groups of P are acidic and their acidity is further enhanced by hydrogen bonding to Al-OH groups, and act as Brønsted acid. When outgassed at high temperatures, the OH groups condense to form Lewis acid sites. The transformation of the Brønsted acid sites into Lewis acid sites can be drawn as follows.

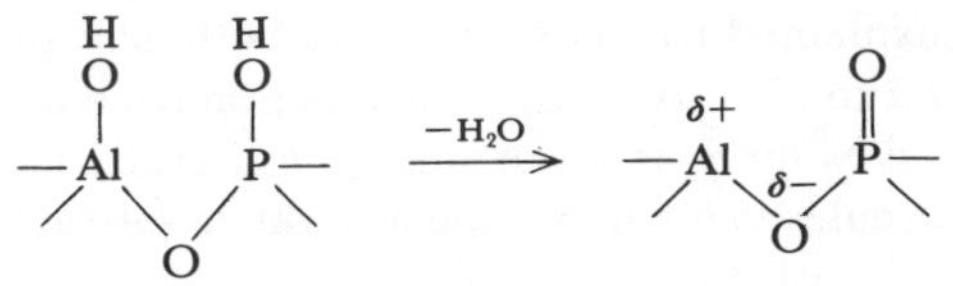

Oxide ions in the dehydrated aluminum phosphorous oxide are held primarily by the P atoms, and the P=O bond is of covalent nature. Therefore, the oxide ions are inadequate to act as base sites.[5] Two types of Lewis acid sites are demonstrated from IR measurement of coadsorption of ammonia and pyridine; one is isolated and the other is the acid site adjacent to basic site.[6]

The OH content in aluminum phsphorous oxide varies with the media in the preparation procedures. Comparing aluminum phosphorous oxides in aqueous ammonia, ethylene oxide, and propylene oxide, both acidic and basic properties are highest for the sample prepared in aqueous ammonia.[7)]

The Al/P ratio affects the acidic properties. The maximum number of acid sites is obtained at the Al/P ratio of 0.3.[8)] The strength of the acid sites also depends on the Al/P ratio. Acid sites of $H_0 \leq -5.6$ are generated for the sample of Al/P < 1, and $H_0 \leq -3.5$ and 4 for samples of Al/P = 2 and 4, respectively.

The acid strength of aluminum phosphorous oxide is enhanced by the addition of SO_4^{2-} up to 3 wt%.[9,10)] Enhancement of acid strength is suggested by high catalytic activities for 1-butanol dehydration[11)] and cyclohexene skeletal isomerization.[9)] However, addition of excess SO_4^{2-} ions reduces the activity for the reactions. The catalytic activity of the oxide prepared from aluminum sulfate is different from those prepared from chloride and nitrate for 1-butanol dehydration; equilibrium mixture of butene isomers is produced over the oxide from sulfate whereas the ratios of 1-butene/2-butenes and *cis*/*trans* are larger than the equilibrium ratios over the oxides from chloride and nitrate. The catalytic behavior of the oxide from sulfate different from the other oxides is caused by the presence of a small amount of SO_4^{2-} ions generating strong acid sites.

The presence of base sites on aluminum phosphorous oxide is confirmed by titration methods[12)] and adsorption of CO_2, HCl[5)] and phenol.[7)] The amounts of base sites are small. The numbers of base sites are about one order of magnitude less than those of acid sites for most cases.[7)]

Dehydration of alcohols proceeds over aluminum phosphorous oxides. Correlation between 1-butanol dehydration activity and amount of acid sites stronger than $H_0 = 1.5$ has been reported.[8)] The dehydration of *cis*- and *trans*-2-methylcyclohexanol to 1-methylcyclohexene and 3-methylcyclohexene is catalyzed by the aluminum phosphorous oxides of different A1/P ratios.[13)] While the *cis* isomer is converted more extensively than the *trans* with all compositions, the relative amount of the two olefinic products (1-/3-methylcyclohexene) increases markedly on increase in the amount of P. The formation of 3-methylcyclohexene from either alcohol takes place on strong acid sites by the *E*1 process in which carbenium intermediates are involved. On the other hand, the formation of 1-methylcyclohexene takes place on pairs of acid and base sites by the *E*2 process. The aluminum atoms and aluminum atoms with hydroxyl groups attached function as Lewis acid sites and Brønsted acid sites, respectively.

Oxidative dehydrogenation of ethylbenzene to styrene is catalyzed by various oxides containing phosphorous oxide.[12)] Aluminum phosphorous oxide shows the highest activity; this is explained by the presence of both acid sites of H_0=1.5 to -5.6 and base sites of H_- =17.2 to 25.6. Although the reaction itself is not an acid-base catalyzed reaction, the acid sites serve as adsorption sites for ethylbenzene while the base sites activate oxygen molecules to form O^- species which abstract the H from the intermediates.

Aluminum phosphorous oxide shows characteristic features in butene isomerization.[14–16)] At low pretreatment temperatures (473 – 773 K), the catalyst behaves like a typical acid catalyst; the reaction proceeds via *s*-butyl cation intermediate and gives *cis*/*trans* ratios close to unity. At high pretreatment temperatures, however, *cis-trnas* interconversion predominates to an overwhelming degree; double bond migration becomes very slow. The activity maximum is observed when aluminum phosphorous oxide is outgassed at 1173 K.[14,15)] The specific catalytic activity is considered to be due to a pair of strong acidic and weak basic sites favoring *cis-trans* isomerization without double bond migration. The reaction scheme is drawn as follows:

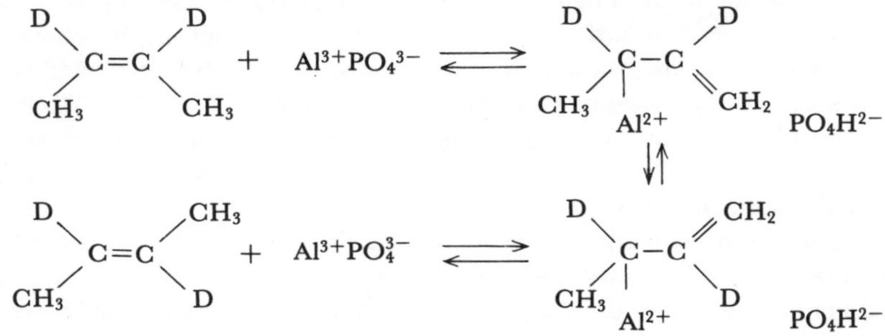

A weak basic site may be able to abstract an allylic H^+ only with assistance from a strong acidic site to interact with carbon atom 3. As a result, an anion that strongly interacts with the acidic site at carbon atom 3 will be formed. The attack of H^+ at the carbon atom 3 would result in production of 1-butene. However, a strong interaction between the acidic site and carbon atom 3 would not allow an H^+ to attack the latter. Therefore, only *cis-trans* isomerization occurs.

Moffat calculated the charge densities of each atoms in various phosphate model clusters by CNDO/2, and the results are shown in Fig. 3.81.[17] The exposed oxygen atoms were found to have maximum electron density and hence function as Lewis base. On the basis of the magnitude of the positive charge on cations, it is inferred that

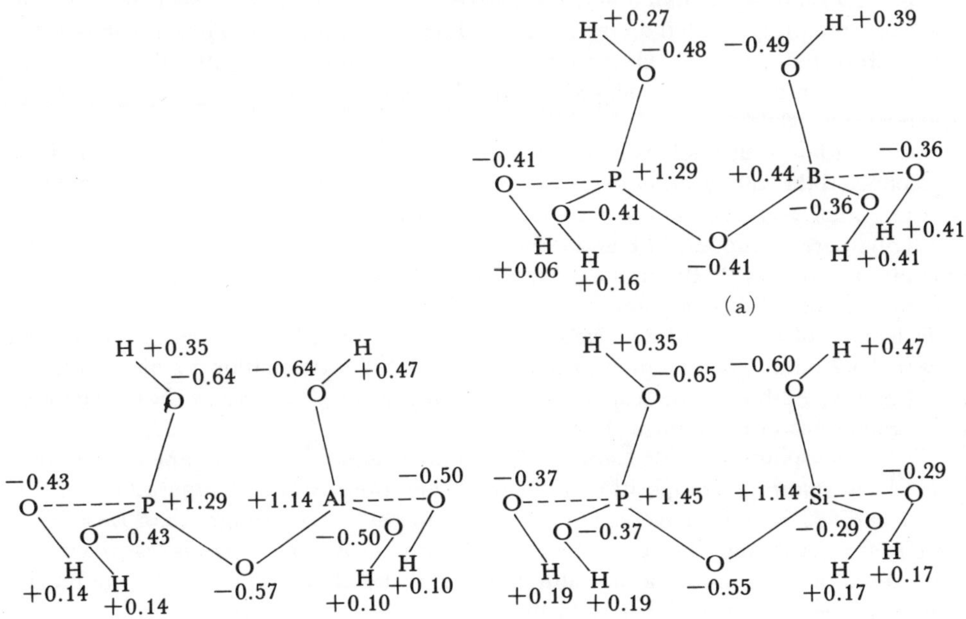

Fig. 3.81 **Calculated charge densities for the clusters representing (a) boron phosphate, (b) aluminum phosphate and (c) silicon phosphate.**
(Reproduced with permission by J. B. Moffat, *Molec, Catal.*, **30**, 174(1985)).

phosphorous atoms function as the acidic site in these phosphates. Brønsted acidity of these catalysts is due to protons attached to exposed oxygen, since these possess maximum positive charge. The order of Lewis acid strength is boron phosphate = aluminum phosphate < silicone phosphate, while the order of Brønsted acidity is boron phosphate < aluminum phosphate = silicon phosphate, and the Lewis basicity order is boron phosphate < aluminum phosphate = silicone phosphate.

B. Boron Phosphorous Oxide

Boron phosphorous oxide has acidic sites on the surface and catalyzes acid-catalyzed reactions. The oxide possesses acid sites of $H_0 = -5.6$ to -8.2, which are comparable to those on aluminum phosphorous oxide. The base sites, however, have not been measured.

Boron phosphorous oxide is usually prepared by heating a mixture of boric acid and phosphoric acid at about 313 K followed by calcination at about 630 K.[18-20] Before use for catalyst, the resulting boron phophorous oxide is further pretreated at higher temperatures in a vacuum or in air, because the acidic properties and consequently catalytic activities are dependent on the pretreatment conditions.

The existence of both Lewis and Brønsted acid sites on boron phosphorous oxides of different compositions is evidenced by IR measurements of adsorbed alkene[21] and substituted pyridines.[22] On adsorption of alkenes, carbenium ions are formed, and on adsorption of substituted pyridines, both pyridinium ions and coordinated pyridines are formed.

The P/B ratio in the oxide and the pretreatment temperature affect the acid site type and strength generated on the surfaces. The oxide with P > B has a high percentage of Brønsted acid sites while that with P < B contains predominantly Lewis acid sites.[22] Pretreatment in a vacuum at 873 K generates the strongest acid sites of $H_0 = -5.6$ to -8.2.[23]

Dehydration of alcohols is catalyzed by boron phosphorous oxide. The reactivity of alcohols in the dehydration decreases as follows: *tert*-amyl alcohol > 3-pentanol > > 2-propanol > 1-pentanol > ethanol.[24] The catalytic activities of the oxides composed of different amounts of B and P for propanol dehydration correlate with the total amount of acid sites.[20] Butanol undergoes dehydration on boron phosphorous oxide. The maximum activity is obtained at the P/B ratio of 0.6.[25] The activity correlates with the sum of Lewis and Brønsted acid sites. 2-Butanol, 2-methyl−2-butanol, and 3-methyl−2-butanol also undergo dehydration.[26] The formation of *trans*-2-butene and 2-methyl−2-butene increases with increasing surface acidity. In these reactions, the carbenium ion mechanism is operating.

Boron phosphorous oxide shows a high catalytic activity for 1-decane oligomerization.[23] The optimum composition of the oxide and pretreatment temperature are P/B = 1.1 and 873 K, respectively. For the reaction, acid sites stronger than $H_0 = -5.6$ are effective. Although the surface area of boron phosphorous oxide is less than one-tenth those of silica alumina and alumina boria, the conversion of 1-decane on the former catalyst is much higher than those on the latter catalysts.

In addition to the reactions mentioned above, boron phosphorous oxide shows catalytic activities for the following reactions; alkylations of phenol[27] and 1,2-benzenediol with methanol,[28] hydrogen transfer of cyclohexane to cyclohexene and methylcyclopentene-1,[29] and conversion of amides to nitriles.[30] The acid sites on boron phosphorous oxide are involved in all the above reactions.

C. Zirconium Phosphorous Oxide (Zirconium bis(monohydrogen phosphate))

Zirconium phosphorous oxide investigates in catalysis is restricted to the stoichiometric oxide. The oxide has the general formula $Zr(HPO_4)_2 \cdot nH_2O$ and is either in amorphous form or crystalline form of various layered structures. One of the crystalline zirconium phosphates is α-layered acid salt, zirconium bis(monohydrogen orthophosphate) of monohydrated form. The structure is schematically drawn in Fig. 3.83[31,32] Each layer consists of planes of Zr atoms bridges through phosphate groups alternating above and below the Zr atom planes. Each layer is considered a planar macro-anion $[Zr_n(PO_4)_{2n}]^{2-}$, whose negatively charged oxygens are balanced by an equivalent number of protons or other cations.

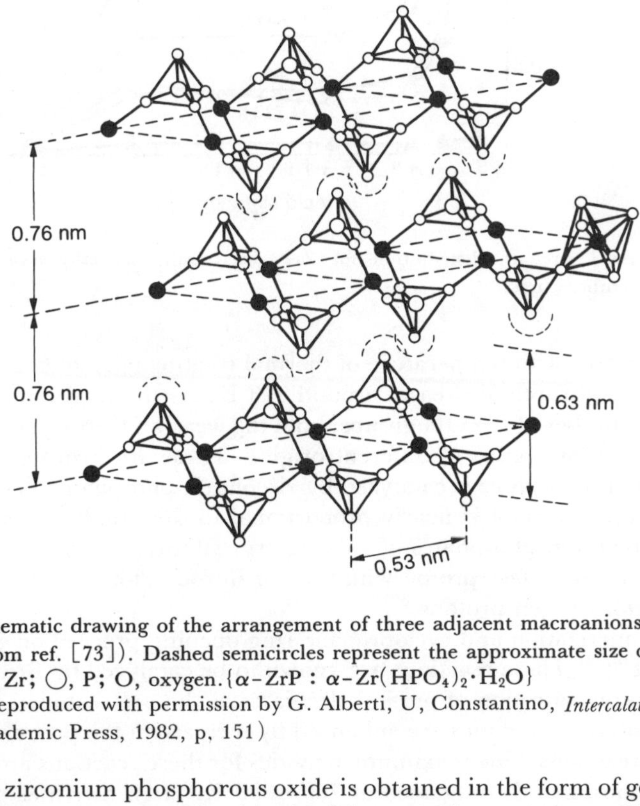

Fig. 3.83 Schematic drawing of the arrangement of three adjacent macroanions in α-ZrP (from ref. [73]). Dashed semicircles represent the approximate size of the O^-; ●, Zr; ○, P; O, oxygen. {α-ZrP : α-$Zr(HPO_4)_2 \cdot H_2O$}
(Reproduced with permission by G. Alberti, U, Constantino, *Intercalation Chemistry*, Academic Press, 1982, p. 151)

Amorphous zirconium phosphorous oxide is obtained in the form of gel by addition of a soluble Zr(VI) salt to phosphoric acid. The stoichiometric crystalline zirconium phosphate is prepared by refluxing the amorphous gel in concentrated phosphoric acid.[33,34] Depending on the reflux conditions, different crystalline forms result.

The acidic properties of crystalline zirconium phosphate vary with the pretreatment temperature. Two types of acid sites different in strength exist on the surfaces: one is $H_0 = 4.8$ to 3.3, the other is $H_0 = -3.0$ to -5.6.[35,36] The acid site distribution of zirconium phosphate is shown in Fig. 3.84. The number of the strong acid sites

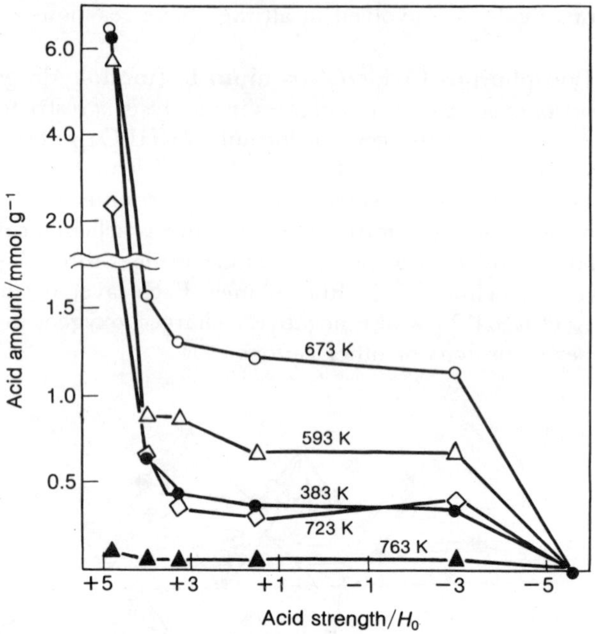

Fig. 3.84 Acid strength distribution for ZrP dried and thermally treated at various temperatures.

gradually increases with temperature of thermal treatment up to 673 K. The number of the acid sites abruptly decreases around 723 K. As the total number of acid sites is close to the number of P−OH groups, both the weak and strong acid sites are considered to be P−OH groups of different origins. These are Brønsted acid sites.

Dehydration of alcohols is catalyzed by zirconium phosphate. The activity for dehydration of cyclohexanol is nearly proportional to the number of surface hydroxyl groups for zirconium phosphates of varying crystallinity.[37] The activity is poisoned by the exchange of surface proton with Cs^+ or introduction of quinoline. The active sites are monohydrogen protons.[36]

Butene isomerization and cyclopropane ring opening proceed on zirconium phosphate catalyst[32,38] These reactions are known to be catalyzed by Brønsted acid sites. Although the concentration of protons decreases as the pretreatment temperature increases, the catalytic activities are enhanced by increase in the pretreatment temperature for both reactions. The maximum activities for these reactions are obtained when ϵ-zirconium phosphate is outgassed at 700−800 K. High activities observed on the catalysts pretreated at high temperatures are caused by the increase in acidity. The origin of the strong acid sites active for the reactions is the terminal phosphate groups (residual phosphate groups) located between Zr atom planes. The ^{31}P-MAS-NMR measurements of the zirconium phosphates of different crystalline forms and outgassed at different temperature indicate that the electron movement from the surface residual phosphate to bulk phosphorous atoms occurs on high temperature treatment.[32] This movement enhances the acidic properties of catalyst. The proposed adsorbed states of

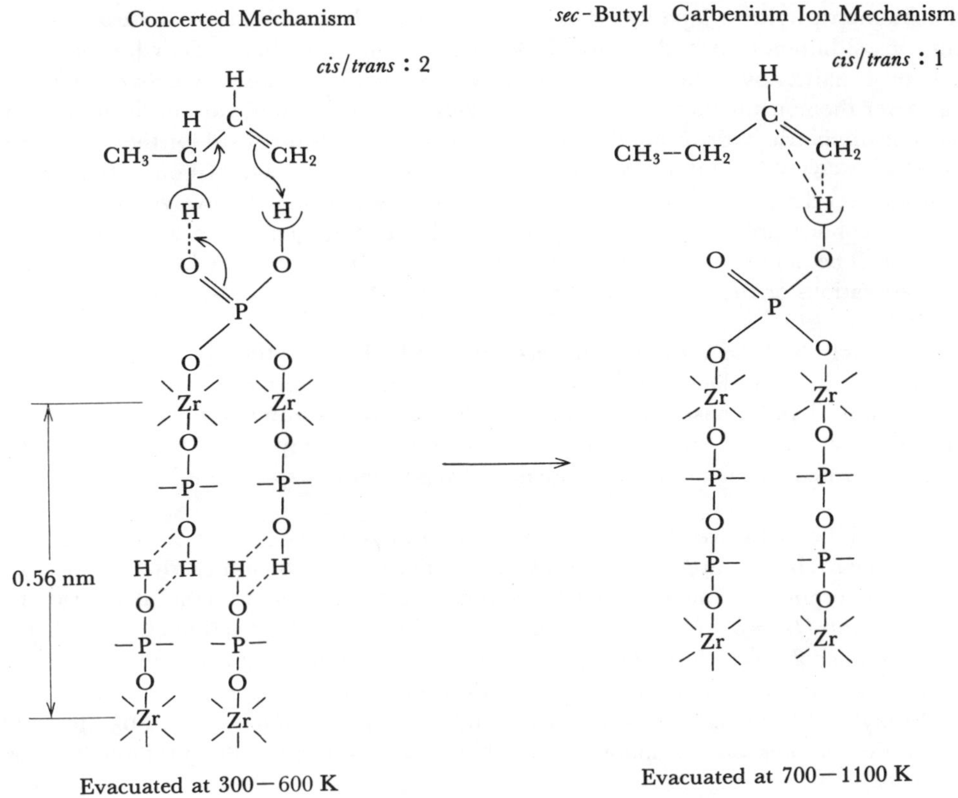

Scheme 1 isomerization of 1-butene on ε-ZrP.

1-butene undergoing isomerization are illustrated below for zirconium phosphate catalysts pretreated at low and high temperatures.[39)]

For the catalyst pretreated at low temperatures, butene interacts both with OH group and O atom, and undergoes isomerization by a switch mechanism. For the catalyst pretreated at high temperatures, the proton on the terminal phosphate interacts with butene to form secondary carbenium ion.

D. Calcium Phosphorous Oxide

Hydroxyapatite, $Ca_{10}(PO_4)_6(OH)_2$, and calcium phosphate, $Ca_3(PO_4)_2$, show both acidic and basic properties. The main characteristic in catalysis involving these materials is the large contribution of the base sites on the surfaces.

Calcium phosphate has about 0.07 mmol g^{-1} weak acid sites at H_0=4.8, and about 0.05 mmol g^{-1} base sites at H_0=7.1.[27,40)]

Alkylation of phenol with methanol is effectively catalyzed by calcium phosphate which is much more active than boron phosphorous oxide. The selectivity to ortho methylation products, *o*-cresol and 2,6-xylenol, is 88%.[27)] The high selectivity is caused by the basic properties on the surface.

Catalytic properties of hydroxyapatite were studied in details by Hall *et al.*[41-44)] Hydroxyapatite of stoichiometric composition P/Ca = 0.6 is less acidic than the nonstoichiometric hydroxylapatite P/Ca = 0.63. The catalytic activities of these are different in 2-butanol dehydration. 2-Butanol undergoes both dehydration and dehydrogenation over the stoichiometric hydroxyapatite, while only dehydration occurs over the nonstoichiometric one. From dehydrogeneation of various alcohols over the stoichiometric hydroxyapatite, a positive value, 1.5, is obtained for the Taft constant ρ_{α}^{*}, suggesting that a negative charge is developed at the α-carbon in the transition state for dehydrogenation. The dehydrogenation mechanism involves the transfer of the alcoholic proton to the base sites and the α-hydrogen to the acid sites of the catalyst. The acid sites acting in the dehydrogenation are assumed to be either the catalyst cations or the protons of HPO_4^{2-} group, while the base sites are assumed to be OH^- or PO_4^{3-}.

Although the base sites on nonstoichiometric hydroxyapatite are not strong compared with those on the stoichiometric hydroxyapatite, the dehydration of alcohols involves an acid and a base site, where a nearly concerted elimination occurs. It should be noted that the acid and base sites participating in the dehydration are not the same as those in dehydrogenation, as evidenced by poisoning experiment.

E. Correlations Between Catalytic Activity and Acidity

Good correlations are observed between acidity or basicity and catalytic activity of various phosphorous metal oxides for several reactions. For propanol dehydration, acid amounts ($H_0 = 3.0$ to $+1.5$) correlate well with the catalytic activities. Phosphates of metal ions of large electronegativity possess large amounts of acid sites and hence show high catalytic activities for the dehydration.

Propylene oxide undergoes isomerization to different products depending upon the properties of sites on phosphorous metal oxides. Over acid sites propionaldehyde

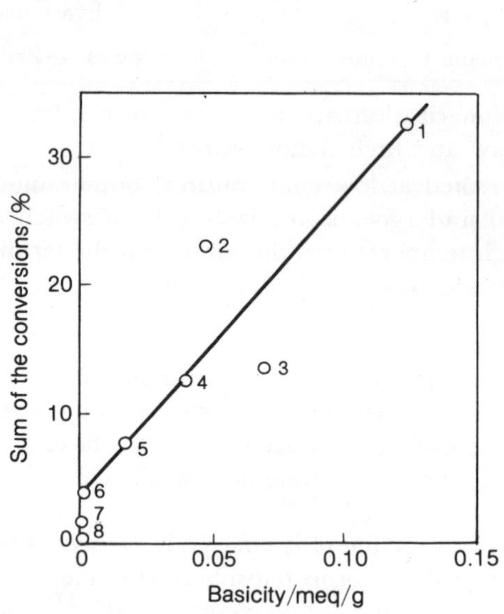

Fig. 3.85 Propylene oxide isomerization conversion *vs.* basicity of catalyst.
1; Li_2NaPO_4, 2; KLi_2PO_4, 3; $Ca_3(PO_4)_2$, 4; Na_2LiPO_4, 5; K_2LiPO_4, 6; $Mg_3(PO_4)_2$, 7; $AlPO_4$, 8; Li_3PO_4

yeilds proportional to the amounts of the acid sites are produced. Base sites participate in the formation of allyl alcohol and 1-proponol formed by acid-base cooperation. Isomerization of propylene oxide to acetone is catalyzed by base sites. The sum of the yields for allyl alcohol, 1-propanol, and acetone correlates well with basicity, as shown in Fig. 3.85.[35)]

In the alkylation of toluene with methanol, the yields of the ring alkylation products and side-chain alkylation products correlate well with the acidity and basicity, respectively, as shown in Fig. 3.86.[46)] Among phosphorous metal oxides, boron phosphorous oxide shows the highest yield of the ring alkylation product xylene and the largest number of acid sites, while potassium phosphorous oxide supported on active carbon shows the highest yield of the side-chain alkylation products ethylbenzene and styrene and the largest number of base sites.

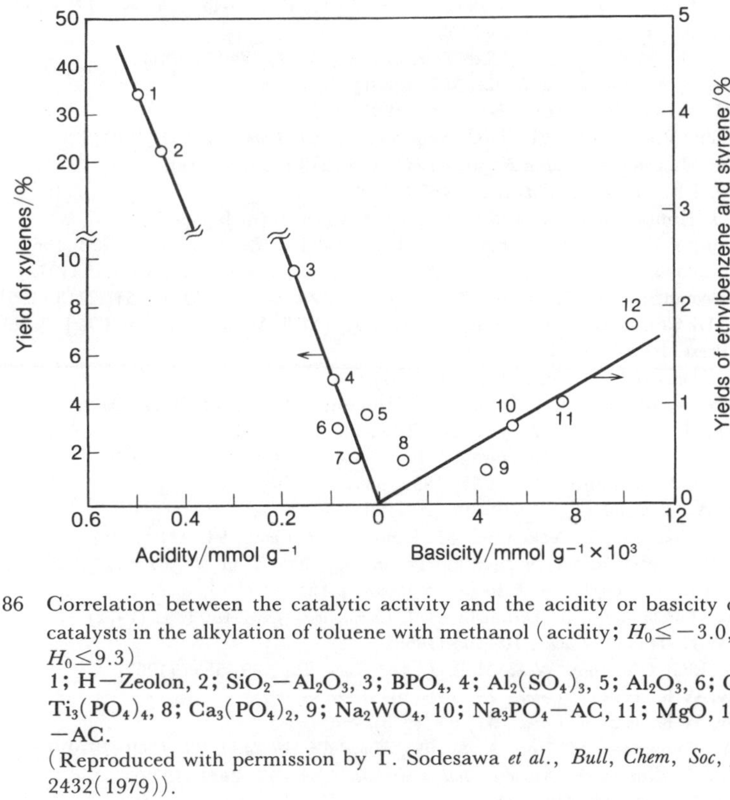

Fig. 3.86 Correlation between the catalytic activity and the acidity or basicity of various catalysts in the alkylation of toluene with methanol (acidity; $H_0 \leq -3.0$, basicity; $H_0 \leq 9.3$)
1; H—Zeolon, 2; SiO_2—Al_2O_3, 3; BPO_4, 4; $Al_2(SO_4)_3$, 5; Al_2O_3, 6; $GrPO_4$, 7; $Ti_3(PO_4)_4$, 8; $Ca_3(PO_4)_2$, 9; Na_2WO_4, 10; Na_3PO_4—AC, 11; MgO, 12; K_3PO_4 —AC.
(Reproduced with permission by T. Sodesawa *et al.*, *Bull, Chem, Soc, Jpn.*, **52**, 2432(1979)).

REFERENCES

1a. H. Itoh, Ph. D. Thesis, Hokkaido University, 1986 (in Japanese).
b. M. Tsuhako, I. Motooka, M. Kobayashi, *Nippon Kagakukaishi*, **92**, 318 (1971) (in Japanese).
2. M. Tsuhako, I. Motooka, M. Kobyashi, *Nippon Kagakukaishi*, **92**, 1131 (1971) (in Japanese).

3. R.F. Vogel, G. Marcelin, *J. Catal.*, **80**, 496 (1983).
4. K. Kearby, Proc. 2nd Intern. Congr. Catal., 1960, Paris, p.2567.
5. J.B. Peri, *Disc Faraday Soc.*, **52**, 55 (1971).
6. H. Itoh, A. Tada, H. Hattori, K. Tanabe, *J. Catal.*, **115**, 244 (1989).
7. J.M. Campelo, J.M. Marinas, S. Mendioroz, J.A. Pajares, *J. Catal.*, **101**, 484 (1986).
8. A. Tada, M. Yoshida, M. Hirai, *Nippon Kagakukaishi,* **1973**, 1739 (in Japanese).
9. J.M. Campelo, A. Garcia, D. Luna, J.M. Marinas, *J. Catal.,* **102**, 447 (1986).
10. H. Itoh, A. Tada, *Nippon Kagakukaishi*, **1976**, 698 (in Japanese).
11. A. Tada, M. Yoshida, *Nippon Kagakukaishi,* **1973**, 856 (in Japanese).
12. T. Tagawa, K. Iwayama, Y. Ishida, T. Hattori, Murakami, *J. Catal.,* **79**, 47 (1983).
13. A. Schmidtmeyer, J.B. Moffat, *J. Catal.*, **96**, 242 (1985).
14. Y. Sakai and H. Hattori, *J. Catal.*, **42**, 37 (1976).
15. A. Tada, *Materials and Chem. Phys.*, **17**, 145 (1987).
16. H. Itoh, A. Tada, H. Hattori, *J. Catal.*, **76**, 235 (1982).
17. J.B Moffat, *J. Mol. Catal.*, **30**, 171 (1985).
18. J.B. Moffat, H.L. Goltz, *Can. J. Chem.*, **43**, 1680 (1965).
19. R. Tartareli, J.F. Neeleman, A. Lucchesi, G. Stoppato, F. Morelli, *J. Catal.*, **17**, 41 (1970).
20. J. B. Moffat, A.S Riggs, *J. Catal.*, **42**, 388 (1976).
21. H. Miyata, J. B. Moffat, *J. Chem. Soc. Trans. Faraday 1,*, **77**, 2493 (1981).
22. H. Miyata, J.B, Moffat, *J. Catal.*, **62**, 357 (1980).
23. A. Tada, H. Suzuka, Y. Imizu, *Chem. Lett.*, **1987**, 423.
24. L. K. Freidlin, V.Z. Sharf, Z.T. Tukhyamuradov, *Kinet. Katal.*, **5**, 347 (1964).
25. A. Tada, K. Mizushima, *Nippon Kagakukaishi,* **1972**, 278 (in Japanese).
26. S.S. Jewuer, J.B. Moffat, *J. Catal.*, **57**, 167 (1979).
27. F. Nozaki, I. Kimura, *Bull. Chem. Sci. Jpn.*, **50**, 614 (1977) (in Japanese).
28. L. K Freidlin, V.Z. Sharf, O.M. Kho'lmer, L.L. Malkina, *Kinet. Katal.*, **2**, 228 (1961).
29. M. G. Mitichenko, H.M. A. Shenari, A.A. Kubasov, *Zh. Prikl. Khim.*, **45**, 455 (1972).
30. E. N. Zil'berman, N. M. Teplyakov, A.E. Kulikova, *Zh. Obschch. Khim.*, **34**, 2713 (1964).
31. G. Alberti, U. Constantino, in *Intercalation Chemistry,* (M.S. Whittingham and A.J. Jacobson., eds.) Academic Press, New York, 1982, p.147.
32. K. Segawa, *Materials Chem. Phys.*, **17**, 181 (1987).
33. Y. Watanabe, Y. Matumura, Y. Izumi, Y. Mizutani, *J. Catal.*, **40**, 76 (1975).
34. A. Clearfield, J. A. Stynes, *J. Inorg. Nucl. Chem.*, **26**, 117 (1964).
35. T. Hattori, A. Ishiguro, Y. Murakami, *J. Inorg. Nucl. Chem.*, **40**, 1107 (1978).
36. A. Clearfield D.S. Thakur, *J. Catal*, **65**, 185 (1980).
37. D. Thakur, A. Clearifield, *J. Catal.*, **69**, 230 (1981).
38. K. Segawa, Y. Kurusu, Y. Nakajuma, M. Kinoshita, *J. Catal.*, **94**, 491 (1985).
39. K. Segawa, Proc. 4th Seminar, Science and Technology, New Type of Catalysis, (Japan-Taiwan Seminar), 1986, Tokyo, Interchange Association Japan, p.121.
40. T. Imanaka, Y. Okamoto, S. Teranishi, *Bull. Chem. Soc., Jpn.,* **45**, 1353 (1972).
41. J.A. Bett, W.K. Hall, *J. Catal.*, **10**, 105 (1968).
42. J.A.S. Bett, L.G. Christner, W.K. Hall, *J. Am. Chem. Soc.*, **89**, 5535 (1967).
43. C.L. Kibby, W.K. Hall, *J. Catal.*, **29**, 144 (1973).
44. C.L. Kibby, W.K. Hall, *J. Catal.*, **31**, 65.(1973).
45. A. Tada, Y. Yamamoto, M. Itoh, A. Suzuki, *Kogyo Kagaku Zashi,* **73**, 1069 (1970) (in Japanese).
46. T. Sodesawa, I. Kimura, F. Nozaki, *Bull. Chem. Soc. Jpn.*, **52**, 2431 (1979).
47. K. Thomke, *J. Catal.*, **44**, 339 (1976).
48. A.Y. Bakaev, T.V. Zamulina, *Kinet. i Katal..*, **16**, 462 (1975).
49. M. Tsuhako, I. Motooka, M. Kobayashi, *Nippon Kagakukaishi*, **92**, 318 (1971) (in Japanese).
50. J.B. Moffat, A.S. Riggs, *J. Catal.*, **28**, 157 (1973).
51. H. Griselbach, J.B. Moffat, *J. Catal.*, **80**, 350 (1983).
52. T.N. Frianeza, A.Clearfield, *J. Catal.*, **85**, 398 (1980).
53. A. Clearfield, D.S. Thakur, *J. Catal.*, **65**, 185 (1980).
54. H. Monma, *J. Catal.*, **75**, 200 (1982).

55. A. Tada, H. Itoh, Y. Kawasaki, J. Nara. *Chem. Lett.*, **1975**, 517.
56. J. B. Moffat, S.S. Jewur, *J. Chem. Soc., Faraday Trans. 1*, **76**, 746 (1980).
57. S. Kikkawa, Y. Shimizu, S. Higuchi, *Chem. Lett.*, **1979**, 849.
58. T. Sodesawa, H. Nakajima, H. Nozaki, *Chem. Lett.*, **1979**, 607.
59. B. Gallance, J.B. Moffat, *J. Catal.*, **76**, 182 (1982).
60. H. E. Sheffer, R. L. Perry, R.J. Thimineur, B.T. Adams, J.L. Simoniau, N.L. Zutty, R.A. Clendenning, *Ind. Eng. Chem. Prod. Res. Dev.*, **10**, 362 (1971).
61. J. Morey, J.M. Marianas, J.V. Sinisterra, *React. Kinet. Catal. Lett.*., **22**, 175 (1983); **52**, 2431 (1979).
62. R. Tartarelli, M. Giogini, A. Jucchesi, G. Stoppato, Morelli, *J. Catal.*, **17**, 41 (1970).
63. M. Giorgini, P.F. Marconi, G. Monzani, R. Simula, R. Tartareni, *J. Catal.*, **24**, 521 (1972).
64. W.T Reichle, *J. Catal.*, **17**, 297 (1970).
65. N.S. Figoli, H.R. Keselman, P.C. L'Argentière, C.L. Lazzaroni, *J. Catal.*, **77**, 64 (1982).
66. N.S. Figoli, C.L. Lazzaroni, H.R. Keselman, and P.C. L'Argentière, *J. Catal.*, **85**, 583 (1984).
67. J.M. Cowley, J.C. Wheatley, W.L. Kehl., *J. Catal.*, **56**, 185 (1979).
68. F.M. Scheidt, *J. Catal.*, **3**, 372 (1964).
69. T.G. Arefeva, Y.A. Gorin, *Kinet. i Katal.*, **11**, 176 (1970).
70. A. Tada, *Bull. Chem. Soc. Jpn.*, **48**, 1391 (1975).
71a) F. Nozaki, H. Ohta, *Bull. Chem. Soc. Jan.*, **47**, 1307 (1974).
b) F. Nozaki, Y. Iimori, *Bull. Chem. Soc. Jpn.*, **49**, 567 (1976).
72. T. Yamanaka, K. Tanabe, *Shokubai (Catalyst)*, **17**, 102 (1975) (in Japanese).
73. G. Alberti, V. Constantino, in *Intercalation Chemistry,* (M.S. Whittingham and A.J. Tacobson, eds.) p.147, Academic Press, New York, 1982.

3.9 SUPERACIDS

3.9.1 $TiO_2-SO_4^{2-}$, $ZrO_2-SO_4^{2-}$, $Fe_2O_3-SO_4^{2-}$

These solid superacids do not contain halogen atoms which are environmentally undesirable, and they are stable even at 773 – 873 K. The discovery of the superacids originated from the study of the effect of anions on the preparation of TiO_2.[1)]

A. Preparation Method

The solid superacids are easily prepared by mounting $(NH_4)_2SO_4$ or H_2SO_4 on $TiO_2 \cdot nH_2O$, $Zr(OH)_4$, and $Fe(OH)_3$, respectively, followed by calcining at 773 – 873 K. As an example, the preparation method of $ZrO_2-SO_4^{2-}$ is described in more detail. A commercially available $ZrOCl_2$, $Zr(NO_3)_4$ or $ZrO(NO_3)_2$ is hydrolyzed with 28% aqueous ammonia and the precipitates formed are washed with distilled water and dried at 373 K overnight to obtain $Zr(OH)_4$. The hydroxide, $Zr(OH)_4$, is immersed into an aqueous solution of $(NH_4)_2SO_4$ or H_2SO_4 and the suspended solution is evaporated to dryness, followed by calcination in air or in vacuum at 773 – 923 K to obtain $ZrO_2-SO_4^{2-}$. The optimum temperature of calcination changes depending on the kind of SO_4^{2-} source ($(NH_4)_2SO_4$ or H_2SO_4). In the acylation of chlorobenzene with chlorobenzoyl chloride, the optimum temperatures are 823 and 873 K for $ZrO_2-H_2SO_4$ and $ZrO_2-(NH_4)_2SO_4$, respectively, as shown in Fig. 3.87.[2)] The optimum content of SO_4^{2-} changes from 1 to 8 wt% depending on the kind of reaction. As a more convenient and simpler method, $ZrO_2-SO_4^{2-}$ is prepared by pouring 30 ml of 0.5 – 1N H_2SO_4 solution onto 2g of $Zr(OH)_4$ placed on a folded filter paper and by allowing to stand in air, followed by calcining in air for 3 h.[3)] An important point

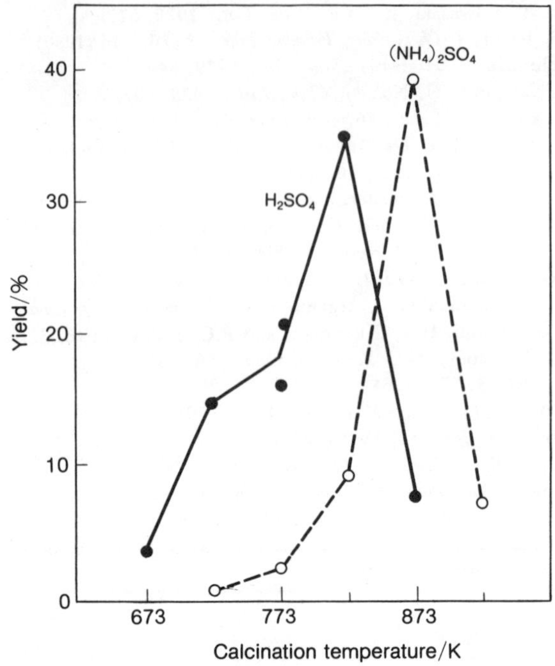

Fig. 3.87 Activities of two kinds of $ZrO_2-SO_4^{2-}$ catalysts prepared by impregnating with H_2SO_4 and $(NH_4)_2SO_4$ followed by calcining at different temperatures. Yield %, in 3 h.

is to use $Zr(OH)_4$ instand of ZrO_2 as an adsorbent of SO_4^{2-}. Since the solid superacid loses super acidity by absorbing moisture when exposed to air, calcination in a pyrex glass tube and storage in sealed tubes are recommended.

As in the case of $(NH_4)_2SO_4$ or H_2SO_4, the addition of SO_3 to ZrO_2, TiO_2 or Fe_2O_3 causes the generation of superacidity.[4)] However, the addition of SO_2 or H_2S (adsorption at 673 K) does not generate superacidity. It is interesting that $Fe_2O_3-SO_2$ or $Fe_2O_3-H_2S$ does not show any acidity, but exhibits strong acidity when oxidized with O_2 at 773 K.[5)] On the other hand, $Fe_2O_3-SO_4^{2-}$ loses its acidity when reduced with H_2 at 773 K. These facts indicate that oxidation and reduction influence the acidity of a sulfur-containing superacid. It should be noted that the solid superacids cannot be used in the presence of reducing reagents such as hydrogen, alcohols, etc. at high temperatures (above 673 K).

B. Morphology and Surface Properties

The exothermic peak of ZrO_2 in a DTA profile was shifted by SO_4^{2-} treatment from 693 K to 813–885 K. Crystallographic phase transformation in various ZrO_2 and $ZrO_2-SO_4^{2-}$ is shown in Table 3.41.[6)] It is evident that a metastable tetragonal phase is stabilized when a small amount of SO_4^{2-} is involved. The development of a monoclinic form in ZrO_2 and the conversion of tetragonal to monoclinic in the SO_4^{2-}-promoted ZrO_2 were almost independent of the preparation methods.

TABLE 3.41 Crystallographic phase transformation in various ZrO_2

Sample	Calcination temperature/K			
	623	773	923	1073
ZrO_2-A	M_-	M	M	M
ZrO_2-A-H_2SO_4 (1.0)	amor.	T_-	T	T, M_-
ZrO_2-A-$(NH_4)_2SO_4$	amor.	T_-	T	T_-, M_-
ZrO_2-U	M_-	M	M	M
ZrO_2-U-H_2SO_4 (1.0)	amor.	T_-	T_-	T, M_-
ZrO_2-U-$(NH_4)_2SO_4$	amor.	T_-	T	T_-, M_-

amor. ; amorphous
M ; monoclinic M_- ; monoclinic (not well developed)
T ; tetragonal T_- ; tetragonal (not well developed)
ZrO_2-A ; prepared by hydrolysis of $ZrO(NO_3)_2 \cdot 2H_2O$ with aqueous ammonia.
ZrO_2-U ; prepared by hydrolysis of $ZrO(NO_3)_2 \cdot 2H_2O$ with urea.
(Reproduced with permissein from *Materials Chem. Phys.*, **16**, 71 (1986)).

The surface areas of $ZrO_2-SO_4^{2-}$ are larger than those of ZrO_2 as shown in Table 3.42.[6] ZrO_2 treated with $(NH_4)_2SO_4$ showed larger surface area than ZrO_2 treated with H_2SO_4.

The acid strengths of $TiO_2-SO_4^{2-}$ and $ZrO_2-SO_4^{2-}$ measured by the indicator method are $H_0 \leq -14.57$[7] and $H_0 \leq -16.04$,[8] respectively. Thus, the acid strength of $ZrO_2-SO_4^{2-}$ is 10,000 times higher than that of 100% H_2SO_4. The acid strength of $Fe_2O_3-SO_4^{2-}$ cannot be measured by color change of indicators because of its dark color, but it is regarded as a superacid, judging from its catalytic activity for skeletal isomerization of *n*-butane at room temperature which is not catalyzed by 100% H_2SO_4.[9] The type of acid sites on the solid superacids evacuated at 773 K is only a Lewis type.[4,10]

According to ESR study, adsorption of perylene gave the typical signal of cation radicals of perylene on $ZrO_2-SO_4^{2-}$, while no signal was found on an unpromoted ZrO_2.[6] This indicates that an oxidizing property appears on the introduction of sulfate ions.

TABLE 3.42 Surface area of the samples prepared

Sample	Surface area ($m^2\,g^{-1}$)
ZrO_2-A	47.1
ZrO_2-U	41.7
ZrO_2-A-$(NH_4)_2SO_4$	119.5
ZrO_2-U-$(NH_4)_2SO_4$	119.2
ZrO_2-A-H_2SO_4	85.2
ZrO_2-U-H_2SO_4	65.1

See Table 3.41 for ZrO_2-A and ZrO_2-U.
(Reaproduced with permission from *Materials Chem., phys.*, **16**, 73(1961)).

C. Structure of Acid Sites[4)]

The structure of acid sites on a solid superacid was studied by infrared and XPS spectroscopy. The infrared spectrum with four absorption bands at 1240, 1140, 1020, and 924 cm^{-1} of a hydrated $Fe_2O_3-SO_4^{2-}$ (Fig. 3.88a) is quite close to that of the inorganic chelating bidentate complexes. This strongly suggests that the analogous sulfur species is formed on the surface of the oxide. The most probable structure of this species is proposed as structure I in Fig. 3.89.[4)] After removal of the molecular water adsorbed, the spectrum changed drastically and the resultant spectrum (1375, 1180, 1025, and 968 cm^{-1} in Fig. 3.88b) became close to that of organic sulfates.[4)] The probable structure of the surface sulfur species at this stage is proposed as structure II. This spectral change indicates that the removal of molecular water has resulted in the structural transformation from a chelating inorganic complex to an organic sulfate. An inferred transformation by desorption of molecular water adsorbed is expressed in Fig. 3.89[4)] An important difference between structures I and II is that structure I has only ionic SO bonds with partial double bond character, while structure II has covalent double bonds.

Since spectrum c in Fig. 3.88 was commonly observed in SO_3/Fe_2O_3, SO_2/Fe_2O_3 oxidized at 723 K, and H_2S/Fe_2O_3 oxidized at 723 K, and since those

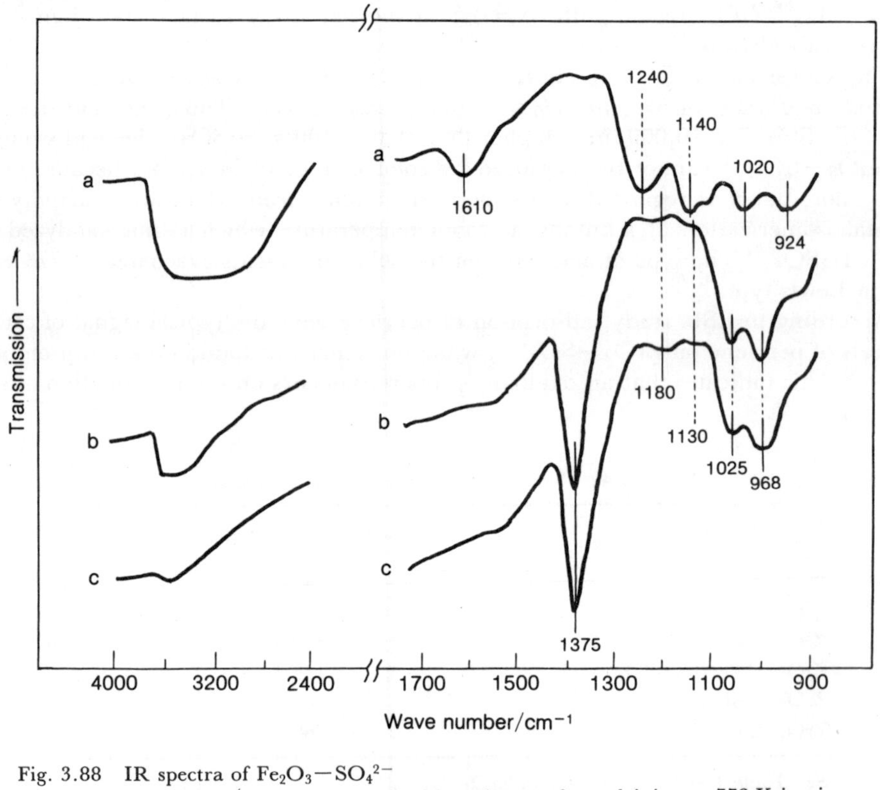

Fig. 3.88 IR spectra of $Fe_2O_3-SO_4^{2-}$
a; original (a sample wetted with moisture after calcining at 773 K in air, b; evacuated at 573 K, c; evacuated at 673 K.

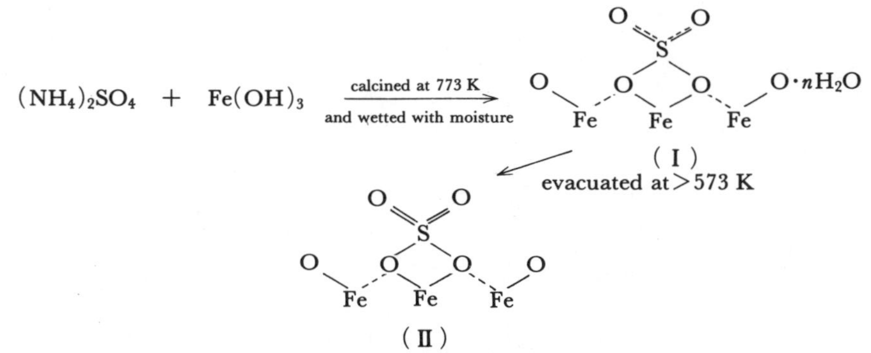

Fig. 3.89 Process of formation of a superacid complex, $Fe_2O_3-SO_4^{2-}$.

samples were active in the isomerization of cyclopropane, it is reasonable to conclude that structure II is essential for the acid-catalyzed reactions as a common active site on the samples described above. Infrared spectroscopic observations of pyridine adsorbed on those catalysts revealed that the catalysts possess solely Lewis acidity; no Brønsted acidity was found. Thus the central Fe ion acts as a Lewis acid site, whose acid strengh can be strongly enhanced by the inductive effect of S=O in the sulfur complex, as shown in Fig. 3.90. The appearance of an intense band at 1375 cm^{-1} which was assigned to the assymmetric stretching vibration of S=O bonds having a high double bond nature is necessary for causing the inductive effect to generate superacidity. Since the S=O stretching vibration of SO_4^{2-} in metal sulfates usually appears around 1100 − 1235 cm^{-1}, the structure of solid superacids which show much higher frequency is different from the structure of metal sulfates.

According to XPS measurements, the S 2p signals of $Fe(OH)_3-SO_4^{2-}$ and FeO_3-H_2S change on oxidation and reduction as shown in Fig. 3.91.[11] When $Fe(OH)_2-SO_4^{2-}$ showing only the signal of S^{6+} was reduced *in situ* by a few torr of H_2 at 723 K, the intensity of the S 2p signal of S^{2-} increased and that of S^{6+} decreased. On the other hand, H_2S treatment of $Fe(OH)_3$ calcined at 773K gave only one signal which indicates the presence of the S^{2-} state. *In situ* oxidation of both samples by a few torr of O_2 at 573 K resulted in the complete oxidation from S^{2-} to S^{6+} state. Since a complex of S^{6+} state shows high activity for acid-catalyzed reactions while a complex of S^{2-} state is inactive, the S^{6+} state is consi-

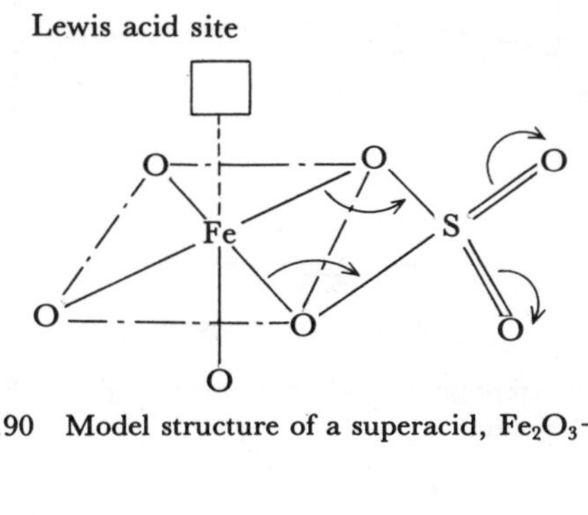

Fig. 3.90 Model structure of a superacid, $Fe_2O_3-SO_4^{2-}$.

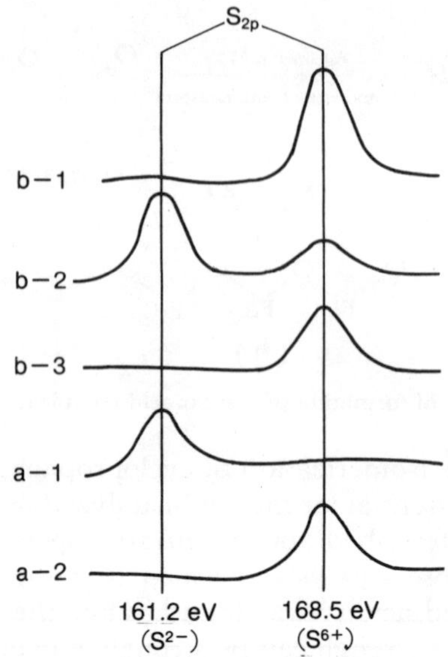

Fig. 3.91 Changes in S_{2p} signals by reduction and oxidation for the following samples : (a-1) $Fe(OH)_3$ treated by a few torr of H_2S at 773 K; (a-2) sample a-1 oxidized *in situ* by a few torr of O_2 at 773 K; (b-1) AS/$Fe(OH)_3$ calcined at 773 K, followed by evacuation *in situ* at 773 K; (b-2) sample b-1 reduced *in situ* by a few torr of H_2 at 723 K; (b-3) sample b-1 oxidized *in situ* by a few torr of O_2 at 573 K.

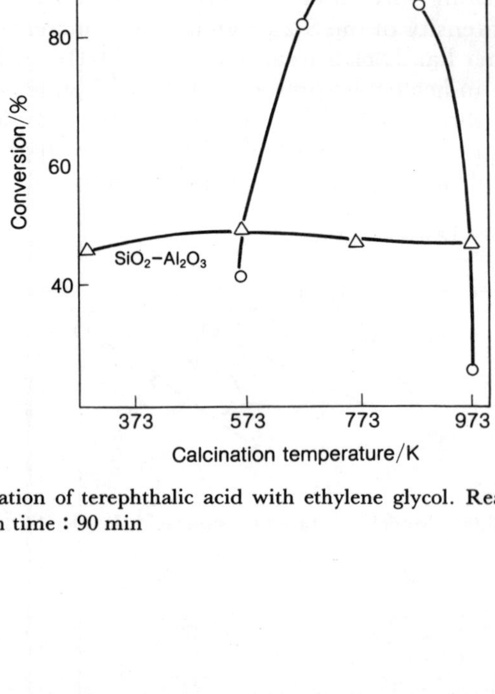

Fig. 3.92 Esterification of terephthalic acid with ethylene glycol. Reaction temp., 473 K, Reaction time : 90 min

dered to be necessary for the generation of superacidity.[4,11]

The solid superacids which were obtained by the introduction of a sulfur compound to ZrO_2, TiO_2, and Fe_2O_3 has a strong tendency to reduce the bond order of SO from a highly covalent double-bond character to a lesser double-bond character when a basic molecule is adsorbed on its central metal cation as evidenced by the shift of 1375 cm^{-1} band to lower frequency.[12] The change of electronic structure caused by pyridine adsorption is understood by Fig. 3.90[12] where the coordination number of a surface metal cation of a metal oxide was taken as 5. The strong ability of a sulfur complex to absorb electrons from a basic molecule is a driving force to generate superacidity.

D. Catalytic Activity

Skeletal isomerization of paraffins such as butane, pentane, etc. is not catalyzed even by 100% H_2SO_4. It was found, however, that $ZrO_2-SO_4^{2-}$, $TiO_2-SO_4^{2-}$, and $Fe_2O_3-SO_4^{2-}$ catalyzed the skeletal isomerization of butane at 293 − 323 K, the main products being isobutane.[7−9] The activity of the solid superacids is lowered as the reaction proceeds probably due to coke formation. To prevent the catalyst from its deactivation, a catalyst on which a small amount of Pt, Ni etc. was added was developed. Over a $Pt-ZrO_2-SO_4^{2-}$ catalyst, no deactivation was observed for more than 100 h for the skeletal isomerization of pentane at 413 K under 20 kg/cm^2 of hydrogen pressure.[13]

A $ZrO_2-SO_4^{2-}$ catalyst is also active for the acylation of aromatics which has been known to be catalyzed only by $AlCl_3$ and can be used as a promising heterogeneous catalyst instead of a homogeneous catalyst (cf. Section 4.3). The solid superacids were found to exhibit extremely high activities for the reactions such as dehydration of alcohol,[14,15] double-bond isomerization of 1-butene,[14,15] isomerization of cyclopropane to propylene,[14,15] esterification of terephthalic acid,[16] and polymerization of ethers.[17] As an example, $TiO_2-SO_4^{2-}$ calcined at 773 K is much more active than $SiO_2-Al_2O_3$ for the esterification of terephthalic acid with ethylene glycol at 473 K, as shown in Fig. 3.92[16]

For further information on the topic, the volume entitled *Superacids* by Olah *et al.*[18] is recommended.

REFERENCES

1. K. Tanabe, M. Itoh, K. Morishige, H. Hattori, in: *Preparation of Catalysts*, (B. Delmon, P.A. Jacobs, G. Poncelet, eds), Elsevier, Amsterdam, 1976, p.65.
2. K. Tanabe, T. Yamaguchi, K. Akiyama, A. Mitoh, K. Iwabuchi, K. Isogai, Proc. 8th Intern. Congr. Catal., Berlin, 1984, Verlag Chemie, Weinheim, Vol.5, p.601.
3. K. Arata, M. Hino, *Hyomen,* **19**, 75 (1981).
4. T. Yamaguchi, T. Jin, K. Tanabe, *J. Phys. Chem.*, **90**, 3148 (1986).
5. Y. Nagase, T. Jin, H. Hattori, T. Yamaguchi, K. Tanabe, *Bull. Chem. Soc. Jpn.*, **58**, 916 (1985).
6. T. Yamaguchi, K. Tanabe, Y.C. Kung, *Materials Chem. Phys.*, **16**, 67 (1986).
7. M. Hino, K. Arata, *J. Chem. Soc., Chem. Commun.*, **1979**. 1148.
8. M. Hino, K. Arata, *J. Chem. Soc., Chem. Commun.*, **1980**, 851.
9. K. Arata, M. Hino, *Shokubai*, **21**, 217 (1979).

10. T. Jin, Thesis for the degree D. Sc., Hokkaido Univ., 1985 (in Japanese).
11. T. Jin, M. Machida, T. Yamaguchi, K. Tanabe, *Inorg. Chem.*, **23**, 4396 (1984).
12. T. Jin, T. Yamaguchi, K. Tanabe, *J. Phys. Chem.*, **90**, 4794 (1986).
13. S. Baba, T. Shimizu, H. Takaoka, T. Imai, S. Yokoyama, Disc. Meeting, Petrol. Chem., Preprint No. 2-I-17, 1986.
14. K. Tanabe, A. Kayo, T. Yamaguchi, *J. Chem. Soc., Chem. Commun.*, **1981**, 602.
15. A. Kayo, T. Yamaguchi, K. Tanabe, *J. Catal.*, **83**, 99 (1983).
16. K. Tanabe, H. Hattori, Y. Ban'i, A. Mitsutani, Japanese Patent, 55−115570 (1980).
17. M. Hino, K. Arata, *Chem. Lett.*, **1980**, 963.
18. G.A. Olah, G.K. Surya Prakash, J. Sommer, *Superacids*, John Wiley & Sons, New York, 1985.

3.9.2 Complex Metal Halides and Mounted Superacids

Metal halides complexed with certain compounds exhibit superacidic character as evidenced by conversion of saturated hydrocarbons at room temperature or below. Among these materials are $AlCl_3$ complexed with other halides or sulfates, and SbF_5 mounted on metal oxides.

A. Aluminum Chloride-based Solid Superacids

Aluminum chloride, when combined with proper cocatalyst, shows high catalytic activity at low temperatures for acid catalyzed reactions such as alkanes cracking and skeletal isomerization. Hydrogen chloride is the most common cocatalyst.

Some attempts have been made to prepare solid superacids by combining $AlCl_3$ with solid cocatalysts. A complex of $AlCl_3$ with crosslinked polystyrene sulfonic acid was prepared and its catalystic activity examined in alkanes reactions.[1] The catalyst showed activity to catalyze cracking and isomerization of hexane at 353 K.

A solid superacid is prepared from $AlCl_3$ and $CuCl_2$. A mixture of $AlCl_3$ and $CuCl_2$ is kneaded under nitrogen atmosphere. The resulting material catalyzes pen-

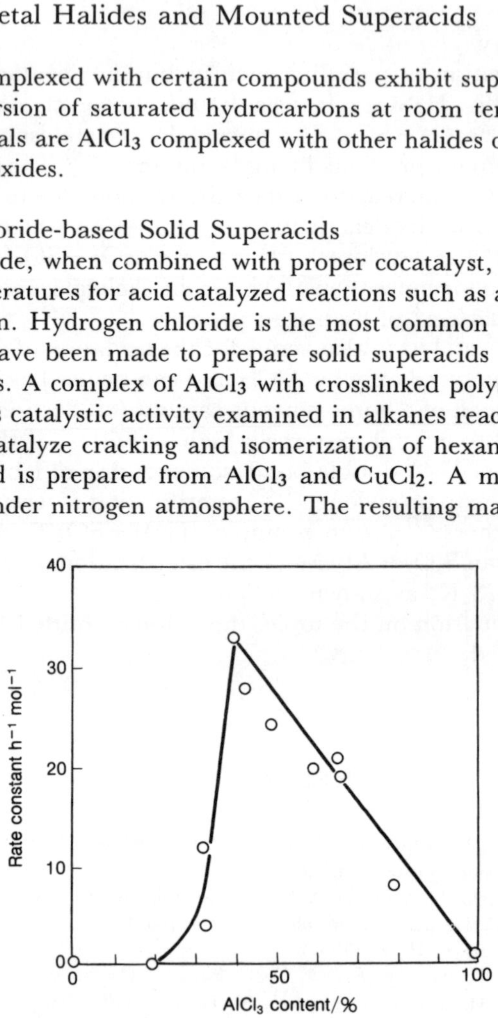

Fig. 3.93 Dependence of the rate constant on the composition of $AlCl_3-CuCl_2$ mixture; 317.7 K.
(Reproduced with permission from *J. Catal.*, **56**, 49 (1979)).

tane isomerization at room temperatures.[2)] The catalytic activities vary with the composition as shown in Fig. 3.93. The maximum activity is observed at the composition of $AlCl_3$ 40% and $CuCl_2$ 60%.

By combining $AlCl_3$ with different metal chlorides, solid superacids are also prepared. The catalytic activities of the superacids for pentane conversion are summarized in Table 3.43. The most active catalyst is obtained by a combination of $AlCl_3$ and $TiCl_3$ 1/3 $AlCl_3$.

TABLE 3.43 Pentane conversion with $AlCl_3$-metal chloride mixtures†

Cocatalyst	Total conversion	Liquid-phase composition				
		n-C_5	*iso*-C_5	C_6	*iso*-C_4	*n*-C_4
$MnCl_2$	11.8	88.2	10.7	0.6	0.5	Trace
$CuCl_2$	11.4	88.6	8.9	1.1	1.4	Trace
$CoCl_2$	8.8	91.2	7.7	0.6	0.5	Trace
$NiCl_2$	6.7	93.3	5.6	0.7	0.4	Trace
$TiCl_3$(AA)	31.3	68.7	29.7	0.8	0.8	0
$TiCl_3$(HA)	0.5	99.1	0.5	0.3	0.1	0
VCl_3	6.9	93.1	5.6	0.5	0.8	Trace
$BiCl_3$	5.8	94.2	5.0	0.4	0.4	Trace
$FeCl_3$	3.2	96.8	2.6	0.4	0.2	0
$ZrCl_4$	0.8	99.2	0.2	0.6	Trace	0
$HfCl_5$	0.2	99.8	0.1	0.1	Trace	0
$TaCl_5$	2.4	97.6	2.1	0.2	0.1	0
$MoCl_5$	0.6	99.4	0.4	0.2	Trace	0
—	2.4	97.6	1.7	0.6	0.1	0

† Reaction time, 3 h; temperature, 301 k; pentanec, 10 ml; catalyst, $AlCl_3$ (7.5 mmol)+metal chloride (7.5 mmol) except $HfCl_4$ (5.4 mmol) and $TaCl_5$ (6.1 mmol).

B. Antimony Pentafluoride Mounted on Metal Oxides and Intercalated Graphite

Certain metal oxides treated with SbF_5 exhibit superacidic character. The SbF_5-treated metal oxides can catalyze skeletal isomerization of saturated hydrocarbons at room temperatures.[3)] The catalysts were prepared by repeated exposure of the heat-treated metal oxides to SbF_5 vapor followed by outgassing to remove excess SbF_5.

Time dependence of composition in the reaction of butane over SbF_5-treated $SiO_2-Al_2O_3$ at 291 K is shown in Fig. 3.94.[3)] Besides $SiO_2-Al_2O_3$, TiO_2, SiO_2, SnO_2, MgO, TiO_2-ZrO_2, and 13X molecular sieves became highly active catalysts on treatment with SbF_5.[4)]

In addition to butane isomerization, SbF_5-treated metal oxides can catalyze the conversions of propane, 2-methylpropane (isobutane), pentane, 2-methylbutane (isopentane), hexane, cyclohexane, and methylcyclopentane at room temperatures.[3)] Methane, ethane and 2,2-dimethylpropane (neopentane), however, do not undergo any reactions over SbF_5-treated metal oxides.

Acid strengths of the SbF_5-treated metal oxides in H_0 scale are summarized in Table 3.44.[3)] The strongest acid sites of $SbF_5-SiO_2-Al_2O_3$ are in the H_0 range -13.75 to -14.52, while those of $SbF_5-TiO_2-SiO_2$ and of $SbF_5-Al_2O_3$ are in the

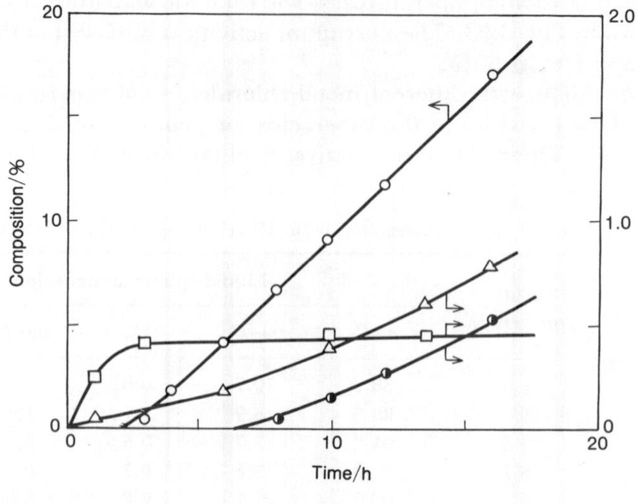

Fig. 3.94 Time dependence of composition in the reaction of butane at 291 K. Catalyst; 0.51 g, initial pressure; 96 Torr, ○; 2-methylpropane, □; 2-methane, △; propane, ◑; 2-methylbutane.

TABLE 3.44 Acid strength of SbF_5-treated catalyst†

Catalyst	H_0			
	−12.70	−13.16	−13.75	−14.52
SiO_2-Al_2O_3 (I)	+	+	−	−
SbF_5-SiO_2-Al_2O_3 (I)	+	+	+	−
TiO_2-SiO_2	−	−	−	−
SbF_5-TiO_2-SiO_2	+	+	−	−
SbF_5-Al_2O_3 (II)	+	+	−	−

† +, present; −, absent.

range − 13.6 to − 13.75. Compared with $SiO_2-Al_2O_3$, Al_2O_3, Al_2O_3, and the other metal oxides, treatment with SbF_5 greatly enhances acid strength.

The types of acid sites are dependent on the SbF_5 treatment conditions as measured by IR spectroscopy of adsorbed pyridine. Both Brønsted and Lewis acid sites are present when the treatment with SbF_5 is carried out below 373 K, while only Lewis acid sites are detected for the catalyst treated at 573 K. The structures of acid sites of SbF_5-treated $SiO_2-Al_2O_3$ are suggested as shown in Fig. 3.95. At low temperature treatment, both surface OH groups and Al cations exist. Their acid strengths are enhanced by adsorption of SbF_5. At temperatures higher than 573 K, SbF_5 reacts with the OH groups to give $-OSbF_4$ and F.

Besides treatment with SbF_5, treatment with NH_4F, FSO_3H, $SbCl_5$, and FSO_3H-SbF_5 enhanced the activities of metal oxides for acid-catalyzed reactions. The FSO_3H-SbF_5-treated catalyst catalyzed the reaction of butane, although the ac-

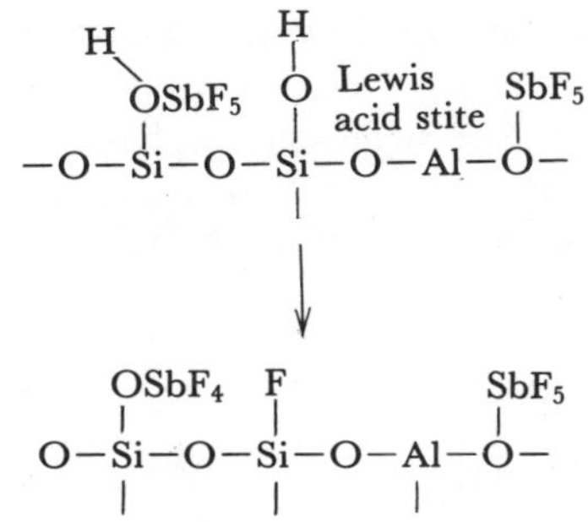

Fig. 3.95

tivity is not as high as those of SbF_5-treated catalysts. The $SbCl_5$-treated catalyst catalyzes the reaction of 2-methylpropane and 2-methylbutane, but it cannot catalyze the reactions of butane and pentane. Neither FSO_3H-treated catalyst nor NH_4F-treated catalyst catalyzes the reactions of alkane at room temperature, but their activities for 1-butane isomerization are much higher than those before treatment. Therefore, it is evident that SbF_5 is the most effective among the reagents with which metal oxides are treated. The catalytic activities of these catalysts for reactions of different types of hydrocarbons are qualitatively classified in Table 3.45. The catalytic activities of liquid superacids and conc. H_2SO_4 are included for comparison.

The reaction mechanisms of hydrocarbon conversions over $SbF_5-SiO_2-Al_2O_3$ were studied by coisomerization of perdeuterio and nondeuterio compounds; coisomerization of pentane-d_0/d_{12}, 2-methylbutane-d_0/d_{12}, 2-methylpropane-d_0/d_{10}, and cyclohexane-d_0/d_{12} being carried out.[5)] For all skeletal isomerizations, an intramolecular H (or D) transfer is involved in the rearrangements of the carbon skeletons, though the methyne H atom of 2-methylbutane and all H atoms of 2-methylpropane rapidly exchange among the molecules before yielding isomerized products. It is suggested that all the reactions proceed by the carbenium ion mechan-

TABLE 3.45 Classification of catalytic activities of various catalysts

Catalyst	Reaction of alkane with			Reaction alkene
	Primary C-H	Secondary C-H	Tertiary C-H	
FSO_3H-SbF_5 $HF-SbF_5$ (liquid superacid)	+	+	+	+
SbF_5-metal oxide FSO_3H-SbF_5-metal oxide	−	+	+	+
$SbCl_5$-metal oxide conc. H_2SO_4	−	−	+	+
FSO_3H-metal oxide NH_4F-metal oxide	−	−	−	+

† +, active; −, inactive at room temperature.

ism in which the reactions are initiated by abstraction of an H^- from the reactants. No indications were observed for the formation of carbonium ions as observed for the reactions in liquid superacids. The suggested mechanisms for cyclohexane isomerization and 2-methylbutane isomerization are shown as Scheme III in Fig. 3.96. Scheme I and II, in which the reaction is initiated by the addition of H^+ to the reactant, are not plausible.

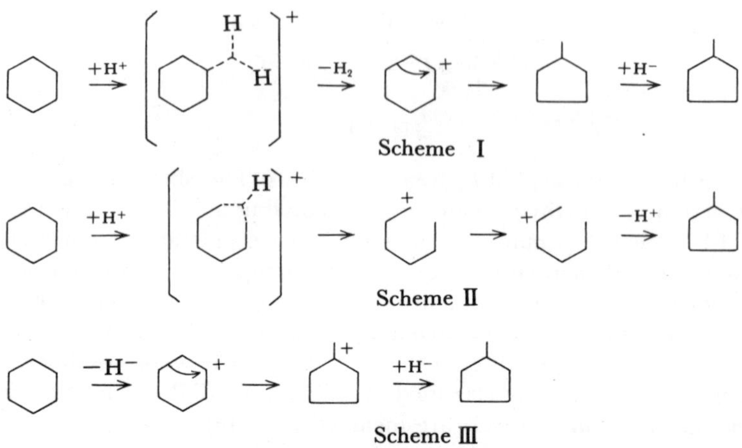

Fig. 3.96 Possible scheme for cyclohexane isomerization

The catalytic behavior of the $SbF_5-SiO_2-Al_2O_3$ catalyst for the reactions of alkanes has much in common with those of metal halides. However, the $SbF_5-SiO_2-Al_2O_3$ catalyst differs from metal halide on a few points. One difference is that no promoters are required for the $SbF_5-SiO_2-Al_2O_3$ catalyst, whereas certain promoters necessary for metal halides. In the latter case, the presence of promoters makes it possible for metal halides to abstract an H^- from alkanes by the chain transfer reaction to form carbenium ions. On the other hand, the surface Lewis acid sites of $SbF_5-SiO_2-Al_2O_3$ catalysts can directly abstract an H^- from alkanes in an initial step. Another difference is observed in the relative reactivities of hydrocarbons. With metal halides, the hydrocarbons having a methyne H react faster than the hydrocarbons having methylene H. Over the $SbF_5-SiO_2-Al_2O_3$ catalyst, the general tendency does not hold as shown by a faster isomerization of cyclohexane than 2-methylbutane. Probably because abstraction of an H^- from alkanes may not be a slow step for the $SbF_5-SiO_2-Al_2O_3$ catalyst.

Antimony pentafluoride intercalated graphite shows very high activity for isomerization and cracking of methylpentanes.[6,7] The catalyst has the molar formula $C_{6.5}SbF_5$. Under hydrogen pressure, methylpentane undergoes isomerization at 243 K, and cracking at room temperature. The use of ^{13}C-labeled reactants (hexane isomer) shows that the isomerization, which involves only intramolecular rearrangements of the hexyl cations, is described by 1,2 alkyl shifts of methyl and ethyl groups and rearrangements via protonated cyclopropane rings. Although the acid strength is not measured, the catalytic behavior demonstrates the catalyst to be superacid.

REFERENCES

1. V.L. Magnotta, B.C. Gates, *J. Catal.*, **46**, 266 (1977).
2. Y. Ono, T. Tanabe, N. Kitajima, *J. Catal.*, **56**, 47 (1979).
3. H. Hattori, O. Takashashi, M. Takagi, K. Tanabe, *J. Catal.*, **68**, 132 (1981).
4. K. Tanabe, H. Hattori, *Chem. Lett.*, **1976**, 625.
5. O. Takahashi, H. Hattori, *J. Catal* **68**, 144 (1981).
6. F. Le Normand. F. Fajula, F. Gault, J. Sommer, *Nouv. J. Chim.*, **6**, 411 (1982).
7. F. Le Normand, F. Fajula, F. Gault, J. Sommer, *Nouv. J. Chim.*, **6**, 417 (1982).

3.10 SUPERBASES

Addition of alkali metals to certain types of oxides resulted in the formation of very strong base sites. Materials which possess base sites stronger than $H_- = -26$ are called superbases. The H_- value 26 proposed to be set in conformity with the definition of superacid.[1] The critical H_0 value for superacid is *ca.* -12, which differs by 19 H_0 units from $H_0=7$, the neutral acid-base strength. The H_- value 26 differs from $H_- = 7$ for neutral acid-base strength by 19 H_- units. Although alkali and alkaline earth oxides show superbasicity without the addition of alkali metals, as described in Sections 3.1.1 and 3.1.2, this section deals only with alkali metal-added materials showing strong basicity. The materials which show superbasicity by the addition of alkali metals are limited to alkaline earth oxides and alumina.

A. Preparation of Solid Superbases

Alkali metals are added by exposing the surfaces of alkaline earth oxides pretreated at high temperatures to the vapor of alkali metals produced by heating the metals or by decomposition of alkali azides.[2] More complex procedures were taken for superbasic alumina as follows.[3] To the calcined γ-alumina, NaOH was added at $583-593$ K with stirring. Water generated in this process was removed by flowing nitrogen. The sample was stirred continuously for three hours and Na metal was added at the same temperature. The sample was stirred one more hour to become pale blue.

B. Basicity

Malinowski measured the basicity of alkali-added MgO in H_- scale and given in Table 3.46.[4,5] Among alkali metals, K is the most effective in creating strong basic sites on MgO. Base sites stronger than $H_- = 35$ increased if two kinds of alkali metals were added on MgO surface.[6]

The appearance of base sites stronger than $H_- = 37$ was observed for the above mentioned $Na-NaOH-Al_2O_3$.[3]

TABLE 3.46 Amounts of alkali metals deposited on magnesia surface and concentrations of superbasic sites on catalyst surfaces[3)]

Catalyst system	Ionization energy of evaporated metal (eV)	Amount of deposited metal (mmol g^{-1})	Concentration of superbasic centers (mmol g^{-1})	
			$27<H_-<33$	$H_->35$
MgO† - Na	5.17	0.42	0	0.154
MgO - K	4.37	0.63	0.013	0.517
MgO - Cs	3.98	1.2	0.446	0.189

† MgO pretreatment temperature 823 K

C. Catalytic Activity

Malinowski *et al.* measured the catalytic activities of alkali metals doped MgO for dehydrogenation of isopropylbenzene and hydrogenation of alkenes.[5,7,8)] For isopropylbenzene dehydrogenation, the catalyst prepared by the addition of both Na and K to MgO shows the highest yield for α-methylstylene. They reported that the reaction takes place on very strong one-electron donor centers and has a free radical character.

Methylformate decomposition to CO and H_2 is effectively catalyzed by solid superbase.[2)] The Na – MgO catalyst prepared by heating a mixture of MgO and sodium azide shows much higher activity than MgO alone. Although the basic strength was not measured, the high activity of Na – MgO is considered to be due to superbasicity of the catalyst.

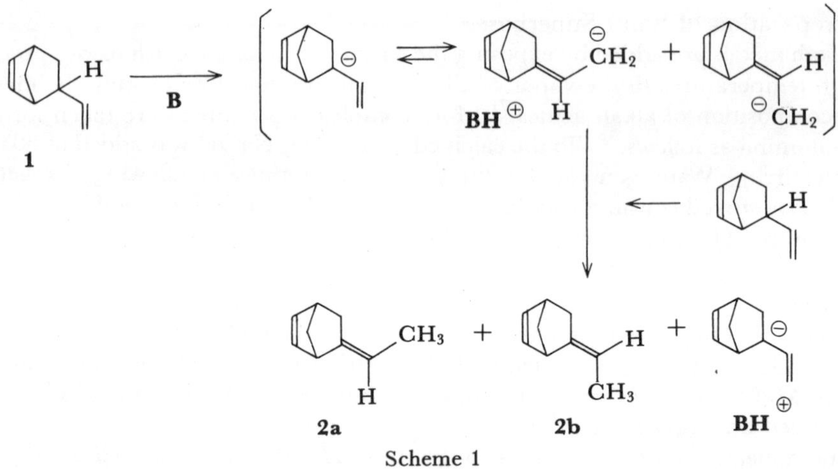

Scheme 1

The above mentioned Na – NaOH – Al_2O_3 was prepared for the catalyst for the isomerization of 5-vinylbicyclo [2,2,1] heptene **1** to 5-ethylidenebicyclo-[2,2,1] hept – 2-ene **2** as shown in Scheme 1.[3)] The reaction takes place at the temperature of 243 K which meets the reaction condition to avoid thermal rearrangement of the reactant. This reaction does not occur over the $NaOH-Al_2O_3$ catalyst possessing basic sites of H_- 27 without Na metal treatment. This reaction, too, is catalyzed by superbasic sites at 243 K.

D. Structure of Superbasic Sites

Malinowski *et al.* proposed the generation of superbasic sites on MgO as follows.[4,5,7,9–12)]

$$[\ \ ]\text{ anionic vacancy} + Me^0 \longrightarrow [e]\ F^+\text{ center} + Me^+ \quad (1)$$

$$O^-[\ \ ]\text{ hole trapped on oxygen anion} + Me^0 \quad O^{2-}[\ \ ] + Me^+ \quad (2)$$

$$2OH_S + Me^0 \longrightarrow OMe_S + H_2O \quad (3)$$

$$OH_S + Me^0 \longrightarrow OMe_S + 1/2H_2 \quad (4)$$

The color center formed according to eq. (1) is of strong one-electron donor character, while the other sites formed by eqs. (2)−(4) are of strong electron-pair donating character. The increases in basic strength are caused by the introduction of an electron from the alkali metal to the hole trapped on the O^2-anion for eq. (2), and by replacement of an H atom by a more electropositive alkali metal atom for eqs. (3) and (4).

REFERENCES

1. K. Tanabe, in: *Catalysis by Acids and Bases* (B. Imerik, C. Naccache, G. Coudurier, Y. Beu Tararit, J.C. Vedrine, eds.) Elsevier, Amsterdam, 1985, p.1.
2. T. Ushikubo, H. Hattori, K. Tanabe, *Chem. Lett.*, **1984**, 649.
3. G. Suzukamo, M. Fukao, M. Minobe, *Chem. Lett.*, **1987**, 585.
4. S. Malinowski, J. Kijenski, in: *Catalysis* (C. Kemball and D.A. Dawden, eds.) Vol. 4, p.130, Royal Soc. Chem. (1981).
5. J. Kijenski, S. Malinowski, *J. Chem. Soc., Trans. 1*, **74**, 250 (1978).
6. J. Kijenski, K. Brzozka, S. Malinowksi, *Bull. Acad. Pol. Sci. Ser. Sci. Chim.*, **26**, 271 (1978).
7. J. Kijenski, S. Malinowski, *J. Res. Inst. Catal.* Hokkaido Univ., **28**, 97 (1980).
8. J. Kijenski, S. Malinowski, *Bull. Acad. Pol. Sci, Ser. Sci. Chim.*, **25**, 749 (1977).
9. J. Kijenski, M. Marczewski, S. Malinowski, *React. Kinet. Catal. Lett.*, **7**, 151 (1977).
10. J. Kijenski, M. Marczewski, S. Malinowski, *React. Kinet. Catal. Lett.*, **7**, 157 (1977).
11. J. Kijenski, S. Malinowski, *Bull. Acad. Pol. Sci. Ser. Sci. Chim.*, **25**, 501 (1977).
12. J. Kijenski, K. Brzozka, S. Malinowski, *Bull. Acad. Pol. Sci. Ser. Sci. Chim.*, **26**, 271 (1978).

4
Catalytic Activity and Selectivity

4.1 ISOMERIZATION

4.1.1 General Remarks

Isomerization is a reaction in which a molecule is transformed into a molecule having the same molecular formula but a different structure, i.e., isomers. Industrially important isomerization reactions are rearrangements of the carbon skeleton of C_4-C_8 hydrocarbons, and isomerization among alkyl benzene isomers such as xylenes and ethylbenzene.[1,2] Isomerization including heteroatoms, such as propylene oxide to allyl alcohol and Beckmann rearrangement of cyclohexanone oxime to ϵ-caprolactam, are also significant industrial processes.

Isomerization with respect to the double bond of olefinic compounds, that is, double-bond shift and *cis*−*trans* isomerization, has been extensively studied to elucidate the reaction mechanisms and for the characterization of catalysts.

Most of the isomerizations mentioned above are catalyzed by solid acids and bases. In the case of solid acids the reactions proceed via classical or non-classical carbocations and the stabilities of the reaction intermediates are generally very important in determining rate and selectivity. For base catalysis the stability of anionic intermediates is important.

4.1.2 Double-Bond Isomerization

The double-bond isomerization proceeds by the breaking and formation of C−H bonds on or next to the double bonds.[2] The mechanism may be classified by the charge of hydrogen atom added or removed, H^+, H (neutral) or H^-, and the timing of the C−H bond breaking and formation, that is, (a) breaking followed by formation or (b) formation followed by breaking. Simultaneous occurrence of bond breaking and bond formation is formally possible, but no firm evidence of this mechanism has been reported. Examples of those mechanisms are illustrated below for the case of *n*-butene. Eq. (1) is for solid protonic acids (proton addition followed elimination of proton) and eq. (2) for solid bases (the first steps is proton abstraction to form allyl anion).

$$CH_2{=}CH-CH_2-CH_3$$

$$\xrightarrow{+H^+} \quad CH_3-CH^+-CH_2-CH_3$$

$$\xrightarrow{-H^+} CH_3-CH=CH-CH_3 \quad (1)$$

$$\xrightarrow{-H^+} [CH_2 \overset{...}{=} CH \overset{...}{=} CH-CH_3]^- \xrightarrow{+H^+} CH_3-CH=CH-CH_3 \quad (2)$$

The presence of a variety of reaction intermediates described above can be clearly demonstrated by detailed tracer studies in the case of deuterium exchange of propylene, which is closely related to double-bond isomerization of olefins.[3)]

A. Double-bond Isomerization of Butene

Double-bond isomerization of *n*-butene is described in more detail in relation to the acid-base properties of catalysts. The rate usually increases either with increasing acid strength and acid amount or with increasing base strength and base amount. As for the selectivity there are apparent correlations with the acidity or basicity of catalysts as summarized in Table 4.1.[4)] However, since the correlation is not one-to-one cor-

TABLE 4.1 Selectivity and mechanism of *n*-butene isomerization

Catalyst	Selectivity: 1-Butene →*cis*/*trans*	Selectivity: *cis*-2-Butene →trans/1	Reaction intermediate	Type of 1, 3— hydrogen shift
Acids				
H_2SO_4	0.5	2	2-Butyl cation	H^+, intermolecular
$H_3PW_{12}O_{40}$	1	*ca.*10		
$SiO_2 \cdot Al_2O_3$	1	7		
Metal sulfates/SiO_2	1	1-10		
Ion-exchanged zeolites	1-2	—		
Bases or acid-bases				
KO-*t*-Bu	26	—	*π*-Allyl anion	H^+, intramolecular
CdO	20	0.1		
MgO	11-16	1		
CaO	1-7	1-5		
ZnO	11	1		
Na/Al_2O_3	4	0.8		
ZrO_2	5	0.5		
TiO_2	2-7	0.1		
La_2O_3	5	*ca.*0		
ThO_2	3	0.3		
CeO_2	*ca.*2	—		
Al_2O_3	3-6	*ca.*4		
Others				
MoS_2	1	*ca.*20	2-Butyl	H, intermolecular
Co_3O_4	2	-15		
$AlPO_4$	—	∞		
SO_2/SiO_2	—	∞	Radical	No shift
MoO_x/TiO_2	—	∞	Carbene	

respondence, caution must be paid when speculation of the mechanism is made based on the selectivity.

The isomerization between the three isomers of *n*-butene may be characterized by six rate constants in the scheme given below. Provided that the reaction orders are the same, the following relations hold for these rate constants.

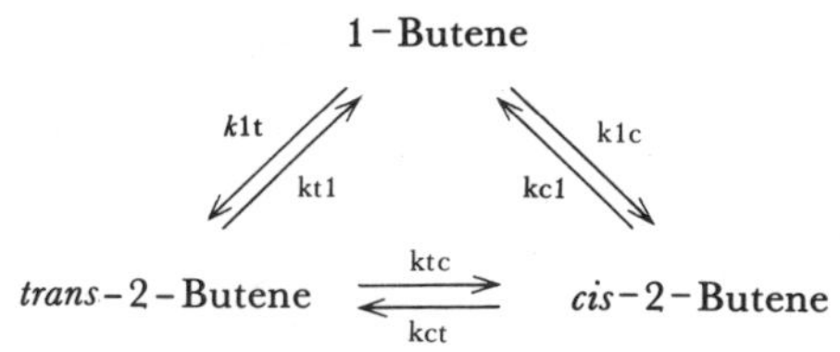

Scheme 1

$$(k_{ct}/k_{cl})(k_{tl}/k_{tc})(k_{lc}/k_{lt})=1 \tag{3a}$$

$$(k_{ij}/k_{ji})=K_{ij} \quad (K_{ij}=\text{equilibrium constant between i and j isomers}) \tag{3b}$$

Solid acids Over Brønsted (protonic) solid acids, the reaction intermediate is a *sec*-butyl cations formed by the addition of a proton from the solid surface to butene. Hence, when solid acids are deuterated the deuterium of the catalyst is incorporated into both reactant and isomerized butenes.[5,6] The same intermediate is often involved in the case of Lewis solid acids. In the latter case, protonic acid is induced by the reaction of butene on the Lewis acid site. For example, $CH_3CH=CH_2 + L \rightarrow CH_3CH^+ - CH_2 - L$ (L: Lewis site), where H at CH_3 or CH_2L acts as acidic proton. The intermediate is illustrated in Fig. 4.1 A. Fig. 4.1B shows the stereochemical reaction scheme for the case in which the deuterium atom (D_A) is attached to *cis*-2-butene from below. If the H_B atom is removed downward from **1** or **2**, the intermediate is transformed to *trans*-2-butene. The removal of D can produce *trans*-2-butene if D is removed upward from **1** by a certain mechanism or removed downward after the "rollover" of the intermediate takes place (**3**). 1-Butene is produced if one of three protons at C-1 of the intermediate is removed. Analysis of the tracer experiments based on this model reveals detailed information on the dynamic behavior of the intermediate.[6]

Over silica-alumina, butene isomerizes on protonic acid sites which are originally present in small quantity on the surface or on protonic sites induced on Lewis acid sites. This conclusion was deduced from the tracer studies of Ozaki and Kimura[5] and Hightower and Hall.[7] The latter group carried out coisomerization of butene-d_8 and butene-d_0 and found that mixing of hydrogen isotope between the d_0 and d_8 species took place in starting as well as in isomerized butene isomers to comparable extents.[7] Their analysis demonstrated that the *sec*-butyl cations is the intermediate. The analysis has been further advanced by Misono *et al.*,[6] taking into account the stereochemistry of proton addition and abstraction shown by the model in Fig. 4.1B.

It has also been demonstrated that not only the catalytic activity but the selectivity is also strongly dependent on the acid strength; the *trans*/1 ratio from *cis*-2-butene and *cis*/1 ratio from *trans*-2-butene markedly increased with the acid strength, while the

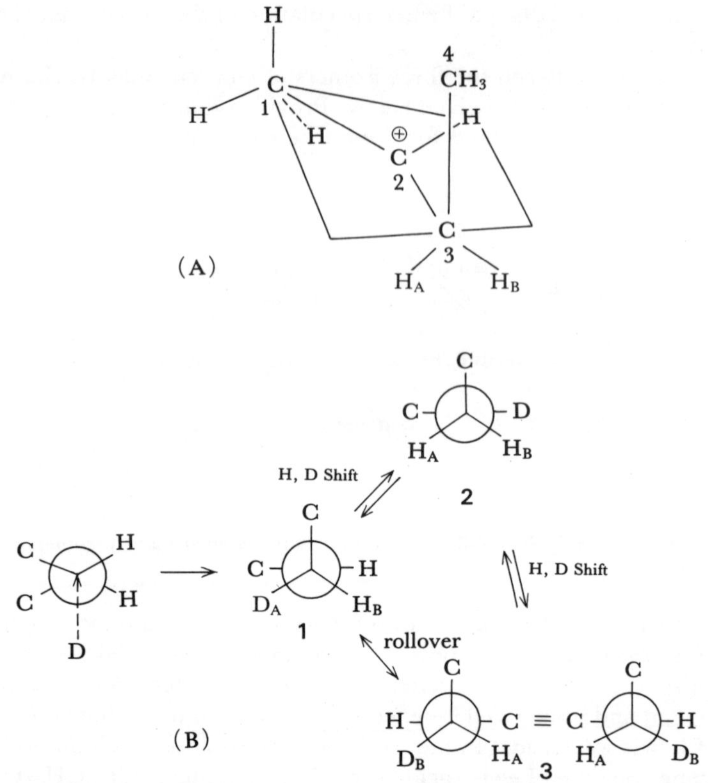

Fig. 4.1 2-Butylcation (A) and stereochemistry of *n*-butene isomerization via 2-butyl cation (B).
Suffixes A, B indicate two diastereomeric positions of carbon-3

cis/trans ratio from 1-butene was never far from unity for metal sulfates,[8)] ion-exchange zeolites[9)], and ion-exchange resins.[10)] Results of metal sulfates are shown in Fig.4.2. The electronegativity of metal ion represents the acid strength, as indicated by the indicator test.

The variations in rate and selectivity with increasing acid strength were explained by considering the linear free energy relationships and the stability of *sec*-butyl cation in the data of kinetic[11)] and tracer studies.[6a)] Rate of isomerization is highly accelerated if alkyl groups are substituted at the carbocation center because the carbocation is stable in the order tertiary > secondary > primary. For example, 3-methyl-1-butene isomerizes more than 100 times faster than 1-butene over $MgSO_4-SiO_2$.[12)]

Solid bases As shown in Table 4.1, the *cis/trans* ratio from 1-butene is usually very large in the case of solid bases. This is because the allyl intermediate formed by the abstraction of proton from butene (eq. 2) is more stable in the anti (*cis*) form than in the syn (*trans*) form (eq. 4). Superbases such as Cs_2O and Rb_2O are reported to catalyze butene isomerization in a similar way.[13)]

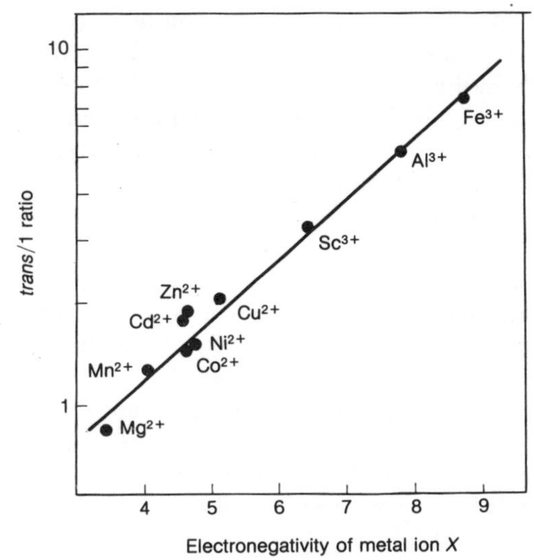

Fig. 4.2 *Trans*/1 ratio from isomerization of *cis*-2-butene over metal sulfates on silica gel plotted against the electronegativity of metal ion.
Cf. Ref. (8b) for the electronegativity of metal ion.

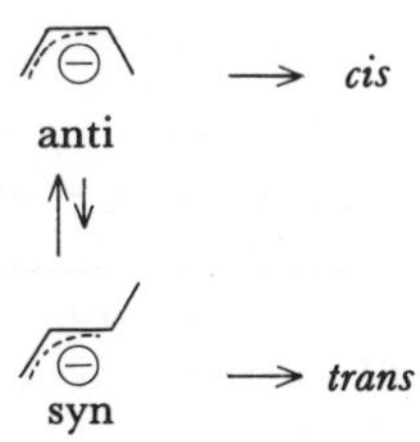

(4)

Tracer studies demonstrated that the hydrogen is shifted intramolecularly.[14)] For example[14a)]

$$CH_2{=}CH{-}CHD{-}CH_3 \longrightarrow \begin{cases} CH_2D{-}CH{=}CH{-}CH_3 \\ CH_3{-}CH{=}CD{-}CH_3 \end{cases} \quad (5)$$

Coisomerization of d_0- and d_8-butene over MgO exhibited little isotopic mixing in accordance with intramolecular mechanism.[14b)] It is not known whether the hydrogen shift takes place in one step or in two steps. The presence of a slight amount of isotopic mixing[14b)] and quantum chemical considerations[15)] favor the two-step mechanism. Isotopic mixing did not occur to a significant degree, probably because the strong basic sites which can abstract proton from butene are scarce on the surface. The allyl mechanism has been demonstrated also by *in situ* IR study in the case of ZnO.[16)]

It has been indicated that the rate increases with the basicity.[17)] Alkyl substituents at the allyl position accelerate the isomerization rate but to a much smaller extent than in the case of solid acids; the ratio was about three for 3-methyl-1-butene to 1-butene over CaO.[12)]

Other solid catalysts Specific *cis-trans* isomerization takes place over $AlPO_4$ although the mechanism is not very clear (see Section 3.8.2).[18] Over MoS_2 and Co_3O_4, isomerization proceeds rapidly in the presence of hydrogen. The intermediate is a butyl group covalently bonded to metal ion. Isomerization accompanied by metathesis is also possible with catalysts containing low valent Mo, W, and Re. These catalysts may be regarded to be soft Lewis acids. Radical mechanisms in which *cis-trans* isomerization takes place exclusively are known.[37]

B. Isomerization of Other Olefins

Isomerization of olefins in general proceeds in a way similar to that of butene. Effects of substituents on the stability and reactivity of intermediates and steric effect of bulky substituents are the factors to be considered.

Usually isomerization proceeds more cleanly over solid bases than over solid acids because of the absence of polymerization for the former. The turnover frequency is probably greater over solid bases. For example, the isomerization of α-pinene and related cyclic compounds having exo-cyclic double bonds efficiently catalyzed by solid bases such as SrO and MgO.[19]

As for olefins containing heteroatoms, the following isomerization reactions have been reported.

$$C{=}C{-}C{-}O{-}C{-}C \longrightarrow \textit{cis and trans-}\ C{-}C{=}C{-}O{-}C{-}C \quad (6)$$

$$C{=}C{-}C{-}N\langle \longrightarrow \textit{cis and trans-}\ C{-}C{=}C{-}N\langle \quad (7)$$

For the isomerization of 2-propenyl ether the catalytic activity is in the order; $CaO > La_2O_3$, SrO, $MgO \gg ZnO$, Al_2O_3, $SiO_2-Al_2O_3 \approx 0$.[20] The initial isomerized product was exclusively *cis* form, indicating the intermediacy of an anionic allyl species as in the case of butene isomerization over solid bases. In the case of isomerization of 2-propenylamine, MgO and CaO were very active, while ZrO_2 and ZnO were inactive.[21] By the isomerization of *N,N*-dimethyl-2-propenylamine, 100% *cis-N,N*-1-propenylamine was initially formed, so that an anionic allyl intermediate was proposed.

Allyl halides and acetates isomerize easily with acid catalysts via cationic allylic species.[22] Alkyl substitution at allylic position increases the rate to an intermediate extent between alkyl cation (solid acid) and allylic anion (solid base).[12]

4.1.3 Isomerization of Paraffins

Skeletal isomerization of *n*-C_5, C_6 paraffins to corresponding isoparaffins is important for improving the octane number as they are mixed in gasoline. Since low temperature is favored for the equilibrium of this reaction, catalysts active at low temperatures are desirable. Noble metals loaded on zeolites such as Pt−Y zeolite with low Na content are effective and used at about 520 K.[23] Fig 4.3 shows the effect of Na content of zeolite on the catalytic activity for hexane isomerization. As the acidity increases with decreasing Na content the optimum temperature of operation is greatly suppressed.

The isomerization over noble metal-solid acid bifunctional catalysts proceeds by the combination of two functions: The dehydrogenation-hydrogenation on metallic sites and the isomerization of olefin on acid sites.[24,25] It has been pointed out that no-

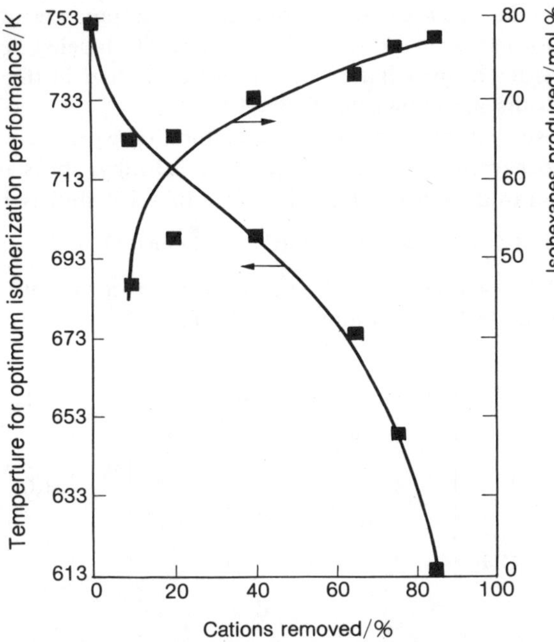

Fig. 4.3 Effect of sodium removal of Y-zeolite in hexane isomerization.
(Reproduced with permission by J. A. Rabo *et al.*, *Actes Congn. Int. Catal.*, 2nd, Paris, **2**, 2063 (1961)).

ble metals alone can isomerize paraffins.[26] However, the rate of isomerization over bifunctional catalysts is much faster than over metal catalysts, so that isomerization over the former proceeds mainly on acidic sites of solid acids.[24] Further, skeletal

TABLE 4.2 Isomerization of ^{13}C-labeled 3-methylpentane, 2-methylpentane, and 2,3-dimethylbutane

Run number	Starting hydrocarbon†1	2,3-Dimethylbutanes %†2			2-Methylpentanes %†2			†3	3-Methylpentanes %†2		†3	
1		0.03	4.7	95.3	0.22	73.2	0.3	27.5	99.75	97.9	2	0.1
2		0.13	53.2	46.8	0.77	2	68.3	29.7	99.1	0.8	5.5	93.7
3		0.13	67.8	32.2	99.25	0.2	• 1.6	98.2	0.62	2.5	78.8	18.7
4		0.62	69.4	30.6	97.81	0.6	96.4	3.6	5.4	4.3	71.3	24.4
5		0.07	0.4	99.6	99.35	98.9	0.1	1	0.58	26.8	71.4	1.8
6		98.65	0.5	99.5	0.75	44.9	27.6	27.5	0.6	20.3	66.6	13
7		94.84	98.7	1.3	5.1	5.1	51	43.9	2.51	2.8	66.7	30.5

†1 Labeled carbons are indicated by solid circles.
†2 Percentage among the monolabeled isomers
†3 Most probably in run numbers 1 and 5.

isomerization can take place easily without metal components in the case of strong acids such as H mordenite.[27] The isomerization of ^{13}C-labeled hexanes over H mordenite at 443K has been investigated with a pulse method in the presence of hydrogen.[27] Typical results are shown in Table 4.2.

It was suggested that the interconversion between 2-methylpentane and 3-methylpentane occurs mainly by the 1,2- and 1,3-alkyl shifts in the hexyl cations which accompany rapid hydride shift. The rate of alkyl shift is in the order,

$$1,2\text{-ethyl} > 1,3\text{-methyl} > 1,2\text{-methyl shift}$$

and the relative value is about 10:4:1. The difference in the rate reflects the relative stability of intermediate hexyl cations: I>II>III.

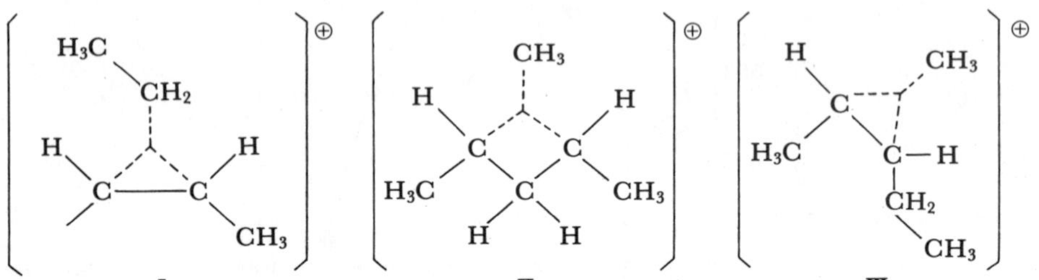

I II III

Interconversions of 2-methylpentane ⇄ 2,3-dimethylpentane as well as 3-methylpentane ⇄ 2,3-dimethylpentane proceed probably via protonated cyclopropane-type intermediates (IV);

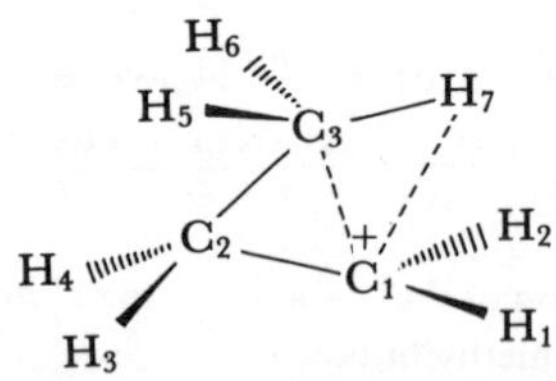

IV

In addition to the products explained by the above mechanism, there is a relatively minor mechanism which causes random distribution of tracer. It was suggested that these products are probably formed by polymerization-decomposition of hydrocarbons.

Reforming is the transformation of naphtha ($\geq C_7$) into alkylaromatics in the presence of hydrogen by using bifunctional catalysts, e.g. Pt–Re, Ir/Al_2O_3. The process includes several kinds of reactions: (a) dehydrogenation of cyclohexanes to armatics, (b) isomerization of *n*-alkanes to branched alkanes, (c) dehydroisomerization of alkylcyclopentanes to aromatics, (d) dehydrocyclization of alkanes to aromatics, and (e) hydrocracking of alkanes and cycloalkanes to low molecular weight alkanes. The acidity of the catalyst acts bifunctionally in reactions with metal components (see Section 4.14).

Dehydrogenation and isomerization of cyclohexane was studied on $TiO_2-ZrO_2-V_2O_5$.[27] Both reactions correlated well with the surface acidity. The conversion of *n*-butane to isobutane is catalyzed in a similar way.

4.1.4 Isomerization of Alkylbenzenes

Transformation of *m*-xylene to *p*-, *o*-xylene or ethylbenzene as well as isomerization of *o*-xylene or ethylbenzene to *p*-xylene are significant in industrial processes. The reactions are catalyzed by acid sites and the mechanism has been suggested to be follows:

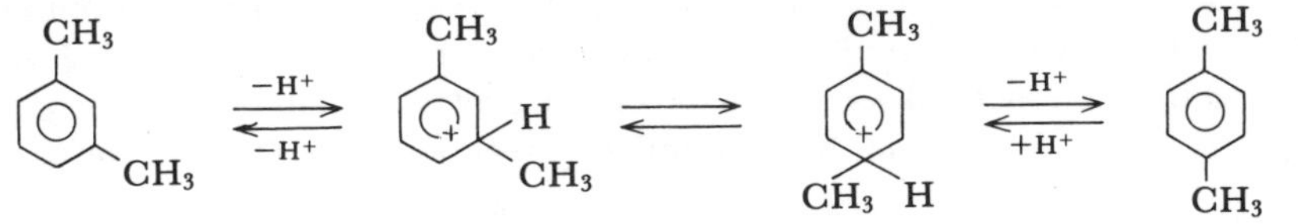

(8)

In accordance with this, the rate of xylene isomerization rapidly decreased upon Na treatment of silica-alumina in parallel with the rate of cumene cracking.[29] The rate of 1,2-alkyl shift is in the order

$$t\text{-butyl} > \text{isopropyl} > \text{ethyl} > \text{methyl}$$

reflecting the stability of alkyl cation. Cracking which accompanies the isomerization also increases in the same order. Over strong acids the conversion of ethylbenzene to xylene via alkylcyclopentane can take place.[30] Generally, the cracking and disproportionation, the main side reactions, tend to increase over strong acids. In these reactions shape selectivity due to the micropores of zeolites is significant (see Section 3.4).

4.1.5 Isomerization Including Heteroatoms

A. Isomerization of Epoxide

Production of allyl alcohol by the isomerization of propylene oxide is an industrialized process. Lithium phosphate is specifically selective for this reaction. It is believed that the appropriate balance of the acidity and basicity of catalyst is essential for high selectivity. If the acidity is dominant, isomerization to aldehyde becomes the main reaction (eq. 9a), and a path to acetone is favored on basic catalysts (eq. 9c), as shown below.

propylene oxide →
- → acetaldehyde (acid) (9a)
- → allylic alcohol (acid+base) (9b)
- → acetone (base) (9c)

The above relationships have been confirmed between the acid-base properties of several metal phosphates and their catalytic activities and selectivities.[31]

According to the patent literature,[32] the method of preparation of Li_3PO_4 is crucial for obtaining selectivity higher than 90%. A correlation exists between the line broadening of XRD (002) line and catalytic performance of Li_3PO_4 prepared by several methods.[33] This indicates the importance of a crystal plane as well as high surface area for efficient isomerization.

Isomerization of ethylene oxide to acetaldehyde is an undesirable reaction in the oxidation of ethylene over Ag catalyst. Addition of Cs has been reported to suppress the isomerization by weakening the Lewis acidity of Ag (electronic effect).[34)]

B. Beckmann Rearrangement

Beckmann rearangement of cyclohexanone oxime to ε-caprolactam is catalyzed by sulfuric acid in the industrial process. Several attempts have been reported to substitute sulfuric acid by suitable solid acids,[35)] but it is rather difficult to obtain high yields. Recently, it was reported the silica-supported boria catalyst prepared by vapor phase decomposition method was very efficient (oxime conversion: 98%, lactam selectivity 96% at 523K), with slight deactivation with reaction time. (see Section 3.1.11)[36)]

REFERENCES

1. J.H. Sinfelt, in *Catalysis*, (J.R. Anderson, M. Boudart, eds.) Vol. 1, Springer, Berlin 1981, p.257; J.W. Ward, in: *Applied Industrial Catalysis* (B.E. Leach, ed.), Vol. 3, Academic Press, Orlando, 1981, p.272.
2. T. Uematsu, *Shokubai Koza,* Vol. 8, Kodansha, 1985, p.85; M. **Misono**, *Kagaku no Ryoiki,* **27**, 437 (1973) (in Japanese).
3. T. Kondo, S. Saito, K. Tamaru, *J. Am. Chem. Soc.,* **96**, 6857 (1974).
4a) T. Okuhara, M. Misono, *Shokubai* (Catalyst), **25**, 280 (1983) (in Japanese);
b) N.F. Foster T. Cvetanovic, *J. Am. Chem. Soc.*, **82** (1960).
5. A. Ozaki, K. Kimura., *J. Catal.*, **3**, 395 (1964).
6a) M. Misono, N. Tani, Y. Yoneda, *J. Catal.*, **55**, 314 (1978).
b) J.L. Lemberton, G. Perot, M. Guisnet, Proc. 7th Intern. Congr. Catal., Tokyo, 1980, Kodansha, Tokyo and Elsevier, Amsterdam, 1981, p.993.
7. J. Hightower, W.K. Hall, *J. Am. Chem. Soc.*, **89**, 778 (1967); *J. Phys. Chem.*, **71**, 1014 (1967).
8a) M. Misono, Y. Saito, Y. Yoneda, *J. Catal.*, **9**, 135 (1967); *ibid.*, **10**, 88 (1968); *Bull. Chem. Soc. Jpn.*, **44**, 3236 (1971).
b) M. Misono, E. Ochiai, Y. Saito, Y. Yoneda, *J. Inorg. Nucl. Chem.*, **29**, 2685 (1968). See also ref. (7) of Section 4.8.
9. E. Lombardo, W.K. Hall, *J. Catal.*, **22**, 54 (1971).
10. T. Uematsu, K. Tsukada, M. Fujishima, H. Hashimoto, *J. Catal.*, **32**, 369 (1974).
11. M. Misono, Y. Yoneda, *J. Phys. Chem.*, **76**, 44 (1972).
12. M. Misono, K. Sakata, F. Ueda, Y. Nozawa, Y. Yoneda, *Bull. Chem. Soc. Jpn.*, **53**, 648 (1980).
13. S. Tsuchiya, S. Takase, H. Imamura, *Chem. Lett.*, **1984**, 661.
14a. N. Tani, M. Misono, Y. Yoneda, *Chem. Lett.*, **1973**, 591.
b. I.R. Shannon, C. Kemball, H.F. Leach, Symp. Chemisorption and Catalysis, Inst. Petrol., London, 1970.
15. M. Misono, W. Grabowski, Y. Yoneda, *J. Catal.*, **49**, 363 (1977).
16. R.J. Kokes, A.L. Dent, *Advan. Catal. Relat. Subj.*, **22**, 40 (1972).
17. H. Hattori, N. Yoshii, K. Tanabe, Proc. 5th Intern. Congr. Catal., Palm Beach, 1972, North-Holland, Amsterdam, 1973, p.233.
18. H. Itoh, A. Tada, H. Hattori, *J. Catal.*, **76**, 235 (1982).
19. Y. Fukuda, H. Hattori, K. Tanabe, *Bull. Chem. Soc. Jpn.*, **51**, 3150 (1978); H. Hattori, K. Tanabe, K. Hayano, H. Shirakawa, T. Matsumoto, *Chem. Lett.*, **1979**, 133; T Yamaguchi, N. Ikeda, H. Hattori, K. Tanabe, *J. Catal.*, **67**, 324 (1981).
20. H. Matsuhashi, H. Hattori, *J. Catal.*, **85**, 457 (1984).
21. A. Hattori, H. Hattori, K. Tanabe, *J. Catal.*, **65**, 246 (1980).
22. W.G. Young, H.E. Green, A.F. Diatz, *J. Am. Chem. Soc. Jpn.*, **93**, 4782 (1971).

23. J.A. Rabo, P.E. Pickert, D. Stamires, J.E. Boyle, Actes Congr. Int. Catal., 2nd, Paris, **1960**, 2055.
24. G.A. Mills, H. Heinemann, T.H. Milliken, A.G. Oblad, *Ind. Eng. Chem.*, **45**, 134 (1953).
25. J.H. Sinfelt, in *Catalysis* (J.R. Anderson, M. Boundart, eds.), Springer, Berlin, 1981, Vol. 1, p.257.
26. F.G. Gault, *Advan. Catal. Relat. Subj.*, **30**, 1 (1981).
27. M. Daage, F. Fajula, *J. Catal.*, **81**, 394, 405 (1983).
28. R.-C. Chang, I. Wang, *J. Catal.*, **107**, 195 (1987).
29. I. Mochida, Y. Yoneda, *J. Catal.*, **7**, 393 (1967); H. Matsumoto, Y. Saito, Y. Yoneda, *J. Catal.*, **11**, 211 (1968).
30. M. Nitta, P.A. Jacobs, in *Catalysis by Zeolites* (B. Imelik *et al.*, eds.), Studies Surf. Sci. Catal., Vol. 5, Elsevier, Amsterdam, 1980, p.251.
31. T. Imanaka, Y. Okamoto, S. Teranishi, *Bull. Chem. Soc. Jpn.*, **45**, 1353 (1972).
32. Fr. Pat. 1496221; Ger. Offen 1810120.
33. T. Mochizuki, T. Okuhara, M. Misono, 54th National Meeting of the Chem. Soc. Jpn., April, 1987.
34. S.A. Tan, R.B. Grant, R.M. Lambert, *J. Catal.*, **106**, 54 (1987).
35. T. Yashima, S. Horie, J. Saito, N. Hara, *Nippon Kagaku Kaishi*, **1977**, 77.
36. S. Sato, K. Urabe, Y. Izumi, *J. Catal.*, **102**, 99 (1986).
37. Y. Sendoda, Y. Ono, *J.Chem. Soc, Faraday I*, **76**, 435 (1980).

4.2 ALKYLATION

4.2.1 Alkylation of Aromatics with Alcohols

A considerable body of literature exists concerning the Friedel-Crafts alkylation using conventional protic acids[1], proton donor-promoted Lewis acids such as aluminum chloride-hydrogen chloride.

The synthetic zeolites, whose application to catalysis was developed in the early 1960's, attracted attention as alkylation catalysts because of high acidity, easy separation of catalysts from products, regenerability, absence of corrosive substances such as halogen and volatile acids, and lack of environmentally hazardous streams such as spent aluminum chloride.

Early studies on X- and Y-type zeolites, especially rare earth exchanged varieties, revealed effective performance in alkylation of benzene or toluene with olefins or alcohols.[2,3]

p-Xylene is a valuable aromatic compound because of the demand for oxidation to terephthalic acid, a major component in polyester fibers. Though toluene is the major single component produced in catalytic reformers, the demand for toluene is limited compared to that for benzene or xylenes. Therefore, it is very desirable to convert toluene to xylenes, especially *p*-xylene by alkylation of toluene. The equilibrium amount of *para* isomer in xylenes is only about 24% of the total, and the separation of these isomers is not easy because of the closeness of their boiling points. Therefore, it is of great industrial importance to alkylate toluene directly to *p*-xylene.

In 1970, Yashima and co-workers focused attention on the distribution of xylene isomers produced by alkylation of toluene with methanol over a variety of cation-exchanged Y-zeolites.[4] The relatively high amount of *para* isomers (45 – 50% selectivity) was obtained with certain catalysts. They attributed this to the preferential formation of *para* isomer and the suppression of the isomerization of the *para* isomer thus formed in the supercage of the zeolites.

The selectivity for *para* isomer was greatly improved by using modified ZSM-5 zeolites.[5,6] The shape selectivity of ZSM-5 is modified significantly by treatment with a variety of chemical reagents. For example, modification with phosphorus or boron was made by impregnating the zeolite crystals with aqueous phosphoric acid or orthoboric acid, followed by calcination in air to convert the acid into the oxides. Selected results are summarized in Table 4.3. Though the selectivity for *para* isomer in alkylation with ordinary ZSM-5 is close to that expected from the thermal equilibrium, selectivity as high as 97% is achieved with the modified ZSM-5.

The chemical treatments are assumed to reduce the effective pore openings or channel dimensions of ZSM-5. This results in discrimination based on differences in the dimension of xylene isomers. The selectivity is proposed to be determined by following factors.[4-8].

i) A bulky species such as phosphorus partially blocking pore openings would greatly favor outward diffusion of the *para* isomer relative to the *ortho* and *meta* isomers. Diffusion of *p*-xylene is $> 10^3$ times faster than that of *o*- and *m*-xylenes.[5]

ii) Alkylation at the *para* position is favored within the more confined cylindrical pore of the modified catalysts and the isomerization is hindered.

iii) Phosphorus or boron compounds on the external surface cover strong acid sites located there and prevent rapid isomerization of the *p*-xylene which has emerged from the pore. The poisoning of the strong acid sites inside the pore may also serve to inhibit the isomerization.

Yashima *et al.* carried out alkylation of toluene with C_2-C_4 aliphatic alcohol over

TABLE 4.3 Alkylation of toluene with methanol over modified ZSM-5 catalyst

Modification element	—	P(8.51%)	B	
Temperature/K	873	873	873	
W H S V†1	5.3	10	3.8	
Toluene/Methanol (mole ratio)	2	2	2	
Conversion/%				
toluene	30	21	20	
methanol	100	92	—	
Product distribution/wt %				
C_6^-		1.7	—	
Benzene	1.4	0.1	1.6	
Toluene	67.1	74.1	77.8	
Xylene				
para	9.0	20.7	17.1	
meta	12.9	0.4	1.7	
ortho	5.8	0.2	0.7	
Others	3.8	2.2	1.1	
% Xylene				equilibrium
para	26	97	88	23
meta	50	2	9	51
ortho	24	1	3	26

†1: Weight of toluene and methanol mixture per h per unit weight of catalyst

HY zeolites and discussed the geometric effect of the zeolite structure on the selectivity of alkylation and the subsequent isomerization of alkylation products.[9] No *ortho* isomers were produced in the alkylation with isobutyl and *t*-butyl alcohols.

Alkylation of xylene isomers with methanol over H – ZSM-5 was reported by Namba *et al.*[10] The main products were xylene isomers and 1,2,4-trimethylbenzene, the latter constituting more than 99% of the trimethylbenzene fraction.

In the alkylation of trimethylbenzene with methanol over H-ZSM-5, the 1,2,3,4-isomer fraction in tetramethylbenzenes are very high (to 98%).[11] The selectivity for the isomer is further improved by selective dealumination of the external surface of the zeolite crystals or by selective poisoning of the active sites on the external surface.[12]

Alkylation of chlorobenzene with methanol over ZSM-5 at 523 K gives a mixture of *o*- and *p*-chlorotoluene, the amount of the *meta* isomer being less than a few percent.[13] The selectivity depends on the crystal size of the zeolite, high *para* selectivity (90%) being obtained over large crystals (220 μm). Dealuminated mordenite gave a selectivity of about 40% for *p*-chlorobenzene.

High β-selectivity in alkylation of naphthalene and methylnaphthalene with methanol over H – ZSM-5 was reported by Fraenkel *et al.*[14] Thus, 76% selectivity for 2-methylnaphtalene and 2,6/2,7-dimethylnaphthalene was achieved with 15% naphthalene conversion at 623 K.[14]

4.2.2 Alkylation of Aromatics with Olefins

Ethylbenzene is the key intermediate in the manufacture of styrene, one of the most important industrial monomers. Almost all ethylbenzene is synthesized from benzene and ethylene.

In the conventional ethylbenzene technology, an aluminum chloride – hydrogen chloride combination is the most widely used catalyst. The highly corrosive nature of aluminum chloride requires special resistance materials in the construction of the reaction vessel and product handling equipment. The polluting nature of aluminum chloride further necessitates treatment of the product for disposal of spent catalysts.

The Alkar process using boron trifluoride supported on alumina introduced in 1958 was a high pressure fixed-bed process.[15] The process permitted the utilization of the light olefins (ethylene + propylene) of the refiners' gas, which had been burnt as a fuel. The quality of ethylbenzene and isopropyl benzene was excellent. However, commerical experience showed that corrosion problems were still substantial and product pretreatment was necessary to remove boron trifluoride.[16]

With the introduction of faujasite zeolite into petroleum cracking, interest in vapor phase alkylation was renewed. There were several reported studies on the use of faujasite or mordenite to ethylate benzene.[2,3,17,18] They were very active, but associated with rapid aging attributed to coke formation. Therefore, a feasible commercial alkylation using faujasite as a catalyst never evolved.

In 1976, the Mobil/Badger ethylbenzene process was announced.[16,19] This is a vapor-phase, fixed-bed process that utilizes ZSM-5. Because of the unique characteristics of the catalysts, aging rate is low and yields of nonselective byproducts are also low. The first commercial unit with a capacity of 50,000 t/y was streamed by the American Höechst Corp. in 1980. Alkylation is carried out in the gas-phase at about

680 K and 20 bar. The molar benzene/ethylene ratio in the feed is 6 to 7. The conversion of ethylene is 100%.

Poly (*p*-methylstyrene) has been claimed to be superior to conventional polystyrene at least for some applications.[20] *p*-Ethyltoluene can be readily dehydrogenated to *p*-methylstyrene.[20] Recently, selective production of *p*-ethyltoluene using modified ZSM-5 zeolites was established.[20,21] Modification of ZSM-5 was made by impregnating with a solution of inorganic reagents such as diammonium hydrogen phosphate or manganese acetate. After removal of solvents by evaporation, the catalyst was dried and calcined in air to decompose the salts and convert the metal component to the corresponding oxides.

The reaction data are summarized in Table 4.4. With unmodified H–ZSM-5 zeolite catalyst, near equilibrium ratio of the *meta* and *para* isomers was observed. However, lower than equilibrium amounts of *o*-ethyltoluene were produced. A dramatic increase in selectivity compared with unmodified catalysts was obtained for the *para* isomer (up to 98%). A corresponding decrease in *meta* isomer and virtual elimination of *o*-ethyltoluene was also observed. Thus, a completely different isomeric mixture of ethyltoluenes was obtained with modified ZSM-5 zeolites compared with $HCl/AlCl_3$ catalyst presently used for commercial vinyltoluene production.

A new process for the production of 97% *p*-methylstyrene monomer and properties of the corresponding polymers has been described.[20] Selective formation of *p*-diethylbenzene from ethyltoluene and ethylene using modified ZSM-5 catalyst has also been reported.[22] *Para*-selectivity of >99% was observed with ZSM-5 modified with

TABLE 4.4 Selective formation of *p*-ethyltoluene from toluene and ethylene over modified ZSM-5 zeolites

Catalyst Modification/wt %†1	H-ZSM-5 —	P-ZSM-5 P(5.6)	PB-ZSM-5 B(4), P(1)	CaP-ZSM-5 Ca(6),P(3.6)	MnP-ZSM-5 Mn(6.4),P(3.5)
Conditions					
Temp/K	673	625	625	673	673
WHSV toluene	6.9	6.9	6.9	7	7
toluene/C_2H_4 C_2H_4	0.5	0.5	0.5	0.5	0.5
	4.5	4.5	4.5	4.2	4.2
Conversion/%					
toluene	20.5	18.9	13.3	22.8	18.7
C_2H_4	91.3	92.4	63.0	—	—
Selectivity to products (wt %)					
Ethyltoluenes	84.0	90.0	95.4	89.5	94.5
Other liquid products	14.5	5.8	2.3	11.5	5.5
Gas (C_1–C_4)	1.5	4.2	2.3	†2	†2
Ethyltoluene					
para	26.2	37.9	82.0	91.9	98.3
mata	59.6	60.9	17.9	8.1	1.7
ortho	14.2	1.2	0.1	0	0

†1 Present as the oxide

†2 Not measured and not included

both Mg and P oxides at 673 – 798 K (see Table 4.5).

Isopropylbenzene is an important intermediate for the production of phenol. There are two main industrial methods for the synthesis of isopropylbenzene from benzene and propylene. One is the liquid-phase process using sulfuric acid as catalyst and the other is the vapor-phase process using phosphoric acid (or P_2O_5) supported on silica or kieselguhr, the content of P_2O_5 and SiO_2 being 62 – 65% and 25%, respectively. In the latter process, UOP process,[23)] the reaction is carried out at 570 – 640 K and 17 – 30 bar with a benzene-to-propylene ratio of 5 – 7. Isopropylbenzene yield of 96 – 97% of the theoretical value based on benzene and 91 – 92% based on propylene is typical. The formation of isopropylbenzene in the alkylated products is more than 90%. This process has also been applied to the alkylation of benzene with ethylene.

Alkylation of aromatic hydrocarbons such as toluene or xylenes with styrene over Nafion-H (perfluorinated resin-sulfonic acid) and Amberlyst 15 has been reported.[24)]

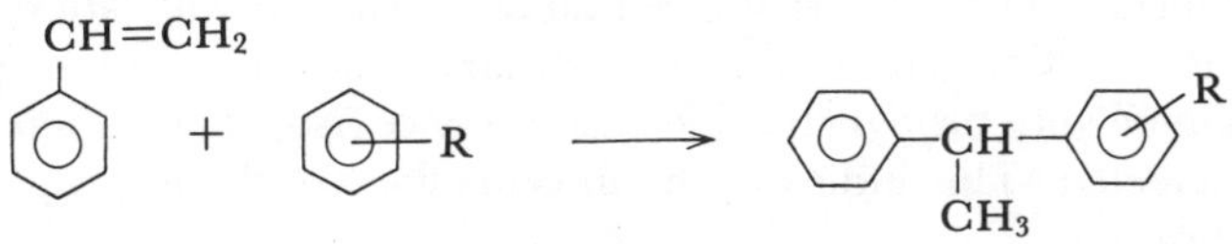

TABLE 4.5 Alkylation of ethylbenzene (EB) with ethylene: large-crystal Mg-P-ZSM-5 catalyst

	Run no.		
	1	2	3
Conditions			
Temp./K	673	698	798
Pressure/kg cm^{-2}	7.0	7.0	7.0
WHSV			
EB	29.7	30.31	30.21
C_2H_4	1.16	1.16	1.16
H_2	0.24	0.24	0.24
$EB/C_2H_4/H_2$ (mole)	6.77/1/3	6.9/1/3	6.9/1/3
Conversion			
EB	11.4	12.6	16.6
C_2H_4	51.1	37.4	28.8
Selectivity to Producks (wt %)			
Benzene	8.0	15.2	58.0
Toluene	1.0	2.1	1.6
Xylene	0.4	0	0.4
Ethyltoluene	1.1	0.8	3.8
Diethylbenzene	88.4	79.7	32.7
Other aromatics	0.6	1.2	0.4
Light gas	0.5	1.0	3.1
Diethylbenzene			
para	99.2	99.9	99.6
meta	0.8	0.1	0.4
ortho	0	0	0

(Reproduced with permission by W. W. Kaeding *et al.*, *J. Catal.*, **95**, 516(1985)).

The reaction was carried out at 343 K. The reaction proceeded selectively with virtually no competing styrene dimerization. Nafion-H® is more effective than Amberlyst 15, but both are better catalysts than soluble protonic acids such as trifluoro-sulfonic acid or *p*-toluenesulfonic acid.

Naphthalene can be alkylated by olefins. Alkylation of naphthalene with propylene with solid phosphoric acid,[25,26] perfluorinated alkane-sulfonic acids ($C_{10}F_{21}SO_3H$ and $C_{12}F_{25}SO_3H$)[27] and Nafion-H®[27] is patented.

4.2.3. Alkylation of Aromatics with Alkyl Halides

Vapor-phase alkylation of benzene, toluene, *o*-, *m*-, *p*-xylenes and fluorobenzene with alkyl halides was studied with Nafion-H as a catalyst.[28] Conversion of as high as 87% (based on isopropyl halide) was obtained at 353 K from a 5:2 mixture of benzene and the chloride. The catalyst showed no deactivation. Alkylation ability follows the order RF > RCl > RBr and secondary > primary. The only product obtained in the alkylation of toluene with propyl chloride was cymenes (isopropyltoluenes); no propyltoluenes were detected. This indicates the intermediacy of the isopropyl cation in the alkylation reaction.

Arata and coworkers studied extensively the alkylation of toluene with alkyl halides. They found that iron (II, or III) sulfate, when calcined at 973 K, became good catalysts for benzylation, benzoylation of toluene[29,30] and also for polycondensation of benzyl chloride.[31] Later, they found that the activity is significantly enhanced by treating the catalyst with hydrogen chloride.[32,33]

Arata and Hino[33] found that better catalysts could be obtained by calcining $Fe(OH)_3$ at 573 – 873 K. The hydroxide was prepared by hydrolyzing $FeCl_3$ or $Fe(NO_3)_3 \cdot 9H_2O$. The alkylation reactions were carried out at room temperature with 50 cm^3 of toluene solution (0.5 mol l^{-1}) of benzyl chloride, *t*-butyl chloride or acetyl chloride and 0.1 g (for benzylation or *t*-butylation) or 0.5g (for acetylation) of catalyst. Benzylation and *t*-butylation was completed within 2 min and 10 min, respectively. For acetylation with acetyl chloride, the reaction was slow, the conversion being 28% after 6 h of reaction. The reaction with acetyl bromide is slightly faster; conversion of 30% was obtained after 4 h. The isomer distribution of alkyltoluenes was 42% *ortho*, 6% *meta* and 52% *para* for benzylation and 3% *meta* and 97% *para* for butylation with *t*-butyl chloride. It was presumed that iron chloride formed on the surface of amorphous iron oxide by its reaction with hydrogen chloride is a catalytically active species for alkylation.[34]

The same catalyst was also very active for polycondensation of benzyl chloride.[35] Thus, when 0.1 g of the catalyst was added to 5 cm^3 of benzyl chloride at room temperature, polymerization occurred immediately with violent evolution of hydrogen chloride and completed in less than 10 s. The yield of methanol – insoluble polymer was about 70%. Elemental and NMR analyses indicated that the product is predominantly linear *para*-substituted polybenzyl. The molecular weight as determined by vapor pressure osmometry was 8175, the degree of polymerization being *ca.* 90.

4.2.4 Alkylation of Aromatics with Alkyl Chloroformates and Oxalates

Alkylation of toluene and phenol with alkyl esters of carboxylic acids and alkyl chloroformate over Nafion-H was studied by Olah *et al.* in both liquid and vapor phase.[36)]

$$ArH + R^1COOR^2 \longrightarrow ArR^2 + R^1COOH$$

$$ArH + ClCOOR^2 \longrightarrow ArR^2 + HCl + CO_2$$

Diethyl oxalate shows particularly good alkylating ability even at milder conditions. Thus, its reaction with toluene under reflux (383 K) for 12 h gave up to 50% ethyltoluene. The advantage of alkyl chloroformate in liquid-phase alkylation lies primarily in their volatile byproducts.

Gas-phase alkylation of toluene with alkyl chloroformate was also reported to be efficient. Due to the high reactivity of alkyl chloroformates, as compared to that of alcohols, higher yields of alkylation of toluene were obtained under the same reaction conditions. A 59% conversion of methyl chloroformate was observed in the alkylation of toluene at 573 K, as compared to about 10% conversion using methanol.

4.2.5 Alkylation of Phenols with Alcohols and Olefins

A number of works have been reported on the alkylation of phenol with methanol over metal oxides as catalyst. Generally, alkylation over acidic oxides such as silica – alumina,[37,38)] phosphoric acid[39)] and Nafion-H®[40)] give mainly anisole and a mixture of three isomers of cresol. On the other hand, basic metal oxides such as MgO[38)] and Mg-containing mixed oxides[38,41,42)] favor alkylation at *ortho* positions. This reaction is industrially important since the reaction product, 2,6-xylenol, is a monomer for good heat-resisting resin.

Kotanigawa and coworkers found that mixed oxides containing Fe_2O_3 are selective catalysts for *ortho*-alkylation of phenol.[43)] Table 4.6 shows the activities and the selectivities in the alkylation at 623 K over mixed metal oxides, where the composition of

TABLE 4.6 Reaction products from phenol and methanol over $MO-Fe_2O_3$ catalyst

M of $MO-Fe_2O_3$	Cu	Mg	Ca	Ba	Zn	Mn	Co	Ni[†1]
Phenol converted, mol %	95.3	8.8	68.7	82.5	88.4	24.0	63.9	67.5
Selectivity, %[†2]								
o-Cresol	41.0	75.3	79.3	64.3	43.5	83.9	82.6	53.2
2,6-Xylenol	59.0	24.7	20.3	35.6	56.5	13.1	17.3	18.6
Methanol converted, mol %	42.3	5.1	23.1	28.7	66.5	2.2	23.8	98.3
Selectivity, %[†3]								
Methylation	31.5	22.1	41.9	38.7	21.0	100	32.5	6.6
Gasification	68.5	77.9	58.0	61.3	79.0	—	67.5	93.3

†1 Selectivity for benzene, toluene, xylene, and carbonization are 12.4, 5.0, 1.0, and 9.8, respectively.

†2 Given by (moles of *o*-cresol or 2,6-xylenol per moles of phenol converted).

†3 Given by (moles of methyl group in products or gaseous products per moles of methanol converted).

Reaction conditions: 623 K; phenol+methanol=63 kPa; phenol/methanol=1/10; contact time 1.6s.

(Reproduced with permission by T. Kotanigawa *et al.*, *Bull. chem. Soc. Jpn.*,**44**, 1962 (1971)).

oxides is M/Fe ratio of 2, M standing for the second metal component. As shown in the table, phenol is selectively methylated to the *ortho* position; except for $NiO-Fe_2O_3$ anisol and cresols are not produced at all. The mixed oxide $CuO-Fe_2O_3$ and $ZnO-Fe_2O_3$ catalysts are also active for *ortho* alkylation of phenol with ethanol, 1-propanol and 2-propanol.[43)]

Nozaki and Kimura found that $Ca_3(PO_4)_2$ is more active than MgO or CaO for *ortho*-alkylation on phenol with methanol.[44)] At 773 K, the selectivity for *ortho*-alkylation on phenol basis was 88%, while the selectivity on methanol basis was 93%. Thus, the selectivity on methanol basis is much higher than that with $ZnO-Fe_2O_3$, though the activity is lower than the latter catalyst.

Tanabe and Nishizaki studied the infrared spectra of phenol adsorbed on MgO and $SiO_2-Al_2O_3$.[38)] Phenol molecules are dissociatively adsorbed on both catalysts to form the surface phenolate. However, the ratio of the intensity of the band at 1496 cm^{-1} to that around 1600 cm^{-1} was quite different in the two catalysts, though both bands are due to the in-plane skeletal vibrations of the benzene ring. With MgO, the ratio was the same as that of phenol in the liquid phase, while it was quite different from that of liquid phenol with $SiO_2-Al_2O_3$. From these observations, they suggested the cause of the selectivity difference in the two catalysts: On the acidic oxide, the interaction of the aromatic ring of the phenolate and the surface is strong and the aromatic ring of the phenolate lies close to the surface. This facilitates the ring alkylation at *meta* and *para* positions and also *o*-alkylation. On the other hand, the interaction of the phenolate and the surface is weak on the basic oxide and the aromatic ring of the phenolate is in a more or less upright posture. This inhibits alkylation at positions other than the *ortho* position.

The surface phenolates are also suggested to be the intermediates for *ortho*-alkylation of phenol with methanol over $ZnO-Fe_2O_3$[45)] and $Ca_3(PO_4)_2$[44)] from infrared spectroscopic studies. Kapsi and Olah studied the methylation of phenol and the rearrangement of anisole and methyl anisole over Nafion-H®, and concluded that *o*-methylation forming anisole is followed by intermolecular O→C methyl transfer leading to the formation of cresols.[40)]

Namba *et al.* studied the alkylation of phenol with methanol over H, K−Y zeolites with varying ratio of H/K and examined the dependence of the product selectivity on the acid strength of the zeolites.[46)] Zeolites with acid strength of $-3.0 \geq H_0 \geq 8.2$ gave *o*- and *p*-cresols selectively. Zeolites with weaker acid sites favored anisole, while those with stronger acid sites favored the formation of xylenols and *m*-cresol. Over H, K−Y zeolite (K, 85%), a 35% yield of cresols was obtained with 65% *para* and 35% *ortho* isomers.

Alkylation of phenol with 2-propanol over H−ZSM-5 catalysts gives more than 50% selectivity for *para* isomer in isopropylphenols at 523−573 K, while the selectivity is 20−25% with amorphous $SiO_2-Al_2O_3$.[47)]

Phenols can also be alkylated with olefins. The alkylation of *m*-cresol with propylene to produce thymol (2-isopropyl−5-methylphenol) was studied with metal sulfates and alumina as catalysts.[48,49)] A 90% selectivity to thymol was obtained at the phenol conversion of 63% from a 1:1 mixture of phenol and propylene at 673 K.

Kijiya *et al.* reported the alkylation of phenol with isobutane with $SiO_2-Al_2O_3$, Zn or Ca-exchanged X-type zeolites at 573 K.[50)] High selectivity to *p*-isobutylphenol (to 90%) was obtained. Metal oxides such as MgO, CaO, and TiO_2 showed no activity for the reaction.

Ion exchange resins are effective catalysts for alkylation of phenol. The reaction of of phenol with nonene, dodecene and 2-methylpentene-1 proceeds under reflux in mixtures of reactant and water. The *ortho*-to-*para* ratios of the product phenol depends on the amount of water in the mixtures.[51]

Bisphenol A (2,2′-bis (4′-hydroxyphenyl)-propane) can be prepared from a 10:1 mixture of acetone and phenol over cation-exchange resin of which sulfo-groups are partially esterified with mercaptoethanol.[52]

4.2.6 Side-chain Alkylation of Aromatics

While alkylation of aromatics with olefins or alcohols occurs at the aromatic ring over acid catalysts, alkylation of the alkyl groups proceeds over basic catalysts. Pines and coworkers reported that the side-chain alkylation of toluene with ethylene is effectively catalyzed by the use of a mixture of sodium and a promotor such as anthracene or *o*-chlorotoluene.[53]

Podall and Foster reported that the reaction of toluene with olefins with KC_8, a graphite inclusion compound, gave the alkylation of the side chain.[54] A 50% conversion of toluene to 3-phenylpentane was obtained together with higher alkylbenzenes at 298 K from toluene and ethylene. At 323 K, the main product was propylbenzene (48%) together with 3-phenylpentane and a small quantity of higher alkylated products. Similarly, the reaction of isopropylbenzene with ethylene gave a 42% yield of *t*-amylbenzene at 473 K.

In recent years, particular attention has been paid to the side-chain alkylation of toluene with methanol to styrene and ethylbenzene. The commercial incentive stems from using toluene, instead of more expensive benzene, as the raw material for the production of styrene.

Sidorenko *et al.*[55] found that the alkylation of toluene with methanol over alkali metal-exchanged zeolites gives a mixture of xylenes, styrene and ethylbenzene at 678 and 728 K. In particular, KX (K^+ exchanged X-type zeolite) and RbX gave predominantly styrene and ethylbenzene. Yashima *et al.* studied the reaction in more detail.[56] Over LiX or LiY zeolite, xylenes were the sole products, while over Na^+, K^+, Rb^+, and Cs^+-exchanged zeolites, styrene and ethylbenzene were produced selectively. The activity for side-chain alkylation has a tendency to be greater with X-type zeolites than the corresponding Y-type zeolites, and also depends on the size of the alkali metal cations, that is, Na < K < Rb < Cs. These trends were also found in the alkylation of toluene with formaldehyde. Addition of hydrogen chloride to the reaction system promoted the ring alkylation and inhibited the side-chain alkylation. On the other hand, addition of aniline inhibited xylene formation over LiY, but promoted the side-chain alkylation. From these facts, Yashima *et al.*[56] stressed the importance of basic sites in side-chain alkylation. The basicity of KX and KY was confirmed by the color change of adsorbed indicators, cresol red and thymolphtalein.

Unland *et al.*[57] made an extensive study of the side-chain alkylation of toluene with methanol. They confirmed the general features of the alkylation, which had been reported by other investigators. In addition, they found that the addition of certain inorganic materials such as phosphoric acid or boric acid to the ion-exchange solution improved the selectivity for the side-chain alkylation.[57–59] The borate-promoted CsX zeolite was the most favorable and a selectivity of >50% on the methanol basis for the side-chain alkylation was obtained, as shown in Fig. 4.4.[58] Here, toluene and methanol at a mole ratio of 5.2/1.0 was fed at space velocity of 950 h^{-1} at 683, 703 and 673 K. From the IR, Raman and NMR studies,[58-61] Unland and coworkers[57] suggest that high selectivity with CsX is based on the adsorption of a toluene molecule between two (or more) large cations in an overcrowded supercage of X-type zeolite in such a way that (1) the electrostatic potential at the molecule is higher than one would normally expect; (2) only the methyl group is exposed for alkylation; (3) because of the strong interaction of cations with aromatic molecules, the protons of the methyl group become more acidic and susceptible to attack. They also suggest that incorporation of borate in the supercage is slowing down the decomposition of formaldehyde, a real alkylating agent.

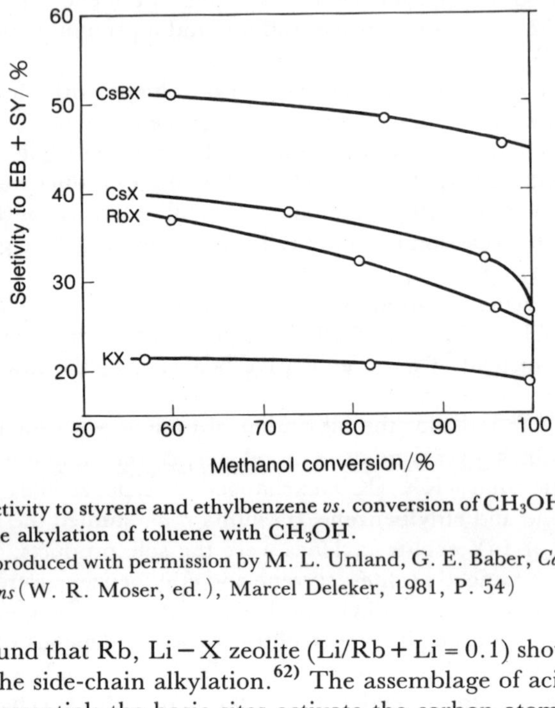

Fig. 4.4 Selectivity to styrene and ethylbenzene *vs*. conversion of CH_3OH for various zeolites in the alkylation of toluene with CH_3OH.
(Reproduced with permission by M. L. Unland, G. E. Baber, *Catalysis of Orgenic Reactions* (W. R. Moser, ed.), Marcel Deleker, 1981, P. 54)

Itoh *et al*. found that Rb, Li−X zeolite (Li/Rb + Li = 0.1) showed a higher activity than RbX for the side-chain alkylation.[62] The assemblage of acid and base sites was assumed to be essential; the basic sites activate the carbon atom of the side chain of toluene and the acid sites adsorb and stabilize toluene molecules.[63] They further suggest that weakly acidic sites are generated by incorporation of Li cations and this serves also to suppress the decomposition of formaldehyde.[64]

The reactions of xylenes and ethylbenzene with methanol over RbX also give side-chain alkylation products.[56,64] Alkylation of toluene with ethylene over RbX gives isopropylbenzene and α-methylstyrene.[56] Similarly, the reactions of α- and β-methylnaphthalene with methanol give the corresponding ethylnaphtalenes with traces of vinylnaphthalenes over KX and RbX at 670−720 K.[65]

Side-chain alkylation of toluene with methanol also proceeds over alkali metal oxides supported on active carbon.[66]

4.2.7 *N*-Alkylation of Aniline with Methanol or Dimethyl Ether

Alumina is one of the best catalysts for *N*-methylation of aniline with methanol to form *N,N*-dimethylaniline.[67,68] Evans and Bourns[68] reported that under optimum conditions, 558 K and a molar ratio methanol to aniline of 10:1, a 95.5% yield of *N,N*-dimethylaniline was obtained, the remainder being mainly *N*-methylaniline with trace amounts of rearrangement products (ring-methylated products). A considerable amount of methanol was converted to dimethyl ether.

Aniline can also be *N*-methylated with dimethyl ether. With alumina as catalyst, dimethylaniline yield of 98.5% to 99% was obtained at 548 to 573 K using a dimethyl ether-to-aniline ratio of 5:1 at LHSV of 0.08.[68]

Takamiya *et al.*[69] reported aniline alkylation with methanol over MgO catalysts. The products consisted exclusively of *N*-methylaniline. The optimum reaction temperature was 753 K. The activity depends on the type of MgO catalyst. The MgO containing 2 wt% SO_4^{2-} showed the highest activity, followed by the MgO prepared from hydroxide, the MgO containing 2 wt% PO_4^{3-}, and the MgO prepared from carbonate. The reaction was retarded by introduction of both pyridine and carbon dioxide, indicating that both acid and base sites are required for the reaction to proceed.

The Hammett plot for substituted anilines gave $\rho = -1.73$. The negative values of ρ indicate the reaction as being electrophilic. The rate determining step was suggested to be the attack of methyl cation to anilino group (step III) in the following scheme.

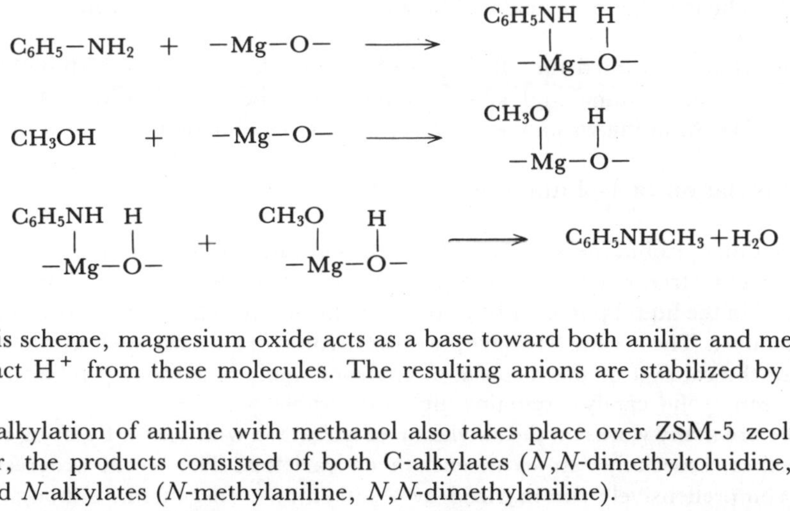

I

II

III

In this scheme, magnesium oxide acts as a base toward both aniline and methanol to abstract H^+ from these molecules. The resulting anions are stabilized by Mg^{2+} cation.

The alkylation of aniline with methanol also takes place over ZSM-5 zeolites.[70] However, the products consisted of both C-alkylates (*N,N*-dimethyltoluidine, toluidine) and *N*-alkylates (*N*-methylaniline, *N,N*-dimethylaniline).

The modification of ZSM-5 with metal oxides and variation of Si/Al ratio in ZSM-5 catalysts result in ZSM-5 catalysts of different acid-base properties. In Fig. 4.5, the correlation between the aniline conversion and the acid amount is shown for Na−ZSM-5 of different Si/Al ratios. Correlation is clearly observed between the conversion and the acid amount, indicating the presence of weak acid sites being required. Besides acid sites, the presence of base sites is required. The ZSM-5 catalysts modified with MgO or Cs_2O which possess large quantity of basic sites show high conversions

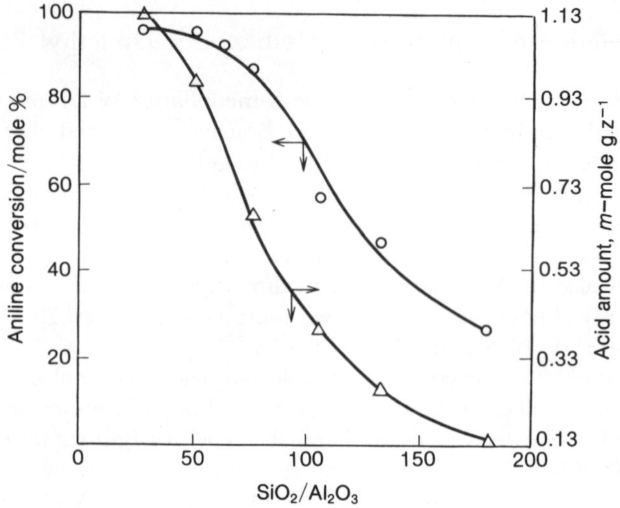

Fig. 4.5 Effect of NaZSM-5 with various SiO_2/Al_2O_3 ratio on aniline conversion., Conditions :
calcination temp. 823 K; calcination time, 3h; reaction temp., 693 K;
whsv, 0.8 h^{-1}; time on stream, 4 h; MeOH/Aniline=3.
(Reproduced with permission by P. Y. Chem, *et al.*, Proc. 7th Intern. Zeolite Conf., 1986,(Y. Murakami, ed.), 1986, p. 741).

of aniline. The fact that both acid and base sites are required is supported by poisoning experiments.

Parera *et al.*[71] studied the alkylation of *N*-methylaniline to *N*,*N*-dimethylaniline over a series of alumina and silica-alumina. The best catalyst was the synthetic $SiO_2-Al_2O_3$. Alumina was a good catalyst, but gave dimethyl ether as a byproduct.

4.2.8 Alkylation of Isobutane with Olefins

Alkylation of isobutane with C_3 to C_5 olefins to form C_7 to C_9 isoparaffins is a very important industrial reaction for the production of high octane fuels. The reaction is performed in the liquid phase with sulfuric acid or nearly anhydrous hydrogen fluoride as the catalyst. These catalysts have, however, some drawbacks such as the corrosive nature of the catalyst or the disposal of enviromentally hazardous products. A clean process using solid catalyst remains highly desirable.

There are many works on the alkylation of isobutane with olefins by using faujasite-type zeolites, especially rare-earth exchanged varieties. The recent progress in this field is comprehensively summarized by Weitkemp.[72] In general, although zeolites are initially very active, they undergo rapid deactivation.

Weitkemp[72-74] studied the time course of alkylation of isobutane with *n*-butenes over CaX and CaY zeolites. At low times on stream, alkylation is extremely selective; no olefins, naphthenes or aromatics are formed. During this initial "alkylation stage," the conversion of the feed olefins is 100%. Carbon number distribution of the products at this stage is given in Fig. 4.6. A complex mixture of C_5 to C_{12} isoparaffins is formed. In all cases, isooctanes predominate, though the distribution changes with time on

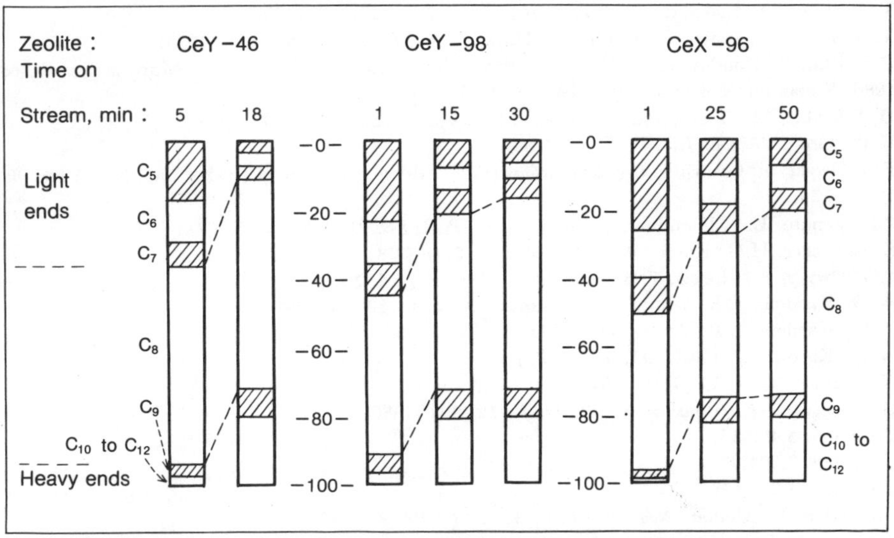

Fig. 4.6 Alkylation of 150 butane with *n*-butenes on cerium-exchanged faujasites (the numbers stand for %-exchange) isobutane/butene=11/1, fixed bed reactor, $m_{CeY-46}=1.1$ g, $m_{CeY-98}=1.4$ g, $m_{CeX-96}=1.5$ g, T=353 K, P=3.1 MPa, liquid feed rate=7.5 cm^3/h. Carbon number distributions in wt.−%.
(Reproduced with permission by J. Weitkemp, Proc. Inter. Symp. on Zeoite Catalysis, 1985, p. 280).

stream. After a certain time, the alkylation stage ends and butenes begin to appear After this point, the C_8 fraction mostly consists of octanes, indicating that butenes ar consumed mainly by dimerization or oligomerization. This change indicates that th strong acid sites required for hydride-transfer are deactivated by carbonaceou deposits.

Because of the rapid deactivation, the alkylation of isobutane with butenes is eco nomically unattractive at the present time.

REFERENCES

1. G.A. Olah, *Friedel-Crafts and Related Reactions*, Vol. I-IV Interscience, New York, 1964.
2. P.B. Venuto, L.A. Hamilton, P.S. Landis, J.J. Wise, *J. Catal.*, **5**, 81 (1966)
3. P.B. Venuto, L.A. Hamilton, P.S. Landis, *J. Catal.*, **5**, 484 (1966).
4. T. Yashima, H. Amhad, K. Yamasaki, M. Katsuta, N. Hara, *J. Catal.*, **16**, 273 (1970).
5. N.Y. Chen, W.W. Kaeding, F.G. Dwyer, *J. Am. Chem. Soc.*, **101**, 6784 (1979).
6. W.W. Kaeding, C. Chu, L.B. Young, B. Weinstein, S.A. Butter, *J. Catal.*, **67**, 159 (1981).
7. T. Yashima, Y. Sakaguchi, S. Namba, Proc. 7th Intern. Congr. Catalysis (T. Seiyama, K. Tanabe eds.), 1981, Kodansha, Tokyo, p.739.
8. L.B. Young, S.A. Butter, W.W. Kaeding, *J. Catal.*, **76**, 418 (1982).
9. T. Yashima, N. Yokoi, N. Hara, *Bull. J. Jpn. Petrol. Inst.*, **13**, 215 (1971).
10. S. Namba, A. Inaka, T. Yashima, *Zeolites*, **3**, 106 (1983).

11. T. Yashima, A. Inaka, S. Namba, *J. Jpn. Petrol. Inst.*, **28**, 13 (1985).
12. T. Yashima, A. Inaka, S. Namba, N. Hara, *J. Jpn. Petrol Inst.*, **28**, 498 (1985).
13. C.F. Ren, C. Condurier, C. Naccache, Proc. 7th Intern. Zeolite Conf. (Y. Murakami *et al.*, eds.), 1986, Kodansha, Tokyo and Elsevier, Amsterdam, p.733.
14. D. Fraenkel, M. Cherniavsky, B. Ittah, M. Levy, *J. Catal.*, **101**, 273 (1986).
15. H.W. Grote, *Oil Gas J.*, **56** (13); **73** (1956).
16. F.G. Dwyer, in: *Catalysis of Organic Reactions* (W.R. Moser, ed.) Marcel Dekker Inc., New York, Basel, 1981, p.39.
17. P.B. Venuto, L.A. Hamilton, *Ind. Eng. Chem. Prod. Res. Dev.*, **6**, 190 (1967).
18. K.A. Becker, H.G. Karge, W.D. Streubel, *J. Catal.*, **28**, 403 (1973).
19. F.G. Dwyer, P.J. Lewis, F.H. Schneider, *Chem. Eng.*, **1976** (1) 55.
20. W.W. Kaeding, L.B. Young, A.G. Prapas, *Chemtech,*, **12**, 556 (1982).
21. W.W. Kaeding, L.B. Young, C-C. Chu, *J. Catal.*, **89**, 267 (1984).
22. W.W. Kaeding, *J. Catal.*, **95**, 512 (1985).
23. *Hydrocarbon Process*, **55** (3) 91 (1976).
24. H. Hasegawa, T. Higashimura, *Polym. J.*, **12**, 407 (1980).
25. US Patent 3,458,587
26. US Patent 3,504,046
27. US Patent 4,288,646
28. G.A. Olah, D. Meidar, *Nouv. J. Chim.*, **3**, 269 (1979).
29. K. Arata, I. Toyoshima, *Chem. Lett.*, **1974**, 929.
30. K. Arata, K. Yabe, I. Toyoshima, *J. Catal.*, **44**, 385 (1976).
31. K. Arata, A. Fukui, I. Toyoshima, *J. Chem. Soc., Chem. Comm.*, **1978**, 121.
32. M. Hino, K. Arata, *Chem. Lett.*, **1977**, 277.
33. K. Arata, M. Hino, K. Yabe, *Bull. Chem. Soc. Jpn.*, **53**, 6 (1980).
34. K. Arata, M. Hino. *Chem. Lett.*, **1980**, 1479.
35. M. Hino, K. Arata, Chem. *Lett.* **1979**, 1141.
36. G.A. Olah, D. Meider, P. Malhotra, J.A. Olah, *J. Catal.*, **61**, 97 (1980).
37. M. Inoue, S. Enomoto, *Chem. Pharm. Bull. (Tokyo),* **20**, 232 (1972).
38. K. Tanabe, T. Nishizaki, Proc. 6th Intern. Congr. Catal. 1956, London, 1977, p.863.
39. M. Inoue, S. Enomoto, *Chem. Pharm. Bull. (Tokyo),* **19**, 2518 (1971).
40. J. Kapsi, G.A. Olah, *J. Org. Chem.*, **43**, 3142 (1978).
41. Y. Fukuda, T. Nishizaki, K. Tanabe, *Nippon Kagaku Zasshi*, **1972**, 1754 (in Japanese).
42. Jpn. Kokai Tokkyo Koho, 48-97825, 99128, 99129, 49-7235, 13128, 14432, and 18834
43. T. Kotanigawa, M. Yamamoto, K. Shimokawa, Y. Yoshida, *Bull. Chem. Soc. Jpn.*, **44**, 1961 (1971); T. Kotanigawa, K. Shimokawa, *Bull. Chem. Soc. Jpn.*, **47**, 1555 (1974).
44. F. Nozaki, I. Kimura, *Bull. Chem. Soc. Jpn.*, **50**, 614 (1977).
45. T. Kotanigawa, *Bull. Chem. Soc. Jpn.*, **47**, 950 (1974).
46. S. Namba, T. Yashima, Y. Itaba, N. Hara, in: *Catalysis by Zeolites*, (B. Imelik *et al.*, eds.) 1980, Elsevier, Amsterdam, p.105.
47. US Patent 4,391,998.
48. M. Nitta, K. Yamaguchi, K. Aomura, *Bull. Chem. Soc. Jpn.*, **47**, 2897 (1974).
49. M. Nitta, K. Aomura, K. Yamaguchi, *Bull. Chem. Soc. Jpn.*, **47**, 2760 (1974).
50. M. Kijiya, S. Okazaki, *Nippon Kagaku Zasshi*, **1978**, 1071 (in Japanese).
51. Japan Patent 1962-18182
52. Japan Patent 1962-14721
53. H. Pines, J.A. Vasely, V.N. Ipatieff, *J. Am. Chem. Soc.*, **77**, 554 (1955).
54. H. Podall, W.E. Foster, *J. Org. Chem.*, **23**, 401 (1958).
55. Y.N. Sidorenko, P.N. Galich, V.S. Gutyrya, V.G. I1'in, I.E. Niemark, *Dokl. Akad. Nauk SSSR*, **173**, 132 (1967).
56. T. Yashima, K. Sato, T. Hayasaka, N. Hara, *J. Catal.*, **26**, 303 (1972).
57. M.L. Unland, G.E. Baker, in: *Catalysis in Organic Reactions* (W.R. Moser, ed.), Marcel Dekker, New York, Basel, 1981, p.51.
58. US Patent 4,115,424; 4,140,726

59. J.J. Freeman, M.L. Unland, *J. Catal.*, **54**, 183 (1978).
60. M.L. Unland, *J. Phys. Chem.*, **82**, 580 (1978).
61. M.D. Sefcik, *J. Am. Chem. Soc.*, **101**, 2164 (1979).
62. H. Itoh, T. Hattori, K. Suzuki, A. Miyamoto, Y. Murakami, *J. Catal.*, **72**, 170 (1981).
63. H. Itoh, A. Miyamoto, *J. Catal.*, **64**, 284 (1980).
64. H. Itoh, T. Hattori, K. Suzuki, Y. Murakami, *J. Catal.*, **79**, 21 (1983).
65. O.D. Konoval'chikov, P.N. Galich, V.S. Gutyrva, G.P. Lugovskaya, *Kinet. Katal.*, **9**, 1387 (1968).
66. Jpn. Kokai Tokkyo Koho, 45-133932.
67. A.G. Hill, J.H. Shipp, A.J. Hill, *Ind. Eng. Chem.*, **43**, 1579 (1981).
68. T.H. Evans, A.N. Bourns, *Can. J. Tech.*, **29**, 1 (1951).
69. N. Takamiya, Y. Koinuma, K. Ando, S. Murai, *Nippon Kagakukaishi*, **1979**, 1452 (in Japanese).
70. P.Y. Chen, M.C. Chen, H.Y. Chu, N.S. Chang, T.K. Chuang, Proc. 7th Intern. Zeolite Conf., 1986, Kodansha, Tokyo and Elsevier, Amsterdam, p.739.
71. J.M. Parera, A. González, M.M. Barral, *Ind. Eng. Chem. Prod. Res. Dev.*, **7**, 259 (1968).
72. J. Weitkemp, Proc. Intern. Symp. Zeolite Catalysis, Siofok, 1985, p.271.
73. J. Weitkemp, Proc. 5th Intern. Conf. Zeolites (L.V.C. Rees, ed.) Heydon, London, 1980, p.858.
74. J. Weitkemp, *in: Catalysis by Zeolites* (B. Imelik *et al.*, eds.) Elsevier, Amsterdam, 1980, p.65.

4.3 ACYLATION

Acylation reactions using Lewis acid catalysts such as $AlCl_3$, $TiCl_4$, $SnCl_4$, $FeCl_3$, etc. and Brønsted acid catalysts such as CF_3SO_3H, FSO_3H, etc. are important in organic synthesis and the chemical industry (manufacture of weed killers, etc.). However, the process has several disadvantages; i.e. waste of large amounts of catalyst, corrosion of reactor, water pollution by acidic waste water, and difficulty of catalyst recovery. In order to eliminate these disadvantages of the homogeneous reaction, the use of solid acid catalysts such as heteropoly acids,[1)] activated iron sulfate,[2)] and iron oxide[3)] has been attempted, but it was found that these catalysts dissolve into the reaction mixture during the reaction and do not act as heterogeneous catalysts.[4)]

Ordinary solid acids such as $SiO_2-Al_2O_3$ and zeolites are almost or completely inactive for acylation reactions. However, a solid superacid, $ZrO_2-SO_4^{2-}$ (cf. Section 3.9) was recently found to exhibit high activity for the acylation of chlorobenzene or toluene with benzoyl chloride or *o*-chlorobenzoyl chloride in liquid phase.[5)] This was ascertained by separation of the solid superacid from the reaction mixture during a reaction in which the solid superacid acted as the perfect heterogeneous catalyst.

Catalytic activities of various solid acids for the acylation of chlorobenzene with *o*-chlorobenzoyl chloride are shown in Table 4.7.[5)] The yield in the case of $ZrO_2-SO_4^{2-}$ was 100% at 406 K for 10 h, whereas yield in the case of ZSM-5 and $SiO_2-Al_2O_3$ was 0 and 0.17%. In the acylation of toluene with *o*-chlorobenzoyl chloride, the yield of substituted benzophenone derivatives was 93% at 373 K for 1 h over ZrO_2 containing 7 wt% of SO_4^{2-} (Table 4.8), while ZSM-5 and $SiO_2-Al_2O_3$ did not show any activity under the same reaction condition.

$FeSO_4$ calcined at 973 − 1073 K and activated with benzyl chloride was reported to be active for the acylation of toluene and benzene with acetyl halides or acetic acid anhydride. The catalytic activity of $FeSO_4$ is higher than that of $AlCl_3$ and $FeCl_3$ as shown in Table 4.9.[2)] However, it is highly probable that $FeSO_4$ reacts with benzyl chloride as an activator or acetyl chloride to form $FeCl_3$ which acts as a homogeneous catalyst. To examine whether the active species is on the solid surface or in the liquid, the solid catalyst was separated from a reaction mixture by filtration during the reaction and the reaction was continued without solid catalyst, but the time variation of

TABLE 4.7 Activities of solid acid catalysts for acylation

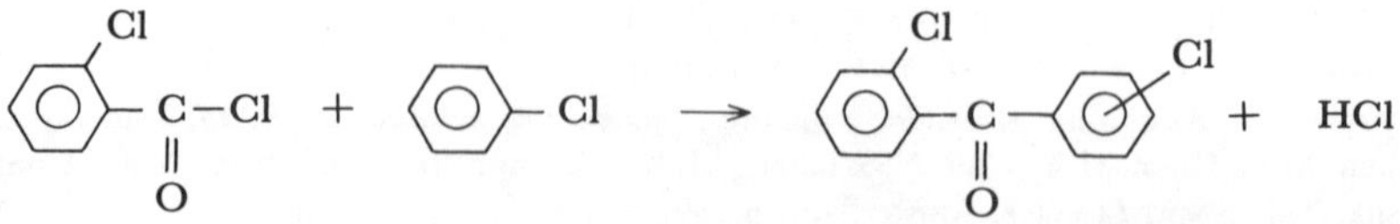

Catalyst	Amount g	Reaction temp., K	Reaction time, h	Product: Yield %	Composition, % 2,4′-	2,2-	4,4′-
ZrO_2	3	408	3	0			
$ZrO_2-SO_4^{2-}$	3	406	1	26.2	88.1	11.6	0.3
$ZrO_2-SO_4^{2-}$	6.1	406	10	100			
ZrO_2-NH_4F	3	406	3	0			
$ZrO_2-SnO_2-SO_4^{2-}$	3	406	3	0.6			
$SiO_2-Al_2O_3$	2.5	406	1	0.17	90.4	9.6	0
$SiO_2-Al_2O_3-NH_4F$	3	406	3	0			
$TiO_2-SO_4^{2-}$	3	406	1	6.3	87.6	12.4	0
ZSM-5	3	408	3	0			
Mg-Y	3	408	3	0			
H_2SO_4	0.2 ml	408	3	0.17	91.0	6.6	2.4

TABLE 4.8 Acylation of substituted benzene derivatives with benzoyl chlorides over $ZrO_2-SO_4^{2-}$[†1] Reaction time: 60 min

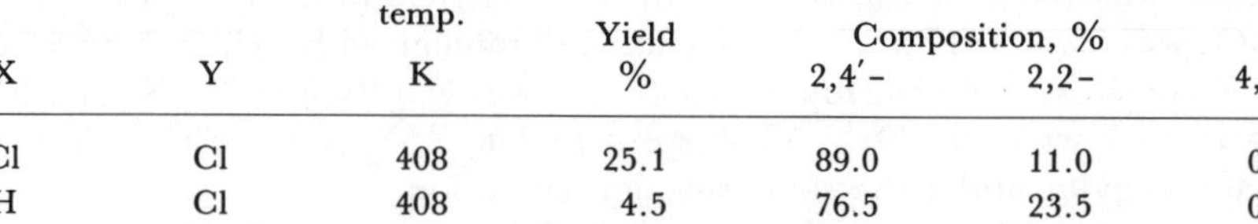

Substituent X	Substituent Y	Reaction temp. K	Product: Yield %	Composition, % 2,4′-	2,2-	4,4′-
Cl	Cl	408	25.1	89.0	11.0	0
H	Cl	408	4.5	76.5	23.5	0
Cl	CH_3	383	93.0	84.4	14.2	1.4
H	CH_3	383	13.8	69.1	27.3	3.6

[†1] SO_4^{2-} content: 7 wt %.

TABLE 4.9 Acylation of toluene with acetic acid anhydride

Catalyst	Calcination temp. K	Reaction temp. K	Yield, %		Isomer fraction		
			3 h	5 h	*ortho*	*meta*	*para*
$FeSO_4$	973	353	38	44			
	973	373	39		15	3	82
	1073	353		33			
	1073	373	48	55	14	2	84
$FeCl_3$		353	15		8	2	90
$AlCl_3$		353	11		4	6	90

the reaction of the liquid portion is almost the same as that of a continuous run where solid catalyst is included. Therefore, the true active species in the case of activated $FeSO_4$ is most likely in liquid phase and the reaction proceeds homogeneously. In the case of $ZrO_2-SO_4^{2-}$, the acylation reaction completely stopped when solid catalyst was separated from the reaction mixture. Thus $ZrO_2-SO_4^{2-}$ is concluded to be a true heterogeneous acid catalyst.[5)]

Besides $ZrO_2-SO_4^{2-}$, a superacidic perfluororesin sulfonic acid (Nafion-H®) is reported to be catalytically active for several types of acylation reactions.[6)]

REFERENCES

1. K. Nomiya, Y. Sugaya, S. Sasa, M. Miwa, *Bull. Chem. Soc. Jpn.*, **53**, 2089 (1980).
2. K. Arata, M. Hino, *Bull. Chem. Soc. Jpn.*, **53**, 446 (1980).
3. K. Arata, M. Hino, *Chem. Lett.*, **1980**, 1479.
4. T. Yamaguchi, A. Mitoh, K. Tanabe, *Chem. Lett.*., **1982**, 1229.
5. K. Tanabe, T. Yamaguchi, K. Akiyama, A. Mitoh, K. Iwabuchi, K. Isogai, Proc. 8th Intern. Congr. Catal., Berlin, 1984, Vcrlag Chemie, Weinheim, Vol.5, p.601.
6. G.A. Olah, P.S. Iyer, G.K. Surya Prakash, *Synthesis*, **1986**, 513.

4.4 TRANSALKYLATION OF ALKYLAROMATICS

4.4.1 General Mechanism

In transalkylation, one of the alkyl groups is transferred from one alkylaromatic molecule to another aromatic molecule. The mechanism of transalkylation was studied extensively in Friedel–Crafts chemistry. Though the reaction conditions are quite different from those of Friedel–Crafts catalysts, it seems quite probable that an essentially same mechanism is operative also in transalkylation with solid-acid catalysts. Thus, Kaeding *et al.* proposed the following mechanism for disproportionation of toluene over zeolites.[1)]

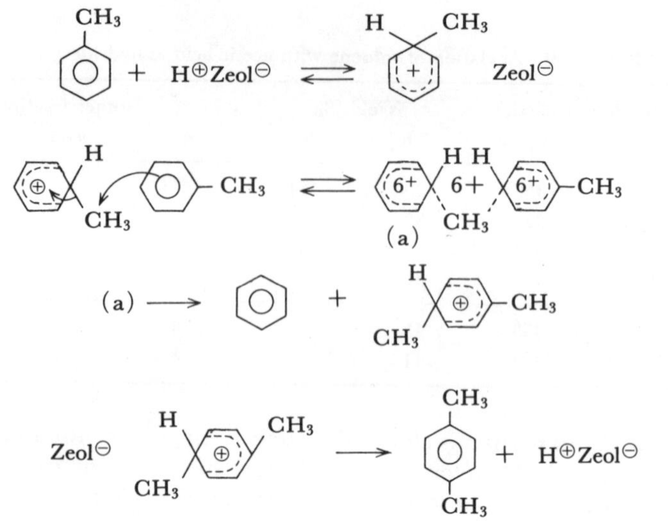

(1)

(2)

(3)

(4)

The protonation of an alkylaromatic molecule occurs at its ipso position (eq. 1). This weakens the carbon – methyl bond and initiates transfer to a second aromatic molecule (eqs. 2 and 3). Transfer of a proton back to the zeolite from the protonated xylene gives the xylene product and regenerates the acid site in the catalyst (eq. 4).

4.4.2 Disproportionation of Toluene

The most important transalkylation from the industrial standpoint is the disproportionation of toluene into benzene and xylenes, especially *p*-xylene, since *p*-xylene is a starting material for terephthalic acid, a major component in polyester fibers.

Amorphous silica-alumina was the first catalyst used for this disproportionation. The first industrial process using $SiO_2-Al_2O_3$ was the Xylene – Plus process established by Atlantic Richfield Co.[2]

Since the development of zeolite, chemistry transalkylation has been studied mainly using zeolite catalysts. Frilette used natural mordenite treated by acid.[3] The activity was much higher than amorophous $SiO_2-Al_2O_3$, but the activity could not be maintained. Benesi reported that mordenite was about 8 times more active than Y-type zeolites and that the active centers were Brønsted acid sites.[4] Various efforts including dealumination and cation exchange have been made to improve the aging.

In 1969, a commercial process (Tatray process) using a mordenite-based catalyst was announced by Toray Industries.[5] The reaction conditions are 620 – 720 K, 20 – 30 bar, H_2/hydrocarbon molar ratio of 6 – 10.

ZSM-5 zeolites are also active catalysts for transalkylation reactions. A high concentration of *para* isomer is attained by modifying the zeolite with inorganic oxides.[6,7] The reaction over ordinary H – ZSM-5 gives a near-equilibrium mixture of xylene isomers. Modification of ZSM-5 with phosphorus, boron, or magnesium compounds reduces the catalytic activity for the disproportionation. However, the concentration of *para* isomer in the xylene product increases significantly. Results with a catalyst containing about 11wt% magnesium, present as an oxide, are shown in Fig. 4.7.[7] Here, toluene conversion at each temperature was varied by changing the space velocity. Para-selectivity of 80 – 90% is obtained. The selectivity decreased with toluene conversion due to the isomerization of the primary products. The higher reaction temperature favors the *para* selectivity. Approximately equimolar amounts of benzene and xylenes are formed, indicating the absence of dealkylation reaction.

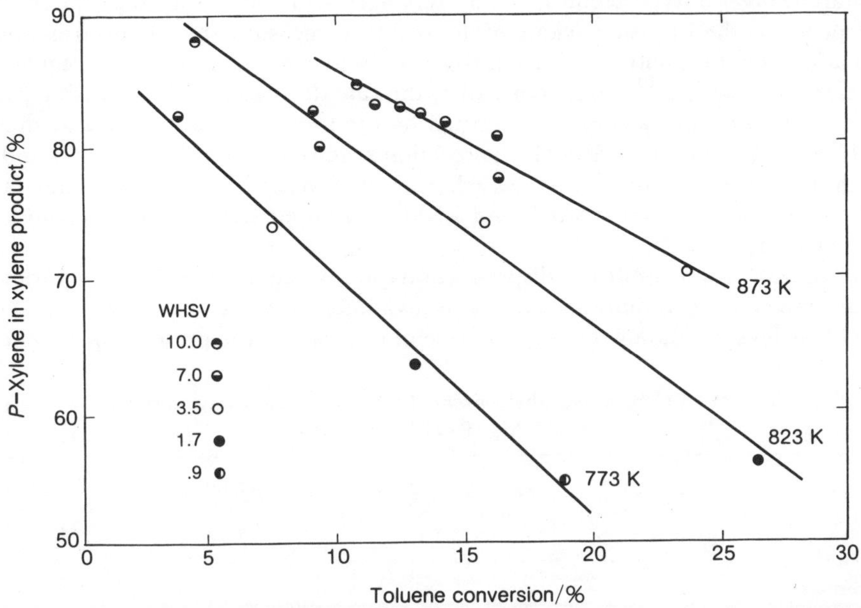

Fig. 4.7 *Para*-selectivity in toluene disproportionation over ZSM-5 modified with magnesium oxide (Mg=11 wt %).
(Reproduced with permission by W. W.Kaeding *et al.*, *J. Catal.*, 69, 396(1981)).

Treatment with inorganic compounds is considered to reduce the dimensions of pore openings and channels sufficiently to favor outward diffusion of *p*-xylene, the isomer with the smallest molecular dimension.

Young *et al.*[8] suggested the following kinetic situation under the toluene disproportionation conditions. The transalkylation reaction to form benzene and xylenes within the pores is relatively slow. Benzene diffuses out of the pores rapidly. The xylenes isomerize rapidly within the pores. (Xylene isomerization is about 1000 times faster than toluene disproportionation.) *para*-Xylene diffuses out moderately fast while the *ortho* and *meta* isomers move within the pores relatively slowly and further convert to *para* isomer before escaping from the channel system.

It was also suggested that zeolite-mediated steric effects in the xylene-forming transition state may contribute to enhancing the amount of *para*-isomer formed initially in the pores.[8]

Further evidence for diffusion control of *para*-xylene selectivity in toluene disproportionation over ZSM-5 catalysts has been described by Haag and Olson, who noted a good correlation between the sorption rate of *o*-xylene and the *para*-selectivity in toluene disproportionation for several ZSM-5 catalysts including large crystallite and inorganic-modified ZSM-5.[9]

4.4.3 Transalkylation of Alkylaromatics Other Than Toluene

Karge and coworkers[10-12] studied disproportionation of ethylbenzene over various catalyst. Over Y-type zeolites cation-exchanged with various cations, the reaction rate depends on the Brønsted acidity of the zeolite as measured by IR spectroscopy.[10] It was also noted that only very strong Brønsted acid sites ($H_0 \leq -8.2$) are capable of catalyzing the reaction.[10] Over mordenite, the rate decreased with increasing size of alkaline earth ions and was again governed by the number of Brønsted acid sites.[11] Over H−ZSM-5 and H−ZSM-11, the reaction required higher temperatures and did not exhibit any induction period, which was observed with mordenite and faujasites.[12] The reaction over ZSM-5 and ZSM-11 showed shape-selectivity, no *ortho*-isomer being formed.[12]

Weitkemp[13] carried out the disproportionation of ethylbenzene over a variety of catalysts and concluded that this reaction is a valuable test reaction for the characterization of zeolites of unknown structure. Useful criteria are the presence or absence of

TABLE 4.10 Selective ethylbenzene disproportionaion: large-crystal Mg-P-ZSM-5

	Run no.		
	1	2	3
Conditions			
Temp./K	698	748	798
Pressure/kg cm^{-2}	7.0	7.0	7.0
EB	30.2	30.2	30.2
H_2	0.24	0.24	0.24
TOS/h	429-430	520-521	538-539
Conversion			
EB	14.7	18.9	22.5
Selecivity for Products/wt %			
Benzene	42.7	50.4	62.4
Toluene	2.5	1.6	1.3
Xylene	0	0	0
Ethyltoluene	0.8	0.9	0.6
Diethylbenzene	46.1	31.7	15.4
Other aromatics	1.2	2.1	2.7
Light gas	6.7	13.3	17.6
Total	100.0	100.0	100.0
Diethylbenzene			
para	99.8	99.6	99.3
meta	0.2	0.4	0.7
ortho	0	0	0

(Reproduced with permission by W. W. Kaeding, *J. Catal.*, **95**, 518(1985)).

an induction period, rate of deactivation, yield ratio of diethylbenzene to benzene, and distribution of the diethylbenzene isomers.

A very high *para*-selectivity of inorganic-modified ZSM-5 was also manifested in the disproportionation of ethylbenzene. The results with ZSM-5 modified with magnesium and phosphorus compounds are shown in Table 4.10. The concentration of *para*-isomer in the diethylbenzene products was over 99%.[14)]

The importance of transition state-type selectivity was first demonstrated by Csicsery.[15,16)] In the reaction of 1-methyl – 2-ethylbenzene over mordenite, the amounts of 1,3-dimethyl – 5-ethylbenzene and 1-methyl – 3,5-dimethylbenzene were very small where the 1,3,5-trialkylbenzenes are the main components at equilibrium. It was concluded that symmetrical trialkylbenzenes cannot form in the pores of H-mordenite; too little space is available for diphenylmethane-type intermediates in transition states leading to symmetrical isomers. The other trialkylbenzene isomers can form because their transition states are smaller.

Similar behavior was observed in disproportionation of toluene over mordenite. Among trimethylbenzene isomers, 1,3,5-trimethylbenzene has the largest molecular size, and 1,2,4-trimethylbenzene has the smallest molecular size. Namba *et al.* carried out the disproportionation of *m*-xylene over a variety of cation-exchanged mordenites.[17)] In the trimethylbenzene products, the concentration of 1,2,4-trimethylbenzene is higher than the equilibrium value, at the expense of the concentration of 1,3,5-trimethylbenzene. The selectivities for 1,2,4-trimethylbenzene over various zeolites are given in Table 4.11. H-mordenite shows the highest activity, but it shows a selectivity only slightly higher than equilibrium. The selectivity increased progressively with increasing cation size though the activity is reduced at the same time. Ion-exchange with Cu^{2+} or Zn^{2+} is the most effective. This is ascribed

TABLE 4.11 Activities and selectivites of cation-exchanged mordenite in the disproportionation of *m*-xylene

Cation	Ionic radius/nm	Conversion/%^{†1}	Selectivity/%^{†2}
H		49	8
Be	0.033	29	8
Mg	0.062	24	17
Ni	0.065	25	28
Co	0.068	23	25
Zn	0.074	21	31
Cu (10 %, slow exchange)	0.096^{†3}	41	19
Cu (20 %, slow exchange)	0.096	26	33
Cu (5.5 % rapid exchange)	0.096	34	39
Ca	0.099	9	8
Sr	0.116	8	8
Ba	0.136	8	8

†1 Reaction conditions: 573 K, W/F=100 g h $mole^{-1}$

†2 Selectivity is defined as $(f-f_e)/(1-f_e)\times 100$, where f and f_e are the fraction of 1,2,4-isomer in trimethylbenzene products at 20% conversion, and at equilibrium, respectively.

†3 Ionic radius of Cu^+ since Cu^{2+} ions are assumed to be reduced to Cu^+ under the reaction conditions.

to the narrowing of the effective channel size by the presence of the cations in the mordenite pores. The larger cations such as Sr^{2+} or Ba^{2+} have no effect. With these cations, adsorption of reactant itself is probably hindered as judged from their low activity. Thus the reaction proceeds on the external surface of the crystallites. Namba *et al.* introduced "rapid" copper exchange, which deposits most of the copper near the external surfaces of the zeolite.[17] Thus, less copper (3 – 6% exchange) is needed than in slow exchange, which distributes the copper more evenly. By slow exchange, a high selectivity for 1,2,4-trimethylbenzene is obtained without reducing the catalytic activity of mordenite.

Transalkylation between isopropylbenzene and benzene to yield *n*-propylbenzene over ZSM-5 zeolites was reported by Beyer and Borbely.[18]

(5)

Since the transformation of isopropylbenzene to propylbenzene is not effected in the absence of benzene, it is obvious that this side-chain isomerization proceeds via intermolecular alkyl transfer. The transfer occurs below 530 K; at LHSV of 11.4, the 50% conversion level is reached at 560 K.

In analogy with eq. 5, propyltoluene isomers are formed when benzene is replaced by toluene in the reaction mixture. At LHSV of 11.4, 50% conversion level is reached at about 525 K; no side reactions were observed at this temperature. There is a pronounced shape selectivity effect, inhibiting the formation of *o*-propyltoluene.

Transalkylation between ethylbenzene-toluene proceeds over the same catalyst, again with negligible formation of *o*-methylethylbenzene. The [Ga] – ZSM-5 (ZSM-5 type zeolite containing gallium instead of aluminum) is also active for the reaction, though slightly less active than ZSM-5. There is no substantial difference in the selectivity of the two zeolites.

Nafion-H is a very useful catalyst for transalkylation reactions. Transfer of a *t*-butyl group occurs very easily over Nafion-H at temperatures as low as 330 K. For example, 2,6-di-*t*-butyl – *p*-cresol is dealkylated in 0.5 h to *p*-cresol. Toluene acts as a better acceptor than benzene.[19]

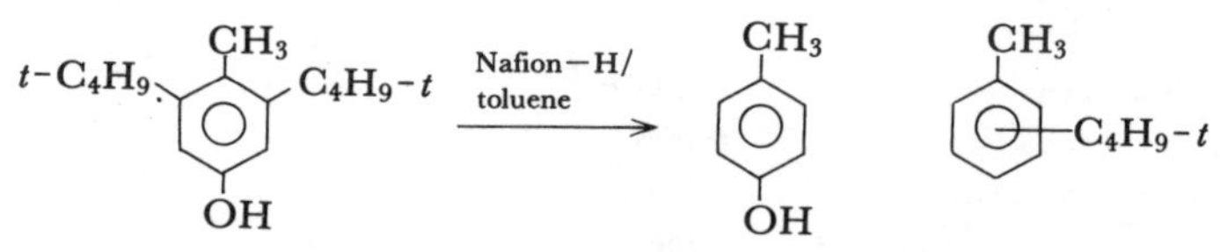

The synthesis of an industrially important intermediate, bisphenol, has been patented.[20,21]

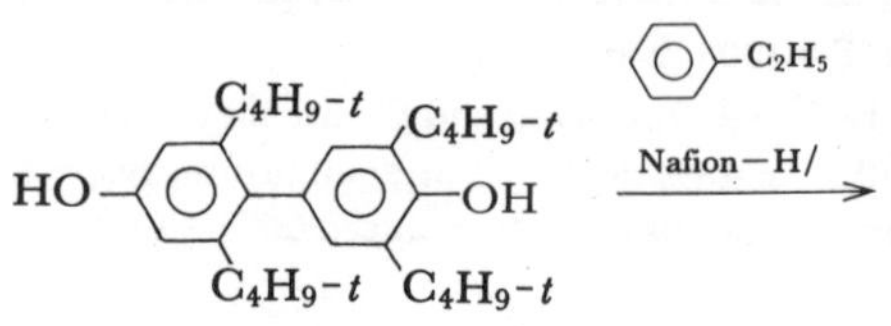

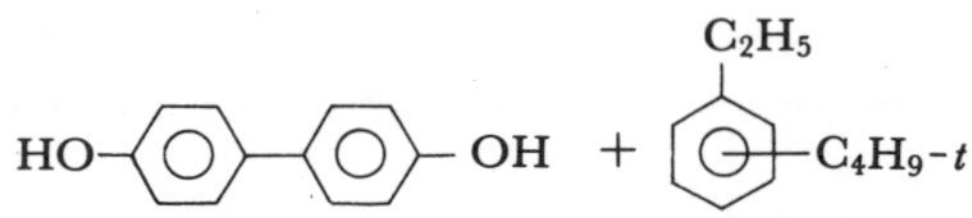

REFERENCES

1. W. W. Kaeding, C. Chu, L. B. Young, S. A. Butter, *J. Catal.*, **69**, 392 (1981).
2. J. A. Verdol, *Oil Gas J.*, **67** (23), 63 (1969).
3. US Patent 3,506,731.
4. H. A. Benesi, *J.Catal.*, **8**, 368 (1967).
5. *Hydrocarbon Process*, **58** (11), 140 (1979).
6. N. Y. Chen, W. W. Kaeding, F. G. Dwyer, *J. Am. Chem. Soc.*, **101**, 6783 (1979).
7. W. W. Kaeding, C. Chu, L. B. Young, S. D. Butter, *J. Catal.*, **69**, 392 (1981).
8. L. B. Young, S. A. Butter, W. W. Kaeding, *J. Catal.*, **76**, 418 (1982).
9. W. O. Haag, D. H. Olson, US Patent 4,117,026 (1978), cited in : P. B. Weisz, Proc. 7th Intern. Congr. Catal. (T. Seiyama, K. Tanabe eds.) Kodansha, Tokyo and Elsevier, Amsterdam, 1980, p.3.
10. H. G. Karge, K. Hatada, Y. Zhang, R. Fiedorow, *Zeolites*, **3**, 13 (1983).
11. H. G. Karge, J. Ladebeck, Z. Sarbak, K Hatada, *Zeolites*, **2**, 94 (1982).
12. H. G. Karge, Y. Wada, J. Weikemp, S. Ernst, U. Girrbach, H. K. Beyer, *Catalysis on the Energy Scene*, (S. Kaliaguine, A. Mahey, eds.) Elsevier, Amsterdam, 1984, p.101.
13. J. W. Weitkemp, *Erdol Kohle*, **39**, 13 (1986).
14. W. W. Kaeding, *J. Catal.*, **95**, 512 (1985).
15. S. M. Csicsery, *J. Catal.*, **19**, 394 (1970).
16. S. M. Csicsery, *J. Catal.*, **23**, 124 (1971).
17. S. Namba, O. Iwase, N. Takahashi, T. Yashima, N. Hara, *J. Catal.*, **56**, 445 (1975); T. Yashima, O. Iwase, N. Hara, *Chem. Lett.*, **1975**, 1215.
18. H. K. Beyer, G. Borbely, Proc. 7th Intern. Zeolite Conf. 1986, Kodansha, Tokyo and Elsevier, Amsterdam, p.867.
19. G. Olah, P. S. Iyer, G. K. S. Prakash, *Synthesis*, **1986**, 513.
20. US Patents 4487978; 4482755.
21. Eur. Patent 45959.

4.5 HYDRATION OF OLEFINS

The hydration of olefins is important for the direct synthesis of alcohols from olefins in the petroleum industry and has been extensively studied over various solid acid catalysts. In the case of ethanol synthesis from ethylene and water, silicotungstic acids,[1-5] silicophosphoric acids,[6] solid phosphoric acids,[5-7], metal sulfates,[8-10] and metal oxides[1,11,12] have been studied as solid acid catalysts. In its industrial process, a solid phosphoric acid catalyst (Shell patent) is widely used throughout the world. The nature of the active (acidic) sites which exhibit high catalytic activity and selectivity is discussed below together with the hydration mechanism involving the catalytic behavior.

4.5.1. Acidic Property *vs.* Catalytic Activity and Selectivity

A. Correlation Between Acidic Property and Catalytic Activity

When metal sulfates were used as catalysts for hydration of ethylene at 463 K, only ethanol was formed, no by-products such as ethylene polymer, diethyl ether or acetaldehyde being detected. The acid amounts of nickel sulfates preheated at various temperatures and their catalytic activities for hydration of ethylene are shown in Fig. 4.8.[10] The activities correlate well with the acid amounts at acid strength $H_0 \leq -3$,

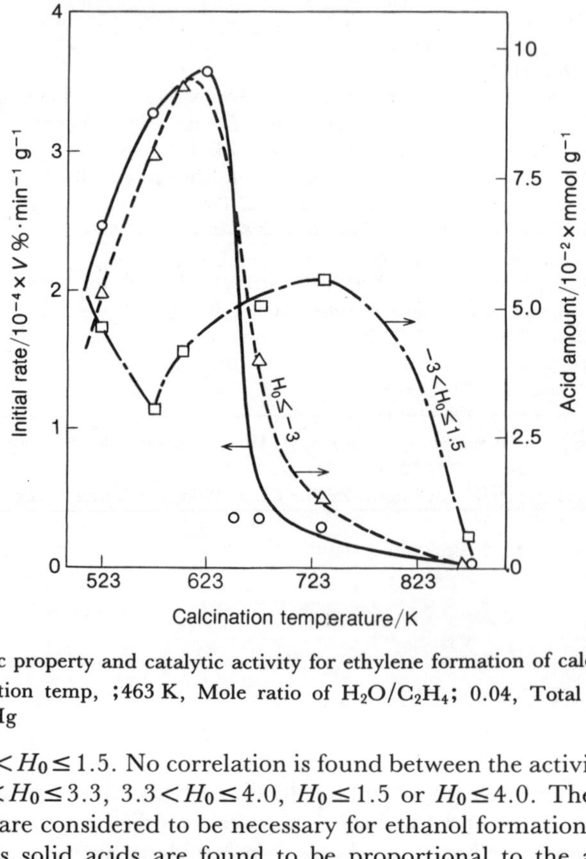

Fig. 4.8 Acidic property and catalytic activity for ethylene formation of calcined $NiSO_4$. Reaction temp, ;463 K, Mole ratio of H_2O/C_2H_4; 0.04, Total pressure; 620 mmHg

but not with $-3 < H_0 \leq 1.5$. No correlation is found between the activities and the acid amounts at $1.5 < H_0 \leq 3.3$, $3.3 < H_0 \leq 4.0$, $H_0 \leq 1.5$ or $H_0 \leq 4.0$. Therefore, the acid sites of $H_0 \leq -3$ are considered to be necessary for ethanol formation. In fact, the activities of various solid acids are found to be proportional to the acid amounts at $H_0 \leq -3$, as shown in Fig. 4.9.[10]

It is known that both Brønsted and Lewis acid sites are formed on the surface of heat-treated nickel sulfate and that the maximum of Brønsted acidity appears when heat-treated at 523 K and the maximum of Lewis acidity at 673 K,[13] while the sum of both acidities shows the maximum at 623 K.[14] Since the maximum activity of nickel sulfate for ethanol formation was observed when heat-treated at 623 K and the activity curve correlated well with the Brønsted plus Lewis acidity curve (Fig.4.9), the ethanol formation is considered to be catalyzed by both Brønsted and Lewis acids. It should be noticed that the Lewis acid sites on the dehydrated nickel sulfate is converted to Brønsted acid sites when water vapor is present during the reaction,[13] but the acid strength of $H_0 \leq -3$ on the surface is not affected by water vapor, if temperature is higher than 353 K.[15]

It has been reported that ion-exchange resins are catalytically active for hydration of propylene,[16,17] isobutene,[18] and isopentene.[19]

B. Acidic Property and Selectivity

The activity for ethanol formation of $SiO_2-Al_2O_3$, which has a comparatively large acid amount at $H_0 \leq -3$, was much lower than that expected from the linear relation shown in Fig. 4.9. Since ethylene polymer and acetaldehyde formed as by-products, the ethylene formation was decreased. The decrease in the selectivity for ethylene formation is considered due to the existence of too strong acid sites of $H_0 \leq -8.2$ on the surface of $SiO_2-Al_2O_3$. In fact, ethylene polymer and acetaldehyde was also formed over the alumina and aluminum phosphate catalysts which have strong acid sites of $H_0 \leq -8.2$, but not over solid phosphoric acid and boron phosphate which have no such strong acids.[10] These results combined with those mentioned in Section 4.5.1A indicate that the effective acid strength for ethanol formation is $-8.2 < H_0 \leq -3$ from the view point of both activity and selectivity.

In the case of hydration of propylene, the effective acid strength of catalysts was reported to be $-3 < H_0 \leq 1.5$ under the reaction conditions of 433 – 493 K and 5 atm. The weaker effective acid strength can be attributed to the higher basic strength of propylene.[20]

As for cation exchanged zeolites A and Y, the condition for selectivity is different.[9] For the zeolites A exchanged with Mg, Ca, Cd, Zn etc. whose pore size is 3 – 5 Å, large molecules such as ethylene polymer cannot be formed in hydration of ethylene despite the fact that the zeolites possess strong acid sites of $H_0 \leq -8.2$ active for polymerization of ethylene. However, with zeolites Y which have large pore size (*ca.* 9Å),

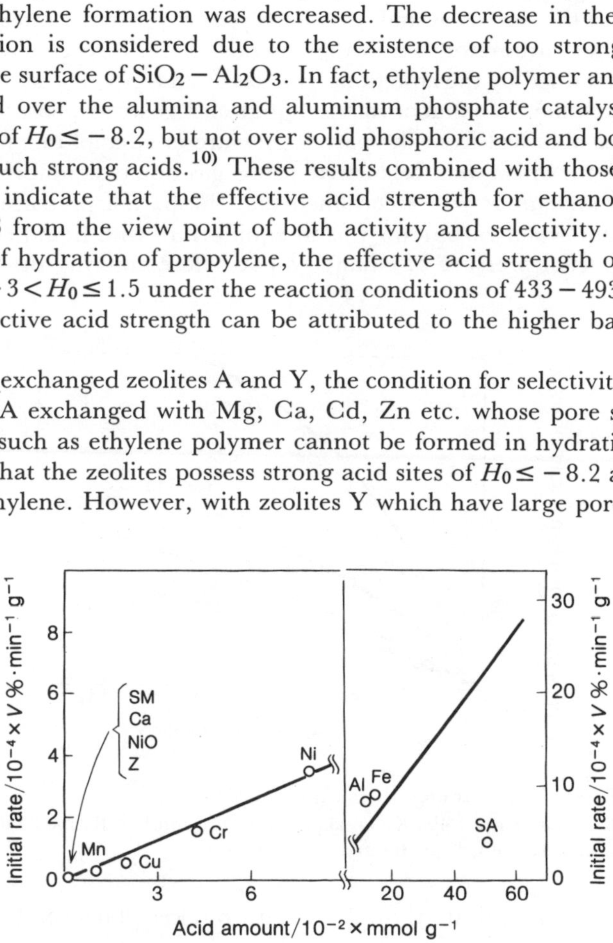

Fig. 4.9 Acid amount at $H_0 \leq -3.0$ and catalytic activity for ethylene formation of various solid acid.
Reaction conditions : the same as those in Fig. 4.8. Ni; $NiSO_4$, Cu; $CuSO_4$, Mn; $MnSO_4$, Ca, $CaSO_4$; Cr, $Cr_2(SO_4)_3$; Al, $Al_2(SO_4)_3$; Fe, $Fe_2(SO_4)_3$; SA, $SiO_2-Al_2O_3$; SM, SiO_2-MgO; NiO, NiO; Z, ZnS.

by-products of large molecular size become possible.

4.5.2 Mechanism of Hydration

The mechanism of the hydration of ethylene is discussed here based on experimental results of the effect of the molar ratio of H_2O/C_2H_4 on the reaction rate and of the deuterium exchange reaction of ethylene with heavy water.

Figure 4.10 shows the plot of the concentration of ethanol in aqueous solution against reaction time. The concentration of ethanol formed increases as the molar ratio R of H_2O/C_2H_4 is decreased from 0.27 to 0.04. The data of Fig. 4.10 were analyzed as described below by applying Hougen-Watson's rate equations[21) to determine the rate-determining step of the hydration reaction. For the analysis, the reaction rate γ was obtained as the sum of γ_1 and γ_2, where γ_1 and γ_2 are the rates of ethanol formation in the water carburetor and in the gaseous phase, respectively.[10) Values for γ_1 and γ_2 were obtained by the equations $\gamma_1 = dn_l/dt$ and $\gamma_2 = dn_g/dt$, where n_l and n_g are molar concentrations of ethanol in the liquid and gaseous phase, respectively. The value of n_g was obtained from the partial pressure of ethanol in equilibrium with the ethanol solution, which was calculated using gas-liquid equilibrium data[22) of the water-ethanol system. It was ascertained that the partial pressure of ethanol vapor coming from the water carburetor was in equilibrium with liquid ethanol and almost equal to the partial pressure of ethanol vapor before entering the carburetor.

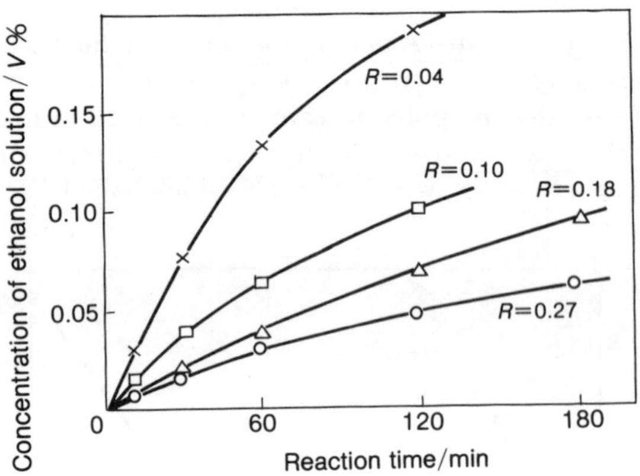

Fig. 4.10 Ethanol concentration *vs.* reaction time.
Reaction temp. ; 493 K, Total pressure; 620 mmHg, R; mole ratio of H_2O/C_2H_4, Catalyst; NiO_4 heat-treated at 573 K.

Now, if we assume that ethanol formation takes place on the catalyst surface and that the reaction rate is controlled by a surface reaction between adsorbed ethylene and water molecules, the reaction rate γ is expressed by the quation[21)

$$\gamma = \frac{ZK_E K_W (P_E P_W - P_A/K_P)}{(1 + K_E P_E + K_W P_W + K_A P_A)^2} \tag{1}$$

where P_E, P_W and P_A are partial pressures of ethylene, water and ethanol, K_E, K_W and K_A adsorption equilibrium constants of ethylene, water and ethanol, and K_p and Z equilibrium constant of reaction and constant, respectively.

If the total pressure, P, (the sum of P_E and P_W is kept constant, $P_E = P/(1 + R)$ and $P_W = PR/(R + 1)$. Assuming $P_A = 0$ at the initial stage of reaction, the initial rate γ_0 is expressed by the equation

$$\gamma_0 = \frac{ZK_EK_WRP^2}{(1+R+K_EP+K_EPR)^2} \tag{2}$$

where γ_0 is obtained by extrapolation of γ to $t = 0$.

Equation (2) can be written as follows.

$$\left(\frac{R}{\gamma_0}\right)^{1/2} = \frac{1+K_wP}{(ZK_EK_W)^{1/2}\cdot P}\cdot R + \frac{1+K_EP}{(ZK_EK_W)^{1/2}\cdot P} \tag{3}$$

If the surface reaction (step III in scheme (4)) is the rate-determining step, the plots of $(R/\gamma_0)^{1/2}$ against R should lie in a straight line. Fig. 4.11 shows that the plots give a good straight line. However, the data of Fig. 4.10 did not fit any equations derived by assuming that the rate-determining step is adsorption of ethylene or water (step I or II), desorption of adsorbed ethanol (step IV), reaction of adsorbed ethylene with free water, or reaction of adsorbed water with free ethylene.

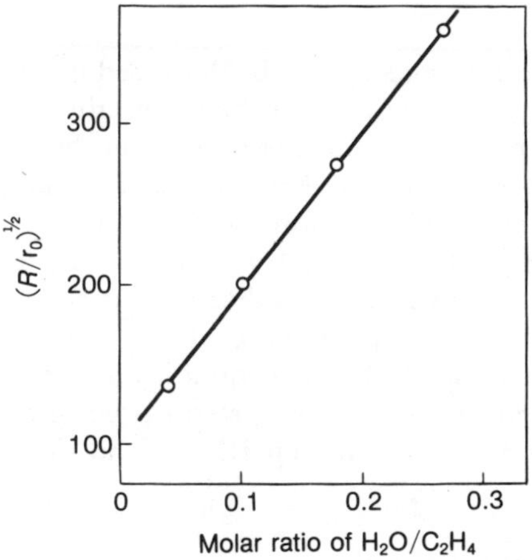

Fig. 4.11 Plot of equation (3).
Reaction temp., 493 K; Total pressure; 620 mmHg, Catalyst; $NiSO_4$ heat-treated at 573 K

$$\left.\begin{array}{l} C_2H_4 \overset{\text{I}}{\rightleftharpoons} C_2H_4(a) \\ \\ H_2O \overset{\text{II}}{\rightleftharpoons} H_2O(a) \end{array}\right\} \overset{\text{III}}{\rightleftharpoons} C_2H_5OH(a) \overset{\text{IV}}{\rightleftharpoons} C_2H_5OH \qquad (4)$$

It was concluded from these results that the rate-determining step was the surface reaction of adsorbed ethylene and water molecules (step III). Since it is generally accepted that hydration of olefin catalyzed by acids in a homogeneous liquid phase proceeds by nucleophilic attack of the hydroxyl ion to carbenium ion formed by addition of a proton to the olefin,[23] hydration on a solid surface may also be considered to proceed via the ethyl carbenium ion. Thus, step III in eq. (4) may be written more in detail as follows.

$$\left.\begin{array}{l} C_2H_4(a) \\ H^+(a) \end{array}\right\} \overset{\text{III-1}}{\rightleftharpoons} C_2H_5^+(a)$$

$$H_2O(a) \overset{\text{III-2}}{\rightleftharpoons} \left\{\begin{array}{l} H^+(a) \\ OH^-(a) \end{array}\right.$$

$$\left.\begin{array}{l} C_2H_5^+(a) \\ OH^-(a) \end{array}\right\} \overset{\text{III-3}}{\rightleftharpoons} C_2H_5OH(a) \qquad (5)$$

The detailed rate-determining step will be discussed below in light of the observed results of the deuterium exchange reaction between ethylene and heavy water. The mass spectra of ethylene remaining during the reaction showed no peak at mass numbers 30, 31 and 32 and a very small peak at 29. The ratio of peak height at 29 to that at 28 was the same as that of ethylene before use for the reaction. These results indicate that no C_2H_3D, $C_2H_2D_2$, C_2HD_3 or C_2D_4 is formed. The fact that the deuterium in heavy water did not transfer to ethylene indicates that either step I in (4) or III-1 in (5) must be a rate-determining step, because, if any of steps II, IV, III-2 and III-3 is rate-determining, steps I and III-1 are in equilibrium and, therefore, deuterium exchange should be observed. Since, however, step I is not the rate-determining step as described above, it is concluded that step III-1 is the rate-determining step of the hydration.

4.5.3 Design of Hydration Catalyst

As mentioned in Section 4.5.1, the optimum acid strength of the catalyst which gives high activity as well as high selectivity for hydration of ethylene must be in the range of $-8.2 < H_0 \leq -3.0$. Solid phosphoric acid, calcined nickel sulfate, and TiO_2-ZnO which possess the optimum acid strength showed 100% selectivity for formation of ethylene. Those catalytic activities are compared in Fig. 4.12, the difference in the activities being mainly due to the difference in their acid amounts. The TiO_2-ZnO catalyst containing a small amount of SO_4^{2-} which showed the highest

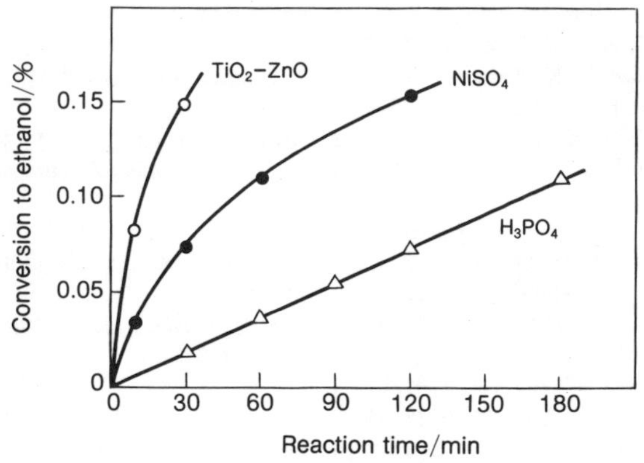

Fig. 4.12 Hydration of ethylene over solid acid catalysts.
○; TiO_2-ZnO (ZnO : 7 wt %) 2 g calcined at 673 K, Reaction temp.; 473 K
●; $NiSO_4$(2 g)calcined at 573 K, Reaction temp.; 493 K
△; Solid phosphoric acid (support : Celite)17 g, Reaction temp.; 493 K

activity as well as 100% selectivity [24] is, however, gradually deactivated by the loss of SO_4^{2-} due to its dissolution into water during the hydration reaction. The activity of TiO_2-ZnO which does not contain SO_4^{2-} is low due to the decrease of the acid sites of $H_0 \leq -3$. The nickel sulfate calcined at 623 K has a disadvantage of weakness in mechanical strength for practical use. The solid phosphoric acid also has the disadvantages of reactor corrosion and necessity of a continuous supply of phosphoric acid under high pressure (50 atm.). Recently, niobic acid ($Nb_2O_5 \cdot nH_2O$) calcined at 573 K whose acid strength is in the range of $-8.2 < H_0 \leq -3$ was reported to exhibit high activity and selectivity without being deactivated,[25] but some problems of catalyst pelleting and regeneration after use for a long time still remain. In order to design an ideal catalyst, development of mixed oxides which do not contain any impurities such as SO_4^{2-}, but possess optimum acid strength of $-8.2 < H_0 \leq -3$ is necessary. For such development, the hypothesis regarding the acidity generation of mixed oxides presented in Section 3.2 may be useful.

REFERENCES

1. C. V. Mace,, C.F. Bonilla, *Chem. Eng. Prog.*, **50**, 385 (1954).
2. J. Muller, H. I. Waterman, *Brenstoff-Chem.*, **38**, 321 (1957).
3. T. Imai, Y. Yoshinaga, *Koatsu Gasu,* **27**, 212 (1963) (in Japanese).
4. M. Kurita, T. Hosoya, H. Uchida, T. Imai, Y. Yoshinaga, *Tokyo Kogyo Shikensho Hokoku*, **61**, 218 (1966) (in Japanese).
5. H. Kuribayashi, M. Kugo, *J. Chem. Soc. Japan, Ind. Chem. Sect.*, **69**, 1930, 1935 (1966).
6. C. Wagner, U.S. Patent 2,876,266 (1959).
7. O. Uemaki, I. Yanai, M. Fujikawa, M. Kugo, *J. Chem. Soc. Japan, Ind. Chem. Sect.*, **73**, 2142 (1970).
8. M. Nitta, K. Isa, I. Matsuzaki, K. Tanabe, *J. Jpn. Petrol. Inst*, **14**, 779 (1971).
9. M. Nitta, K. Tanabe, H. Hattori, *J. Jpn. Petrol. Inst.*, **15**, 113 (1972).
10. K. Tanabe, M. Nitta, *Bull. Jpn. Petrol. Inst.*, **14**, 47 (1972).

11. F. J. Sanders, B. F. Dodge, *Ind. Eng. Chem.*, **26**, 208 (1934).
12. R. H. Bliss, B. F. Dodge, *ibid.*, **29**, 19 (1937).
13. H. Hattori, S. Miyashita, K. Tanabe, *Bull. Chem. Soc. Japan,* **44**, 893 (1971).
14. K. Tanabe, T. Takeshita. *Advances in Catalysis*, **17**, 315 (1967).
15. I. Matsuzaki, M. Nitta, K. Tanabe, *J. Res. Inst. Catalysis, Hokkaido Univ.*, **17**,46 (1969).
16. J. R. Kaiser, H. Beuther, L. D. Moor, R. C. Odioso, *Ind. Eng. Chem., Research and Development,* **1,** 296 (1962).
17. W. Neier, J. Woellner, *Hydrocarbon Processing*, 113 (1972).
18. A. Delion, B. Torck, M. Hellin, *Ind. Eng. Chem., Process Des. Dev.*,**25**, 889 (1986).
19. A. Delion, B. Torck, M. Hellin, *J. Catal.*, **103**, 177 (1987).
20. Y. Ogino, *Shokubai* (Tokyo), **4**, 73 (1962) (in Japanese); *J. Catalysis,* **8**, 64 (1967).
21. O. A. Hougen, K. M. Watson, *Ind. Eng. Chem.*, **35**, 529 (1943).
22. M. Wremsky, *Z. Physik. Chem.*, **81**,1 (1912).
23. R. W. Traft, *J. Am. Chem. Soc.*, **74**, 5372 (1952).
24. K. Tanabe, C. Ishiya, I. Matsuzaki, I. Ichikawa, H. Hattori, *Bull. Chem. Soc. Jpn.*, **45**, 47 (1972).
25. K. Ogasawara, T. Iizuka, K. Tanabe, *Chem. Lett.*, **1984**, 645.

4.6 CONVERSION OF METHANOL INTO HYDROCARBONS

4.6.1 Methanol to Gasoline Process (MTG process)

The announcement of a new process for converting methanol into gasoline by Mobil Oil Corp.[1] greatly influenced the strategy of obtaining syn-fuel from coal. By employing the Mobil MTG process, along with methanol from coal technology, a new route for the conversion of coal to gasoline became available. The first MTG process was commercialized in 1984 in New Zealand using methanol from natural gas. The MTG process and the chemistry of methanol conversion have been reviewed and summarized by Chang.[2]

The process can be operated in either a fixed bed or a fluidized bed reactor. Typical process conditions and products yield from fixed-bed and fluidized-bed pilot plants are shown in Table 4.12.[3] In either case, methanol conversion is virtually complete. Gasoline yield is high, and hydrocarbons with boiling points above that of gasoline are not produced in significant amount. On a hydrocarbon basis, the $C_5{}^+$ synthesized gasoline yield is 60 wt%. Additional gasoline can be obtained by alkylating the propylene and butenes with isobutane by conventional technology. For a typical fixed bed operation, the yield of gasoline is 85wt% and the gasoline has an unleaded octane number of 93.

The key to the MTG process is the unusual selectivity and activity maintenance of H−ZSM-5 zeolite catalyst. Although many acidic catalysts such as heteropolyacids are known to convert methanol into hydrocarbons,[2,4,5] none has ever exhibited the selectivity to aromatics and activity maintenance of ZSM-5 zeolites. To a large extent, these characteristics are due to the unusual pore structure of these zeolites.

The dimension of the pores (to 6Å) inhibits the growth of hydrocarbons beyond a C_{10} aromatic molecule and therefore controls the product distribution. This pore system also minimizes the formation of polynuclear aromatic coke precursors because such structures do not fit easily in the channel system. Furthermore, the diffusion coefficient in the ZSM-5 channels for molecules larger than C_{10} aromatics decreases

TABLE 4.12 Yields from methanol for two reactor systems

	Typical fixed-bed	4 bpd pilot fluid-bed
Inlet temp., K	633	—
Outlet temp., K	688	—
Avg.bed temp., K	—	686
Pressure, *kPa*	2170	275
Recycle ratio (mole)	9.0	—
Space velocity, WHSV	2.0	1.0
Yield, wt % of methanol charge		
Methanol + ether	0.0	0.2
Hydrocarbons	43.4	43.5
Water	56.0	56.0
$CO+CO_2$	0.4	0.1
Coke+other	0.2	0.2
Total	100.0	100.0
Hydrocarbon distribution, wt %		
Light gas	1.4	5.6
Propane	5.5	5.9
Propylene	0.2	5.0
Isobutane	*8.6*	*14.5*
Butane	*3.3*	*1.7*
Butenes	1.1	7.3
C_5^+ gasoline	79.9	60.0
Total	100.0	100.0
Product distribution, wt %		
Casoline incl. alky	85.0	88.0
LPG	13.6	6.4
Fuel gas	1.4	5.6
Total	100.0	100.0

(Reproduced with permission by B. M. Harney, G. A.Mills, *Hydrocorbon processing*, **59**(2), 70(1980)).

exponentially with carbon number. Durene, a symmetrical tetramethylbenzene, is the heaviest molecule produced.

4.6.2 Reaction Mechanism

The conversion of methanol over ZSM-5 zeolites proceeds according to the following general reaction path.

$$2\,CH_3OH \underset{}{\overset{-H_2O}{\rightleftarrows}} CH_3OCH_3 \xrightarrow{-H_2O} C_2'{\sim}C_5' \longrightarrow \begin{array}{l}\text{paraffins}\\ \text{aromatics}\\ \text{cycloparaffins}\\ C_6^+\ \text{olefins}\end{array}$$

This was established by monitoring changes in product distribution as a function of varying contact time.[6)]

One of the most intriguing problems, still the subject of much controversy, is the mechanism of formation of incipient olefins from methanol. Among the mechanisms proposed, three are described below.

A. Carbene Mechanism

An α-elimination involving a carbenoid intermediate was proposed by Chang and Silvestri.[6] This mechanism was originally speculated by Schwabb and Gates[7] for the formation of traces of olefins in the dehydration of methanol to dimethyl ether over mordenite. It was considered unlikely, however, that the olefins were formed by dimerization of two free carbenes in view of the high reactivity of carbenes. Rather, a concerted reaction between methylene donor and acceptor was proposed involving simultaneous α-elimination and sp^3 insertion into methanol or dimethyl ether.

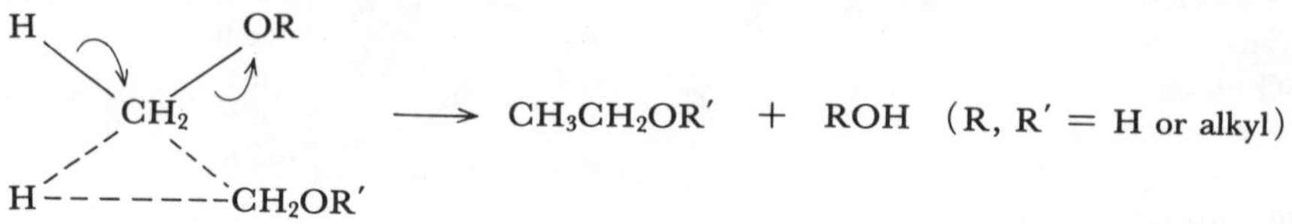

Here, bond scission is facilitated by the cooperative action of acidic and basic sites in the zeolite lattice.

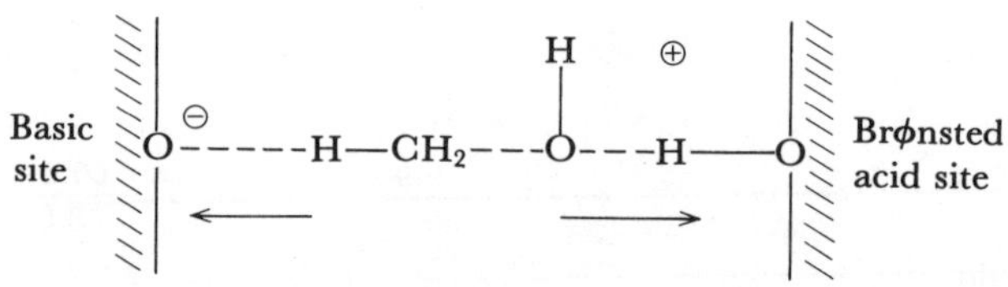

B. Oxonium Ions Mechanism

Van der Berg *et al.*[8] proposed a mechanism involving trimethyl oxonium ions (I), $(CH_3)_3O^+$, which rearrange to methyl ethyl oxonium (II) ions by the action of the basic sites of zeolites.

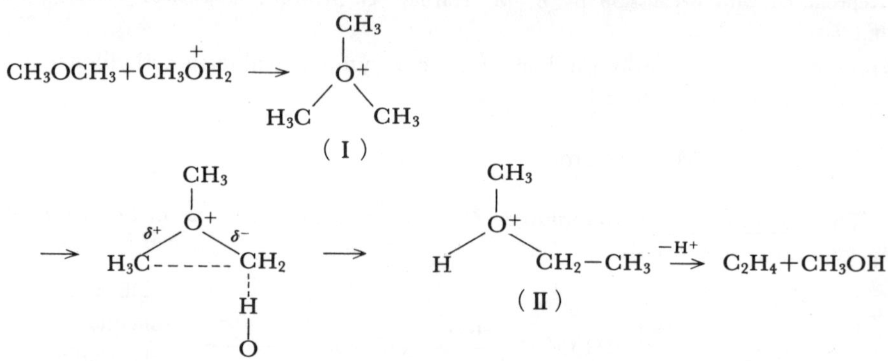

C. Carbonium Ion Mechanism

Ono and Mori[9] proposed the mechanism involving methyl cations.

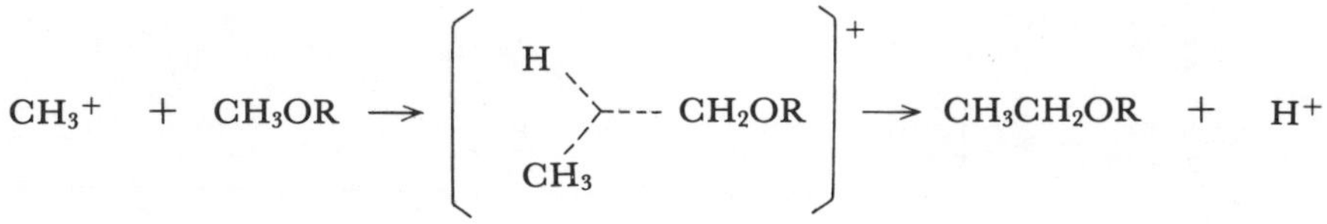

A mechanism of electrophilic attack by methyl cations was put forward by Olah and coworkers for reactions in superacid media.[10] The transition state is believed to be a penta coordinate carbonium ion. Generation of the methyl cation was considered to occur by polarization of surface methoxyl species.[9]

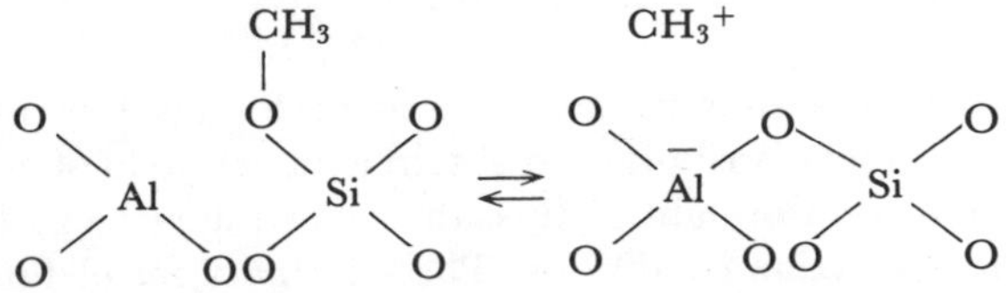

The superacid mechanism was also favored by Kagi.[11] The fact that the product distributions of catalytic conversion of methanol over silver salt of heteropoly acid is very similar to that of the stoichiometric reaction of methyl iodide may be one of the supporting evidence for the mechanism.[12]

Though the solution of initial C – C bond problem remains an open challenge at present,[2] it is generally accepted that the reaction proceeds auto-catalytically.[9,13,14] Thus, the reaction proceeds faster as the extent of conversion increases,[9,13,14] or the rate shows an abrupt increase in a very narrow temperature range.[14] This happens because the reaction between methanol to produce olefins (1) is a slow process, and the subsequent reactions of olefins with methanol (2) is a much faster process.

$$CH_3OH \longrightarrow \text{olefins} \qquad (1)$$

$$RC{=}CH{-}R' + CH_3OH \xrightarrow{\text{fast}} RCH(CH_3){-}C(OH){-}R' \xrightarrow{\text{fast}} RC{=}C(CH_3){-}R' \qquad (2)$$

Thus, under ordinary reaction conditions, methanol is almost exclusively consumed by the reaction with olefins. Once certain amounts of olefins are formed in the zeolite cavities, they undergo various reactions such as dimerization and isomerization of olefins, and cracking of higher hydrocarbons through carbenium intermediates. For example, the formation of aromatic hydrocarbons from lower olefins can be expressed by the following scheme.[15]

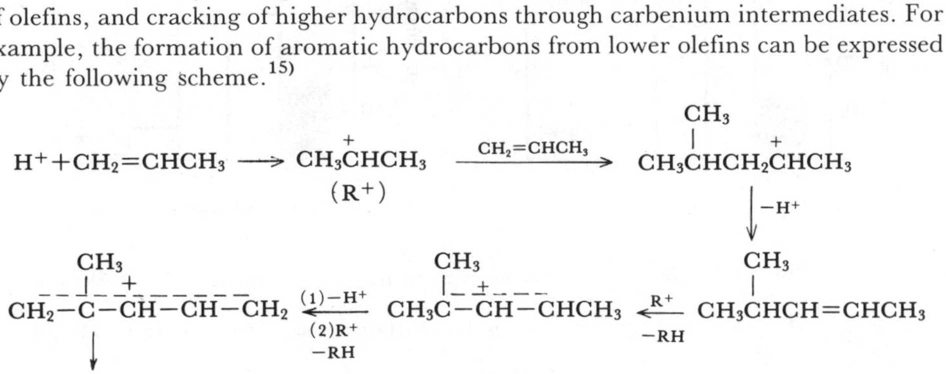

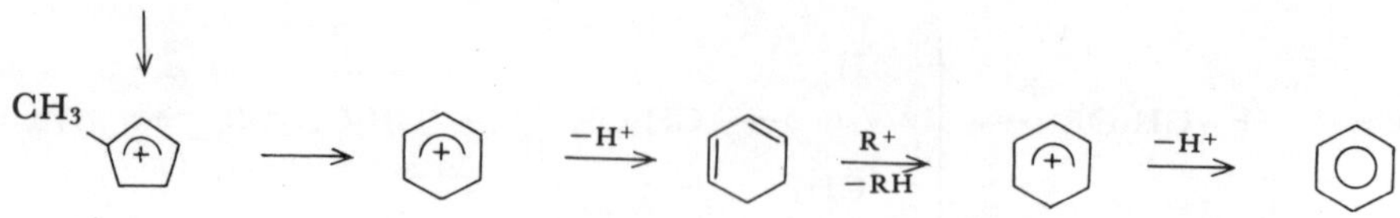

As is evident from the scheme, the formation of one molecule of aromatic hydrocarbons is always accompanied by the formation of three molecules of alkanes. Aromatics is easily alkylated with methanol, and the largest size of the products is limited by the channel system of ZSM-5 zeolites as described earlier.

4.6.3 Modification of Product Distribution

Although the MTG process is an excellent process for gasoline production, it is possible to modify the product distribution to obtain a higher yield of olefins or aromatics. To enhance the production of light olefins, the following strategy has been applied.

i) Modification of reaction conditions. The selectivity for olefins can be increased under the conditions of lower methanol pressures[6)] and high space velocities.[6)]

ii) Use of small pore zeolites. To prevent the formation of large-size molecules, the use of small-pore zeolites such as erionite,[17,18)] chabasite,[17,18)] and ZSM-34[19,20)] increases the selectivity to lower olefins appreciably. Fig. 4.13 shows the product distribution in the methanol conversion over La^{3+}-exchanged chabasite and erionite.[18)] In each case, hydrocarbons with five or more carbon atoms are formed only in small quantities. Small pore zeolite catalysts are, however, generally short lived because of pore plugging with oligomerized products.

iii) Use of zeolites with weak acidity. The selectivity for olefins can be enhanced by suppressing the formation of aromatics and alkanes. Since aromatics or alkanes are formed by the successive hydride transfer between carbenium ion and olefins and the hydride-transfer requires a stronger acid,[15)] the use of zeolites with weak acid strength is effective for obtaining higher selectivity for olefins. The acid strength of

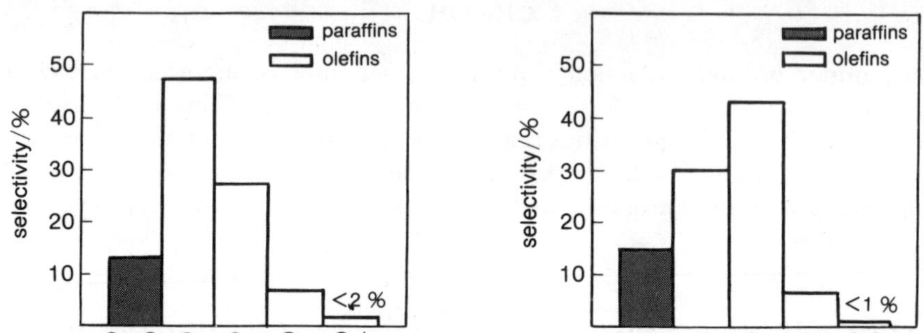

Fig. 4.13 Product distributions in methanol conversion over La^{3+}-exchanged chabasite and erionite.
(Reproduced with permission by U. Dettmeier *et al.*, *Chem, Ing Tech.*, **54**, 591 (1982)).

[B] − ZSM-5 (see Section 3.4) is much weaker than ordinary ZSM-5.[21−23] Holderich *et al.* obtained high selectivity for olefins with [B] − ZSM-5, which was extruded with amorphous silica − alumina then treated with HF followed by HCl.[24] Thus, at 773 K, the $C_2 - C_4$ olefin selectivity was 80%, 47% of propylene being formed. The proportion of aromatics was only 2%. Recently, Inui *et al.* reported that ferrosilicate, which gives an X-ray diffraction pattern similar to that of ZSM-5, gave a high yield of lower olefins from methanol.[25]

Modification of ZSM-5 zeolites with various compounds is also effective.[26,27] Kaeding and Butter[26] reported that ZSM-5 modified by treatment with phosphorus compounds showed high (70%) selectivity to $C_2 - C_4$ olefin formation from methanol. This could also be ascribed to decrease in stronger acid sites by the treatment.

Okado *et al.*[27−29] prepared zeolites with ZSM-5 type structure containing calcium. Though the state of the calcium species is not clear, the zeolites have weak acidity and give a high selectivity for lower olefins at 773 − 873 K in the methanol conversion.

The yield of aromatics can be increased by introducing metal cations having dehydrogenation activity.[30,31] Since the formation of aromatics over acidic catalysts always accompanies the formation of alkanes, it is necessary to introduce another way for producing aromatics. This is accomplished by the direct dehydrogenation of olefinic intermediates. Thus, zinc-exchanged ZSM-5 gave about 70% selectivity to aromatics with a small production of alkanes, as shown in Table 4.13.[30] Zinc species introduced by ion-exchange are considered to act as active sites for dehydrogenation of olefinic intermediates. Other elements such as Ga, or Re seem also effective for this purpose.[31]

TABLE 4.13 Product distributions in methanol conversion over H-ZSM-5 and Zn-ZSM-5

Catalyst	H-ZSM-5	Zn-ZSM-5
Products/C %		
$CO+CO_2$	0.4	0.7
C_1	2.2	1.5
C_2-C_4	46.7	13.0
C_2'-C_4'	6.7	13.1
C_5^+	3.7	4.3
Aromatics	40.3	67.4
Aromatics/C %		
Benzene	9.3	4.5
Toluene	31.5	25.1
C_8	35.0	46.7
C_9^+	24.2	23.7

700 K, CH_3OH=40 kPa, W/F=9.0 g h mol^{-1}

REFERENCES

1. S. L. Meisel, J. P. McCullough, C. H. Lechthaler, P. B. Weisz, *Chemtech,* **6**, 86 (1976).

2. C. D. Chang, *Hydrocarbons from Methanol,* Marcel Dekker, New York (1983).
3. B. M. Harney, G. A. Mills, *Hydrocarbon Process.*, **59** (2), 67 (1980).
4. Y. Ono, T. Baba, J. Sakai, T. Keii, *J. Chem. Soc., Chem. Comm.*, **1981**, 400; T. Baba, J. Sakai, H. Watanabe, Y. Ono, *Bull. Chem. Soc. Jpn.*, **55**, 2557 (1982); T. Baba, J. Sakai, Y. Ono, *Bull. Chem. Soc. Jpn.*, **55**, 2635 (1982); T. Baba, Y. Ono, *Appl. Catal.*, **8**, (1983).
5. T. Hibi, K. Takahashi, T. Okuhara, M. Misono, *Appl. Catal.*, **24**, 69 (1986).
6. C. D. Chang, A. J. Silvestri, *J. Catal.*, **56**, 169 (1979).
7. F. A. Schwabb, B. C. Gates, *Ind. Eng. Chem. Fundam.*, **11**, 540 (1972).
8. J. P. van der Berg, J. P. Wolthulzen, J. H. C. van Hooff, Proc. 5th Conf. Zeolites Naples, Heydon, London, 1980, p.649.
9. Y. Ono, T. Mori, *J. Chem. Soc., Faraday Trans. 1*, **77**, 2209 (1981).
10. G. A. Olah, J. R. De Member, J. Shen., *J. Am. Chem. Soc.*, **95**, 4592 (1973); G. A. Olah, G. Klopman, R. H. Schlosberg, *J. Am. Chem. Soc.*, **91**, 3261 (1969).
11. D. Kagi, *J. Catal.*, **69**, 242 (1981).
12. M. Kogai, Y. Ono, *Bull. Chem. Soc. Jpn.*, **57**, 2987 (1984).
13. N. Y. Chen, W. J. Reagan, *J. Catal.*, **59**, 123 (1979).
14. Y. Ono, E. Imai, T. Mori, *Z. Phys. Chem. N. F.*, **115**, 99 (1979).
15. M. L. Poustma, *Zeolite Chemistry and Catalysis* (J. A. Rabo, ed.) ACS monograph *(7)*, p. 437, American Chemical Society, Washington, D.C. (1976).
16. C. D. Chang, W. H. Lang, R. L. Smith, *J. Catal.*, **56**, 169 (1979).
17. F. A. Wunder, E. I. Leupolk, *Angew. Chem., Int. Ed. Engl.*, **19**, 126 (1980).
18. U. Dettmeier, H. Litterer, H. Baltes, W. Herzog, E. I. Leupold, F. A. Wunder, *Chem. Ing. Tech.*, **54**, 590 (1982).
19. Japan Kokai 58497 (1978).
20. T. Inui, T. Ishihara, Y. Takegami, *J. Chem. Soc., Chem. Commun.*, **1961**, 936.
21. K. F. M. G. J. Acholle, A. P. M. Kentgens, W. S. Veeman, P. Frenken, G. P. M. V. Valdin, *J. Phys. Chem.*, **88**, 5 (1984).
22. C. T-W. Chu, C. D. Chang, *J. Catal.*, **93**, 451 (1985).
23. P. Ratnasamy, S. G. Hedge, A. J. Chandwadkar, *J. Catal.*, **102**, 467 (1986).
24. W. Holderich, H. Eichhorn, R. Lehnert, L. Marosi, W. D. Mross, R. Reinke, W. Ruppel, H. Schlinper, Proc. 6th Intern. Zeolite Conf. (D. H. Olson, A. Bisio, eds.) Butterworth, Guildford, 1984, p. 545.
25. T. Inui, A. Miyamoto, H. Matsuda, H. Nagata, Y. Makino, K. Fukuda, F. Okazumi, Proc. 7th Intern. Zeolite Conf. (Y. Murakami *et al.*, eds.) Kodansha, Tokyo and Elsevier, Amsterdam, 1986, p. 859.
26. W. W. Kaeding, S. A. Butter, *J. Catal.*, **61**, 155 (1980).
27. H. Okado, H. Shoji, K. Kawamura, Y. Kohtoku, Y. Yamazaki, T. Sano, H. Takaya, *Nippon Kagaku Kaishi*, **1987**, 25 (in Japanese).
28. H. Okado, T. Sano, K. Matsuzaki, K. Kawamura, K. Hashimoto, H. Watanabe, H. Takaya, *Nippon Kagaku Kaishi*, **1987**, 791 (in Japanese).
29. H. Okado, H. Shoji, K. Kawamura, Y. Shiomi, K. Fujisawa, H. Hagiwara, H. Takaya, Nippon Kagaku Kaishi, **1987**, 962, (in Japanese).
30. H. Adachi, Y. Sendoda, Y. Ono, *J. Chem. Soc., Faraday Trans. 1*, **84**, 1091 (1988).
31. US Patents 3856872, 3855115, 3953366, 4128504, 4105541

4.7 DEHYDRATION

4.7.1 Dehydration of Alcohols

Dehydration of alcohols to olefins or ethers can be effected with most solid acid catalysts as well as with solid base catalysts. Solid acids are usually more active than bases. Among acid catalysts, alumina is the most versatile. Metal phosphates, metal oxides, and cation-exchange resins are also used for industrial dehydration.

For metal oxides, Batta *et al.* noted that dehydration increases with the covalent character of metal-oxygen bond of the catalyst, whereas dehydrogenation is enhanced by increasing the ionic character.[1] Thus, the catalysts can be placed in the following selectivity sequence.[1]

WO_3 SiO_2 Al_2O_3 TiO_2 Cr_2O_3 FeO ZnO MgO CaO

$\longleftarrow$ dehydration

$\longrightarrow$ dehydrogenation

As described below, dehydration over acidic catalysts generally yields Saytzeff products, while dehydration over basic oxides such as ThO_2 and ZrO_2 yields Hofmann elimination products. Dehydration over strongly basic catalysts such as MgO and CaO is always accompanied by appreciable dehydrogenation.

Heteropoly acids are highly active for dehydration. The activity of dodecatungstophosphoric acid is much higher than that of Y-type zeolites.[2] The dehydration with heteropoly acids is unique, since the reaction proceeds not only on the surface of the solid, but also in the bulk of the solid.[3-5] Because of the ''pseudo-liquid'' nature of the dehydration, every proton in the solid heteropoly acids can participate in the reaction. This explains the very high activity of the acids.

Dehydration is very often accompanied by the subsequent isomerization of primary products. Isomerization may be avoided by poisoning acidic sites with alkali metal ions, ammonia, or organic bases. Cyclohexanol dehydrates to cyclohexene over alumina containing 0.4% sodium or potassium ions, but gives a large amount of methylcyclopentenes over pure alumina.[6] The cyclopentenes do not arise directly from cyclohexanol, but by the isomerization of cyclohexene, the primary product. The selectivity to 3-methyl-1-butene is significantly improved by adding small amounts of base to γ-alumina in the dehydration of 3-methylbutanol.[7]

4.7.2 Mechanisms and Selectivities of Alcohol Dehydration

Catalytic dehydration can proceed according to the following types of mechanisms, where A and B stand for acidic and basic centers of catalysts, respectively.[8]

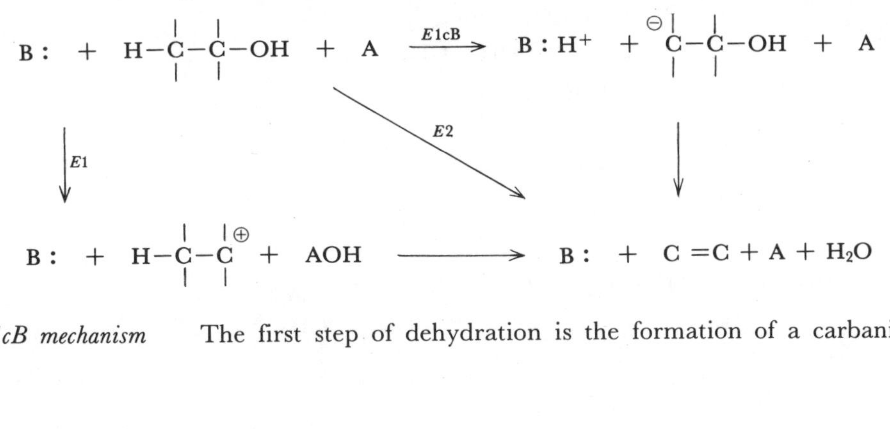

E1cB mechanism The first step of dehydration is the formation of a carbanion, meaning that a C−H bond is loosened or broken in the first step. This mechanism occurs with strongly basic catalysts such as La_2O_3, ThO_2, and alkaline earth oxides.

E1 mechanism The first step of dehydration is the formation of a carbenium ion by abstraction of an OH group. This mechanism occurs with strongly acidic catalysts such as aluminosilicate. The acidic centers A may be either a Brønsted or Lewis type. In the former case, the carbenium ions may be produced with the intermediacy of the oxonium ions.[9)]

$$-\overset{|}{\underset{|}{C}}-OH + H^+ \longrightarrow -\overset{|}{\underset{|}{C}}-\overset{+}{\underset{H}{O}}H \longrightarrow -\overset{|}{\underset{|}{C}}{}^+ + H_2O$$

E2 mechanism The elimination of a proton and a hydroxyl group from alcohols are concerted without formation of ionic intermediates. Alumina is a typical *E*2 oxide.

The three mechanisms can be distinguished in various ways, though, unlike the case of liquid phase reactions, the kinetic method cannot be used. The product distribution is one of the most significant clues. With the *E*1 mechanism, isomerization takes place in the carbenium ion stage. Thus, the formation of 2-butene from 1-butanol is indicative of the *E*1 mechanism. High selectivity for 1-butene (Hofmann orientation) from butan-2-ol is indicative of *E*1cB, whereas *E*1 and *E*2 give mainly 2-butene (Saytzeff orientation).

The dehydration of isobutyl alcohol over $SiO_2-Al_2O_3$ yields a mixture of butenes in which the fraction of *n*-butene is 33%.[10)] Since the rate of skeletal isomerization of isobutene to *n*-butenes is significantly lower than the rate of formation of *n*-butenes in dehydration, *n*-butenes must be primary products.[10)] This indicates that the reaction proceeds via the *E*1 mechanism. Formation of *n*-butenes is associated with the formation of the least stable isopropyl carbenium ions, which are readily rearranged by hydride or methyl transfer to form more stable tertiary or secondary carbenium ions.

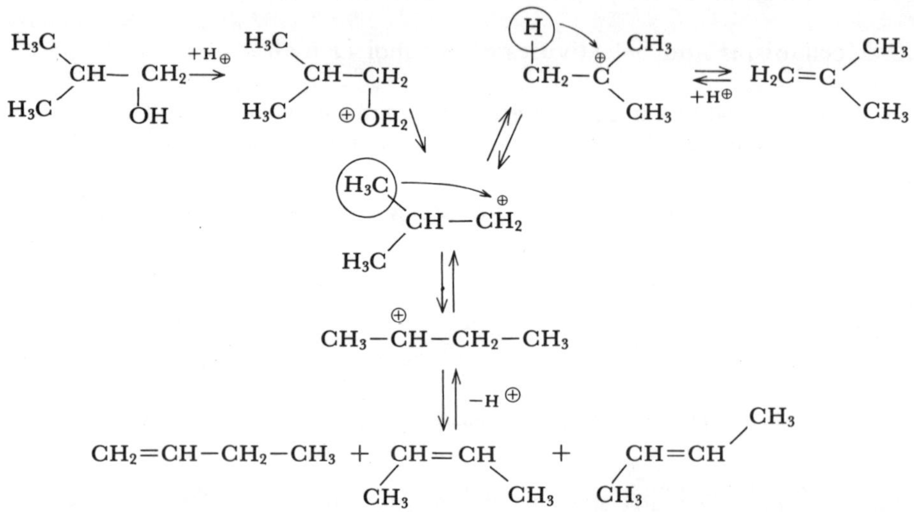

A linear free energy relationship was found between the activation energy of heterolytic dissociation of the C – O bond in the dehydration of aliphatic alcohols over $SiO_2 - Al_2O_3$, which is indicative of the *E*1 mechanism.[11)]

Molybdenum oxide is also an E1-type oxide. The dehydration of 2,2-dimethyl – 3-pentanol over MoO_3 at 453 K gives primarily products requiring methyl migration, 2,3-dimethylpent – 1-ene, 2,3-dimethylpent – 2-ene and 3,4-dimethylpent – 2-ene.[12)] On the other hand, only β-elimination products, *cis*- and *trans*-4,4-dimethylpent – 1-ene are obtained by the dehydration over alumina at the same temperature.[12)]

Noller and Thomke[13)] recommend the use of deuterated reactants or deuterated catalysts (OD group on the surface of oxide instead of OH groups), and give the following criteria for the three mechanisms.

i) With the *E*1cB mechanism, exchange is found in both the alcohol and the olefins, but only in β-positions, indicating that the carbanion undergoes reconversion to the alcohol.

ii) With the *E*1 mechanism, exchange in all positions of the olefins is observed, but no exchange at all in the alcohol (this indicates that the carbenium ion formed is not reconverted to the alcohol).

iii) With the *E*2 mechanism, exchange does not occur, both alcohols and olefins retaining their original isotope composition.

The kinetic isotope effect also provides valuable information.

Some typical examples of mechanistic studies are given below.

ThO_2 is a typical *E*1cB oxide. Dehydration of 2-alcohols gives mainly 1-olefins (Hofmann elimination). The products of dehydration of some alcohols over a ThO_2 catalyst at 600 – 690 K are given in Table 4.14,[14)] though the selectivity depends on the method of catalyst preparation.[15)] Zirconium oxide[16)] and rare earth oxides such as La_2O_3[14)] give also Hofmann dehydration products.

In the dehydration of 2-butene – 2-d_1, there is no loss of D atoms in the unreacted alcohol.[17 – 19)] However, loss of D atoms in the unreacted alcohol are observed in the reaction of 2-butanol-d_6. This indicates that H – D exchange occurs only in the β-position, and reversibility between the alcohol and the carbanion.[17 – 19)] There is no loss of D-atom in 1-butene produced in 2-butanol-2-d_1[17 – 19)]

TABLE 4.14 Thoria-catalyzed dehydration of secondary 2-alkanols

Alcohol	Products, %†		Ratio,
	1-Alkene	2-Alkene	*trans/cis*
2-Butanol	93-94	6-7	1.1
2-Hexanol	95-97	3-5	...
2-Octanol	95-97	3-5	...
4-Methyl-2-pentanol	96-98	2-4	5.0
1-Cyclohexyl-1-ethanol	96-98	2-4	...

† These values do not include 1-2 % ketone formed by competing dehydrogenation.

(Reproduced with permission by A. J. Lindeen, R. van Hoozer, *J. Org. Chem.*, **32**, 3387 (1979)).

The dehydration of three hexanol isomers, 1-hexanol, 2-hexanol and 2-methyl – 2-pentanol over La_2O_3 and ThO_2 was studied.[8] Though the product selectivity clearly indicated the E1cB mechanism, the order of reactivity was found to be primary < secondary < tertiary alcohol, as in the *E*1 and *E*2 mechanism, indicating that the rupture of C – O bond is the rate-determining step. This result is in conformity with the reversibility between the reactant alcohol and the carbanion, as evidenced by H – D exchange experiments.

BPO_4 is a typical *E*1 oxide.[18,20] An appreciable amount of 2-butene is formed from 1-butanol. There is a loss of D atom in 1-butene formed from 2-butanol-3*d*, and also from 2-butanol-2d_1. However, there is no loss of D atoms in the reactant alcohol. All these features are indicative of intermediacy of carbenium ions.[18,20]

Over $Ca_3(PO_4)_2$, dehydration of butanol proceeds with *E*2 mechanism.[20] There is no loss of D atoms in 1-butene formed from *d,l*-erythro-2-butanol-3d_1, and only a slight loss of D atoms in 1-butene formed from 2-butanol-2d_1. There is no uptake of D atom of 2-butanol unreacted from the deuterated catalysts.[20]

Köchloefl and Knözinger[21] studied also the kinetic isotope effect of the dehydration of 2-propanol(I), 2-propanol-d_1(II), and 2-propanol-1,3-d_6(III) over Al_2O_3, ZrO_2, TiO_2 and SiO_2. Over alumina, the value of the isotope effect α_1 (k_I/k_{II}), and α_2 (k_I/k_{III}) are 1.0 and 1.46 respectively, in agreement with an *E*2-like mechanism. The α_1-value increased in the sequence Al_2O_3, ZrO_2, TiO_2, SiO_2, whereas the reverse trend was observed for the deuterium substitution at C_β. These effects indicate a change in the elimination mechanism from *E*2-like to *E*1-like in the above series of catalysts.

Knözinger and Schengllia[22] studied the kinetic isotope effect of the dehydration of *t*-butanol, *sec*-butanol and isobutyl alcohol over alumina and found that the deuteration of the hydroxyl group does not give rise to an isotope effect, whereas substitution of β-proton by deuterium produces an appreciable effect. From the dependence of the isotope effects on substrate structure and temperature, it was concluded that at temperatures below 573 K alcohols are dehydrated via *E*2-like reaction intermediates over alumina. With increasing temperature, the contribution of the ionic structure increased so that at elevated temperatures – depending on the reactant structure – the reaction may proceed via an *E*1-mechanism.

The effect of the substituents on the dehydration of secondary alcohols of the general structure over alumina was studied by Dautzenberg and Knozinger and the ratios of 2-olefins and 1-olefins in the products (S_{21}) determined.[23]

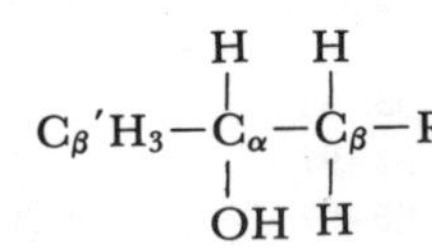

The rario S_{21} increased in the sequence of R.

$$CH_3 > C_2H_5 > n\text{-}C_3H_7 > iso\text{-}C_3H_7H > t\text{-}C_4H_9$$

Since the inductive effect of the substituent R will alter only the reactivity of the β-hydrogens without influencing the reactivity of the β'-hydrogens, the S_{21} values decrease in the observed sequence because of the reduced rate of formation of the 2-olefins with increasing inductive power of the substituent R. The S_{21}-selectivity is predominantly governed by inductive effects, the rupture of C_β-H bonds being involved in the rate-determining step (*E*2 mechanism) in accordance with deuteruim-isotope effect.

The mechanism of dehydration may differ from alcohol to alcohol, even if the same catalyst is used. Thomke carried out the dehydration of a series of alcohols over BPO_4, $Ca_3(PO_4)_2$ and Sm_2O_3, and determined the mechanism by two criteria: uptake of deuterium from deuterated catalysts into produced olefin and unreacted alcohol.[20] The results are given in Table 4.15. Most of the examined alcohols showed *E*1 on BPO_4, *E*2 (sometimes mingled with *E*1) on $Ca_3(PO_4)_2$ and *E*1cB on Sm_2O_3. There are two exceptions: dehydration of ethanol over BPO_4 and 2-methyl − 2-propanol over Sm_2O_3. Both proceed with the *E*2 mechanism.

TABLE 4.15 Mechanism of dehydration of α- and β-substituted ethanol

Substance	BPO_4	$Ca_3(PO_4)_2$	Sm_2O_3
2-methyl-2-propanol	*E*1	*E*2(+*E*1)	(*E*2)
3-pentanol	*E*1	*E*2	*E*1cB
2-butanol	*E*1	*E*2	*E*1cB
2-propanol	E1	(E2)	E1cB
1-butanol	*E*1	*E*2(+*E*1)	*E*1cB
1-propanol	*E*1	*E*2(+*E*1)	*E*1cB
ethanol	(*E*2)	*E*2	*E*1cB

(Reproduced with permission by K. Thomke, *Z. Phys. Chem. Neue Folge*, **106**, 302 (1977)).

The dehydrogenation of alcohol accompanied the dehydration of alcohols over basic oxides. Thomke studied the dehydration (and dehydrogenation) of 2-butane-2-d_1 and erythro − 2-butanol-3-d_1 over La_2O_3, ThO_2, Sm_2O_3, MgO and CaO and concluded that the formation of butenes and 2-butanones occurs through the same carbanion intermediate.[24]

In the E_2 mechanism, there is another kind of selectivity, anti and syn elimination. Syn elimination products are formed when the groups are removed from the same side of the molecule and anti elimination products formed when the groups are removed from the opposite side. Over alumina, menthol yields mainly 2-menthene, and neomenthol eliminates mainly to 3-menthene.[25]

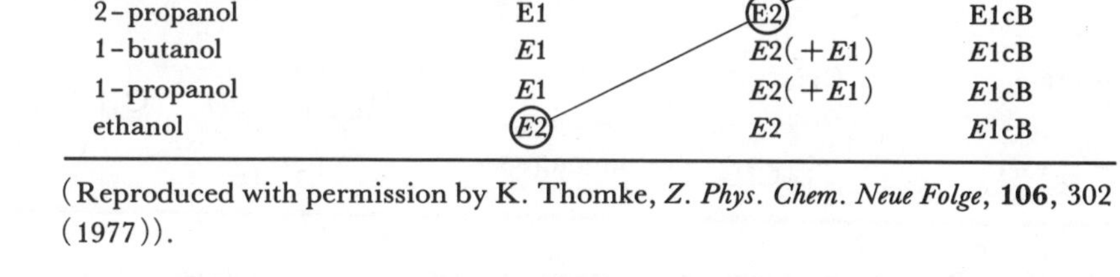

Thus, the dehydration of menthol and neomenthol exhibits anti rather than syn

stereospecificity. The products of dehydration of *cis*- and *trans*-2-*R*-cyclohexanol (R = alkyl or phenyl) differ from each.[26] The *cis*-2-substituted cyclohexanols gave 1-substituted cyclohexenes as the major product, whereas the 3-substituted cyclohexenes were the major products from the *trans*-2-substituted cyclohexanols. These results are also in accord with anti elimination. Anti-elimination itself is good evidence for a concerted ($E2$) rather than a carbenium ion mechanism.

Dehydration of *threo*- and *erythro*-2-butanol-3-d_1 is a good test for preferred mode of elimination.[27]

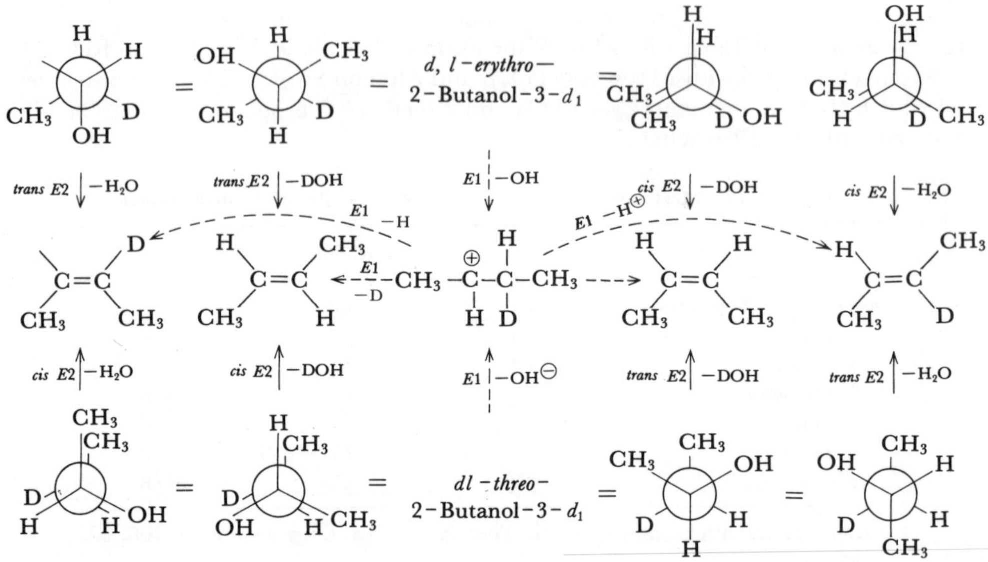

Kibby *et al.* studied the reactions of *threo*- and *erythro*-2-butanol-3-d_1 over alumina and hydroxyapatite catalysts.[28] The percentages of the monodeuterated molecules in 2-butenes are: 80% of the *trans*-2-butene, 28% of the *cis*-2-butene from the *threo* stereoisomer and 28% and 84%, respectively, from the *erythro* compound. Over hydroxyapatite, the corresponding results were 29% and 74% for the *threo*, and 80% and 39% for the *erythro* stereoisomer. Thus, anti elimination was the preferred mode over alumina and syn elimination over hydroxyapatite. Thomke and Noller determined the ratio of syn/anti ratio of elimination over various phosphates from the same reactants.[27,29] The ratio was calculated to be 85/15 with $Ca_3(PO_4)_2$, 70/30 with $CaHPO_4$ and $Ba_3(PO_4)_2$ and 45/55 with $AlPO_4$.

The dehydration of 1-butanol over alumina gives mainly the β-elimination product, 1-butene (97%), but also small amounts of the 2-butenes of which *cis*-isomer predominates by a factor of two.[6] To account for the 2-enes, a γ-elimination with a concerted migration of a β-hydrogen was invoked.[6]

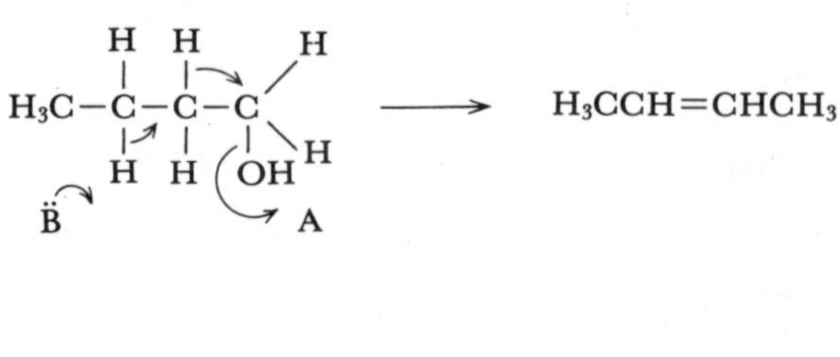

The bimolecular ether formation from alcohols proceeds at much lower temperatures than the olefin formation.[30] Over alumina, it occurs even at 400 – 410 K.[30] The mechanism involving surface alkoxide groups has been discussed.[30,31]

4.7.3 Dehydration of Alcohol with Ring Transformation

Dehydration of cycloalkylmethyl alcohol gives ring-expanded products via Wanger-Meewein rearrangement. Pines and Brown performed dehydration of cycloalkylmethanol with 4 to 7-membered rings.[32] The dehydration at 573 K over fresh alumina of cyclobutane-, cyclopentane-, cyclohexane-, and cycloheptane-methanol yielded 99, 55, 7 and 6% of ring-expanded product.

Dehydration of tetrahydrofurfuryl alcohol (tetrahydrofuran-2-methanol) over silica – alumina at 623 K gives dihydropyrane.[33] The dehydration at higher temperatures (723 – 763 K) gives acrolein and ethylene as the main products.[33]

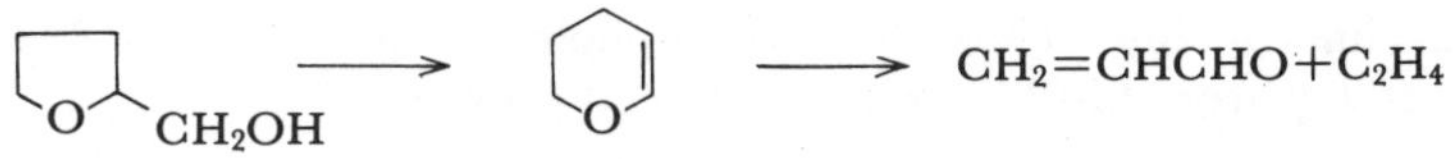

Dehydration of tetrahydropyran-2-methanol over high purity η-alumina without any carrier gas produced cyclopentane carbaldehyde in a yield of 71% at 603 K, while dehydration over copper-chromia oxide supported on kieselguhr with hydrogen carrier gas at 703 K yielded 2,3,4,5-tetrahydrooxepine and cyclopentanecarbaldehyde (44.8% and 4.7%, respectively).[34]

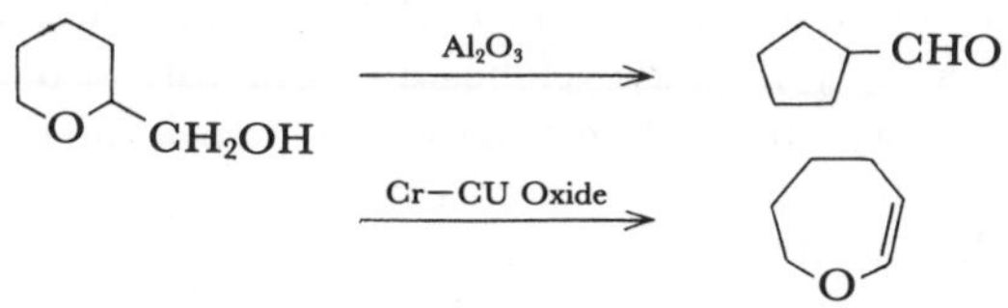

4.7.4 Dehydration of Heterocyclic Alcohols

Dehydration of allylic heterocyclic alcohols containing 1-metallacyclopent – 4-ene-3-ols (M = Ge, Si) over Al_2O_3 or ThO_2 gives the corresponding C-methylated silole or germole (minor product) and the exocyclic isomeric dienes (major product).[35] The latter may be produced by isomerization of the former via 1,3-sigmatropic hydrogen migration.

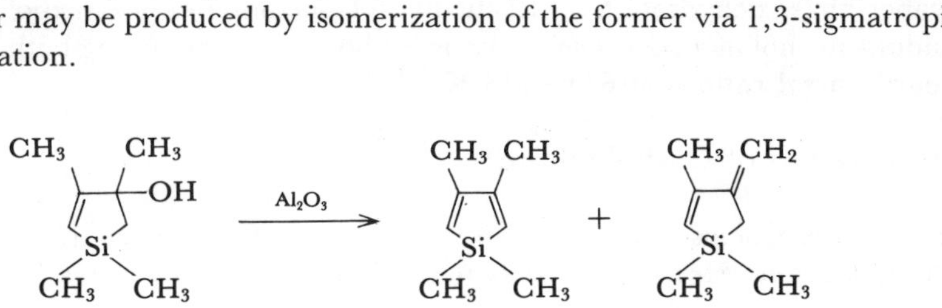

4.7.5 Dehydration of Diols

The dehydration of 1,4-, 1,5-, or 1,6-alkanediols gives the corresponding 5- to 7-membered oxyacyclanes. Tetrahydrofuran is obtained from 1,4-butanediol by dehydration over Al_2O_3[36] or silica – alumina.[37]

Distillation from 1,5-pentandiol with a catalytic amount of Amberlyst 15 yields 94% tetrahydropyran.[38] The distillation from 1,6-hexandiol with alumina gives oxepane in a 53% yield.[36] Similarly, the distillation from diethylene glycol with alumina results in the formation of 1,4-dioxane in a 49% yield.[36]

The vapor-phase dehydration of *cis*-2-butene 1,4-diol over alumina yields 2,5-dihydrofuran with 100% selectivity, while the dehydration over silica – alumina or tungsten oxide produces an appreciable amount of crotonaldehyde as a by-product.[39]

The reaction of bis(2-hydroxyethyl) – alkylamine over $SiO_2 - Al_2O_3$ (silica-rich) yields *N*-alkylmorpholine. Thus, *N*-ethylmorpholine is obtained in a 87% yield with 87.3% selectivity from bis(2-hydroxyethyl) – ethylamine.[40]

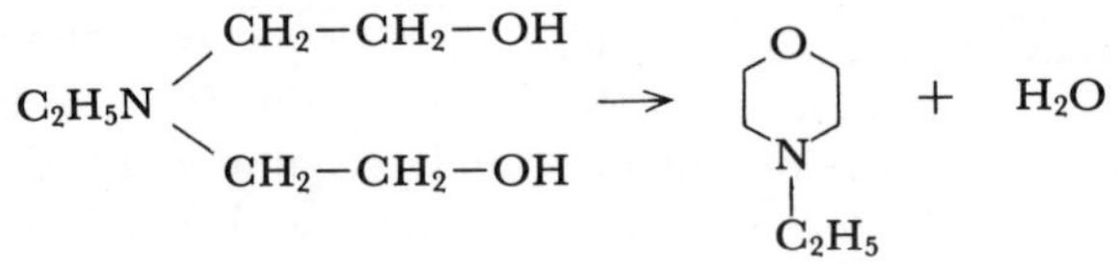

Cis- and *trans*-1,4-cyclohexanediol were dehydrated at about 523 K over alumina.[41,42] The *cis*-diol eliminates to give the cyclohexenol as the major product, but the *trans* isomer yields mainly 1,4-epoxycyclohexane. The rate of dehydration of the *trans* isomer is about fifty times faster than than of the *trans*-isomers. The intramolecular concerted ring closure through a boat conformation has been suggested for the formation of 1,4-expoxycyclohexane.[42]

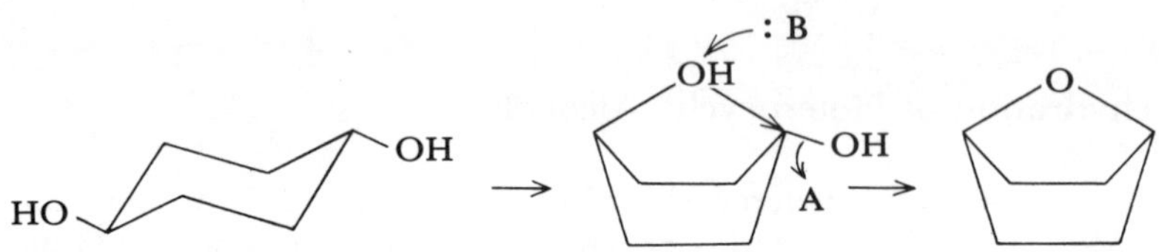

The vapor-phase dehydration of decanediol, octanediol and hexanediol yields the corresponding alkenol in a good selectivity over phosphates containing both alkali and alkaline earth metal cations at 673 – 713 K.[43]

4.7.6 Dehydration of Carbohydrates

The reaction of fructose in alcohols in the presence of Amberlyst 15 at 373 K for 20h gives the laevulic ester and hydroxymethylfurfuryl ether mixtures.[44]

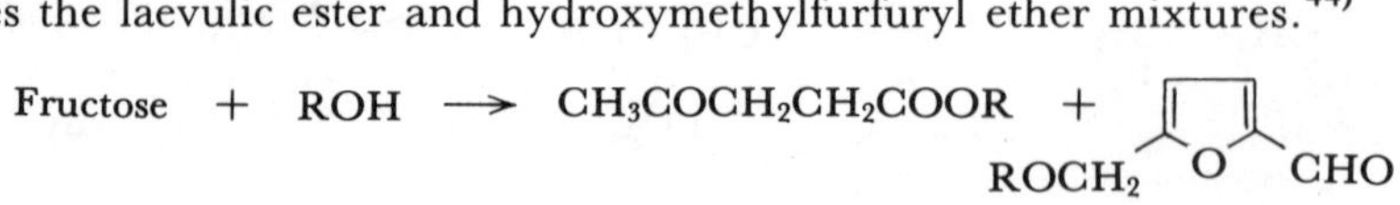

The dehydration of sorbitol and oligosaccharide can be effected by cation-exchange resins.[45)]

4.7.7 Dehydration of Cyclic Ethers and Epoxides

The dehydration of oxolanes yields dienes. 2,5-Dimethylcyclohexadienes are selectively formed by ring-opening dehydration of 2,2,5,5-oxolane over platinum/Al_2O_3, Al_2O_3 or NaY zeolite.[46)] Bartok and Molnar found that (±)-2,2,3,4,5,5-hexamethyl oxolane (I) is converted mainly into 2,3,4,5-tetramethylhexa−1,5-diene (IV), while its meso-isomer (II) reacts at a higher rate than (I) to give 2,3,4,5-tetramethylhexa−2,4-deiene (III) with high selectivity on γ-alumina.[47)] Mutual isomerization of the dienes is not observed.

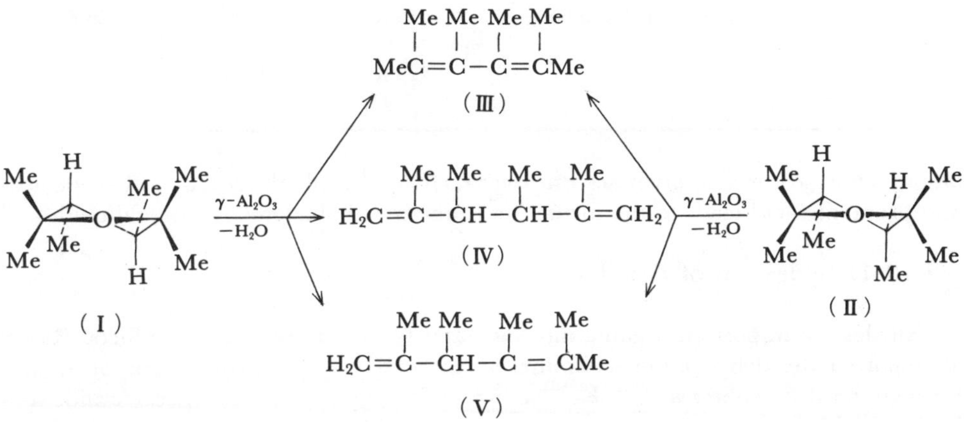

Cyclohexeneoxide is dehydrated to 1,3-cyclohexadiene with SiO_2-TiO_2-MgO in liquid phase at 381 K.[48)]

4.7.8 Dehydration of Aldehydes

The preparation of aldehydes by dehydration is useful because the aldehydes are easily accessible via hydroformylation of olefins.

Catalysts used for dehydration of aldehyde are phosphoric acid[49)] or borophosphates.[50)] Recently, it was found that borosilicates with the structure of a pentasil zeolite are very useful for dehydration of aldehyde.[51,52)] The results are summarized in Table 4.16.[51)]

4.7.9 Dehydration of Carboxylic Acids

The intramolecular and intermolecular dehydration of carboxylic acids leads to the formation of ketenes and acid anhydrides, respectively. Methylketene is formed in a 65.5% yield by the dehydration of propionic acid over $Bi_2O_3-SiO_2$ at 973 K.[53)] Maleic anhydride is produced by the dehydration of maleic acid over alumina or metal phosphates at 503−513 K.[54)]

Intermolecular dehydration of carboxylic acid may be accompanied by decarboxy-

TABLE 4.16 Dehydration of aldehydes to dienes over borosilicates

Feed	product	Condition	Conversion/%	Selectivity/%
pivaleraldehyde	isoprene	723 K WHSV=1.8 h^{-1}	99.6	51.0
2-methylbutanol	isoprene	673 K 2 h^{-1}	51.2	95.0
isovaleraaldehyde	isoprene	773 K 2 h^{-1}	50.7	85.5
2-methylpentanol	2-methyl-1,3-pentadiene	623 K 2 h^{-1}	21.0	91.5
cyclohexanealdehyde	methylenecyclohexene	623 K 2 h^{-1}	63.9	90.8
	+ methylhexadienes	673 K 2 h^{-1}	85.9	95.6
		723 K 2 h^{-1}	95.5	89.6

lation. Thus, acetic acid gives acetone over Cr_2O_3,[55] while the reaction of acetic acid and propionic acid leads to the formation of methylethylketone over metal oxides.[56]

4.7.10 Dehydration of Amides

Nitriles are important organic intermediates for a variety of ene products. Rao *et al.* reported the dehydration of amides to nitriles by passing the vapor of amides through ZSM-5 zeolites at 673 K.[57] Benzamide, phenacetamide, nicotinamide and isovaleramide gave the corresponding nitriles in 90%, 85%, 89% and 85% yields, respectively.

Dehydration of amides is often carried out in the presence of ammonia. Acetonitrile,[58] adiponitrile,[59] and ε-aminocaprolactam[60] are formed from acetamide, adipic acid and ε-aminocaprolactam over $Mg_2P_2O_7$, phosphoric acid supported on silica, and $MgHPO_4$, respectively in high yields. *o*-Chloronitrile can be produced from *o*-chlorobenzoic acid or its amide over alumina around 623 K.[61]

The dehydration of formamides leads to the formation of nitriles. This reaction can be classified as a homologation reaction, i.e., a reaction bringing about an increase in the hydrocarbon chain of an organic molecule. The transformation corresponds to dehydration to isonitrile, followed by rearrangement of the latter to nitrile.

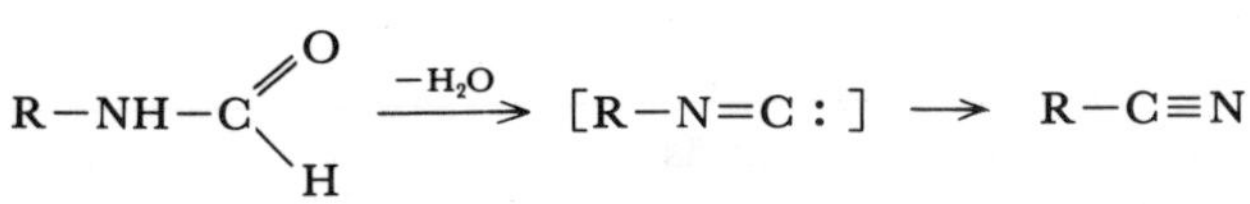

Rodrigues and Delmon reported the transformation of *N*-ethylformamide to propiononitrile with a 90% yield over mixed oxides containing Bi, P and Mo.[62] The most active catalysts contained essentially MoO_3 and $BiPO_4$. The origin of the synergy of the two phases has been discussed in detail.[62–65] For this transformation, oxygen is absolutely necessary for maintaining high selectivity and avoiding deactivation, though it does not react to any appreciable extent. It was proposed that molecular oxygen was dissociated on $BiPO_4$ to a mobil oxygen species, which migrate onto the surface of MoO_3, where it creates the catalytic sites for dehydration.[63,64)]

Ishida and Chono describe similar reactions over MoO_3 or WO_3 supported on silica.[66)] Thus, an *n*-valeronitrile yield of 73% was obtained with a selectivity of 84% over MoO_3/SiO_2 at 703 K. Similarly, *N*-hexylformamide, formanilide, *N*-cyclohexylformamide, *N,N'*-hexamethyleneformamide and *N*-benzylformamide gave nitriles in 50 – 80% selectivity.

REFERENCES

1. I. Batta, S. Boresok, F. Solymosi, Z. G. Szabo, Proc. 3rd Int. Congr. Catalysis, Amsterdam 1964, North Holland, Amsterdam, 1965, p.1340.
2. M. Otake, T. Onoda, *Shokubai*, **17**, 13 (1975) (in Japanese).
3. T. Okuhara, A. Kasai, N. Hayakawa, *Chem. Lett.*, **1981**, 391.
4. T. Okuhara, T. Hashimoto, M. Misono, Y. Yoneda, *Chem. Lett.*, **1983**, 576.
5. T. Baba, J. Sakai, H. Watanabe, Y. Ono, *Bull. Chem. Soc Jpn.*, **55**, 2657 (1982).
6. H. Pines, W.O. Haag, *J. Am. Chem. Soc.*, **83**, 2847 (1961).
7. C.A. Drake, M.M. Martinovich, S.J. Marwil, *Chem. Eng. Prog.*, **81**, 52 (1984).
8. I. Halasz, H. Vinek, K. Thomke, H. Noller, *Z. Phys. Chem. Neue Folge*, **144**, 157 (1985).
9. T. Yamaguchi, K. Tanabe, *Bull. Chem. Soc. Jpn.*, **47**, 424 (1974).
10. S. Kotsarenko, L.V. Malysheva, *Kinetai Katal.*, **24**, 877 (1983).
11. J. Take, T. Matsumoto, Y. Yoneda, *Bull. Chem. Soc. Jpn.*, **31**, 1612 (1978).
12. B.H. Davis, *J. Catal.*, **79**, 58 (1983).
13. H. Noller, K. Thomke, *J. Mol. Catal.*, **6**, 376 (1979).
14. A.J. Lundeen, R. van Hoozer, *J. Org. Chem.*, **32**, 3386 (1979).
15. B.H. Davis, W.S. Brey, Jr., *J. Catal.*, **25**, 81 (1972).
16. EP 150832
17. K. Thomke, Proc. 6th Intern. Congr. Catalysis, London, The Chemical Society, London, 1977, Vol. 1, p.303.
18. K. Thomke, *Z. Phys. Chem. Neue Folge*, **105**, 75 (1977).
19. K. Thomke, *Z. Phys. Chem. Neue Folge*, **105**, 87 (1977).
20. K. Thomke, *Z. Phys. Chem. Neue Folge*, **106**, 295 (1977).
21. K. Kochloefl, H. Knözinger, Proc. 5th Intern. Congr. Catalysis, Palm Beach, North Holland, Amsterdam, 1973, Vol.2, p.1171.
22. H. Knözinger, A. Schengllia, *J. Catal.*, **17**, 252 (1970).
23. D. Dautzenberg, H. Knözinger, *J. Catal.*, **33**, 142 (1974).
24. K. Thomke, *Z. Phys. Chem. Neue Folge*, **106**, 225 (1977).
25. H. Pines, C.N. Pillai, *J. Am. Chem. Soc.*, **83**, 3270 (1961).
26. K. Kochloefl, M. Mraus, C. -S. Chou, L. Beranek, V. Bazant, *Collect. Czeh. Chem. Comm.*, **27**, 1199 (1962).
27. K. Thomke, H. Noller, Proc. 5th Intern. Congr. Catalysis, Palm Beach, 1972, North Holland, Amsterdam, 1973, Vol. 2, 1183.
28. C.L. Kibby, S.S. Lande, W.K. Hall, *J. Am. Chem. Soc.*, **94**, 214 (1972).
29. K. Thomke, *J. Catal.*, **44**, 339 (1976).
30. H. Knözinger, *Angew. Chem. Intern. Ed.*, **7**, 791 (1968).
31. H. Arai, J. Take, Y. Saito, Y. Yoneda, *J. Catal.*, **9**, 140 (1967).
32. H. Pines, S.M. Brown, *J. Catal.*, **20**, 74 (1971).
33. C.L. Wilson, *J. Am. Chem. Soc.*, **69**, 3004 (1947).
34. A. Misono, T. Osa, Y. Sanami, *Bull. Chem. Soc. Jpn.*, **41**, 2447 (1968).
35. A. Laporterie, G. Manuel, H. Iloughmanne, J. Dubac, *Nouv. J. Chem.*, **8**, 437 (1984).

36. Y. Inoue, S. Deguchi, T. Hakushi, *Bull. Chem. Soc. Jpn.*, **53**, 3031 (1980).
37. DE 2,461,922
38. L.T. Scott, J.O. Naples, *Synthesis,* **1973**, 209.
39. H.B. Singh, G.E. Klinzing, J. Coull., *Ind. Eng. Chem., Prod. Res. Dev.*, **12**, 184 (1973).
40. Japan Kokai 1984-05174.
41. R.C. Olberg, H. Pines, V.N. Ipatieff, *J. Am. Chem. Soc.*, **66**, 1096 (1944).
42. J. Manassen, H. Pines, Proc. 3rd Congr. Catalysis (Schatler *et al.* eds.) 1964, North Holland, Amsterdam, 1965, p. 845.
43. DE 3,510,568
44. D.W. Brown, A.J. Floyd, G. Kinsman, Y. Roshan-Ali, *J. Chem. Tech. Biotechnol.*, **32**, 920 (1982).
45. J.K. Klein, B. Fleischer, J.C. Janssen, H. Widdecke, *Chem. -Ing. -Tech.*, **58**, 436 (1986).
46. US 3,692,743
47. M. Bartok, A. Molnar, *J. Chem. Soc., Chem. Comm.*, **1985**, 89.
48. Japan Kokai 1982-2662845; DT 2,163,396
49. DE 2,163,396
50. US 4,587,372; 4,628,140
51. DE 3,419,379
52. W. Holderich, Proc. 7th Intern. Zeolite Conf. (Y. Murakami *et al.*, eds.) Kodansha, Tokyo and Elsevier, Amsterdam, 1986, p. 827.
53. US 3,336,689
54. DE 1,346,189
55. R. Swaminathan, J.C. Kuriacose, *J. Catal.*, **16**, 357 (1970).
56. T. Imanaka, K. Adachi, Y. Okamoto, S. Teranishi, *Nippon Kagaku Zasshi*, **1974**, 357 (in Japanese).
57. A.V.R. Rao, M.N. Rao, K. Garyali, P. Kumar, *Chem. Ind.*, **1984**, 270.
58. Japan Kokai 1971-0265
59. Brit. Patent 1,206,839 (1970)
60. Japan Kokai 1971-26847
61. DD 235,449 (1986) CA **105**, 210,775u
62. M.V.E. Rodriguez, B. Delmon, Proc. 7th Intern. Congr. Catalysis, 1980, Tokyo (T. Seiyama, K. Tanabe, eds.), Kodansha, Tokyo and Elsevier, Amsterdam, 1980, p.1141.
63. J.M.D. Tascon, P. Grange, B. Delmon, *J. Catal.*, **97**, 300 (1986).
64. J.M.D. Tascon, P. Bartrand, M. Genet, B. Delmon, *J. Catal.*, **97**, 300 (1986).
65. J.M.D. Tascon, M.M. Mestdagh, B. Delmon, *J. Catal.*, **97**, 312 (1986).
66. Japan Kokai 1981-115759

4.8 DEHYDROHALOGENATION

Elimination of hydrogen halide from haloalkanes to form a carbon-carbon double bond in which the hydrogen atom to be removed is located at the β-position to the halogen atom belongs to a group of reactions called β-elimination.

$$\begin{array}{cc} | & | \\ -\mathrm{C}- & \mathrm{C}- \\ | & | \\ \mathrm{H} & \mathrm{X} \end{array} \longrightarrow \;\; \rangle\mathrm{C}{=}\mathrm{C}\langle \;\; + \;\; \mathrm{HX} \quad (\mathrm{X : Halogen}) \qquad (1)$$

This reaction proceeds readily over solid acids and bases. The mechanism, selectivity, and stereochemistry of dehydrochlorination and dehydrobromination have been extensively studied in relation to the acid-base properties of catalysts,[1-4] as have those in homogeneous liquid phase.[5] It has been fairly well established that the

TABLE 4.17 Reaction mechanism of dehydrohalogenation

	Mechanism		
	*E*1cB	*E*2	*E*1
		C^--like C^+-like	Ion pair Isolated ion
Catalyst	more basic (more nucleophilic)	⟷	more acidic (more electrophilic)
Ease of bond breaking			
C-H (acidity of proton)	greater	⟷	smaller
C-X	smaller	⟷	greater
Intermediate	C^-ion	Concerted	C^+ion
Stereochemistry	Syn	Anti	? Non-specific

mechanism, the rate and selectivity in general vary as summarized in Table 4.17, depending on the relative readiness of C – H and C – X bond breakings. For example, the C – X bond tends to dissociate more readily into C^+ and X^-, as the acid strength of the active site increases. Therefore, a carbenium (C^+) ion mechanism (*E*1) or a concerted mechanism (*E*2) with a C^+-ion transition like state is favored (eqs. (2) and (3)).

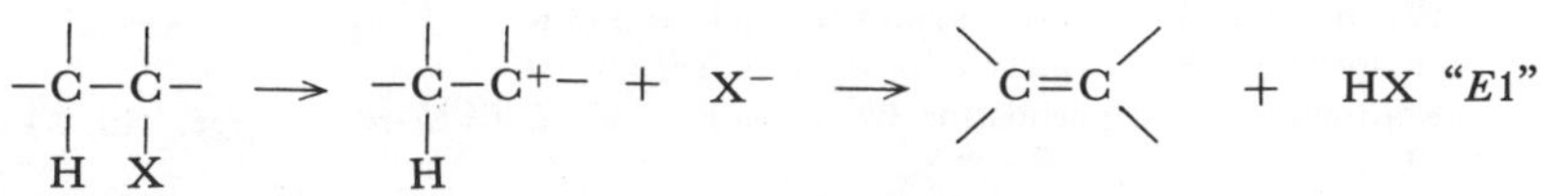 (2)

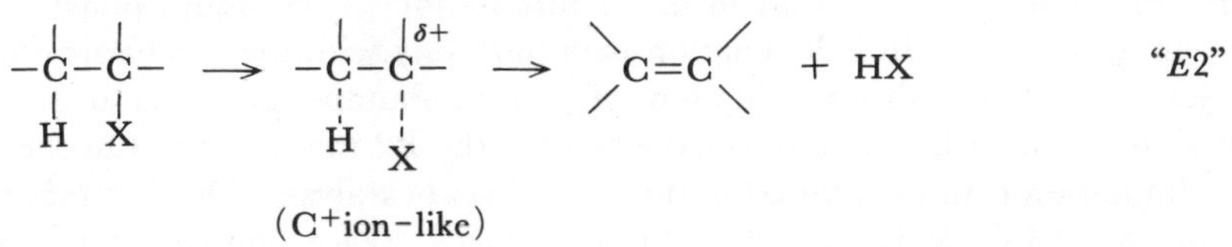 (3)

When the catalyst is strongly basic or β-H is acidic, C – H breaking to form a carbanion intermediate occurs first (*E*1cB) or a concerted mechanism with a C^- ion-like transition state operates.

The reaction rate usually increases with the acid strength of catalyst for *E*1 and C^+-like *E*2 mechanisms, while it increases with the base strength for *E*1cB and C^--like *E*2 mechanisms. Therefore, the catalytic activity often shows an inverse volcano pattern when the activity is plotted as a function of acid-base strength of catalyst. Examples are shown in Fig. 4.14[6] where the log of the rate is plotted against the electronegativity of metal ion which represent the acid strength of metal sulfate catalysts.[7] The difference in the positions of minima reflects the relative ease of the breaking of the C – H and C – X bonds. The β-H of 1,1,2-trichloroethane is more acidic, so the minimum is observed at a more strongly acidic catalyst.

Good correlations are present between the selectivity and the acid-base properties of catalysts as well. Contrasting selectivity has been demostrated in the case of 1,1,2-trichloroethane (see Section 3.1.11).[10] *Trans*/*cis* ratio and 1-/2- ratio of olefins

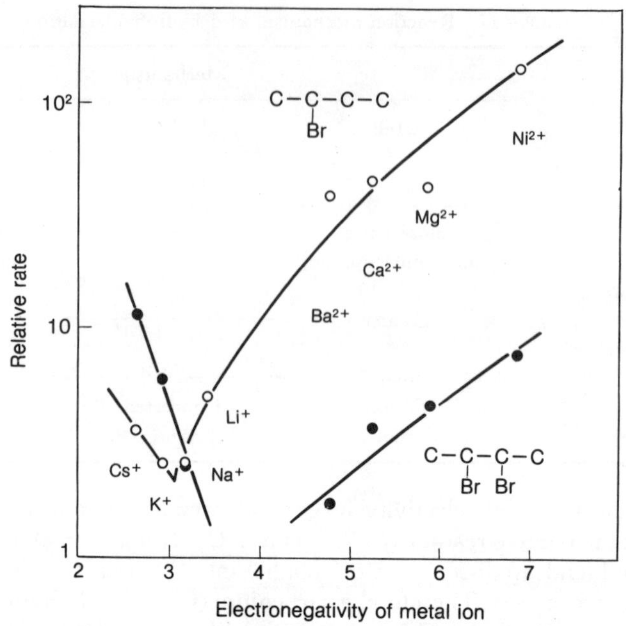

Fig. 4.14 Rates of elimination reactions of 2-bromobutane (○) and 2,3-dibromobutane (●) over metal-ion exchanged silica gels at 433 K
See note for electronegativity of metal ion.[7)]
(Reproduced with permission by Y. Aoki *et al.*, *Bull. Chem. Soc. Jpn.*, **49**, 3438 (1976)).

formed vary in a regular way[3,11)] as in the β-elimination in the liquid phase.[5)]

Stereochemistry of dehydrohalogenation is in harmony with the mechanism given in Table 4.17.[2,4,11)] Dehydrobromination of 2-bromobutane proceeds over basic catalysts mainly by anti elimination as expected for the E2 mechanism. On the contrary, syn elimination is more favored in the case of acid catalysts. The fact that racemization of optically active 2-bromobutane takes place much faster than dehydrobromination indicates the intermediacy of a carbenium ion.[12)] It has been further demonstrated that the steric course of reaction is governed mainly by the acid-base interactions between the reactants and catalysts,[2,11)] although the geometrical factor was originally suggested in earlier studies of dehydration.[13)]

Nearly 100% anti elimination has been observed for dehydrohalogenation of 2,3-dihalobutanes.[1,4)] In this case, the high specificity is due to the cyclic halonium intermediates, as has been shown by the presence of rapid racemization due to the Br shift (eq. (4)).[14)]

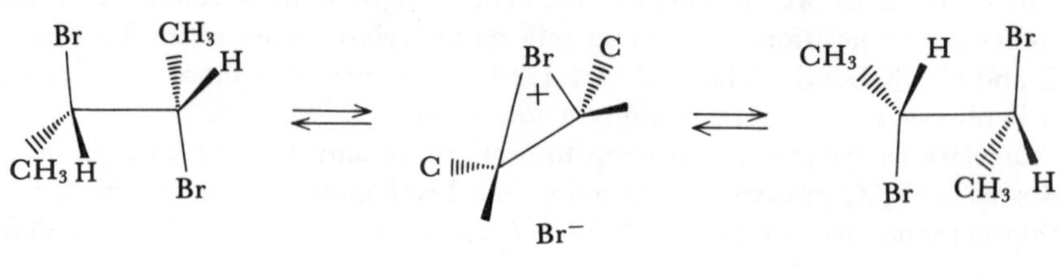

(4)

REFERENCES

1. H. Noller, P. Andreu, M. Hunger, *Angew, Chem. Intern. Ed.*, **10**, 172 (1971).
2. H. Noller, W. Kladig, *Catal. Rev. Sci. Eng.*, **13**, 149 (1976).
3. I. Mochida, Y. Anju, H. Yamamoto, A. Kato, T. Seiyama, *Bull. Chem. Soc. Jpn.*, **44**, 3305 (1971).
4. M. Misono, *J. Catal.*, **30**, 226 (1973).
5. J.F. Bunnett, *Surv. Progr. Chem.* (A.F. Scott, ed.) Vol. 5, Academic Press, New York, 1969, 53.
6. Y. Aoki, M. Misono, Y. Yoneda, *Bull. Chem. Soc. Jpn.*, **49**, 3437 (1976).
7. The electronegativity of metal ion, x_i, was first proposed by Tanaka, Ozaki, and Tamaru (*Shokubai*, **6**, 262 (1964)); the values were calculated by the equation $x_i = (1+Z)\, x_0$ — (1), where Z and x_0 are the oxidation number of metal ion and the electronegativity of metal as element, respectively. Later, Tanaka and Ozaki revised the values by using $x_i = (1+2Z)\, x_0$ — (2).[8] Since x_i from eq. (1) did not fit the data, Misono *et al.*[9] derived an independent equation of $x_i = x_0 + (\Sigma I_n)^{1/2}$ — (3), where I_n is the n-th ionization potential, in order to explain the catalytic activity and selectivity of metal sulfates for butene isomerization. However, there is fairly good correlation between the revised x_i from eq. (2) and x_i from eq. (3). The electronegativity used here is from eq. (3).[9]
8. K. Tanaka, A. Ozaki, *J. Catal.*, **8**, 1 (1967).
9. M. Misono, E. Ochiai, Y. Saito, Y. Yaneda, *J. Inorg, Nucl. Chem.*, **29**, 2685 (1967).
10. I. Mochida, A. Uchino, H. Fujitsu, K. Takeshita, *J. Catal.*, **43**, 264 (1976).
11. M. Misono, Y. Aoki, Y. Yoneda, *Bull. Chem. Soc. Jpn.*, **49**, 627 (1976).
12. M. Ishikawa, M. Misono, Y. Yoneda, *Chem. Lett.*, **1976**, 1229.
13. H. Pines, J. Manassen, *Advan. Catal.*, **16**, 49 (1966).
14. M. Misono, M. Ishikawa, Y. Yoneda. *Chem. Lett.*, **1976**, 69.

4.9 OLIGOMERIZATION AND POLYMERIZATION

4.9.1 Oligomerization of Lower Olefins with Solid Acid Catalysts

The activity of solid acid catalysts for polymerization of olefins to products with higher boiling points has long been known. The polymerization proceeds through the carbenium ion mechanism: proton addition to the double bond followed by a carbenium-ion addition to the double bond.

$$\mathrm{C{-}C{=}C} + \mathrm{H^+} \longrightarrow \mathrm{C{-}\overset{+}{C}{-}C}$$

$$\mathrm{C{-}\overset{+}{C}{-}C} + \mathrm{C{=}C{-}C} \longrightarrow \mathrm{C{-}\underset{\displaystyle C}{\underset{|}{C}}{-}C{-}\underset{\displaystyle C}{\underset{|}{\overset{+}{C}}}} \longrightarrow \longrightarrow \text{polymer}$$

Figure 4.15 shows the correlation between the acid amount ($H_0 < 3.3$) and the activity for propene polymerization at 473 K for a series of $SiO_2-Al_2O_3$ catalysts. The activity is proportional to the amount of acid sites.[1] The effects of the acidity and acid strength of $SiO_2-Al_2O_3$ on propene oligomerization have been demonstrated.[2-4]

Catalytic polymerization of olefins from cracked gases is an important process in the refining industry. Solid phosphoric acid, a mixture of liquid phosphoric acid and kieselguhr, has been used in industrial operations for propene oligomers and polymer

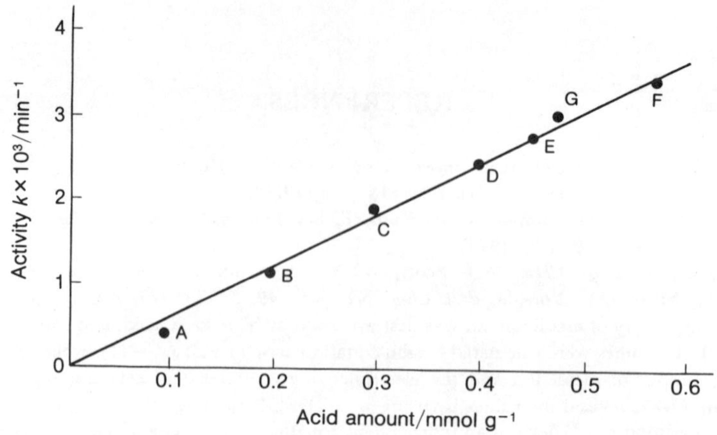

Fig. 4.15 Propene polymerization activity vs. acid amount for a series of $SiO_2 \cdot Al_2O_3$ catalysts.

Catalyst	A	B	C	D	E	F	G
Al_2O_3 (wt %)	0.12	0.32	1.04	2.05	3.56	10.3	25.1

(Reproduced with permission by O. Johnson, *J. Phys., Chem.*, **59**, 829 (1955)).

gasoline. Extensive data have been reviewed by Oblad *et al.*[5] and Jones.[6] The properties of gasoline, obtained by the catalytic polymerization of ethylene in the presence of solid phosphoric acid, have been described by Ipatieff and Corson.[7]

The composition of a sample of commercial propene-oligomer gasoline (after hydrogenation) produced over solid phosphoric acid is illustrated in Fig. 4.16.[8] The main products are liquid oligomers (di-, tri-, and tetramers) indicative of the high rate of the chain-transfer. The C_9-carbons are predominantly doubly branched, consistent with cationic oligomerization. The structure of oligomers does not correspond exactly to a simple reaction scheme due to skeletal isomerization. Thus, the dimers do not have exclusively the 2-methylpentyl skeleton, which is expected from the addition of the 2-propyl cation to propene. 3-Methylpentane and 2,3-dimethylbutane are also formed. The skeletons of these molecules may result from methyl shift in the 2-methyl – 3-pentyl cation.

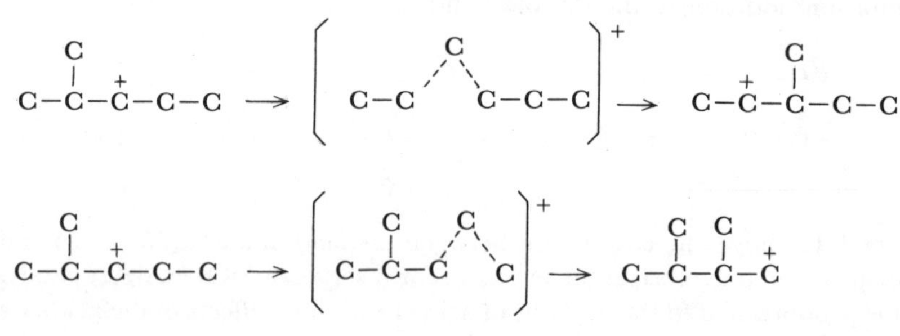

Similarly, the predominant trimers contain 3,5-dimethylheptane and 2,3-dimethylheptane skeletons, which cannot be formed directly from propene and a methylpentyl skeleton.

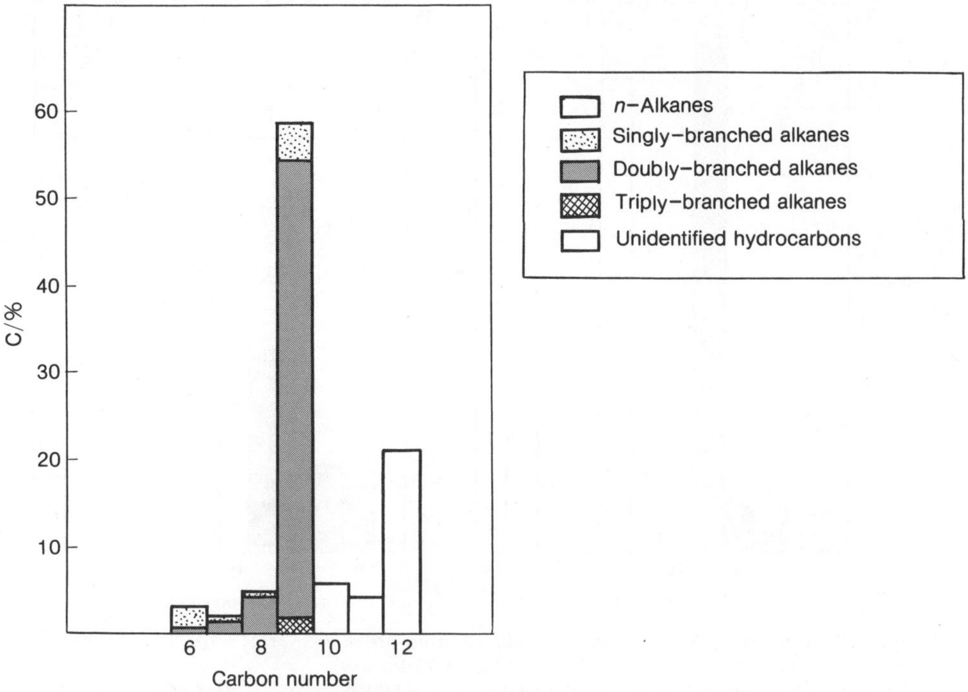

Fig. 4.16 Bar chart showing the composition of (hydrogenated) propene oligomer gasoline. Each bar shows percentage of carbon in the liquid hydrocarbons of that carbon number, and the sections of the bar show how much of this material is straight chain, how much single branched, etc., as detailed in the key.
(Reproduced with permission by K. G. Willshier *et al.*, *Appl. Catal.*, **31**, 344 (1987)).

Wilshier *et al.*[8] studied the oligomerization of propene over ZSM-5 zeolites (1.22% alumina) in a fixed bed reactor at 24 bar and 462 and 557 K. The liquid product contained highly branched olefinic hydrocarbons. The compositions of the products, after hydrogenation over Pd/charcoal, are summarized in Fig. 4.17. The product at 462 K was dominated by trimers, tetramers and, to a lesser extent, pentamers. The C_9 – hydrocarbons are predominantly double-branched, as are the trimer products of the propene-oligomer gasoline (Fig. 4.15), only 3% of the C_9-product being single branched. This shows that the oligomerization proceeds without shape selectivity on the external surface of the zeolite particles. The authors concluded that, at 462 K, the channels of the zeolite must be blocked by oligomerized olefins.[8] This is consistent with the observation that sorption of olefins is hindered at 373 K, and that highly branched products are desorbed at 473 – 573 K.[9]

The distribution of product formed at 557 K differs considerably from that obtained at 462 K. First, a broader distribution of carbon number was obtained, indicating the rapid oligomerization and cracking of oligomer. Second, few of the C_9-products were double-branched, C_7 – C_8 alkenes being produced in a relatively shape selective manner. The product distribution of propene oligomerization was examined in detail also by Quann *et al.*[10]

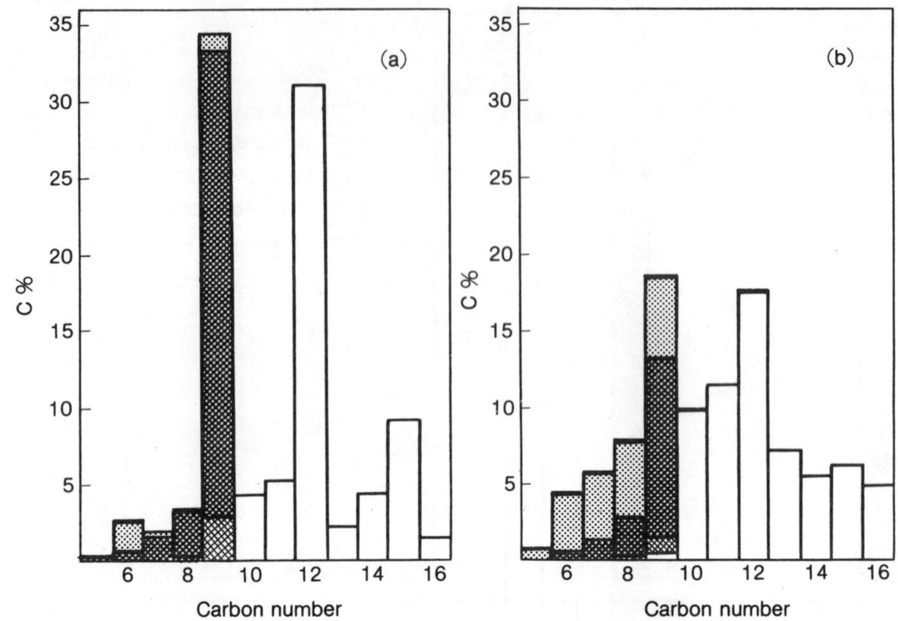

Fig. 4.17 Composition of the liquid produced from propene over H–ZSM–5 zeolite at (a) 462 K, and (b) 557 K. See Fig. 4.15 for key.
(Reproduced with permission by K. G. Willshier *et al.*, *Appl. Catal.*, **31**, 346 (1987)).

Garwood[11)] reported on the shape-selective conversion of C_2-C_{10} alkenes over ZSM-5 zeolites and found that the reaction conditions favoring higher molecular weight products are low temperature (470–520 K) and high pressure (300–1500 psig). The products are 95% alkenes in the conversion of propene at 602 K.

The distribution of C_2-C_6 alkene-products is close to equilibrium in the subatmospheric conversion of C_2, C_3, C_5, C_6, or C_{10} alkene at 544 K. This again indicates the very rapid oligomerization and cracking of oligomers.

Based on these results, a process converting light alkenes to higher molecular weight iso-alkenes using ZSM-5 zeolite has been developed. The product has a high octane number in the gasoline boiling range and a high cetane number in the diesel fuel range after hydrogenation.[12,13)]

Minachev *et al.*[14)] studied the activities of a variety of zeolites for oligomerization of isobutene at atmospheric pressure and found that L-type zeolite manifests the highest activity; oligomer yield greater than 90% was obtained at 373 K.

At higher temperatures, the conversion of olefins over ZSM-5 zeolite produces a mixture of alkanes and aromatics, because of the higher rate of hydrogen transfer.[11,15–18)] The yield of aromatics can be greatly enhanced at the expense of alkanes by incorporating Zn or Ga species into the zeolites.[17,18)]

The polymerization of propene over silica–alumina produces almost exclusively saturated C_4-C_6 hydrocarbons and coke, indicating extensive hydrogen transfer and cracking.[19)]

The catalytic activities of silicoaluminophosphate molecular sieves for propene oligomerization were examined at 643 K and 25 − 50 psig using co-fed hydrogen at a ratio of 2:1.[20]

The large pore SAPO-5 was inactive due to very rapid deactivation. In contrast, high oligomerization activity was observed with medium pore materials, SAPO-11 and SAPO-31, with good selectivity to liquid products (77% and 83% C_5^+ yield) and little gas make. The small pore SAPO-34 was considerably less active. Under the same reaction conditions, a ZSM-5 type zeolite gave considerably lower selectivity to liquid product (37%) with significant quantities of propane and butanes in the product. These results show that SAPO materials have significantly lower hydrogen transfer than ZSM-5 zeolite.

Oligomerization of isobutene was carried out using a silica gel having sulfobenzyl groups as catalyst.[21] At 403 K and atmospheric pressure, the conversion of isobutene was quantitative and yielded 52 wt% of di-, 40 wt% of tri- and 3 wt% of tetraisobutene besides a maximum of 2 wt% of odd carbon-number products.

4.9.2 Dimerization of Olefins with Alkali Metals

Free or supported alkali metals readily dimerize propene to a mixture of 2-methylpentenes at 420 − 470 K and 7.0 MPa.[22] As support materials, silica, α-alumina, oxides or carbonates of alkali and alkaline earth metals are used. The compounds of graphite intercalated by the alkali metals are also used.[22]

The products of propene dimerization using alkali metals dispersed on a variety of supports are given in Table 4.18.[23] The reactions were performed over a support containing 0.1 mole of dispersed alkali metal at 423 K and about 10 MPa pressure. The main product is 4-methyl − 1-pentene, as expected from the allylic carbanion addition to double bond.

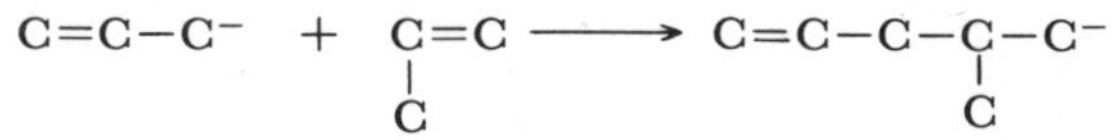

TABLE 4.18 Dimerization of propene over supported alkali metals

Catalyst	K/graphite			K/ K_2CO_3	Na/ graphite (NaC_{64})	Na/ K_2CO_3
	(KC_N)	(KC_{24})	(KC_{60})			
Alkali metal (wt %)	28.8	11.9	5.1	4.4	3.0	3.7
Rate of reaction (gm hexene/ gm atom alkali metal/hr)	63	120	97	150	155	91
Selectivity to dimer (%)	97	98	98	98	89	98
Composition of C_6H_{12} (%)						
4-Methyl-1-pentene	79	62	30	75	21	74
4-Methyl-2-pentene	12	24	50	16	56	18
2-Methyl-1-pentene	4	4	4	—	5	—
2-Methyl-2-pentene	1	6	12	1	15	1
n-Hexene	4	4	4	8	3	7

(Reproduced with permission by H. Pines, W. M. Stalik, *Base Catalyzed Reactions of Hydrocarbons*, Academic Press, 1977, P. 211).

The selectivity to 4-methyl−1-pentene increases with reaction temperature, while the catalyst life decreases with increasing the temperature.

Pis'man *et al.*[24] studied dimerization using dispersed sodium and potassium on various supports and found that sodium deposited on potassium carbonate is the most effective catalyst.

4.9.3 Polymerization of Alkene Oxides

The oxides and hydroxides of alkaline earth metals are active catalysts for the polymerization of ethylene oxide at 340−380 K.[25] As for the hydroxides, the activity decreases in the series $Be(OH)_2$ to $Ba(OH)_2$. Strontium carbonate is several fold more active than calcium or barium oxide.[26] Carbonates and oxalates are also active catalysts. The molecular weight is 10^6-10^7 for the oxides and hydroxides of alkaline earth metals.[25] The presence of protons on the catalysts leads to a lowering of molecular weight because of participation in the breaking reaction.

The oxides of Be, Mg, and Ca are considerably more active than the corresponding hydroxides.[25] It is assumed that the polymerization proceeds by adsorption of a molecule of ethylene oxide on a metal atom which simultaneously retains the growing chain of polymer. The subsequent insertion of the monomer then takes place at the base of the chain.[25]

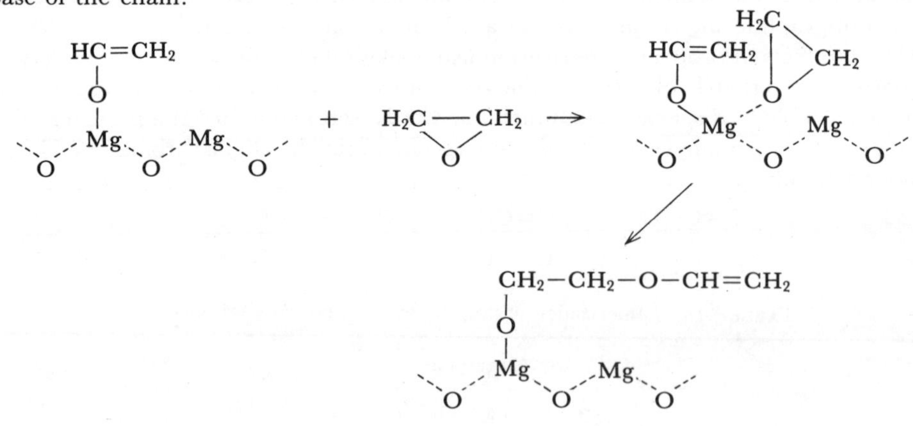

The mechanism of the polymerization of propylene oxide is considered to be similar to that of ethylene oxide; but the rate of the propylene oxide polymerization is 1−2 orders lower than the rate of ethylene oxide polymerization.[25] The polymerization of optically active propylene oxide in the presence of powdered KOH[27] or magnesium oxalate[28] leads to the formation of crystalline optically active polymers.

The double oxides obtained by calcination of hydrotalcite, $Mg_6Al_2(OH)_{16}CO_3 \cdot 4H_2O$ at 723 K, are active catalysts for the polymerization of propylene oxide at 323 K.[29] The anionic mechanism was suggested for the polymerization.

4.9.4 Miscellaneous Polymerization over Solid Acids and Bases

The cationic oligomerization of styrene by a perfluorinated resin sulfonic acid proceeds at 303−343 K in a liquid phase.[30] Complete conversion of the monomer was observed. The effect of temperature on the product distribution is given in Table 4.19. A dimer was the main product irrespective of temperature, though higher oligomers increased at lower temperatures. The dimer yield exceeded 60% at 393 K. It increased with increasing monomer concentration. The product contains three compounds.

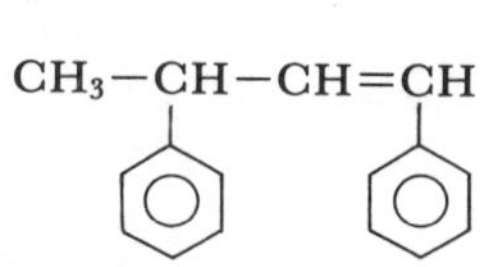

I

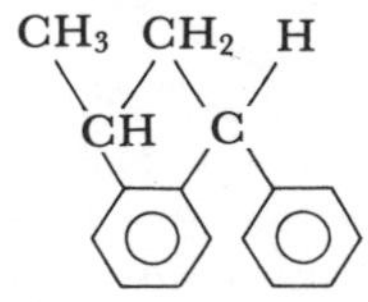

II

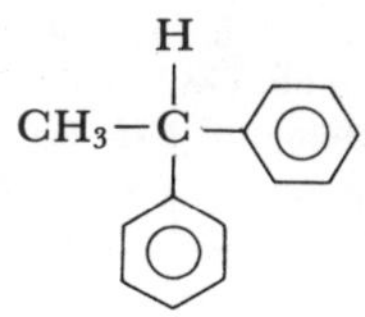

III

Thus, in carbon tetrachloride solvent, the dimer consists of 45% of I and 55% of II. In benzene solvent, the alkylation product, III, was also formed.

TABLE 4.19 Effect of temperature on the distribution of oligomers obtained with Nafion-H®: $[M]_0=0.10$ M; $[C]_0=3.0$ mM

Temp. K	Solvent	Oligomer/wt % Dimer	Trimer	Tetramer	Pentamer	Hexamer	≧Heptamer
303	CCl_4	31	20	16	8	Trace	25
323	CCl_4	48	25	13	4	Trace	10
343	CCl_4	57	25	10	8	0	0
393	Octane	61	25	11	3	0	0

monomer concentration=0.10 M. catalyst$[H^+]$=3.0 mM
(Reproduced with permission by H. Hasegawa, T. Higashimura, *Polymer J.*, **11**, 740(1979)).

Polymerization of gaseous benzyl alcohol was studied over silica−alumina by Olazar *et al.*[31,32] in the temperature range of 523−573 K. The fraction of the polybenzyl soluble in dichloromethane is 30%, 70% being insoluble.

The polymerization of ethylenimine from the gas phase was studied on Al_2O_3-based catalysts at 298 K.[33] The treatment of Al_2O_3 with HCl or BF_3 increased the rate of polymerization. The replacement of surface hydrogen ions of Al_2O_3 with Mg^{2+}, Ca^{2+} or Sr^{2+} decreased the activity. From these results, a mechanism involving protons as active centers was suggested.

Acetaldehyde is converted into high polymer at 203 K in the presence of alumina as catalyst.[34] CoO_3 and MoO_3 also give high polymers in much smaller yields. Polyacetaldehyde is a white, non-sticky and highly elastic material, whose structure is shown to be methylpolyoxymethylene by the infrared spectrum. The cationic mechanism has been suggested for the polymerization.[34]

The double oxide, $MgO-Al_2O_3$, prepared by calcination of hydrotalcite is active for the polymerization of β-propiolactone at 323 K.[35] Strongly basic sites are suggested to be the active centers.

REFERENCES

1. O. Johnson, *J. Phys. Chem.* **59**, 827 (1955).
2. V.C. Holm, G.C. Bailey, A. Clark, *J. Phys. Chem.*, **63**, 129 (1959).
3. R.L. Hodgson, J.H. Raley, *J. Catal.*, **4**, 6 (1965).
4. M. Misono, Y. Yoneda, *Bull. Chem. Soc. Jpn.*, **40**, 42 (1967).
5. A.G. Oblad, G.A. Mills, H. Heinemann, *Catalysis* (P.H. Emmett, ed.) Vol. 6, Reinhold, New York, p. 341.
6. E.K. Jones, *Adv. Catal.*, **8**, 219 (1956).
7. V.N. Ipatieff, B.B. Corson, *Ind. Chem.*, **28**, 860 (1936).
8. K.G. Wilshier, P. Smart, R. Western, T. Mole, T. Behrsing, *Appl. Catal.*, **31**, 339 (1987).
9. J.P. Van der Berg, J.P. Wolthuizen, A.D.H. Clague, G.R. Hays, R. Huis, J.H.C. Van Hooff, *J. Catal.*, **80**, 130 (1983).
10. R.J. Quann, L.A. Green, S.A. Tabak, F.J. Krambeck, *Ind. Eng. Chem. Res.*, **27**, 565 (1988).
11. W.E. Garwood, *Intrazeolite Chemistry*, (G.D. Stucky, F.G. Dwyer, eds.) ACS Symp. Ser. **218**, 383 (1983).
12. S. A. Tabak, F. J. Krambeck, *Hydrocarbon Process.*, **64**, 72 (1985).
13. S. A. Tabak, F. J. Krambeck, W. E. Tabak, *AIChE J.*, **32**, 1526 (1986).
14. Kh. M. Minachev, T. N. Bondaranko, D. A. Kon'drat'ev, *Izv. Akad. Nauk SSSR, Ser. Khim.*, **1987**, 1225.
15. Kh. M. Minachev, D.A. Kondrat'ev, B. K. Nefedov, A. A. Dergachev, T. N. Bondarenko, T. V. Alekseeva, T. B. Borovinskaya, *Izv. Akad. Nauk SSSR, Ser. Khim.*, **1980**, 2509.
16. E. N. Givens, C. J. Plank, E. J. Rosinski, US Patent, 3,827,968 (1974).
17. M. Shibata, H. Kitagawa, Y. Sendoda, Y. Ono, Proc. 7th Intern. Zeolite. Conf. (Y. Murakami, A. Iijima, J. W. Ward eds.) Kodansha, Tokyo and Elsevier, Amsterdam, 1986, p.717.
18. Y. Ono, H. Kitagawa, Y. Sendoda, *J. Chem. Soc., Faraday Trans. 1,* **83**, 2913 (1987).
19. F. E. Shephard, J. J. Rooney, C. Kemball, *J. Catal.*, **1**, 379 (1962).
20. R. J. Pellet, G. N. Long, J. A. Rabo., Proc. 7th Intern. Zeolite Conf. (Y. Murakami, A. Iijima, J. W. Ward, eds.) Kodansha, Tokyo and Elsevier, Amsterdam, 1986, p. 843.
21. A. Saus, E. Schmidl, *J. Catal.*, **94**, 187 (1985).
22. H. Pines, W. M. Stalik, in: *Base Catalyzed Reactions of Hydrocarbons and Related Compounds*, Academic Press, New York, 1977, p.205
23. J. K. Hambling, *Chem. Brit.*, **5**, 354 (1969).
24. I. I. Pis'man, M.A. Dalin, V.R. Ansheles, G.V. Vasil'kovskaya, I.I. Vavilova, *Dokl. Akad. Nauk SSSR*, **179**, 608 (1968).
25. O. V. Krylov, in: *Catalysis by Nonmetals,* Academic Press, New York, 1970, p.87 and p.231.
26. F. N. Hill, F. E. Bailey, Jr., J. T. Fitzpatrick, *Ind. Eng. Chem.*, **50**, 5 (1968).
27. C. C. Price, M. Osgan, *J. Am. Chem. Soc.*, **78**, 4787 (1956).
28. V. S. Livshits, O.V. Krylov, E.I. Klabunovskii, *Dokl. Akad. Nauk SSSR,* **161**, 633 (1965).
29. S. Kohjiya, T. Sato, T. Nakayama, S. Yamashita, *Makromol. Chem. Rapid Commun.*, **2**, 233 (1981).
30. H. Hasegawa, T. Higashimura, *Polymer J.*, **11**, 737 (1979).
31. M. Olazar, J. Bilbao, A. T. Aguayo, A. Romero, *Ind. Eng. Chem. Res.*, **26**, 1956 (1987).
32. J. Bilbao, M. Olazar, J. M. Ardans, A. Romero, *Ind. Eng. Chem. Res.*, **26**, 1960 (1987).
33. O. O. Battan, O. V. Krylov, B. A. Fokina, *Kinet. Katal.*, **7**, 289 (1966).
34. J. Furukawa, T. Saegusa, H. Fujii, A. Kawasaki, T. Tatano, *Makromol. Chem.*, **33**, 32 (1959).
35. T. Nakatsuka, H. Kawasaki, S. Yamashita, S. Kohjiya, *Bull. Chem Soc. Jpn.*, **52**, 2449 (1979).

4.10 ESTERIFICATION

4.10.1 General Remarks

Industrially important esterifications are the reactions of alcohols with saturated and unsaturated aliphatic carboxylic acids, e.g., acetic acid, fatty acids and acrylic acid, and aromatic dicarboxylic acids such as terephthalic acid. These reactions may be carried out in both liquid and vapor phases. Esterification reactions are usually limited by equilibrium particularly in liquid phase. Continuous removal of water produced and/or operation with an excess of one of the reactants are necessary to obtain higher yields of ester. Vapor-phase esterification is favored from this standpoint, and numerous studies have been directed toward the use of solid acid catalysts.[1-5]

Typical catalysts for homogeneous liquid-phase reactions are mineral acids and metal alkoxide.

Ion-exchange resins are representative solid catalysts. In particular sulfonic acid-type cation-exchange resins (sulfonated styrene-divinylbenzene copolymer) are widely used for esterification at about 400 K.[2] Several esterification reactions using fluorinated ion-exchange resins such as Nafion have also been attempted, since they have higher thermal and chemical stability. Various inorganic solid acids such as silica, alumina, zeolites, layered clays, SO_4-treated ZrO_2 and TiO_2, niobic acid, and heteropoly acid have also been applied to vapor-phase and liquid-phase esterifications with good results.[3]

Side reactions are usually dehydration of alcohol to form olefins and ethers and sometimes polymerization. Dissolution of catalytically active sites on the surface (e.g., $-SO_3H$ group in the case of ion-exchange resin) and thermal stability, which shorten the life of catalysts, are other important considerations in the use of solid acids.

Direct formation of ester from C_3, C_4-olefins and carboxylic acids is possible in the liquid phase with cation-exchange resins. Esterification of ethylene in the gas phase was also catalyzed by the resin, but deactivation was considerable.[4]

4.10.2 Reaction Mechanism

Rate equation can be expressed in power form or in terms of the Langmuir-Hinshelwood or Eley-Rideal mechanism.[1,5,6] For example, the reaction of acetic acid and *n*-butanol in the liquid phase catalyzed by HZSM-5 proceeds according to a rate equation which is of the first order with respect to acetic acid and of the zeroth order with respect to *n*-butanol.[6] It has been suggested that vapor-phase esterification of acetic acid with ethanol proceeded on decationized Y zeolites by a reaction between strongly adsorbed acetic acid and ethanol. The reaction rate is expressed by a Rideal-type rate equation, where the influence of the pressures of the acid dimer and products as well as the pressures of the reactants were taken into account.[1]

A plausible mechanism is as follows.

$$RCOOH \underset{}{\overset{+H^+}{\rightleftarrows}} RC^+(OH)_2 \underset{}{\overset{+R'OH}{\rightleftarrows}} RC(OH)(OH_2^+)(OR') \underset{}{\overset{-H_3^+O}{\rightleftarrows}} RCOOR' \quad (1)$$

When porous solid catalysts are used, diffusion processes in intra- and intercrystalline and interparticle pores often play important roles in the kinetics. Santacesaria *et al.* examined the effects of diffusion for both vapor-phase (393 − 427K) and liquid-phase (313 − 343K) esterification of ethanol and acetic acid catalyzed by zeolites.[7] According to them, the vapor-phase reaction proceeded in a condensed phase formed in the pores of catalysts, and intracrystalline diffusion is always a rate-determining step in the case of mordenite and, by contrast, in the case of Y zeolite the influence of intracrystalline diffusion upon reaction rate can be neglected.

The effects of acidic properties of catalysts are described below.

4.10.3 Effects of Chemical Properties of Catalyst

The acid amount of solid catalysts, which was measured by amine titration with an indicator with pK_a = 3.3, was linearly correlated with the catalytic activity.[8] In the case of decationized and metal-ion exchanged zeolites the catalytic activity is related to the surface acidity[1], as shown in Fig. 4.18. It was suggested that there were two mechanisms; one was accelerated by acid and another was minor and independent of the surface acidity. Usually too strong acidity causes unfavorable side reactions such as olefin formation, as in the case of $ZrO_2 - SO_4^{2-}$ [9] and Nafion.

The hydrophobicity of zeolite surface which increases with the Si/Al ratio has also been studied as an important factor controlling the reaction rate.[6]

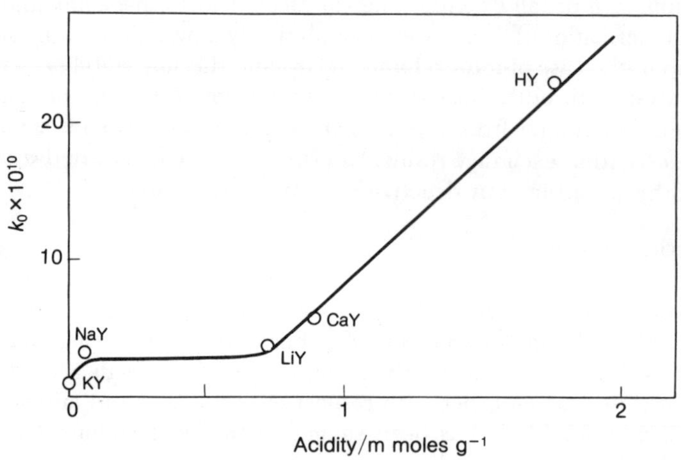

Fig. 4.18 Relationship between the catalytic activities of Y zeolites for esterification and the acid amounts of zeolites. The catalytic activity is expressed by the pre-exponential factor of the rate constant.
Acetic acid/ethanol = 1, reaction temp. = 423 K
(Reproduced with permission by E. Santacesaria *et al.*, *J. Catal.*, **80**, 432 (1983)).

4.10.4 Typical Solid Acid Catalysts

Ion-exchange resins are widely used. Heteropoly acids supported on silica or carbon exhibited higher activity and selectivity than silica – alumina or solid phosphoric acid in the vapor-phase esterification.[11] SO_4^{2-}-treated metal oxides such as TiO_2 and ZrO_2 are much more active than silica – alumina for various esterifications, although slight deactivation upon repeated runs is observed.[9] Excellent performance of niobic acid which were calcined at 393 – 537 K has been reported.[12] The results are compared with other solid acids in Table 4.20. $TiO_2 - SO_4^{2-}$ showed high activity, but the activity rapidly decreased and became much lower than that of niobic acid after 2 h's reaction. The activity of niobic acid did not change even after use for 60 h. The HZSM-5 catalyst also exhibited high activity, but large amounts of diethylether and ethylene were formed.

TABLE 4.20 Catalytic activities of several solid acids for vapor-phase esterification

Catalyst	reaction temp./K	Coversion/% (ethanol basis)	Selectivity/% (ethanol basis)
Niobic acid	393	72	100
	413	86	100
Cation-exchange resin	413	50	98
ZrO_2-SO_4^{2-}	413	56	90
TiO_2-SO_4^{2-}	413	100	95
SiO_2-Al_2O_3	413	14	98
ZSM-5	393	82	92
	413	99	72
$H_3PW_{12}O_{40}$ on Carbon†	393	83	99.7

Acetic acid/ethanol ratio=1 in volume.
†Taken from ref. 11. The other data are from ref. 12.

REFERENCES

1. E. Santacesaria, D. Gelosa, P. Danise, S. Carra, *J. Catal.*, **80**, 427 (1983).
2. M.B. Bochner, S.M. Gerber, W.R. Vieth, A.J. Rodger, *Ind. Eng. Chem. Fundamentals*, **4**, 314 (1965); G.A. Olah, T. Keumi, D. Meidar, *Synthesis*, **1978**, 929.
3. Y. Izumi, *Shokubai Kouza,* Vol. 8, p.285, Kodansha, Tokyo, 1985 (in Japanese).
4. Y. Murakami, T. Hattori, H. Uchida, *Kogyo Kagaku Zasshi,* **72**, 1945, (1969) (in Japanese); Rhone-Poulenc, Jpn. Kokai, 1979 – 151905.
5. J. Giménez, J. Costa, S. Cervera, *Ind. Eng. Chem. Res.*, **26**, 198 (1987).
6. S. Namba, Y. Wakushima, T. Shimizu, H. Masumoto, T. Yashima, in: *Catalysis by Acids and Bases,* (B. Imelik *et al.*, eds.), Elsevier, Amsterdam, 1985, p.205.
7. E. Santacesaria, S. Carra, F. Silva, *J. Catal.*, **85**, 519 (1984).
8. K. Tarama, S. Teranishi, K. Hattori, T. Ishibashi, *Shokubai (Catalyst)*, **4**, 69 (1962) (in Japanese).
9. M. Hino, K. Arata, *Chem. Lett.*, **1981**, 1671; K. Arata, M. Hino, *Shokubai (Catalyst)*, **25**, 124 (1983)

(in Japanese).
10. G.A. Olah, T. Keumi, D. Meidar, *Synthesis*, **1978**, 929.
11. Y. Izumi, K. Urabe, *Chem. Lett.*, **1981**, 663.
12. Z.-H. Chen, T. Iizuka, K. Tanabe, *Chem. Lett.*, **1984**, 1085.

4.11 HYDROLYSIS

4.11.1 Hydrolysis of Esters

The most important heterogeneous catalysts for hydrolysis are strongly acidic ion-exchange resins. Thomas and Davies[1,2] determined the rate constants and apparent activation energies for the hydrolysis of a variety of esters catalyzed by an ion-exchange resin in aqueous solutions. For equivalent amounts, the resin was a more effective catalyst than hydrochloric acid by a factor of 1.7 for methyl acetate, 2.3 for ethyl acetate, and 9.8 for butyl acetate. Bernhard and Hammett[3] found that a lightly cross-linked ion-exchange resin was a better catalyst for the hydrolysis of methyl and ethyl acetates than dilute hydrochloric acid, and that a tightly cross-linked resin was less effective than homogeneous acid.

In 70% aqueous acetone, hydrochloric acid is catalytically more active than the resin by a factor of 2 for methyl acetate, 3 for ethyl acetate and 20 for ethyl hexanoate.[4,5] Hammett and coworkers[4,5] concluded that the resin catalyst imposes a loss in entropy in the transition state.

In industrial hydrolysis processes, distillation through reaction can often be employed for the easy separation of products and to eliminate the equilibrium constraint.[6]

Namba *et al.*[7] carried out the hydrolysis of ethyl acetate in aqueous solution in the presence of a variety of zeolites and found that ZSM-5 zeolites and dealuminated mordenites were the most active. The hydrophobic nature of high-silica zeolites seems to be advantageous for reactions in aqueous solutions.

4.11.2 Hydrolysis of Ethers

The hydrolysis of diethyl ether to ethanol can be effected by using inorganic oxides such as alumina as catalysts.[8] Methyl *t*-butyl ether can be hydrolyzed to obtain *t*-butyl alcohol in the presence of strongly acidic cation-exchange resins at 340 – 400 K.[9]

$$CH_3OC(CH_3)_3 + H_2O \longrightarrow (CH_3)_3COH + CH_3OH$$

Similarly, glycol mono-*t*-alkyl ether can be hydrolyzed to obtain tertiary alcohols with strongly acidic cation-exchange resins.[10] For example, ethyleneglycol mono-*t*-butyl ether gives *t*-butyl alcohol at 350 – 400 K.

$$(CH_3)_3COCH_2CH_2OH + H_2O \longrightarrow (CH_3)_3COH + HOCH_2CH_2OH$$

4.11.3 Hydrolysis of Carbohydrates

The inversion of sucrose can be effected in the presence of strongly ion-exchange resins.[11–13)]

$$C_{12}H_{22}O_{11} \quad + \quad H_2O \quad \longrightarrow \quad C_6H_{12}O_6 \quad + \quad C_6H_{12}O_6$$

(+)-Sucrose D-(+)-Glucose D-(−)-Fructose

The reaction is first order with respect to sucrose concentration for temperatures between 323 and 348 K. The rate is strongly influenced by intraparticle diffusion, with no indication that the external mass transfer is significant.

The hydrolysis of lactose in a continuous fixed-bed column reactor containing strongly acidic resins was reported by Chem and Zall.[14)] The reaction was first order and the activation energy was 154 $kJ\,mol^{-1}$. At 368 K, 99% of the lactose in the whey was hydrolyzed within 3 h of residence time when the substrate was acidified with concentrated nitric acid to a final acid concentration of 0.6 mole l^{-1}.

Hydrolysis of disaccharides — cellobiose, maltose and lactose — was investigated over A, X, and Y type zeolites.[15)] The conversion profiles of the hydrolysis of cellobiose over NaX are shown in Fig. 4.19. The reaction was carried out at 373 K and initial reactant concentration of 1.0 wt%, and a catalyst loading of 0.1 g cm^{-3}. Total conversion is based on the amount of cellobiose consumed by the reaction. A very large fraction of the reaction occurs within the first 0.5 h and the entire reaction terminates after 1 h. The pH profile of the reaction broth is also shown in Fig. 4.19. Upon addition

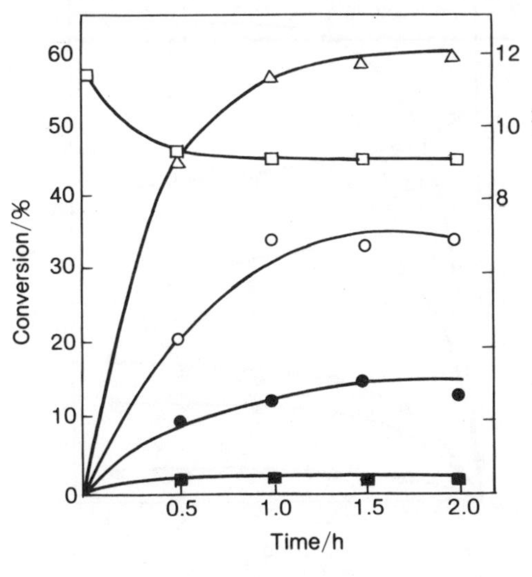

Fig. 4.19 Conversion and pH profiles for hydrolysis of cellobiose over zeolite Na-X. —△—; Total conversion, conversion into : —○—; cellobiulose, —●—; glucose, —■—; fructose, —□—; pH.
(Reproduced with permission by R. Shukla *et al.*, *Carbohydrates Res.*, **143** ,100 (1985)).

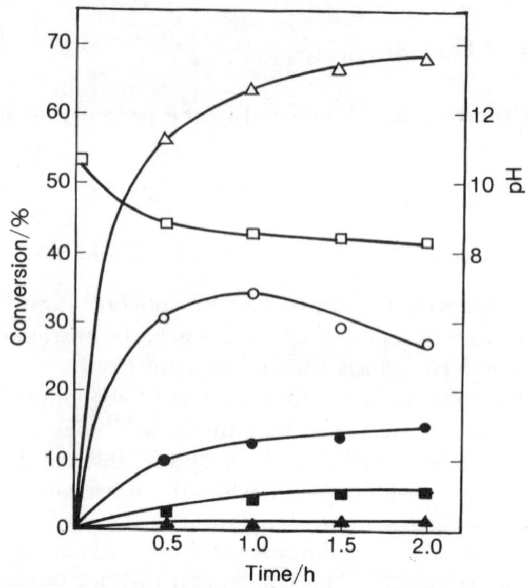

Fig. 4.20 Conversion and pH profiles for hydrolysis of maltose over zeolite Na-X.
—△—; Total conversion, conversion into : —○—; Total conversion, conversion into : meltulose, —●—; glucose, —■—; fructose, —□—; pH.
(Reproduced with permission by R. Shukla *et al.*, *Carbohydrates Res.*, **143**,101 (1985)).

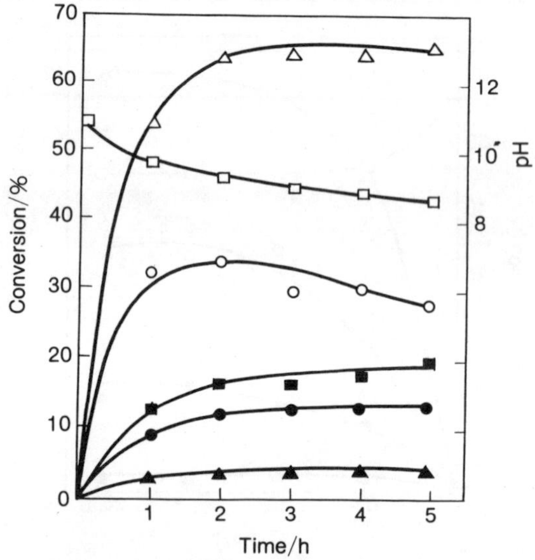

Fig. 4.21 Conversion and pH profiles for hydrolysis of lactose over zeolite 4A.
—△—; Total conversion, conversion into : —○—; lactulose, —■—; galactose, —●—; glucose, —▲—; tagatose, —□—; pH.
(Reproduced with permission by R. Shukla *et al.*, *Carbohydrates Res.*, **143**,101 (1985)).

of the catalyst to the sugar solution, the pH rises to 11.3 then drops rapidly as the reaction proceeds, stabilizing at pH 9.0. The major product formed is cellobiulose followed by glucose and fructose. Approximately 10 − 13% of the disaccharide reacted is not accounted for in the product distribution, indicating that degradation reactions occur.

The time courses of hydrolysis of maltose over NaX and of lactose over NaA at 358 K are shown in Figs. 4.20 and 4.21, respectively.

As shown in Figs. 4.19 − 4.21, the concentration of the ketodisaccharides—cellobiulose, maltulose and lactulose—passes a maximum. On the other hand, the concentration of glucose and fructose (with cellobiose and maltose) and galactose, glucose and tagatose (with lactose) increases monotonically with time. These profiles indicate that zeolite catalyzes two types of reaction, namely hydrolysis and isomerization. Thus reaction routes consist of parallel hydrolysis and isomerization of the disaccharides to their corresponding ketoses, followed by hydrolysis of the ketoses.

4.11.4 Hydrolysis of Nucleosides

Purine nucleosides are known to be susceptible to acid, giving nucleic acid and **D**-ribose. This property is based on the hydrolysis of the glycosidic bond. Treatment with cation-exchange resin in H-form is useful for the industrial isolation or purification of purine nucleosides.[16,17] The rates of hydrolysis of gluconsine, hypoxanthine, and adenosine with cation exchange resins were compared with the hydrolysis rate with hydrochloric acid.[18] The hydrolysis of adenosine with a cation-exchange resin was slower that that of guanosine, while the reverse was true for hydrolysis with hydrochloric acid; this was attributed to the difference in the position of protonation of the two substances.

4.11.5 Hydrolysis of Acetals

The acetals and diacetals of various aldehyde or dialdehydes were reported to be hydrolyzed to the corresponding aldehydes in the presence of strongly acidic cation-exchange resins.[19] Yields of aldehydes are given in Table 4.21. The hydrolysis with the cation-exchange resins takes place in an aqueous medium under milder conditions than in the presence of acids.

The hydrolysis of the diethylacetals of acetaldehyde (I), croton aldehyde (II), benzaldehyde (III), citral (IV) was realized with stirring at 293 K in the presence of the cation-exchange resin. The tetraethyl diacetals of glutaraldehyde (VI), 1,2-diformyl-*gem*-dichlorocyclopropane (VII), and fumaraldehyde (VIII) are hydrolyzed in a similar manner.

The hydrolysis of phenyl propagyl aldehyde diethyl acetal (V) is realized at 363 − 373 K with simultaneous distillation. In the case of the diacetals of unstable dialdehydes, e.g., malon aldehyde (IX), hydrolysis occurs in the presence of aniline hydrochloride, and 3-amilino-propenylanilinium chloride (X) is formed.

$$\underset{(\mathrm{IX})}{(EtO)_2CHCH_2CH(OEt)_2} + C_6H_5NH_2HCl \longrightarrow \underset{(\mathrm{X})}{[C_6H_5NH{=}CH{-}CH{=}CH{-}NHC_6H_5]Cl^-}$$

TABLE 4.21 Yields and characteristics of the aldehydes obtained by acetal hydrolysis

Initial aldehyde No.	Hydrolysis product	Yield, %
I	CH_3CHO	70
II	$CH_3CH{=}CHCHO$	98[†1]
III	PhCHO	71
IV	$(CH_3)_2C{=}CH(CH_2)_2C(CH_3){=}CHCHO$	68
V	$PhC{\equiv}CCHO$	72
VI[†2]	$OHC(CH_2)_3CHO$	40
VII	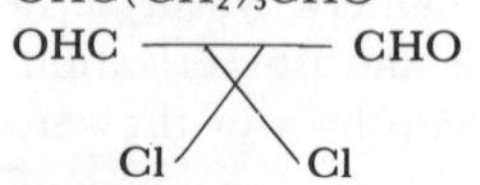	40
VII	$(MeO)_2CHCH{=}CHCHO$	60, 95[†1]
IX[†3]	$[Ph\overset{+}{N}H{=}CH{-}CH{=}CH{-}NHPh]Cl^-$	80

†1 The yield was determined by spectrophotometry.
†2 Analogous results were obtained during hydrolysis of 2-methoxy-3,4-dihydropyran.
†3 Hydrolysis was realized in the presence of $PhNH_2 \cdot HCl$.
(Reproduced with permission by A. Kh. Khusid *et al.*, *Zh. Org. Khim.*, **18**, 2275(1982)).

4.11.6 Hydrolysis of Methylhalides and Methylene Chloride

The gas-phase hydrolysis of methyl halides over γ-alumina to methyl alcohol and dimethyl ether was reported by Olah *et al.*[20] The reactions were carried out in the temperature range of 473 − 673 K while the CH_3Cl to H_2O ratio varied from 1:5 to 1:20. The results are given in Table 4.22. At 648 K with a H_2O/CH_3Cl ratio of 10 and GHSV of 1500 to 1800 $cm^3\ g^{-1}h^{-1}$, the per pass average conversion of methyl chloride was 16%, with the product containing 70% methanol and 30% dimethyl ether, which seems to be formed as a secondary product from the bimolecular dehydration of methyl alcohol. Increasing the H_2O/CH_3Cl ratio to 15 increased the conversion to 18%.

TABLE 4.22 Hydrolysis of methyl halides over γ-alumina

H_2O/CH_3Cl	reaction temp. K	GHSV	conversion, %	% products	
				CH_3OH	CH_3OCH_3
Methyl chloride					
5	648	1500	13	45	55
10	648	1500	16	70	30
15	648	1500	18	80	20
Methyl bromide					
10	648	1000	23	73	27

(Reproduced with permission by G. A. Olah *et al.*, *J. Am. Chem, Soc.*, **107**, 7103(1985)).

Methyl bromide gave comparable results with a conversion of 23% using a H_2O/CH_3Br ratio of 10 at 648 K with GHSV of 1000 $cm^3\ g^{-1}h^{-1}$.

The authors subsequently examined the catalytic activities of the various catalysts comprised of 10 wt% metal oxide, 10 wt% aluminum hydroxide, and 80 wt% γ-alumina, which were prepared by a water slurring procedure followed by calcination. As shown in Table, 4.23, the most effective catalyst was the $ZnO/Al(OH)_3/\alpha$-alumina system which gave 25% conversion. Nickel, bismuth, and magnesium hydroxides gave lower activity, as did other metal oxides when replacing zinc oxide.

The hydrolysis of methylene chloride to formaldehyde was effected by heat-treated nickel sulfate or zinc sulfide.[21] The reaction was suggested to occur through the abstraction of chloride ions from methylene chloride by the Lewis acid sites to form carbenium ions, $^+CH_2Cl$.

TABLE 4.23 Hydrolysis of methyl chloride over γ-alumina-supported metal oxide/metal hydroxide catalysts (H_2O/CH_3Cl ratio 10,648 K, GHSV~1200)

catal, 10 % M_xO_y/ 10 % $M(OH)_x$/ 80 % γ-Al_2O_3	% convers. to CH_3OH/ CH_3OCH_3
$V_2O_5/Al(OH)_3/\gamma$-Al_2O_3	5
$MnO_2/Al(OH)_3/\gamma$-Al_2O_3	13
$Cr_2O_3/Al(OH)_3/\gamma$-Al_2O_3	14
$MgO/Al(OH)_3/\gamma$-Al_2O_3	17
$ZrO_2/Al(OH)_3/\gamma$-Al_2O_3	14
$BaO/Al(OH)_3/\gamma$-Al_2O_3	9
$TiO_2/Al(OH)_3/\gamma$-Al_2O_3	18
$ZnO/Al(OH)_3/\gamma$-Al_2O_3	25
$Fe_2O_3/Al(OH)_3$-$/\gamma$-Al_2O_3	9
$ZnO/Ni(OH)_2/\gamma$-Al_2O_3	11
$ZnO/Bi(OH)_3/\gamma$-Al_2O_3	7
$ZnO/Mg(OH)_2/\gamma$-Al_2O_3	6

(Reproduced with permission by G. A. Olah *et al.*, *J. Am. Chem. Soc.*, **107**,7103(1985)).

REFERENCES

1. G. G. Thomas, C. W. Davies, *Nature*, **159**, 372 (1947).
2. C. W. Davies, C. G. Thomas, *J. Chem. Soc.*, **1952**, 1607.
3. S. A. Bernhard, L. P. Hammett, *J. Am. Chem. Soc.*, **75**, 5834 (1953).
4. V. C. Haskell, L. P. Hammett, *J. Am. Chem. Soc.*, **71**, 1284 (1949).
5. S. A. Bernhard, L. P. Hammett, *J. Am. Chem. Soc.*, **75**, 1798 (1953).
6. Japan Kokai, 68-6602, 82-7259
7. S. Namba, N. Hosonuma, T. Yashima, *J. Catal.*, **72**, 16 (1981).
8. US 2519061
9. Japan Kokai, 74-10535
10. Japan Kokai, 81-138125
11. E. W. Reed, J. S. Dranoff, *Ind. Eng. Chem. Fundam.*, **3**, 304 (1964).

12. N. Lifshutz, J. S. Dranoff, *Ind. Eng. Chem. Process Des. Dev.*, **7**, 266 (1968).
13. T. J. McGovern, J. S. Dranoff, *AIChE J.*, **16**, 536 (1970).
14. H. C. Chem, R. R. Zall, *J. Food Sci.*, **48**, 1741 (1983).
15. R. Shukla, X. E. Verykios, R. Mutharasan, *Carbohydrates Res.*, **143**, 97 (1985).
16. Japan Kokai, 71-3581
17. Japan Kokai, 73-28438
18. Y. Suzuki, *Bull. Chem. Soc. Jpn.*, **47**, 2077 (1974).
19. A. Kh. Khusid, N. V. Chizhova, *Zh. Org. Khim.*, **18**, 2275 (1982).
20. G. A. Olah, B. Gupta, M. Farina, J. D. Felberg, W. M. Ip, A. Husain, R. Karpeles, K. Lammertsma, A. K. Melhorta, N. J. Trivedi, *J. Am. Chem. Soc.*, **107**, 7097 (1985).
21. K. Tanabe, in: *Solid Acids and Bases*, Kodansha, Tokyo and Academic Press, New York, 1970, p.132.

4.12 CATALYTIC CRACKING

4.12.1 Catalytic Cracking and the Catalysts

Cracking reactions are C – C bond ruptures, and thermodynamically favored at high temperatures. Gasoline comprises only about 15 – 25 percent of natural petroleum. Thermal cracking of large molecules was known and utilized to increase the gasoline-fraction of petroleum prior to World War I. Acid-treated montmorillonite catalysts were introduced into a commercial process in 1936 and permitted larger yields of gasoline of high octane number than had been obtainable by thermal cracking of gas oils.

The clay catalysts were sensitive to high-temperature regeneration, hence superseded by amorphous synthetic silica – alumina catalysts. Gasoline yield obtainable from gas oil increased about 20 percent by thermal cracking to over 40 percent with silica – alumina catalysts. In the 1960s, these in turn were replaced by zeolites-containing catalysts. Up to 15 percent Y-type zeolite was incorporated into a silica – alumina matrix. The use of zeolites has resulted in a large increase in gasoline selectivity with low gas and coke yields. Moreover, the gasoline has a higher octane number. The improvement in selectivity is shown in Table 4.24.[1)]

Catalytic cracking in the presence of zeolites results in less alkenes and naphthenes and more aromatics and alkanes than with silica – alumina. This is attributed to increased activity of zeolites for hydrogen-transfer reactions. More C_5 to C_{10} products are formed and less C_3 and C_4 products. This is again attributed to the increasing importance of hydrogen-transfer reactions which stop the cracking at a higher molecular weight.[2)]

Weisz and Miale compared the activity for hexane cracking of a number of zeolites (Table 4.25) with a highly active silica – alumina (10% alumina).[3)] The zeolites are at least 10^4 times as active as amorphous silica – alumina. The catalytic process, however, cannot utilize the activity from a pure zeolite catalyst. The catalyst must be modified to decrease the acid-strength to avoid excessive formation of coke and low molecular weight gases, at the expense of gasoline. Moreover, the catalyst must be able to withstand the thermal and hydrothermal conditions experienced in regeneration. It must also withstand breakup in the mechanical circulation systems. A detailed description of the preparation of industrial catalysts is found in the literature.[1,4)]

TABLE 4.24 Yields of products of commercial cracking reactors with silica-alumina and zeolite catalysts

	Amorphous SiO_2-Al_2O_3	H-Y Catalyst	Difference
Conversion (vol %)	35.6	35.6	—
C_5+gasoline (vol %)	22.1	29.2	+7.1
C_4 (vol %)	8.7	6.2	−2.5
Dry gas (wt %)	5.2	3.5	−1.8
Coke (wt %)	4.3	1.4	−2.9
Gasoline composition†			
Paraffins (%)	13	23	+10
Olefins	17	5	−12
Naphthenes	41	23	−18
Aromatics	29	49	+20

† Resulting from cracking with amorphous and crystalline aluminosilicates.
(Reproduced with permission by J. W. Ward, *Applied Industrial Catalysis*, Vol. 3 (B. E. Leach, ed.) Acedemic Press, 1984, p. 310).

TABLE 4.25 Hydrocarbon cracking activities of various crystalline aluminosilicate catalysts relative to amorphous silica-alumina[3)]

No.	Structure of zeolite	Cations in base exchanging solution	α
1	Amorphous $SiO_2 \cdot Al_2O_3$	—	1.0
2	Faujasite (syn.)	Ca	1.1
3	Faujasite (syn.)	$H(NH_4)$	6400
4	Faujasite (syn.)	La, H	7000
5	Faujasite (syn.)	Ce, La, H	>10000
6	Faujasite (syn.)	Ce, La, $H(NH_4)$	>10000
7	Faujasite (syn.) steam treated	Ce, La, $H(NH_4)$	20
8	A-Zeolite (syn.)	Ca	0.6
9	Zk-5 (syn.)	—	38
10	ZK-5 (syn.)	H	450
11	Mordenite (syn.)	Ca	1.8
12	Mordenite (syn.)	Ca, H	40-200
13	Mordenite (syn.)	H	>10000
14	Mordenite (nat., Nova Scotia)	H	2500
15	Mordenite (nat., Nova Scotia)	H	>10000
16	Gmelinite (nat., Bancroft, Ont.)	$H(NH_4)$	>10000
17	Chabazite (nat., Nova Scotia)	$H(NH_4)$	>10000
18	Stilbite (nat., Halls Harbour, Nova Scotia)	$H(NH_4)$	120

(Reproduced with permission by P. B. Weisz, J. N. Miale, *J. Catal.*, **4** 528 (1965)).

Currently, only Y-type zeolites are of any commercial importance as cracking catalysts. In many cases, rare earth ions are incorporated into Y-type zeolites. So-called ultrastable forms of Y-zeolites are also used. These may be prepared by extracting some of the aluminum from the zeolite framework. The ultrastable Y-zeolites can retain their crystal form at temperatures as high as 1200 K.

4.12.2 Cracking Process

Over the years, many improvements in the cracking process have been made. The initial cyclic operation of the fixed bed units was replaced by designs of moving bed reactors, in which the catalyst moved continuously from a reactor through a purge zone to a regenerator. Fluidized catalytic cracking (FCC) was introduced in 1941, in which the catalyst in the form of fine particles in the 30 − 200 mesh range was maintained in suspension in a stream of vaporized hydrocarbons. The advent of zeolite catalysts led to the further modification of the reactor configuration to achieve shorter residence times and higher temperature operation by taking advantage of the high activities of zeolites. The fluidized bed reactor was replaced or modified by a riser cracker. Fig. 4.22 shows the process diagram of the Kellog FCC units.[5)] The feed is mixed with hot regenerated catalyst at the base of the riser. The slurry of catalyst and oil moves up the riser and most of the reaction occurs in the range of 750 to 790 K. Contact time of 2 − 4 is thus achieved. Regeneration continues to be carried out in a fluidized bed. After separation from the catalyst, the cracked feed stock is sent to the fractionation section.

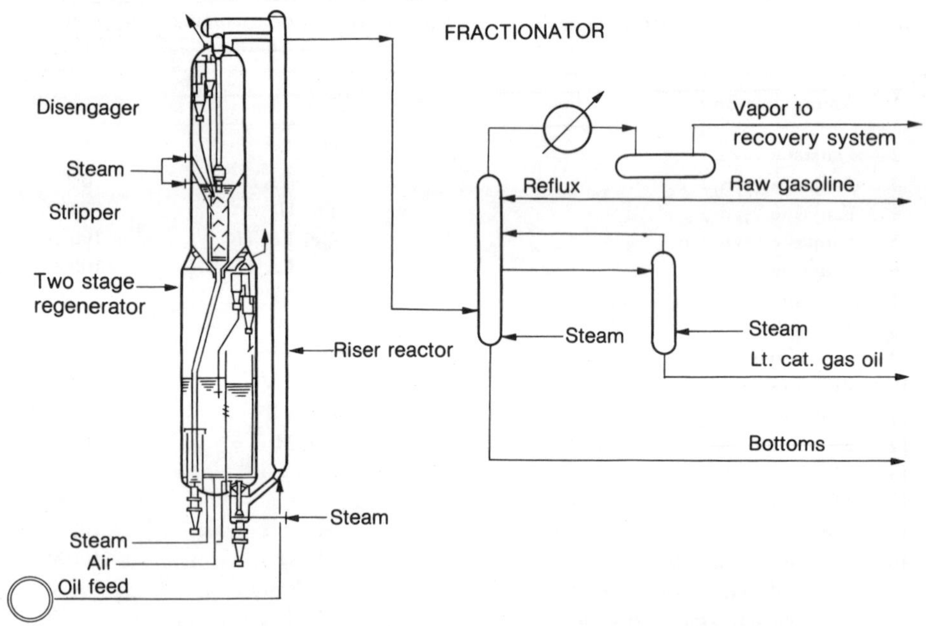

Fig. 4.22 Fluid catalytic cracking.
(Reproduced with permission by *Hydrocarbon Processing,* **58** , 19 (1974)).

The feed to the catalytic cracking reactor may be any distilled fraction, atmospheric or vacuum-distilled, that is to be reduced in molecular weight. Usually, it is a fraction with an initial boiling point above 670 K since more volatile materials can be processed into gasoline.

4.12.3 Mechanism of Catalytic Cracking

Catalytic cracking is essentially carbenium ion chemistry. Thus, the central problem of acid-catalyzed cracking is the mechanism of the generation of carbenium ions.

It is generally accepted that the carbenium ions are formed by a hydride-transfer step.[6,8]

$$R^{+} + H-\overset{|}{\underset{|}{C}}- \longrightarrow RH + {}^{+}\overset{|}{\underset{|}{C}}- \qquad (1)$$

Due to the high temperature, the carbenium ions may split into a smaller carbenium ion and an alkene molecule.

$$-\overset{+}{C}H-CH_2-\overset{|}{\underset{|}{C}}- \longrightarrow -CH{=}CH_2 + {}^{+}\overset{|}{\underset{|}{C}}- \qquad (2)$$

C – C bond scission occurs in the β-position to the carbenium ion atom. The new carbenium ion may either crack or capture a hydride ion from the alkane molecule. The olefin is more easily converted to a carbenium ion than the initial alkane and cracks at a faster rate.

$$-C{=}C- + H^{+} \longrightarrow -\overset{H}{C}-\overset{+}{C}- \qquad (3)$$

Because of the relative instability of primary carbenium ions, small fragments such as $^{+}CH_3$ or $^{+}C_2H_5$ are much more difficult to produce and, in contrast with thermal cracking, catalytic cracking leads to a large amount of C_3-C_4 hydrocarbon gases, and small amounts of methane and ethane.

In feed containing olefins as an impurity, carbenium ion formation occurs readily via reaction (3) and alkanes are converted via hydride transfer reactions (1).

The product distribution of hexadecane was calculated based on the mechanism involving hydride transfer and cracking via β-scission.[7] As shown in Fig. 4.23, the distribution predicted by the theory agrees well with that obtained in hexadecane cracking over $Al_2O_3-ZrO_2-SiO_2$ at 773 K, especially for products with carbon numbers 3 – 14. A large deviation is observed for hydrocarbons of one or two carbon atoms.[7] The product distribution of hexadecane cracking over a rare earth-exchanged Y zeolite (REY) differs from those over $Al_2O_3-ZrO_2$.[2] Thus, the lower yields of C_2-C_4 products and the higher yields of C_5-C_9 products indicate the increased ratio of hydrogen transfer to β-scission rate over REY.[2]

Alkanes are the dominant initial product at 573 K whereas olefins are dominant at 673 K in the cracking of hexadecane over HY zeolite. This was explained as being the result of more extensive hydrogen transfer at the lower temperature.[10]

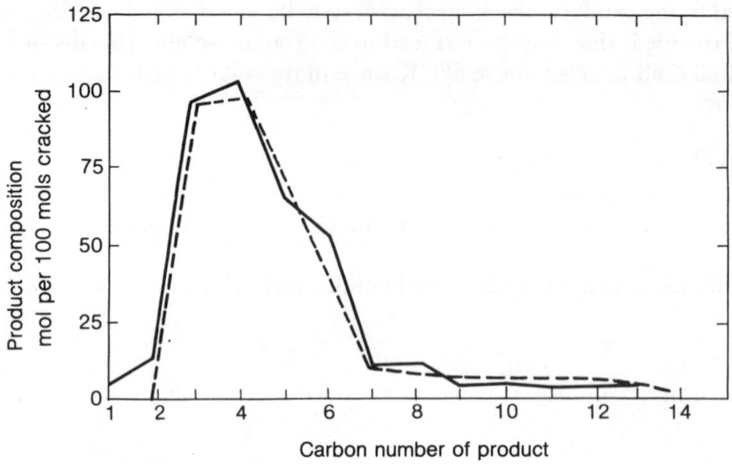

Fig. 4.23 Product distribution in hexadecane cracking over silia-zirconia-alumina at 773 K solid line ; observed, dashed line ; calculated.
(Reproduced with permission by B. S. Greenlsfelder *et al.*, *Ind. Eng. Chem.*, **41**, 2581 (1949)).

The direct formation of carbenium ions from alkane molecules has been the subject of much discussion. Haag and Dessau[11] showed that a monomolecular mechanism via a penta-coordinated carbonium ion intermediate as well as a hydride transfer mechanism make important contributions under certain conditions.

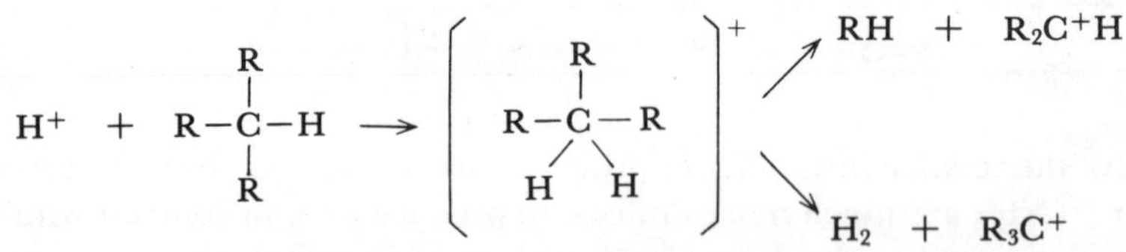

The mechanism can explain the formation of methane, ethane and hydrogen. The carbonium ion mechanism is the main pathway of the cracking of alkanes in superacid media.[12] It was concluded that the carbonium ion mechanism predominates at high pressure, low hydrocarbon pressure and low conversion, and that the opposite applies to the hydride-transfer mechanism.

In the cracking of neopentane over a variety of solid acids, it was demonstrated that neopentane decomposes to form methane and *t*-butyl ion via the protonation of a C–C bond, a carbonium ion mechanism.[13]

For the conversion of propane, it was suggested that the cracking occurs via a carbonium ion mechanism at low conversion levels and that the hydride transfer mechanism prevails at high conversion levels.[14]

Brenner and Emmett[15] examined the cracking of isopentane over silica-alumina catalysts and found the main initial product to be pentenes. This indicates that the first step is dehydrogenation of the alkane, the breaking of carbon–carbon bond being a subsequent step to dehydrogenation.

Some subsequent reactions to the cracking and hydride-transfer may also be significant. Double bond isomerization proceeds so rapidly that products are in chemical equilibrium with respect to this reaction. Because of the higher stability of tertiary carbenium ions over secondary or primary carbenium ions, the latter are easily isomerized to the former. This is the reason for the high fraction of branched isomers in alkane products.

Aromatics may also be formed by dimerization and the cyclization of diolefins. The deprotonation of carbenium ions and hydride transfer are important steps in the formation of aromatics (see Section 4.6.2).[16)]

4.12.4 Shape Selective Cracking

The structure of zeolites often modifies the selectivity of catalytic cracking with respect to both reactants and products, depending on the effective pore size of the zeolites.

Selective cracking of *n*-alkanes in the presence of branched alkanes was first demonstrated by Weisz *et al.*[17)] The cracking of hexane and 3-methylpentane at 773 K were compared (Table 4.26).[17)] No reactions occur over silica, but over an amorphous silica – alumina catalyst, both hexane and 3-methylpentane react at significant rates. NaA (sodium exchanged A-type zeolite) is inactive because its smaller pores severely restrict the diffusion of hexane and probably because it does not have acid sites strong enough for cracking. Over CaA, 3-methylpentane does not react, but hexane reacts quite well. The effective pore diameter of CaA is 0.5 nm, while that of NaA is 0.4 nm. The high selectivity over CaA is attributed to the fact that only hexane can penetrate into the pore system of the zeolite. For the same reason, branched products are essentially absent in the product obtained over CaA while they are the major products over silica-alumina, as expected from carbenium ion chemistry. The principle of shape-selectivity led to Selectforming, a shape selective hydrocracking unit on the product of catalytic reforming.[18)]

The relative cracking rates of heptanes and hexanes over HZSM-5 zeolites are shown in Table 4.27.[19,20)] The rates of cracking are in the following decreasing order.

TABLE 4.26 Comparison of *n*-hexane and 3-methylpentane cracking at 773 K

Catalyst	3-Methylpentane cracking conversion (%)	*n*-Hexane cracking		
		Conversion (%)	$\frac{iC_4}{nC_4}$	$\frac{iC_5}{nC_5}$
Silica	<1	1.1		
Amorphous silica-alumina	28	12.2	1.4	10
Linde Na-A	<1	1.4		
Linde Ca-A	<1	9.2	<0.05	<0.05

(Reproduced with permission by P. B. Weisz *et al.*, *J. Catal.*, **1**, 307 (1962)).

TABLE 4.27 Relative cracking rate constants at 613 K in H–ZSM–5 catalyst

Hexanes

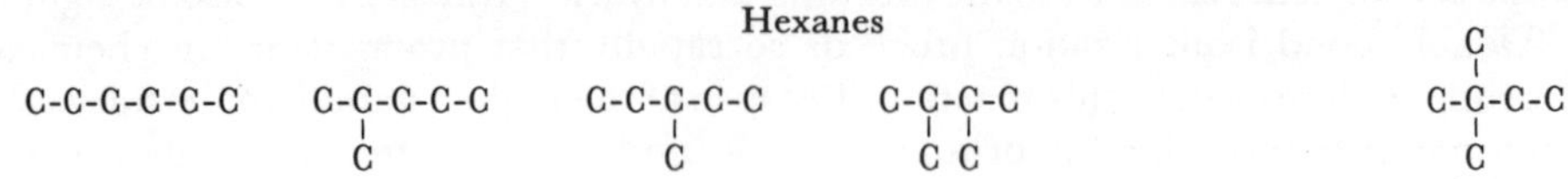

0.71	0.38	0.22	0.09	0.09

Heptanes

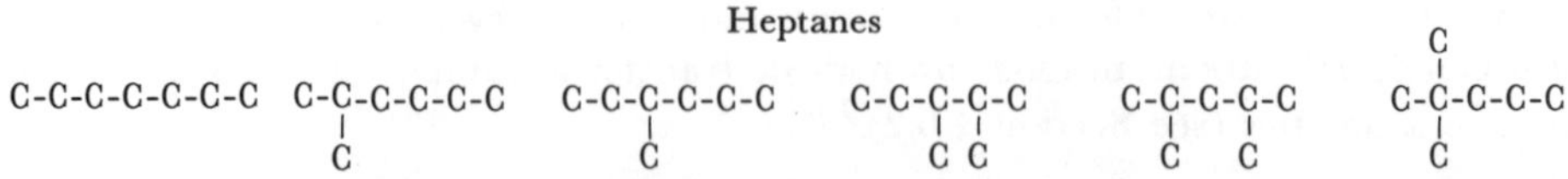

1.0	0.52	0.38	0.09	0.05	0.17

(Reproduced with permission by P. B. Weisz, Proc. 7th International Congr. Catal. 1980, Tokyo (T.Seiyama, K. Tanabe eds.)PartA, 1981, p. 8.

n-alkanes > monomethyl alkanes >> dimethyl alkane

The order of the cracking rates is the reverse of the order of the cracking rate over HY or silica – alumina.[21)] The reasons for shape-selectivity were examined in detail by Lago *et al.*[21)] The selective cracking of *n*-alkanes over monomethyl alkanes is due to a higher intrinsic rate constant of the *n*-alkanes, with diffusional mass-transfer playing no appreciable role. Thus, the methylalkane/*n*-alkane discrimination is a result of steric constraint on the sizable methylalkane-carbenium ion complex in the hydride-transfer reaction. In contrast, dimethyl alkane cracking is strongly diffusion-limited. ZSM-5 zeolites have been utilized as catalysts for dewaxing processes, taking advantage of the shape-selectivity.[22)]

Frilette *et al.* proposed a simple test reaction for estimating the effective pore size of zeolites.[23)] The determination of the "constraint index" is done by continuously passing a mixture of hexane and 3-methylpentane over a zeolite at atmospheric pressure. The constraint index is defined as follows.

$$\text{Constraint Index} = \frac{\log(\text{fraction of hexane remaining})}{\log(\text{fraction of 3-methylpentane remaining})}$$

The constraint index therefore approximates the ratio of the cracking rate constants for the two hydrocarbons. The values of the constraint index of various zeolites are summarized in Table 4.28.

An 8-membered ring zeolite, erionite, gives a constraint index of 38, 10-membered ring zeolites, ZSM-5 and ZSM-11, give values of 8.3 and 8.7, respectively, while 12-membered ring zeolites, mordenite and REY, give values of 0.5 and 0.4. In this way, the constraint index is a very useful tool for estimating the pore size of zeolites of unknown structures.

TABLE 4.28 Constraint index of various zeolites

Zeolite	Constraint index
ZSM-5	8.3
ZSM-11	8.7
ZSM-12	2
ZSM-35	4.5
ZSM-38	2
TMA offretite	3.7
Beta	0.6
ZSM-4	0.5
H-Zeolon (Mordenite)	0.4
REY	0.4
Amorphous silica-alumina	0.6
Erionite	38

(Reproduced with permission by V. J. Frilette *et al.*, *J. Catal.*,**67** 220(1981)).

REFERENCES

1. J. W. Ward, in: *Applied Industrial Catalysis,* Vol. 3, (B. E. Leach, ed.) Academic Press, Orlando, 1984, p. 272.
2. D. M. Nace, *Ind. Eng. Chem., Prod. Res. Dev.*, **8**, 24 (1969).
3. P. B. Weisz, J. N. Miale, *J. Catal.*, **4**, 527 (1965).
4. J. S. Magee, J. J. Blazek, in: *Zeolite Chemistry and Catalysis* (J. A. Rabo, ed.) American Chemical Society, Washington D.C., 1976, p.615.
5. Hydrocarbon Processing **58** (9), (1974).
6. C. T. Thomas, *Ind. Eng. Chem.*, **41**, 2564 (1949).
7. B. S. Greensfelder, H. H. Voge, G. M. Good, *Ind. Eng. Chem.*, **41**, 2573 (1949).
8. H. H. Voge, in: *Catalysis* (P. H. Emmett, ed.) Vol. VI, p.407 (1958).
9. B. C. Gates, J. R. Katzer, G. C. A. Schuit, in: *Chemistry of Catalytic Processes,* McGraw Hill Book Co., New York, 1979, p. 29.
10. J. Abbot, B. W. Wojciechowski, *J. Catal.*, **109**, 274 (1988).
11. W. O. Haag, R. M. Dessau, Proc. 8th Intern. Congr. Catal., Weinheim, 1984, Vol. 2, p. 305.
12. G. Olah, Y. Malpern, Y. Shen, Y. K. Mo, *J. Am. Chem. Soc.*, **93**, 1251 (1971).
13. E. A. Lomberts, R. Pierantozzi, W. K. Hall, *J. Catal.*, **110**, 171 (1988).
14. H. Kitagawa, Y. Sendoda, Y. Ono, *J. Catal.*, **101**, 12 (1986).
15. A. Brenner, P. H. Emmett, *J. Catal.*, **78**, 410 (1982).
16. M. L. Poustma, in: *Zeolite Chemistry and Catalysis* (J. A. Rabo, ed.) American Chemical Society, Washington D.C., 1976, p. 615.
17. P. B. Weisz, V. J. Frilette, R.W. Maatman, E.B. Mower, *J. Catal.*, **1**, 307 (1962).
18. N. Y. Chen, J. Maziuk, A. B. Schwartz, P. B. Weisz, *Oil Gas J.*, **66**, 154 (1968).
19. P. B. Weisz, Proc. 7th Intern. Congr. Catal, 1980, Tokyo (T. Seiyama, K. Tanabe, eds), Part A, p. 3, Kodansha, Tokyo and Elsevier, Amsterdam, 1981, p. 3.
20. N. Y. Chen, W. E. Garwood, *J. Catal.*, **52**, 453 (1978).
21. W. O. Haag, R. M. Lago, P. B. Weisz, *Discussion Faraday Soc.*, No. 72, 317 (1982).
22. N. Y. Chen, R. L. Gorring, H. R. Ireland, T. R. Stein, *Oil Gas J.*, **75**, 165 (1977).
23. V. J. Frilette, W. O. Haag, R. M. Lago, *J. Catal.*, **67**, 218 (1981).

4.13 HYDROCRACKING (HYDROGENOLYSIS)

By hydrocracking is normally meant cracking (C – C bond cleavage) followed by hydrogenation. In a broad sense, C – N, C – S, and C – O bond cleavages followed by hydrogenation of the resulting products are also included. The term hydrogenolysis is often used in this broad sense to indicate hydrocracking.[1)]

In petroleum refineries, hydrocracking processes are used to upgrade low value distillate feedstocks. The hydrocracking processes are classified into three groups according to purpose. The purposes and catalysts are as follows:

i) Removal of sulfur and nitrogen present in various forms from feedstocks. The primary catalysts are $Co{\cdot}Mo/Al_2O_3$, and $Ni{\cdot}Mo/Al_2O_3$.

ii) Hydrogenation of unsaturated compounds to improve the quality of the products. The primary catalysts are supported Ni, Pd, Pt.

iii) Cleavage of C – C bonds followed by hydrogenation to increase the light oil fraction of highly paraffinic nature. The primary catalysts are Pt/acidic zeolite Y for gas oils, and $Co{\cdot}Mo/Al_2O_3$ for residues.

In addition to these processes, dewaxing processes which lower the pouring point by selectively removing straight chained alkanes, and Selectforming which removes $C_5 - C_9$ straight chained alkanes from naphtha and reformates involve cracking and hydrogenation.

The contribution of acid sites in the metal-acid bifunctional catalysis occurring in dewaxing and Selectforming is obvious, and essentially the same as those for cracking and reforming processes. Various contributions of acid and base sites in hydrocracking in the broad sense are described below.

The acidic properties of MoO_3 supported on alumina are measured in conjunction with the activities for hydrodesulfurization. Introduction of Mo to Al_2O_3 produces an increase in acidity through the formation of both extremely weak and strong acid sites.[2)] The acid sites generated contain Brønsted acid sites which are not present on unsupported Al_2O_3.[3)] Sulfidation of MoO_3/Al_2O_3 creates new acid sites stronger than those of the support, Al_2O_3.[4)] Introduction of Co to MoO_3/Al_2O_3 inhibits the formation of strong acid sites.[4)] Incorporation of Ni to MoO_3/Al_2O_3 also increases the acid sites of intermediate strength,[2)] as shown in Fig. 4.24. The acid sites on presulfided Co – and Ni – MoO_3/Al_2O_3 catalysts are of Lewis type. Unlike strong Brønsted acid sites which promote coke formation through carbenium ion reactions, the Lewis acid sites generated by incorporation of Co and Ni are active for hydrodesulfurization.[2)]

The contribution of acid sites was demonstrated in the hydrocracking of diphenylmethane, which was studied as a model reaction for coal liquefaction.[5–7)] Among iron catalysts which were prepared by mixing different metal oxides with iron oxide, $Fe_2O_3 - SiO_2$ showed the highest activity under the reaction condition in which hydrogen pressure is not so high. The catalyst $Fe_2O_3 - SiO_2$ possesses the largest number of acid sites as judged from the highest activity for 2-propanol dehydration. Besides $Fe_2O_3 - SiO_2$ catalyst, a parallel relationship between the activity for the hydrocracking and that for 2-propanol dehydration is observed, as shown in Fig. 4.25. These

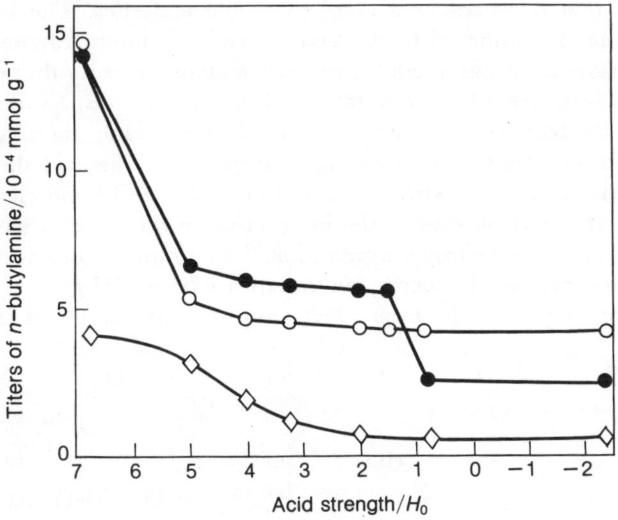

Fig. 4.24 Acid strength distributions.
◇; Al_2O_3, ○; MoO_3/Al_2O_3, ●; $NiO-MoO_3/Al_2O_3$

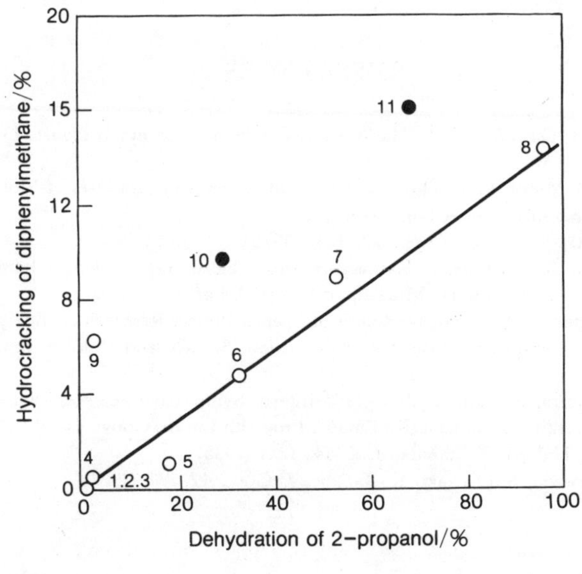

Fig. 4.25 Correlation between diphenylmethane hydrocracking activity and 2-propanol dehydration activity.
1; $Fe_2O_3-SnO_2$, 2; Fe_2O_3 (I), 3; Fe_2O_3-ZnO, 4; $Fe_2O_3-ZrO_2$, 5; $Fe_2O_3-Al_2O_3$, 6; $Fe_2O_3-Nb_2O_5$, 7; Fe_2O_3 (II), 8; $Fe_2O_3-SiO_2$, 9; Fe_2O_3-MgO, 10; Co·Mo/Al_2O_3, 11; Ni·Mo/Al_2O_3

results suggest that hydrocracking takes place on acid sites. The hydrocracking of diphenylmethane is considered to proceed via cationic intermediates formed on the acidic sites. Besides acid sites, metallic iron acts as active sites for the reaction in which the mechanism is different from that taking place on acid sites. As the hydrogen pressure increases, the reduction of catalysts proceeds. As a result, the acidic nature tends to decrease and metallic sites becomes more important. The acid sites do not contribute to the activity for the hydrocracking of ethers (C – O bond cleavage).[8]

There is a report which stresses the importance of the basic properties of the support for MoO_3 used for hydrodesulfurization.[9] The number of effectively supported MoO_3 molecules increases by increasing the surface basic OH groups, because molybdate anions exchange with the basic OH groups on the surface as shown below.

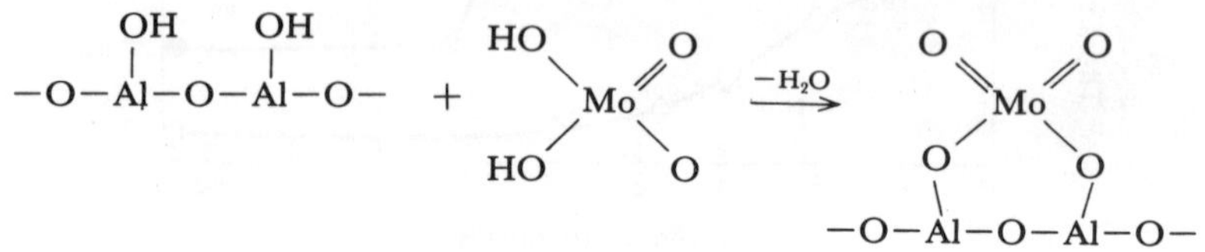

The basic OH groups on Al_2O_3 increase on addition of MgO to Al_2O_3. The resulting catalyst shows higher activity for hydrodesulfurization of thiophene than MgO free catalyst.

REFERENCES

1. P. C. H. Mitchel, in: *Catalysis,* Vol. 4, The Royal Soc. Chem., Burlington House, London, Chapter 7, 1981.
2. J. Laine, J. Brito, S. Yunes, Proc. Third Intern. Conf. Chemistry and Uses of Molybedenum. (H. F. Barry, P.C.H. Mitchel,ed.), Ann Arbor, 1979, p.111.
3. F. E. Kiviat, L. Petrakis, *J. Phys. Chem.,* **77**, 1232 (1973).
4. S. Sivasanker, A. V. Ramaswany, P. Ratnasamy, Proc. Third Intern. Conf. Chemistry and Uses of Molybdenum. (H. F. Barry, P.C.H. Mitchel, ed.), Ann Arbor, 1979, p.98.
5. K. Tanabe, H. Hattori, T. Yamaguchi, Report of Special Project Research on Energy under Grant in Aid of Scientific Research of the Ministry of Education, Science and Culture, Japan, October 1987, SPEY 16, p. 17.
6. H. Hattori, A. Kimura, K. Tanabe, Proc. 1985 Intern. Symp. Coal Science, Sydney, p.165.
7. H. Hattori, K. Yamashita, T. Tanabe, K. Tanabe, Proc. 9th Intern. Congr. Catal., 1988, Calgary, p.27.
8. H. Matsuhashi, H. Hattori, K. Tanabe, *Fuel*, **64**, 1224 (1985).
9. N. Yamagata, Y. Owada, S. Okazaki, K. Tanabe, *J. Catal.*, **47**, 358 (1977).

4.14 CATALYTIC REFORMING

4.14.1 Introduction

Catalytic reforming is one of the most important petroleum refining processes.[1-5] The purpose of catalytic reforming is to process a hydrocarbon fraction boiling within the gasoline range to improve its octane number without changing the carbon number. Catalytic reforming is also the principal source of aromatic chemicals.

The low-octane feedstocks to be reformed contain hydrocarbons with carbon numbers of 5 to 11, most of them having 7, 8 and 9, and consist of large quantities of straight alkanes and naphthenes together with relatively small quantities of branched-chain alkanes, olefins and aromatics. Table 4.29 gives the octane number of pure hydrocarbons with seven carbon atoms.[1] It can be seen that the octane number of alkanes increases with the degree of branching, as well as their dehydrogenation into alkenes or into aromatics. The highest octane number is that of toluene. Thus, improvement in octane number can be achieved by producing aromatics and branched alkanes from linear alkanes and cycloalkanes.

A wide variety of reactions is involved in catalytic reforming. The main reactions are dehydrogenation and isomerization of cycloalkanes (naphthenes), dehydrocyclization of alkanes, alkane isomerization, hydrocracking, and coke formation. A large increase in octane number is achieved through the dehydrocyclization of linear or slightly branched alkanes into aromatics. Hydrocracking and coke formation are undesirable. Hydrocracking produces excessive quantities of light hydrocarbons and coke and con-

TABLE 4.29 Research octane numbers of some C_7 hydrocarbons

Compound	
Alkanes	
Heptane	0
2-methylhexane	42.4
3-methylhexane	56
2,2-dimethylpentane	89
2,3-dimethylpentane	91.4
2,4-dimethylpentane	83.1
3,3-dimethylpentane	83
3-ethylpentane	65
2,2,3-trimethylbutane	112
Alkenes	
1-heptene	54
3-heptene	84
4,4-dimethyl-2-pentene	105.3
Cycloalkanes (naphthenes)	
1,1-dimethylcyclopentane	92.3
methylcyclohexane	74.8
Aromatics	
Toluene	120

sumes hydrogen which could be used elesewhere in the refinery.

4.14.2 Reaction Mechanism

Catalysts for reforming reactions are generally a few tenths percent platinum either alone or in combination with another metal (Re, Ir or Sn) supported on porous promoted alumina. The catalysts are said to be bifunctional since both the metal and oxide components play individual roles in the catalysis. A metal catalyzes dehydrogenation of alkanes into alkenes and cycloalkanes into aromatics; it also catalyzes hydrogenation of isoolefins and contributes to dehydroisomerization and isomerization. An acidic support catalyzes isomerization, cyclization, and hydrocracking through alkenes. Two functions interact through alkenes, which are key intermediates in the reaction network. Fig. 4.26 shows the concept of dual functionality to a variety of reactions of importance in catalytic reforming.[6] Reactions down parallel to the abscissa in this figure occur on the acidic centers on the catalyst, and reactions down parallel to the ordinate occur on the metal center. According to the scheme, alkanes, i.e. hexane, is first dehydrogenated to hexene on the metal centers; there it is protonated to give a carbenium ion, which can then isomerize and desorb as 2-methylpentene and migrate to the metal surface where it can be hydrogenated to give 2-methylpentane. Alternatively, the carbenium ion can react to form methylcyclopentane, which can react further to cyclohexene and to benzene.

A good example of the bifunctionality is seen in Table 4.30,[7] showing the products in the conversion of methylcyclopentane over $SiO_2-Al_2O_3$, Pt/SiO_2 and the mechanical mixture of $SiO_2-Al_2O_3$, and Pt/SiO_2. An acidic oxide is not active for dehydrogenation. Dehydrogenation to methylcyclopentene or to methylcyclopentadiene occurs on the platinum. The acid centers are necessary to isomerize cyclopentene to a six-membered ring. Thus, benzene can be obtained only when both acidic and metal components are present.

In addition to its role in providing sites for hydrogenation-dehydrogenation reac-

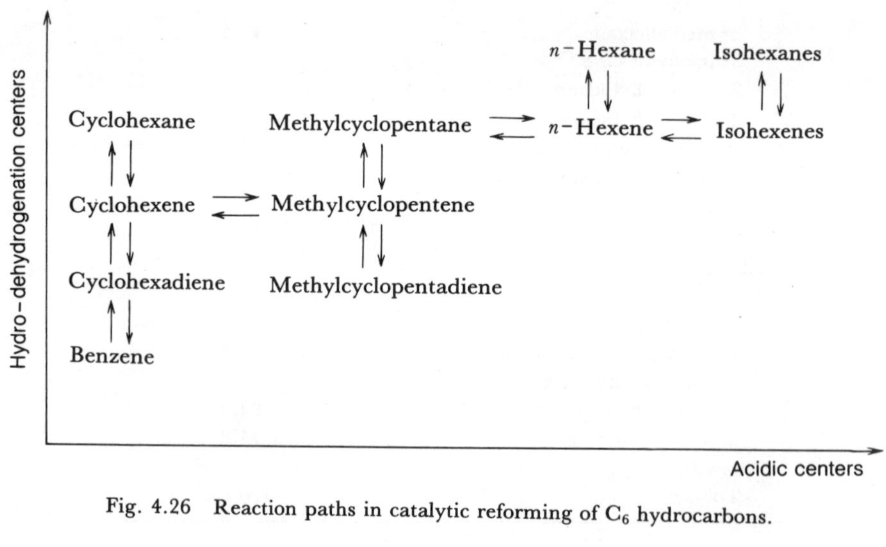

Fig. 4.26 Reaction paths in catalytic reforming of C_6 hydrocarbons.

TABLE 4.30 Conversion of methylcyclopentane catalyzed by acid, metal, and mixed catalysts

Catalyst	Liquid product analysis mol %			
	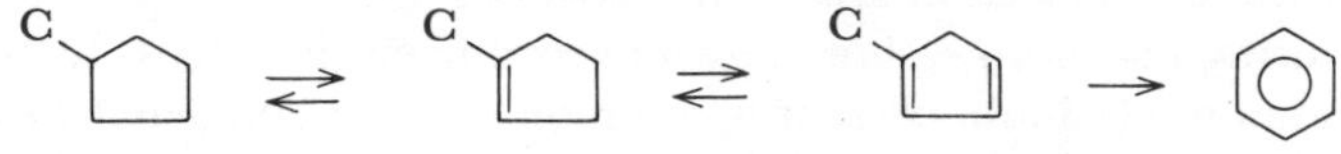			
10 cm^3 of SiO_2-Al_2O_3	98	0	0	0.1
10 cm^3 of Pt/SiO_2	62	20	18	0.8
SiO_2-Al_2O_3+Pt/SiO_2	65	14	10	10.0

† Reaction conditions: 773 K, 82 kPa H_2 partial pressure, 21 kPa methylcyclopentane partial pressure, 2.5 s residence time, and catalysts 0.3 wt % Pt/SiO_2 and SiO_2-Al_2O_3 with 420 m^2 per gram of surface area.

(Reproduced with permission by B. C. Gates *et al.*, *Chemistry of Catalytic Processes*, McGrow Hill, 1979, p.191)

tions to occur, platinum also acts to remove carbonaceous residues. Hydrogen adsorbs and dissociates into atomic species on the platinum and migrates on the surface of the catalyst. Coke precursors are hydrogenated, which improves their desorption from the catalyst. The rate of coke formation is a function of hydrogen pressure. The minimum concentration of platinum in the industrial catalysts is determined by that needed to keep the catalyst clean.

4.14.3 Nature of Reforming Catalysts

A. Alumina Support

In the first bifunctional catalyst, γ-alumina was used as an acidic support. Acidic properties of alumina have been described in Section 3.1.11. Silica – alumina cannot be used, since it catalyzes excessive hydrocracking and undergoes rapid deactivation because of its strong acidity. In general, alumina in its partially hydrated form is not strongly acidic.

The activity of OH groups on the alumina surface can be markedly enhanced by the proximity of Cl^- ions. In the commercial catalysts, γ-alumina is treated with HCl to make it a highly active catalyst. γ-Alumina cannot readily catalyze the skeletal isomerization of alkenes because of its weak acidity. On the other hand, chlorinated alumina is highly active for skeletal isomerization and other strong-acid catalyzed reactions which are desirable in reforming. The strength of acid sites can be controlled by the extent of chlorination.[8,9] If the Cl^- content is too low, the reactions which occur on the acid centers slow down and the octane number of reformate drops. If excess Cl^- ion is present, the extent of hydrocracking increases relative to dehydrocyclization.

Since Cl^- ion is continuously stripped from the surface as HCl by reaction with small amounts of water in the feed, the Cl^- content of the catalyst must be maintained by adding chlorinated organic compounds to the feed.

It was found later that η-alumina was a more stable and regenerative catalyst than γ-alumina. η-Alumina has an inherently stronger acidity than γ-alumina.[2] It can thus be used without necessity for added halogen and provides a high ratio of hydrocyclization relative to hydrocracking.

B. Metallic Component

The first commercial catalytic reforming process using a molybdenum-alumina catalyst was established in 1939. The processes using this type of catalyst were replaced by one using platinum – alumina catalysts during the 1950s. In commercial catalysts, the amount of platinum present is commonly in the range of 0.3 to 0.6 weight percent. One common method of preparing such a catalyst involves impregnation of alumina with chloroplatinic acid, followed by calcination in air at temperatures in the range of 820 ot 870 K.[10)] In commercial reforming units, the catalysts are commonly used in the form of pellets or extrudates with dimensions in the approximate range of 1.5 – 4 mm.

A new generation of bifunctional catalysts was introduced in 1967. The catalyst containing rhenium in addition to platinum provides greater stability.[1,2,11)] In 1975, the process using a catalyst containing platinum and iridium was commercialized. These catalysts are called bimetallic catalysts. The bimetallic catalysts are typically 3 to 4 times more active than the all-platinum catalyst. A bimetallic catalyst with rhenium typically contains about 0.3% platinum and 0.3% rhenium. The reasons for the effectiveness of these bimetallic catalysts are beyond the scope of this volume and the readers should refer to the appropriate monographs or reviews.[2,5)]

4.14.4 Reforming Process

All reforming processes use fixed bed reactors in a series (usually three). Fig. 4.27 shows the process flow diagram of a typical reforming process.[2)] The feed mixed with recycle hydrogen is heated to the reaction temperature (about 750 K) in the first heater and passes into the first reactor. Since the major reaction occurring in the first reactor is the most endothermic of the reactions, dehydrogenation of cyclohexanes to aroma-

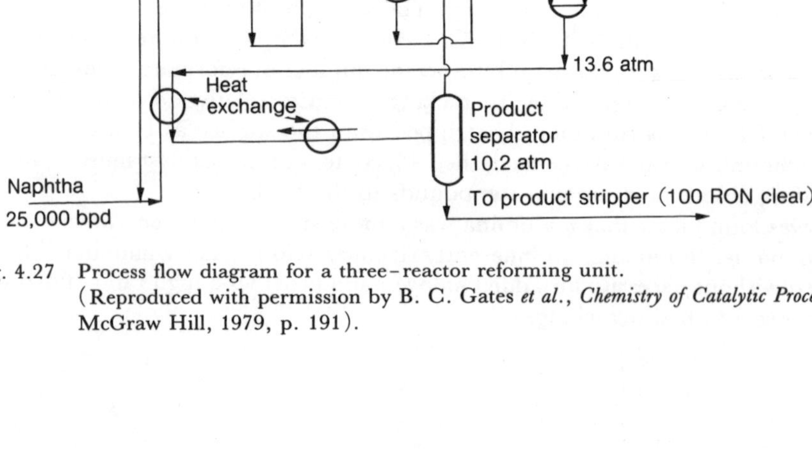

Fig. 4.27 Process flow diagram for a three-reactor reforming unit.
(Reproduced with permission by B. C. Gates *et al.*, *Chemistry of Catalytic Processes*, McGraw Hill, 1979, p. 191).

tics, the temperature of the first reactor falls rapidly. The effluent from the first reactor is reheated to a desired reaction temperature and enters the second reactor, where it undergoes dehydroisomerization of cyclopentanes at a lower rate than that of cyclohexanes. The effluent from the second reactor is passed through an interstage-heater and admitted to the third reactor, where dehydrocyclization and hydrocracking are completed. Because of the difference in endothermicity in the reactors, the amount of catalyst in the upstream reactors is considerably less than in the downstream reactors. Table 4.31 shows typical operating conditions for the three-reactor system.[2)]

Since dehydrogenation and dehydrocyclization are the most important for the octane number gain, operation conditions favoring these reactions are selected. Increasing the reaction temperature favors the thermodynamics, but accelerates the hydrocracking reactions, leading to a loss in yield.

High pressure is harmful to the thermodynamics, but affords better selectivity. Thus a temperature of 770 – 820 K and pressure of 3000 kPa are used for the operation. The molar ratio hydrogen to the hydrocarbon ratio is another important variable. A reduction in the ratio increases the rate of coke make, but enhances dehydrogenation and inhibits hydrocracking. A ratio of 10:1 to 3:1 is employed.

The liquid product, commonly known as reformate, consists essentially of C_5 through about C_{10} hydrocarbons. Compared to the feedstocks, there is a substantial increase in aromatic content (60 – 70 wt%) at the expense of naphthenes. The gaseous product consists of hydrogen and $C_1 - C_4$ hydrocarbons with a hydrogen concentration of commonly 60 – 90 mole percent.

TABLE 4.31 Typical operating conditions for a three-reactor system[1]

	Reactor		
	1	2	3
Inlet temperature, K	775	775	775
Exit temperature, K	708	749	769
Temperature drop, K	69	31	6
Octane number (F-1 clear)[†2]	65.5	79.5	90.0
Octane-number increase	27.0	14.0	10.5
Principal reactions	Dehydrogenation, dehydro-isomerization	Dehydrogenation, dehydro-isomerization, hydrocracking, dehydrocyclization	Hydrocracking, dehydrocyclization
LHSV, h^{-1} per reactor	5.5	2.4	1.7
Percent of total catalyst charge	15	35	50

[†1] Esimated from published information.
[†2] Octane number of feed was 38.5.
(Reproduced with permission by B. C. Gates *et al.*, *Chemistry of Catalytic Processes*, McGraw Hill, 1979, p. 191)

REFERENCES

1. J.F. Le Page, in: *Applied Heterogeneous Catalysis,* Editions Technip, Paris, 1987, p. 468.
2. B.C. Gates, J.R. Katzer, G.A. Schuit, in: *Chemistry of Catalytic Processes*, McGraw-Hill, New York, 1979, p.184.
3. M.D. Edgar, in: *Applied Industrial Catalysis,* Vo. 1 (B.E. Leach, ed.) Academic Press, New York, 1983, p.124.
4. C.N. Satterfield, in: *Heterogeneous Catalysis in Practice*, McGraw-Hill, New York, 1980, p.247.
5. J.H. Sinfelt, in: *Catalysis*, Vol.1, (J.R. Anderson, M. Boudart, eds.) Springer-Verlag Berlin, 1981, p. 257.
6. G.A. Mills, H. Heinemann, T.H. Miliken, A.G. Oblad, *Ind. Eng. Chem.*, **45**, 134 (1953).
7. P.B. Weisz, Actes 2me Cong. Intern. Catal., Editions Technip, Paris, 1961. p.937.
8. R.J. Verderone, C.L. Pieck, M.R. Sad, J.M. Parera, *Appl. Catal.*, **21**, 239 (1986).
9. J.M. Grau, E.L. Jablonski, C.L. Pieck, R.J. Verderone, J.M. Parerea, *Appl. Catal.*, **36**, 109 (1988).
10. H. Sinfelt, *Ann. Rev. Mat. Sci.*, **2**, 64 (1972).
11. J. Beltramini, D.L. Trimm, *Appl. Catal.*, **32**, 71 (1988).

4.15 HYDROGENATION

Hydrogenation catalyzed by alkali metals was first observed in 1930. Ethylene underwent hydrogenation to form ethane in the presence of cesium.[1)] The rate was very slow at room temperature. More active catalyst of alkali metals deposited on Al_2O_3 were found later.[2)] Over $Na-Al_2O_3$, ethylene undergoes hydrogenation at room temperature, and hydrogenation of conjugated dienes to monoolefins takes place much faster than hydrogenation of monoolefins. However, the reaction mechanisms were not extensively studied, and consequently, the nature of the intermediates were not clearly elucidated.

Kokes and his co-workers studied interaction of olefins with hydrogen on ZnO, and proposed heterolytic cleavages of H_2 and C – H bonds.[3)] The negatively charged $_\pi$-allyl ions were intermediate for propylene hydrogenation in which heterolytically dissociated H^+ and H^- were involved. His view was proved to be generally applicable in metal oxide catalyzed hydrogenation. The observation that MgO pretreated at 1273 K exhibited olefin hydrogenation activities was a clear demonstration of solid base-catalyzed hydrogenation.[4)] It was later found that, besides olefins, carbon monixide also undergoes hydrogenation over solid base catalysts.[4)] The hydrogenation occurring on solid base catalysts has characteristic features which distinguish solid base catalysts from conventional hydrogenation catalysts such as transition metals and transition metal oxides. Hydrogenation of olefins and carbon monoxide over solid base catalysts are described in this section.

4.15.1 Hydrogenation of Olefins

The main solid base catalysts active for hydrogenation are alkaline earth oxides, rare earth oxides (from Y_2O_3 to Ln_2O_3 except CeO_2, Pr_6O_{11}, and Tb_4O_7), ThO_2, ZnO and ZrO_2. Alkaline earth oxides must be pretreated at high temperatures to be active for hydrogenation as compared for other base-catalyzed reactions. The variations of the activities of MgO catalyst for hydrogenation of olefins as a function of pretreatment temperature are shown in Fig. 4.28.[6,7] The maximum activities appear on pretreatment of the catalyst at about 1273 K. For other alkaline earth oxides also pretreatment around 1000 – 1300 K is required to reveal maximum activity.

Zirconium oxide exhibits hydrogenation activity following heat treatment at relatively low temperatures; the maximum activity is obtained at the pretreatment temperature of 873 K.[8] It is noteworthy that the transfer hydrogenation of 1,3-butadiene using cyclohexadiene as the hydrogen source occurs at the maximum rate when ZrO_2 is pretreated at 1073K.[8]

All active rare earth oxides show maximum hydrogenation activity following pretreatment at about 923 K.[9,10] This temperature also gives the maximum activity for the other base-catalyzed reactions.

The characteristic features of olefin hydrogenation on solid base catalysts are as follows.[11]

i) Large difference in hydrogenation rate between monoenes and conjugated dienes.

Conjugated dienes undergo hydrogenation much faster than monoenes. For instance, 1,3-butadiene undergoes hydrogenation at 273 K over solid base catalysts, while butenes appreciably react above 473 K over alkaline earth oxides, and above 373 K over rare earth oxides. The products in diene hydrogenation consist exclusively of monoenes, no alkanes being formed at 273 K.

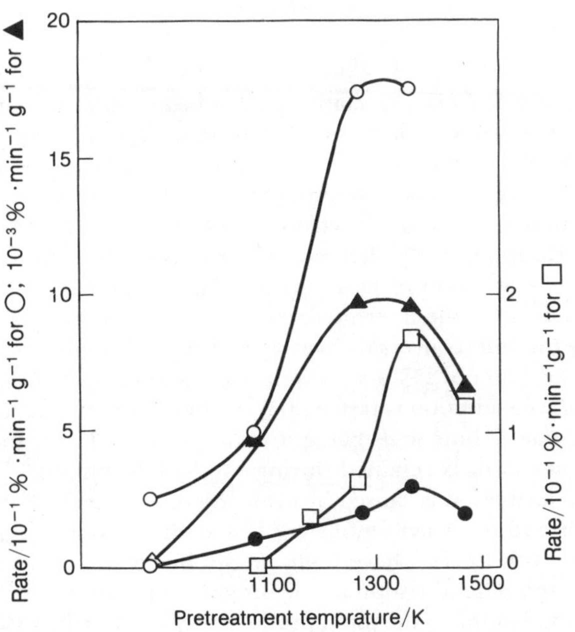

Fig. 4.28 Variation of hydrogenation rates for ethylene(□), propene(▲), 1-butene(●) at 523 K, and 1,3-butadiene(○) at 273 K as a function of pretreatment temperature.

ii) Predominant occurrence of 1,4-addition of H atoms in contrast to 1,2-addition, which is commonly observed for conventional hydrogenation catalysts.

In 1,3-butadiene hydrogenation, 2-butenes are preferentially yielded over solid base catalysts, while 1-butene is the main product over conventional hydrogenation catalysts.

iii) Retention of molecular identity of H atoms during reaction.

While hydrogen molecule dissociates on the catalyst surface, two H atoms used for hydrogenation of one reactant molecule originate from one hydrogen molecule.

Features i) and ii) are characteristic of hydrogenation in which anionic intermediates are involved. The mechanism of 1,3-butadiene hydrogenation is schematically drawn below. Deuterium atoms are used in place of hydrogen atoms for clarity.[11)]

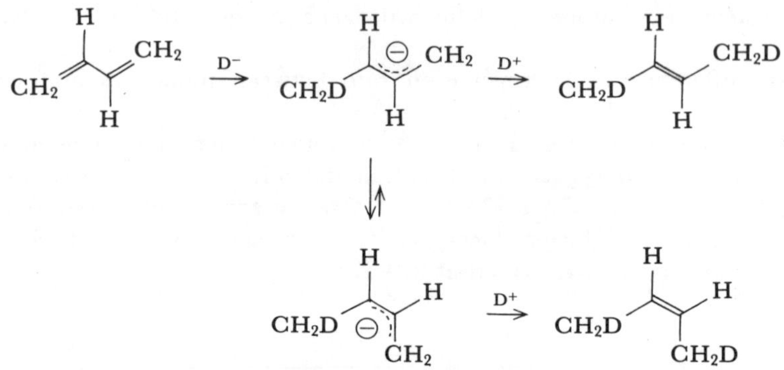

The deuterium molecule is dissociatively adsorbed to form D^+ and D^- on the surface. 1,3-Butadiene consists of 93% s-*trans* conformer and 7% s-*cis* conformer in the gas phase at 273 K. At first, D^- attacks 1,3-butadiene to form the allylic anion of the *trans* form which undergoes either interconversion to form *cis* allylic anion or addition of D^+ to form butenes. Since the electron density of allylic anions is highest on the terminal carbon atom, positively charged D^+ selectively adds to the terminal carbon atom to complete 1,4-addition of D atoms to yield 2-butenes.

The dynamic nature of the intermediates varies with the type of catalyst. On alkaline earth oxides, the interconversion between *cis* allylic anion and *trans* allylic anion is faster than the addition of D^+. As a result, *cis*-2-butene-d_2 is preferentially yielded. On the other hand, the addition is faster than the interconversion on ThO_2, ZrO_2, and rare earth oxides, *trans*-2-butene-d_2 being a main product. Here, the carbon skeleton structure of the *trans* form is retained during the hydrogenation.[8,11,12)]

Lack of butane formation is caused by difficulty of aklyl anion formation compared to allylic anion formation. Alkyl anions are less stable than allylic anions; thus reactions of monoenes with an H^- have high energy barriers.

In 2-methyl-1,3-butadiene (isoprene) hydrogenation, 2-methyl-2-butene resulting from 1,4 addition of H atoms is the main product on solid base hydrogenation catalysts. Deuteration, however, results in the formation of (*E*) and (*Z*) forms of 2-methyl-2-butene-d_2. The ratio of (*E*) form to (*Z*) form varies with catalyst types, and informs us of the dynamic nature of the intermediates in the reaction. The dynamic natures was studied over MgO[11], La_2O_3,[9] ThO_2,[13] and ZrO_2.[8] The reaction scheme is shown below.

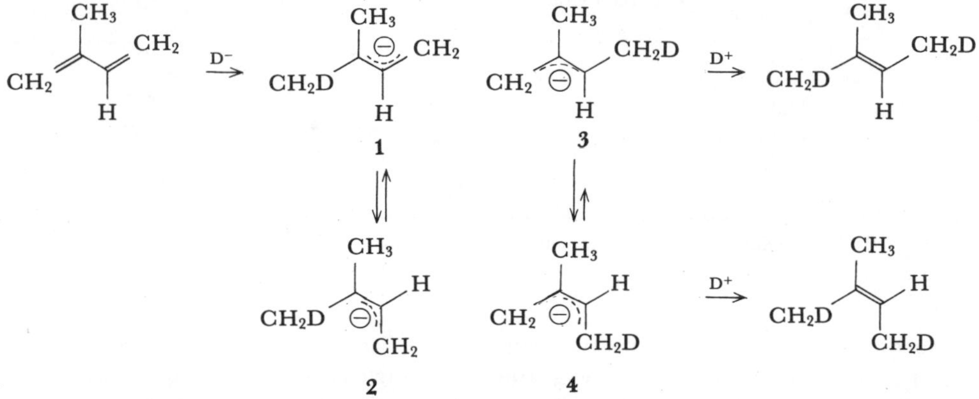

2-Methyl-1,3-butadiene consists of 97% s-*trans* conformer and 3% s-*cis* conformer in the gas phase at 273 K. As shown in the scheme, two possible D-additions may occur. The 1,1-dimethyl allyl anion **1** and 1,2-dimethyl allyl anion **3** are formed by the addition of D^- to carbon atoms 1 and 4 of the reactant, respectively. The anion **1** and **3** can be considered to have two extreme forms.

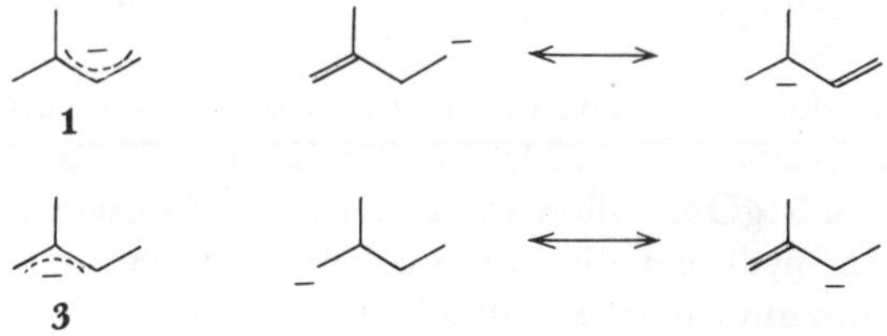

The extreme forms of anion **1** are a primary anion and a tertiary anion, while the extreme forms of anion **3** are a primary anion and a secondary anion. This shows the lower stability of anion **1** as compared to anion **3**.

The products in the deuteration of 2-methyl-1,3-butadiene over different catalysts are summarized in Table 4.32.[12] Over ThO_2 and ZrO_2 89% (*E*) form and 11% (*Z*) form are formed, while over MgO, 60% (*E*) form and 40% (*Z*) form are formed. Although it is not definite yet which is formed first, anion **1** or **3**, the carbon skeleton structure is retained during the reaction over ThO_2, and the interconversion within the allylic anions extends over MgO. The dynamic nature of the allylic anions on La_2O_3 is of an intermediate nature between that for MgO and that for ThO_2.

The retention of molecular indentity observed in both hydrogenations is caused by the isolation of active sites. Two hydrogen ions formed by dissociative adsorption of one hydrogen molecule on one site do not migrate to other sites because each active site is isolated from the others.

The active sites for hydrogenation on alkaline earth oxides are believed to be cation-oxide ion pairs of low coordination as described in section 3.1.2. In the surface model

TABLE 4.32 ^{13}C NMR analysis of 2-methyl-2-butene

Cataslyst	%*E* and *Z* isomers of 2-methylbut-2-ene		% *cis* and *trans* isomers of but-2-ene	
	E	*Z*	*trans*†1	*cis*†1
MgO	60	40	17†2	83†2
La_2O_3	64	36	82†3	18†3
ThO_2	89	11	92†4	8†4
ZrO_2	89	11		

†1, Hydrogenation of 1,3-butadiene. †2, Conversion : 41 %. †3, 21 %. †4 9 %.

structure of MgO, it is plausible that $Mg^{2+}{}_{3C}-O^{2-}{}_{3C}$ pairs act as hydrogenation sites.

1,3-Butadiene hydrogenation was studied on single crystals exposing different crystal planes.[14] The (111) plane of MgO is active but the (100) plane is not. The (111) plane is composed of surface ions of coordination number 3. This supports the hypothesis that the active sites for hydrogenation are ion pairs of low coordination numbers.

TPD results of hydrogen adsorbed on MgO lead us to conclude that hydrogen adsorbed on ion pairs of low coordination numbers do not migrate on the surface and each adsorption site is separated from the others, and that the evolved hydrogen molecule originates from OH and MgH and not from two OH groups or MgH groups.[15] This explains the retention of molecular identity of hydrogen atoms during the reaction.

A key step in the hydrogenation is dissociative adsorption of hydrogen. The amounts of hydrogen adsorbed on MgO at 273 K and 0.75 Pa as a function of pretreatment temperature of MgO are shown in Fig. 4.29.[11] A good correlation of the amount of hydrogen with the hydrogenation rates is found comparing Fig. 4.29 with Fig. 4.28.

In contrast to the information available for alkaline earth oxides, the nature of the active sites on other types of hydrogenation catalysts such as ThO_2, ZrO_2, and rare earth oxides has rarely been investigated.

1,3-Butadiene undergoes transfer hydrogenation over ZrO_2.[16] In transfer hydrogenation, two hydrogen atoms used for the hydrogenation are supplied by hydrogen donor molecules. Transfer hydrogenation of 1,3-butadiene with cyclohexadiene over ZrO_2 mainly yields 1-butene by 1,2 addition of hydrogen atoms. The activity variation of ZrO_2 as a function of pretreatment temperature parallels that of butene isomerization, the activity maximum appearing at 1073 K.[8] The reaction is suggested to occur as follows. The basic sites abstract an H^+ from cyclohexadiene to form a cyclohexadienyl anion. The H^+ attacks 1,3-butadiene at the terminal carbon atom to form an allyl carbenium ion. An H^- from the cyclohexadienyl anion attacks the allyl carbenium ion at the positively charged carbon atom to complete 1,2 addition. Although the half hydrogenated intermediate is cationic, the reaction is base-catalyzed reaction because it is initiated by the abstraction of H^+ by a basic site.

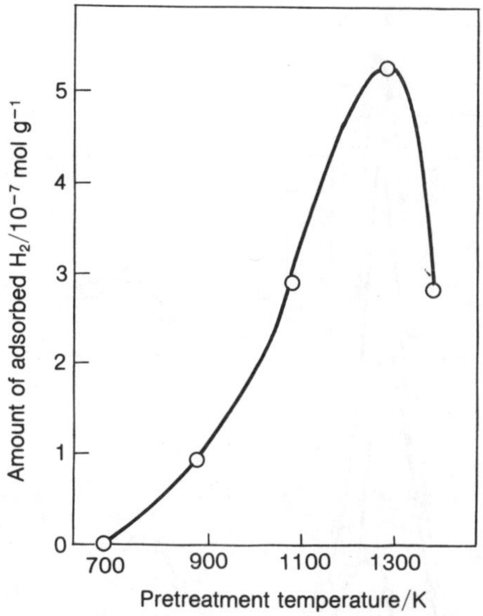

Fig. 4.29 **Adsorption of hydrogen at 273 K and 10^{-2} Torr on MgO pretreated at different temperatures.**

4.15.2 Hydrogenation of Carbon Monoxide

Carbon monoxide hydrogenation catalyzed by a solid base was first observed on MgO in a TPD study of interaction of adsorbed CO with molecular hydrogen. The TPD profiles indicating interaction of adsorbed CO with H_2 are shown in Fig. 4.30.[5,17] Admission of hydrogen to preadsorbed CO on MgO results in a decrease in CO peaks and the appearance of new CO peak at 700 K. The new peak of CO accompanies hydrogen evolution.

IR measurement shows that the surface species resulting from the reaction of adsorbed CO with H_2 are in the form of formate. The reaction is suggested to proceed by the following mechanism.[10]

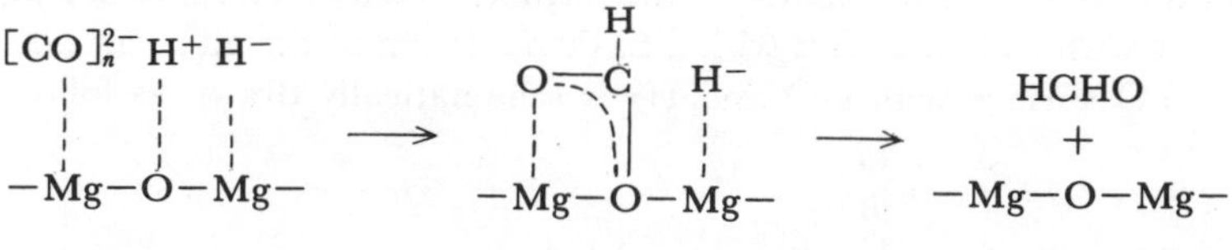

The negatively charged adsorbed species of CO react with H^+ to form [CHO] groups which are adsorbed on surface oxygen ions. The [CHO] group adsorbed on oxygen ion becomes the same structure as the formate group, hence the IR spectra are the same as those for formate.

Essentially the same type of reaction as that of adsorbed CO with H_2 takes place over other types of solid base catalysts such as La_2O_3, ZrO_2, and ThO_2.[18]

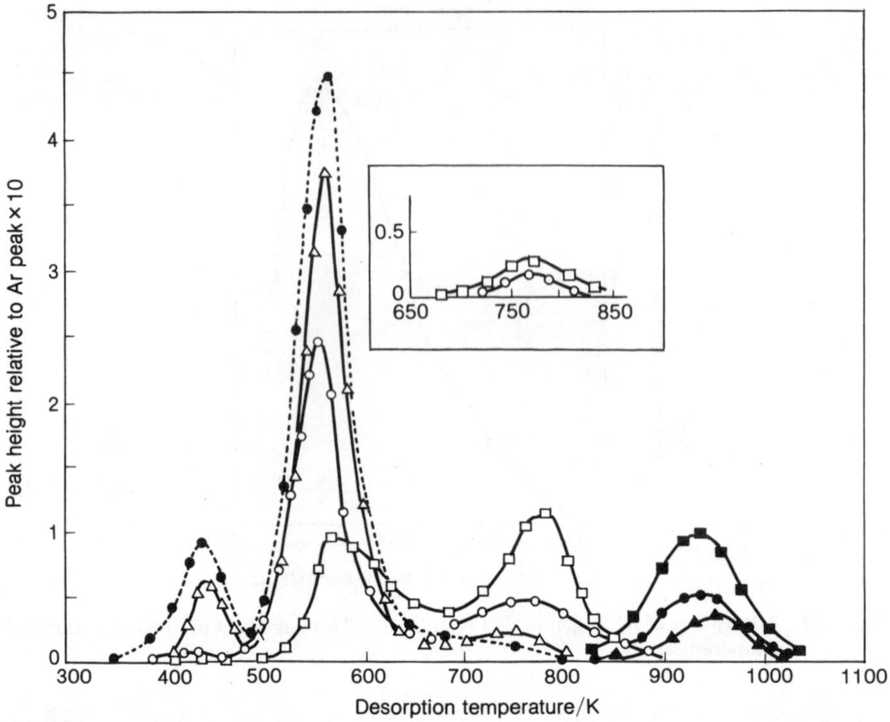

Fig. 4.30 TPD profiles from the surface species formed on MgO by the reaction of adsorbed CO with H_2 at different temperatures. CO : △; 343, ○; 423 and □; 483 K, CO_2 : ▲; 343, ●; 423 and ■; 483 K, H_2 (shown in the inset) : ○; 423 and □; 483 K. The dotted line denotes the evolution of CO when H_2 was not admitted. (Reproduced with permission by H. Hattori, *J. Chem. Soc. Faraday Trans.* I , **80**, 1040 (1984)).

In the catalytic reaction of CO with H_2, formaldehyde is actually formed over MgO at 473 K. At 573 K, however, the products consist mostly of methanol. The reactions over La_2O_3 and ZrO_2 at 473 K give methanol.[18]

Lavalley *et al.* observed the IR spectrum of the formyl group immediately after contact of adsorbed CO with H_2 on ZnO surface.[19-21] The IR bands ascribed to the formyl rapidly diminishes and the bands for the formate develop. The sites of ZnO involve surface Zn ions with two vacancies with a basic oxide ion at the adjacent position. Interaction of ZnO surface with CO and H_2 is schematically drawn as follows.

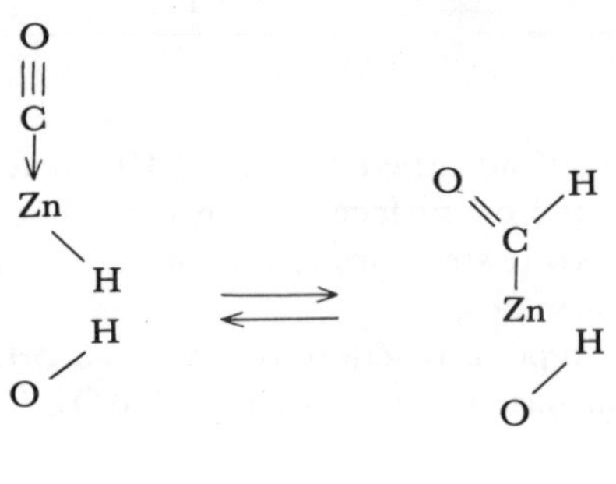

Hydrogen dissociates heterolytically to be adsorbed on Zn and O. The first step for CO hydrogenation is hydride transfer from the metal atom to the carbonyl ligand.

The reaction of CO with H_2 was studied over ZrO_2.[22-24] Although the role of acid – base properties of catalyst was not clearly demonstrated, participation of acid – base sites on the surface was suggested. Around 573 K, CO hydrogenation yields methanol selectively while around 673 K selective formation of isobutane occurs. Hydrogen molecule dissociates heterolytically to be adsorbed on Zr and O to form Zr – H and O – H, as in the case of ZnO. The Zr – H species converts to the O – H species at 525 K. The formation of methanol proceeds via surface bidentate carbonate and methoxide species as shown below.

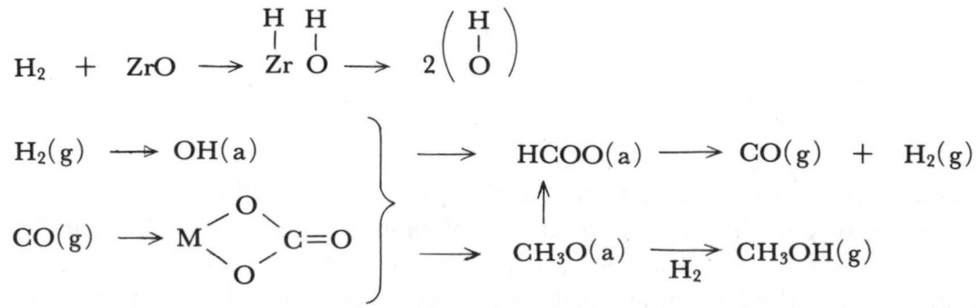

REFERENCES

1. D. G. Hill, G. B. Kistiakowsky, *J. Am. Chem. Soc.*, **52**, 892 (1930).
2. S. E. Voltz, *J. Phys. Chem.*, **61**, 756 (1957).
3. H. Hattori, Y. Tanaka, K. Tanabe, *J. Am. Chem. Soc.*., **98**, 4652 (1976).
4. R. J. Kokes, A. L. Dent, *Advance in Catalysis,* **22**, 1 (1972).
5. G. Wang, H. Hattori, H. Itoh, K. Tanabe, *J. Chem. Commun.*, **1982**, 1256.
6. H. Hattori, Y. Tanaka, K. Tanabe, *Chem. Lett.*, **1975**, 659.
7. Y. Tanaka, H. Hattori, K. Tanabe, *Chem. Lett.*, **1976**, 37.
8. Y. Nakano, T. Yamaguchi, K. Tanabe, *J. Catal.*, **80**, 307 (1983).
9. Y. Imizu, K. Sato, H. Hattori, *J. Catal.*, **76**, 65 (1982).
10. H. Hattori, H. Kumai, K. Tanaka, G. Zhang, K. Tanabe, Proc. National Symp. Catal. India, 1987, Sindri.
11. Y. Tanaka, Y. Imizu, H. Hattori, K. Tanabe, Proc. 7th Intern. Congr. Catal., 1980, Tokyo, p.1254.
12. Y. Imizu, H. Hattori, K. Tanabe, *J. Chem. Commun.*, **1978**, 1091.
13. Y. Imizu, H. Hattori, K. Tanabe, *J. Catal.*, **57**, 35 (1979).
14. K. Miyahara, Y. Murata, I. Toyoshima, Y. Tanaka, T. Yokohama, *J. Catal.*, **68**, 186 (1981).
15. T. Ito, M. Kuramoto, M. Yoshioka, T. Tokuda, *J. Phys.* Chem., **87**, 4411 (1983).
16. T. Yamaguchi, J. W. Hightower, *J. Am. Chem. Soc.*, **99**, 4201 (1977).
17. G. Wang, H. Hattori, *J. Chem. Soc., Faraday Trans. 1*, **80**, 1039 (1984).
18. G. Wang, H. Hattori, Proc. 8th Intern. Congr. Catal., 1984, Berlin, III-219.
19. J. Saussey, J. C. Lavalley, J. Lamotte, T. Rais, *J. Chem. Soc. Chem. Commun.*, **1982**, 278.
20. J. C. Lavalley, J. Sausey, T. Rai, *J. Mol. Catal.*, **17**, 289 (1982).
21. J. Sausey, T. Rais, J. C. Lavalley, *Bull. Soc. Chim. France,* **1985**, 305.
22. T. Maehashi, K. Maruya, K. Domen, K. Aika, T. Ohnishi, *Chem. Lett.*, **1984**, 747.
23. T. Ohnishi, H. Abe, K. Maruya, K. Domen, *J. Chem. Soc. Chem Commun.*, **1985**, 617.
24. T. Ohnishi, K. Maruya, K. Domen, H. Abe, J. Kondo, Proc. 9th Intern. Congr. Catal. Calgary, 1988, p.507.

4.16 DEHYDROGENATION

Dehydrogenation is a reaction in which hydrogen is eliminated from a molecule in the form of dihydrogen or its reacted form by a hydrogen acceptor such as oxygen. The former is called "simple" dehydrogenation (e.g., eq. (1)), and the latter oxidative dehydrogenation (e.g., eq. (2)). Simple dehydrogenation generally requires high reaction temperatures because of the limitation of equilibrium.

$$C_nH_{2n} \longrightarrow C_nH_{2n-2} + H_2 \quad (1)$$

$$C_nH_{2n} + 1/2\,O_2 \longrightarrow C_nH_{2n-2} + H_2O \quad (2)$$

Industrially important processes are dehydrogenation of ethylbenzenes to corresponding styrenes, of paraffins to mono-olefins, of mono-olefins to diolefins, and of alcohols to aldehydes or ketones. For simple dehydrogenation, mixed oxides containing ZnO and Cr_2O_3, which are basic and capable of dissociating H−H and C−H bonds, are commonly used. In the case of oxidative dehydrogenation, since the rate-limiting step is probably the elimination of proton, the basicity of catalyst must play an important role, besides the reduction-oxidation properties of catalysts. However, very few studies have been attempted concerning the quantitative relationship between the basicity and the catalytic activity.

A. Dehydrogenation of Hydrocarbons

Styrene is produced industrially by simple dehydrogenation of ethylbenzene by using catalysts containing Fe oxides as the main component (Fe−Cr−K, Fe−Ce−Mo, Fe−Mg−K oxides) or Ca−Ni phosphate in the presence of steam. According to Lee *et al.*, the promoting effect of alkali was

$$Cs > K > Na > Li$$

and this is the order of increasing basicity.[1] It has been reported that the primary role of added K is to form a basic active phase such as $K_2Fe_2O_3$ by the reaction with iron oxide.[2] In addition, K reduces the deposition of carbon on the catalyst surface and accelerates the desorption of products.

Addition of Ce or Mo to Fe−Cr−K catalysts improves the selectivity. A relation has been reported between the activation energy for styrene formation and the electronegativity of the transition elements added, as shown in Fig. 4.31.[3,4] The acid and base amounts measured correlated well with the electronegativity. According to this relation, Ce is the most effective for promoting dehydrogenation. It was suggested that Mo adjusts the activity at a moderate level to suppress the undesired formation of benzene and toluene.

Recently, the addition of Mg together with K was found to be effective as basic additives.[3,4] It was suggested that Mg increases the number of active sites and thermal

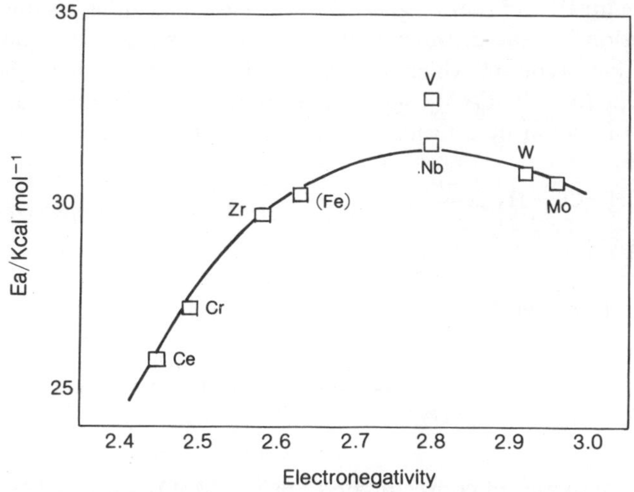

Fig. 4.31 Relation between the activation energies for styrene formation and the electronegativity of various transition metal oxides.
(Reproduced with permission by T. Hirano, *Bull. Chem. Soc. Jpn.*, **59**, 1654 (1986)).

stability by forming a solid solution in Fe_3O_4.

Methylstyrene and divinylbenzene are similarly produced by dehydrogenation. Oxidative dehydrogenation of ethylbenzene is catalyzed by Fe-containing catalysts mentioned above or solid acids such as zeolites and metal phosphates with performance comparable to simple dehydrogenation.[5] Good correlation between the rate of dehydrogenation as well as isomerization of cyclohexane and the acid amount has been observed for $TiO_2-ZrO_2-V_2O_5$.[5] The following scheme in which the abstraction of H^- is rate-limiting has been suggested.

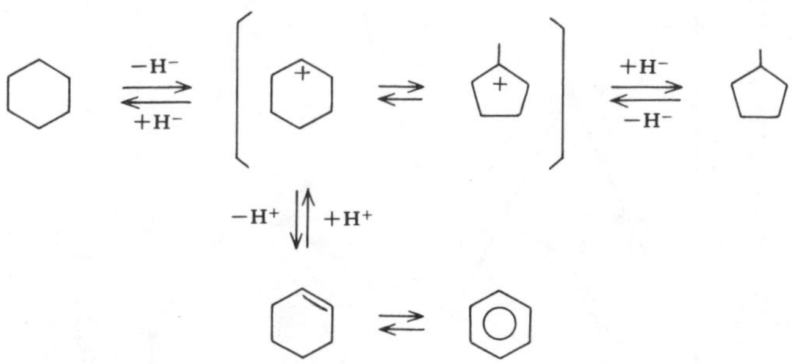

(3)

B. Dehydrogenation of Alcohols

It has been noted that dehydration prevails over metal oxide catalysts which are acidic (e.g., Al_2O_3) and dehydrogenation becomes dominant over basic oxides.[6] Thus, dehydrogenation of alcohols is catalyzed by ZnO, MgO, Cr_2O_3 and CuO. Similar variations of dehydrogenation vs. dehydration can be found for the reaction of formic acid. Increase of dehydrogenation activity with increasing basicity has been reported for alkali-treated zeolites[7] and porous glass.[8] In the case of zeolites, as the pH value during the preparation increased, the activity tended to increase. Two mechanisms have been proposed: keto-type, in which an alkoxide is the intermediate (in most cases[6a, 6b]), and enol-type (over ThO_2[6c]), see eq. (4). In both cases, the reaction proceeds via the abstraction of proton by a basic site (O^{2-}), which is generally rate-determining.

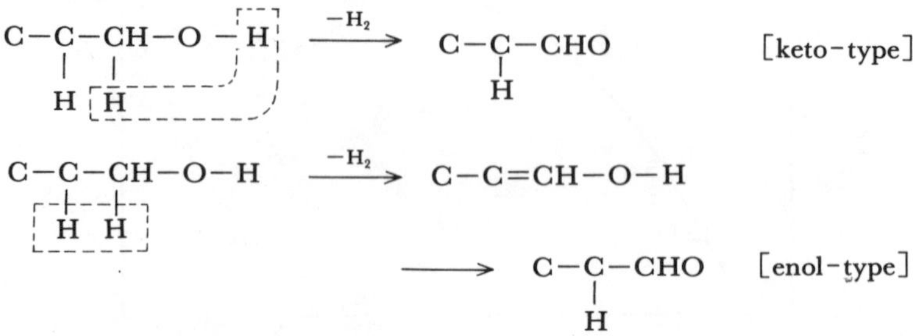

(4)

The catalytic activities of coprecipitated SnO_2-MoO_3 with various Mo/Sn ratios for the dehydration-dehydrogenation of 2-butanol have been studied in relation to the surface composition measured by XPS and the crystallinity estimated by XRD.[9] The results are shown in Fig. 4.32. The SnO_2 phase dissolving Mo in its lattice has been

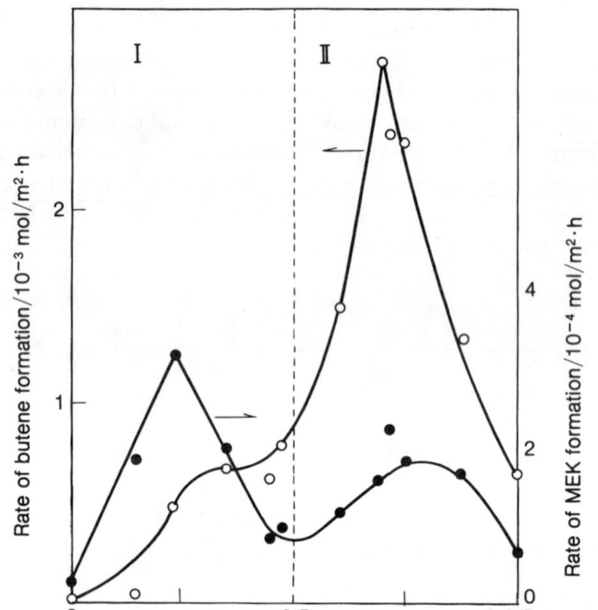

Fig. 4.32 Catalytic activities of SnO_2-MoO_3 catalyst for formation of butenes and methylethylketone (MEK) in the conversion of *sec*-butyl alcohol as a function of the surface composition of the fresh catalyst. Reaction temperature : 463 K, *sec*-butyl alcohol/O_2=1.16
(Reproduced with permission by Y. Okamoto, *et al.*, *J. Catal.*, **71**, 103 (1981)).

proposed to be the active phase for dehydrogenation and accounts for the maximum activity observed at Mo/(Mo + Sn) ratio = 0.25.

Formaldehyde is industrially produced by oxidative dehydrogenation of methanol using either Ag catalyst or Fe – Mo oxide catalyst. Simple dehydrogenation without forming water has been attempted with Cu or Zn compounds,[10] where the addition of phosphoric acid to Cu/SiO_2 is effective to improve the yield of formaldehyde, probably by controlling the oxidation state of Cu.[11] Dehydrogenation of methanol to methyl formate is catalyzed by $CuO-SiO_2$. It was recently reported that Cu exchanged mica showed very high selectivity and durability.[12] The high performance was attributed to the absence of acidity on the surface of the mica. The reaction of 2-butanol was studied with Fe_2O_3-containing mixed oxides.[13]

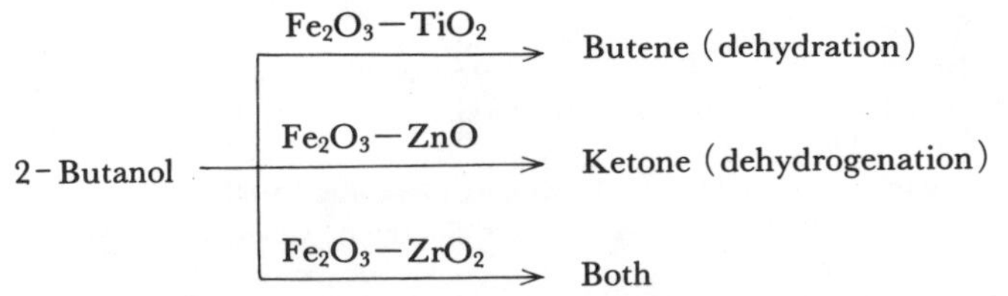

(5)

Reduction of catalysts increased the activity for dehydrogenation and resulted in the formation of butane. It was concluded that butane was formed by the nucleophilic substitution of OH^- by H^- which was liberated in the dehydrogenation of butanol.

Role of acid-base bifunctional catalysis of $MgO-SiO_2$ has been studied for the dehydrogenation of alcohol by acetone.[14] In the case of ethanol, the catalyst basicity plays a predominant role, while for 2-butanol both acidity and bacicity are important.

C. Other Dehydrogenations

Dehydrogenation of propene or isobutene produces benzene or *p*-xylene, respectively, over SnO_2 and ZnO.[15]

$$2\,C_3H_6 + 3/2\,O_2 \longrightarrow C_6H_6 + 3\,H_2O \qquad (6)$$

When acidic catalysts or acidic additives are added to the above catalysts, oxidation to aldehydes becomes dominant. It was suggested that the basic nature of the catalyst surface keeps the allylic intermediate electrically neutral and favors the dimerization of the allyl (see Section 4.17).

Oxidative coupling of methane to form ethane and ethylene proceeds with fairly good selectivity over basic solids such as Li-doped MgO and rare-earth oxides at very high temperatures.[16,17]

$$2\,CH_4 \longrightarrow C_2H_6,\ C_2H_4 \qquad (7)$$

Isobutyric acid can be converted to methacrylic acid by oxidative dehydrogenation using heteropoly compounds.[18,19] In this reaction, suppression of the catalyst acidity which accelerates the decomposition of isobutyric acid to propylene and CO is necessary to improve the selectivity for methacrylic acid.

REFERENCES

1. E. H. Lee, *Catal. Rev.*, **8**, 285 (1973).
2. T. Hirano, *Appl. Catal*, **26**, 65, 81 (1986).
3. T. Hirano, *Bull. Chem. Soc. Jpn.*, **59**, 1653, 2672 (1986).
4. T. Hirano, *Shokubai (Catalyst)*, **29**, 642 (1987) (in Japanese).
5. R-C. Chang, I. Wang, *J. Catal.*, **107**, 195 (1987).

6a) H. Niiyama, E. Echigoya, *Bull. Chem. Soc. Jpn.*, **44**, 1739 (1971).
 b) L. Nodek, J. Sedlacek, *J. Catal.*, **40**, 34 (1975) ; N. Takezawa, C. Hanamaki, H. Kobayashi, *J. Catal.*, **34**, 329 (1974).
 c) K. Thomke, *Z. Phys. Chem. NF*, **106**, 225 (1977).

7. T. Yashima, H. Suzuki, N. Hara, *J. Catal.*, **33**, 486 (1974).
8. T. Imanaka, N. Nakamura, Y. Ido, S. Teranishi, *Nippon Kagaku Kaishi*, **91**, 319 (1970) (in Japanese).
9. Y. Okamoto, K. Oh-Hiraki, T. Imanaka, S. Teranishi, *J. Catal.*, **71**, 99 (1981).
10. Japan Kokai, 1977-215, 1987-22737.
11. T. Yamamoto, A. Shimoda, T. Okuhara, M. Misono, *Chem. Lett.*, **1988**, 273.
12. Y. Morikawa, K. Takagi, Y. Moro-oka, T. Ikawa, Proc. 8th Intern. Congr. Catal., Vol. 5, Verlag Chemie, Weinheim, 1984, p.679.
13. T. Jin, H. Hattori, K. Tanabe, *Bull. Chem. Soc. Jpn.*, **56**, 3206 (1983).
14. H. Niiyama, E. Echigoya, *Bull. Chem. Soc. Jpn.*, **45**, 939 (1972).
15. T. Seiyama, M. Egashira, T. Sakamoto, I. Aso, *J. Catal.*, **24**, 76 (1972).
16. T. Ito, J. X. Wang, C. H. Liu, J. H. Lunsford, *J. Am. Chem. Soc.*, **107**, 5062 (1985).
17. K. Otsuka, K. Jinno, A. Morikawa, *Chem. Lett.*, **1985**, 499.
18. M. Otake, T. Onoda, *Shokubai (Catalyst)*, **18**, 169 (1976) (in Japanese); Japan Kokai 1977-108918, 31018; 1981-15238.
19. M. Akimoto, Y. Tsuchida, K. Sato, E. Echigoya, *J. Catal.*, **72**, 83 (1981).

4.17 OXIDATION

Acid-base properties of catalysts in general play significant roles not only in acid-base catalysis but also in oxidation catalysis. The mechanisms in which the acidity or basicity takes part in oxidation catalysis may be classified into two categories: i) activation of one or more of the reactants, products and intermediates and ii) acceleration of one or more of the reaction paths involved in overall oxidation reactions (parallel and/or consecutive paths).

4.17.1 Activation of Reacting Molecules

Acidic or basic substances such as P_2O_5 or K salts are often added to industrial oxidation catalysts in order to improve catalytic performance. These additives suppress undesirable side reactions and overoxidation by adjusting the acid-base properties of the catalyst surface.

Ai and associates[1,2] reported that for several oxidation reactions catalyzed by binary and ternary metal oxide catalysts, the catalytic activity and selectivity are correlated with the acid-base properties of the catalysts. The correlations were explained

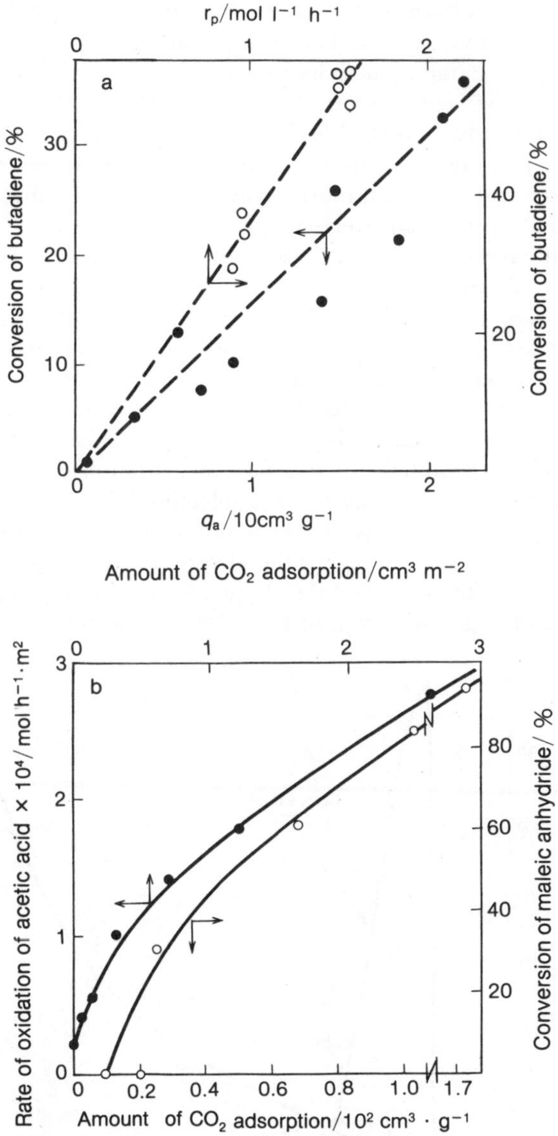

Fig. 4.33 a. Correlation between the catalytic activity for oxidation of butadiene and the acid amount of catalyst.[1)]
●, Mo–Bi–P oxide (P/Mo = 0.2, Bi/(Bi + Mo)) = 0 – 1); ○, V–Mo oxide (Mo = 0 – 30%); r_p, rate of dehydration of isopropanol; q_a, amount of irreversible adsorption of ammonia.

b. Correlation between the catalytic activities for oxidations of acetic acid and maleic anhydride and the base amount of catalysts measured by CO_2 adsorption.[2)]
●, Ti–C–P oxide (V/T_i = 1/9, P/T_i ratio varied), ○, Mo–Bi–P oxide (as in Fig. 4.33a)).

by the strength of acid-base interactions between reacting molecules (reactants and products) and the catalyst surface. For example, in the case of oxidation of butadiene, which is "basic", the catalytic activity of several Mo − Bi − P and V − Mo mixed oxides increased with the acid amount of the catalyst (Fig. 4.33a)). The amount was measured by irreversible adsorpition of NH_3 or by the rate of dehydration of isopropanol. According to the investigators, the more acidic catalysts interact more strongly with "basic" butadiene and activate it more easily. On the other hand, the activities for oxidation of "acidic" acetic acid over Mo − Bi − P and Ti − V − P oxide catalysts were correlated with the base amounts of those catalysts as measured by irreversible adsorption of CO_2 (Fig. 4.33b)).[2)]

It is further claimed that the selectivity of the oxidation is classified to various types according to the acid-base strength (or the ionization potential) of reactants and of the products.[2)] For example, the oxidation of "basic" molecules such as olefins and aromatics to produce "acidic" molecules such as maleic anhydride (Type Base − Acid) increases with increase in the acidity of catalysts. Acid catalysts readily activate the reactants but not the products. In the case of oxidation of basic reactants to basic products, moderately acidic catalysts were selective.

Seiyama and co-workers[3)] reported that the acid-base properties of catalysts controlled the reaction paths by changing the electronic state (cationic or neutral) of the reaction intermediate. They studied the allylic oxidation of propene over various metal oxides and found the relation shown in Fig. 4.34. It is seen in this figure that the selec-

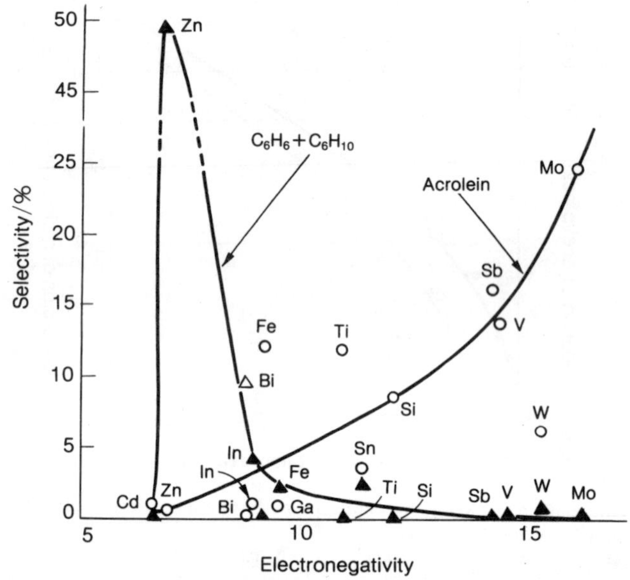

Fig. 4.34 Correlation between the selectivity of allylic oxidation of propene over metal oxide catalysts and the electronegativity of the metal ion.[3)]
C_6H_6 + C_6H_{10}: Dimerization by oxidative dehydrogenation,
Acrolein: Oxidation to acrolein.
(**Reproduced with permission by T. Seiyama *et al.*, Proc. 5th Intern, Congr. Catal., Palm Beach, 2, 1002 (1972)).**

tivity between two competitive paths, that is, dimerization to benzene and oxygen-addition to acrolein is dependent on the acid-base properties of the oxides, which are represented by the electronegativity of metal ion. Essentially the same trend was also found in the same reactions over P- or K-added SnO_2.[4] They explained the results as follows. Propene forms an allyl (π or σ) intermediate, the allylic hydrogen being abstracted by oxide ion of the catalyst, and the allyl intermediate coordinates with metal ion on the surface. If the surface is acidic, the allyl becomes more or less cationic and susceptible to the nucleophillic attack of the oxide ion. In the case of a basic surface, the intermediate may become neutral, which facilitates the dimerization by decreasing the electrostatic repulsion between the two allyls.

$$CH_2{=}CH{-}CH_3 \xrightarrow{-H} CH_2{\overset{...}{=}}CH{\overset{...}{=}}CH_2 \begin{cases} \xrightarrow{\text{dimerization}} \text{Benzene} \\ \xrightarrow[\text{oxygen addition}]{} \text{Acrolein} \end{cases} \quad (1)$$

4.17.2 Acceleration of Certain Reaction Paths

A combination of copper chromite and solid acids such as $SiO_2-Al_2O_3$, WO_3 and MoO_3 (combination of dehydrogenation and hydration) catalyzes the formation of acetone from propene and water at 500 – 600K,[5] although the yield was very low. Acetone was produced in high selectivity from a mixture of propene, oxygen and water on MoO_3-based mixed oxide catalysts.[6,7] A mechanism, in which the reaction proceeds by initial hydration of propene followed by oxidative dehydrogenation, has been proposed.[6]

$$CH_2{=}CH{-}CH_3 + H_2O \longrightarrow CH_3CH(OH)CH_3 \quad (2)$$

$$CH_3CH(OH)CH_3 + 1/2\,O_2 \longrightarrow CH_3C({=}O)CH_3 + H_2O \quad (3)$$

The first step, which is catalyzed by acid, seems to determine the reaction rate, since the overall reaction rate per unit surface area showed a good linear correlation with the density of acid sites which were measured by titration ($H_0 \leq +3.3$), as shown in Fig.4.35 for the case of SnO_2-MoO_3.[8] Formation of ketones from olefins and water probably in a similar manner have been reported for Pt/Al_2O_3,[9a], MoO_3/Al_2O_3,[9a] transition metal-exchanged zeolite,[9b] H – ZSM-5[9c] and heteropoly acid.[10]

The oxidation of acrolein and methacrolein over 12-heteropolymolybdates has been proposed to proceed by the reaction mechanism shown by Eq. (4).[11]

$$RCHO \rightleftharpoons RCH(OM)_2 \xrightarrow{\ /\ } RCOOH \quad (4)$$

$$[M = Mo, H]$$

The first step is catalyzed by acid. This is a pre-equilibrium step and even a weak acid sufficiently catalyzes this step, although the reaction does not proceed on a non-acidic catalyst. The rate-determining step is the second step, that is, the oxidative dehydro-

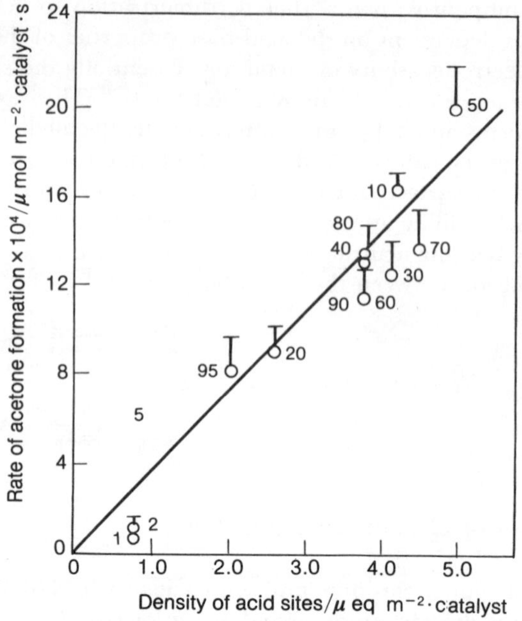

Fig. 4.35 Correlation between the acidity of SnO_2-MoO_3 and the catalytic activity for acetone formation.[8] Numbers show the content of Mo (mol %).
(**Reproduced with permission by Y. Takita, A. Ozaki, Y. Moro-oka** *J. Catal.*, **27**, 190 (1987)).

genation of the ester or diol-type intermediate. Hence, the overall rate is correlated with the oxidizing ability of catalysts.

The main reaction paths of the oxidation of acetaldehyde are the following.[12)]

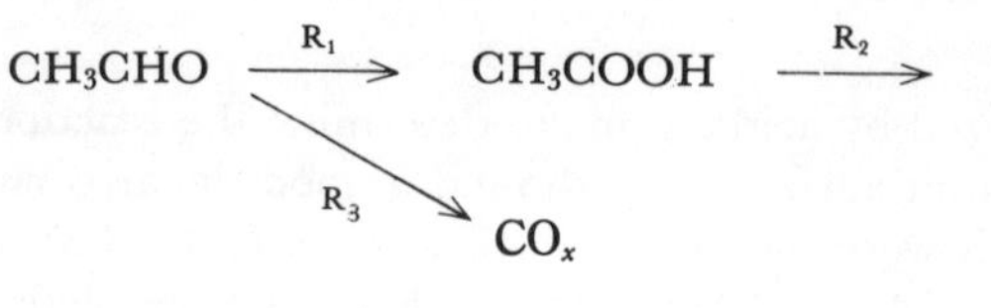

(5)

R_1 and R_3 are accelerated by the oxidizing ability of catalysts, while R_2 is promoted by acid via decomposition to methanol and CO, so that the suppression of the acidity of catalyst, for example by the use of $H_4PMo_{11}VO_{40}$, resulted in higher selectivity.

Synthesis of methacrolein by dehydrogenation of isobutyric acid is also catalyzed in fairly high yield by heteropoly compounds.[13)] In this case the acidic and oxidizing properties of catalyst function competitively, in contrast to the case of oxidation of methacrolein; the acid sites promote the side reactions of decomposition of isobutyric acid to propene and CO (reverse Koch reaction). Therefore, when Mo atoms of 12-phosphomolybdates are substituted by W atoms in increasing degree, the oxidizing ability decreases while the acidity increases, resulting in a change of reaction, as shown in Fig. 4.36.[14)] Acetone, another significant side product, has been suggested to form by the addition of lattice oxygen.[15)] A similar competitive behavior between acidity and

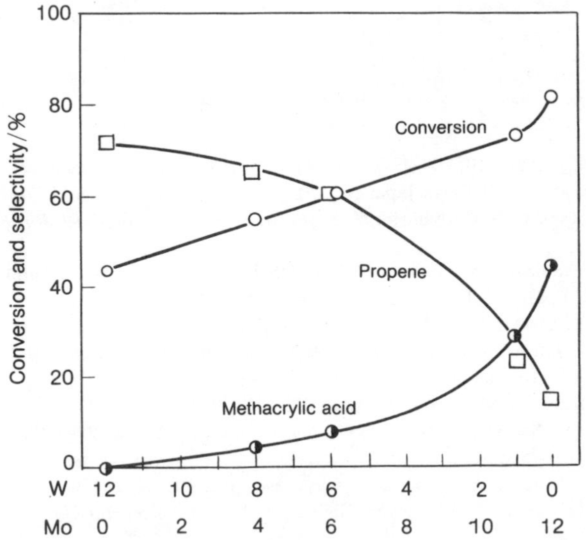

Fig. 4.36. Effects of mixed polyatoms on the reaction of isobutyric acid over $H_3PMo_xW_{12-x}O_{40}$.[14) (Reproduced with permission by M. Otake, T. Onoda, *Shokubai*, **18**, 176 (1976)).

oxidizing ability has also been reported for the reaction of methanol over $H_3PW_{12-x}V_xO_{40}$ dispersed on silica.[16)]

An important role of strong acid sites has also been suggested for the selective oxidation of *n*-butane over $(VO)_2P_2O_7$.[17)]

In the commercial NO removal process, NO is reduced by NH_3 over V_2O_5 catalyst supported on TiO_2, according to eq. (6).[18)]

$$NO + NH_3 + 1/4\,O_2 \longrightarrow N_2 + 3/2\,H_2O \qquad (6)$$

In this reaction acid sites on the catalyst play an important role, that is, the activatation of NH_3 by a protonic site, as shown in the following sequence of reactions.[19)]

$$NH_3 + V{-}OH \longrightarrow V{-}ONH_4 \qquad (7)$$

$$NO + 1/2\,O_2 \longrightarrow NO_2(ads) \qquad (8)$$

$$V{-}ONH_4 + NO_2(ads) \longrightarrow N_2 + 2H_2O + V{-}O \qquad (9)$$

$$2V{-}O + H_2O \longrightarrow 2V{-}OH \qquad (10)$$

A mechanism in which NO and $V{-}ONH_4$ directly react instead of eq. (9) has also been proposed.[20)]

REFERENCES

1. M. Ai, T. Ikawa, *J. Catal.*, **40**, 203 (1975); M. Ai, S. Suzuki. *Nippon Kagaku Kaishi*, **1973**, 21 (in Japanese).
2. M. Ai, *Shokubai*, **18**, 17 (1976) (in Japanese); M. Ai, T. Niikuni, S. Suzuki, *Kogyo Kagaku Zasshi*, **73**, 950 (1979) (in Japanese); D. B. Dadyburjor, S. S. Jewur, E. Ruckenstein, *Catal. Rev. Sci. Eng.*, **19**, 293 (1979). (1979).
3. T. Seiyama, N. Yamazoe, M. Egashira, Proc. 5th Intern. Congr. Catal., Palm Beach, 1972, p. 997.
4. T. Seiyama, M. Egashira, T. Sakamoto. I. Aso, *J. Catal.*, **24**, 76 (1972).
5. T. Yamamoto, A. Ozaki, *Kogyo Kagaku Zasshi*, **70**, 687 (1967) (in Japanese).
6. Y. Moro-oka, S. Tan, Y. Takita, A. Ozaki, *Bull. Chem. Soc. Jpn.*, **41**, 2820 (1968).
7. J. Buiten, *J. Catal.*, **10**, 188 (1968); *ibid*; **13**, 373 (1969).
8. Y. Takita, A. Ozaki, Y. Moro-oka, *J. Catal.*, **27**, 185 (1972).
9a) S. Ogasawara, Y. Nakada, Y. Iwata, Sato, *Kogyo Kagaku Zasshi*, **72**, 2244 (1969) (in Japanese); *ibid*; **73**, 509 (1970) (in Japanese).
b) I. Mochida, A. Kato, T. Seiyama, *Bull. Chem. Soc. Jpn.*, **44**, 2282 (170).
c) M. Iwamoto, H. Ueno, T. Shiozu, M. Tajima, S. Kagawa, Preprint 60th Symp. Catal., 4B08, Fukuoka, 1987.
10. H. Niiyama, Y. Saito, E. Echigoya, Preprint 44th Symp. Catal., Fukuoka, 1979.
11. M. Misono, K. Sakata, Y. Yoneda, W. Y. Lee, Proc. 7th Intern. Congr. Catal., Tokyo, 1980, Kodansha, Tokyo and Elsevier, Amsterdam, 1981, p. 1047; Y. Konishi, K. Sakata, M. Misono, K. Sakata, M. Misono, Y. Yoneda, *J. Catal.*, **77**, 169 (1982).
12. H. Mori, N. Mizumo, M. Misono, Unpublished results cited in *Shokubai*, **30**, 56 (1988) (in Japanese).
13. Japan Kokai 1977 – 138,499; 1977 – 108,918 (Mitsubishi Chem. Ind.).
14. M. Otake, T. Onoda, *Shokubai*, **18**, 169 (1976) (in Japanese).
15. M. Akimoto, Y. Tsuji, K. Sato, E. Echigoya, *J. Catal.*, **72**, 83 (1981).
16. S. M. Sorensen, R. S. Berger, 2nd Japan-China-USA Seminar on Catal., Berkeley, 1985.
17. G. Centi, F. Trifiro, *Chem. Rev.*, **88**, 55 (1988); G. Busca, G. Centi, F. Trifiro, *J. Am. Chem. Soc.*, **107**, 7757 (1985).
18. S. Matsuda, A. Kato, *Appl. Catal.*, **8**, 149 (1983).
19. M. Takagi, T. Kawai, M. Soma, T. Onishi, K. Tamaru, *J. Catal.*, **50**, 441 (1977).
20. M. Inomata, A. Miyamoto, Y. Murakami, *J. Catal.*, **62**, 140 (1980).

4.18 MISCELLANEOUS

4.18.1 Aldol Condensation (Aldol Addition)

Aldol condensation includes reactions of aldehydes or ketone producing β-hydroxyaldehydes or β-hydroxyketones by self-condensation (dimerization) or mixed condensation. A general formula of the reaction may be drawn as follows.

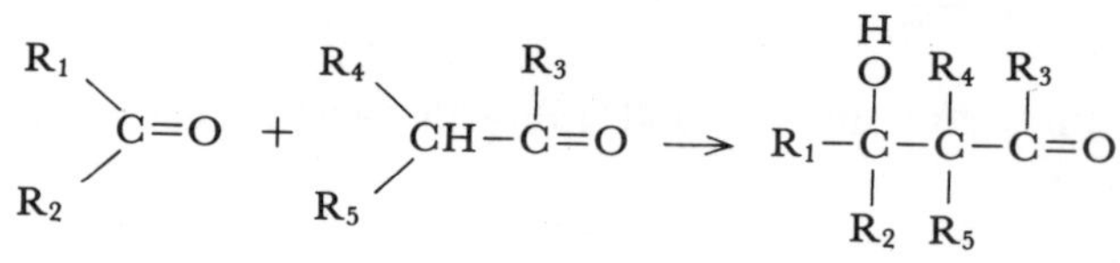

The reaction is essentially the addition of a C − H bond dissociated to the C = O bond of the other molecules.

Catalysts for aldol condensation may be either acidic or basic, but basic catalysts are much more common. The most common catalyst is $Ba(OH)_2$. Besides $Ba(OH)_2$, alkali and alkaline earth hydroxides or phosphates, and anion exchange resins are examples of solid base catalysts for the reactions.[1,2)]

The importance of catalyst basic properties was emphasized by Malinowski *et al.*[2 - 10)] They studied aldol condensation of formaldehyde with acetaldehyde, acetone, and acetonitrile. The rate constants for these reactions on SiO_2 mounted NaOH catalyst show correlation with NaOH content in the catalysts as shown in Fig. 4.37. Essentially the same linear relationship was observed for aldol condensation of acetaldehyde, and acetaldehyde with benzaldehyde. The linear relations support the view that basic properties are actually the cause of the catalytic activities. On SiO_2-supported NaOH catalysts, the groups − Si − ONa are assumed to be the active sites.

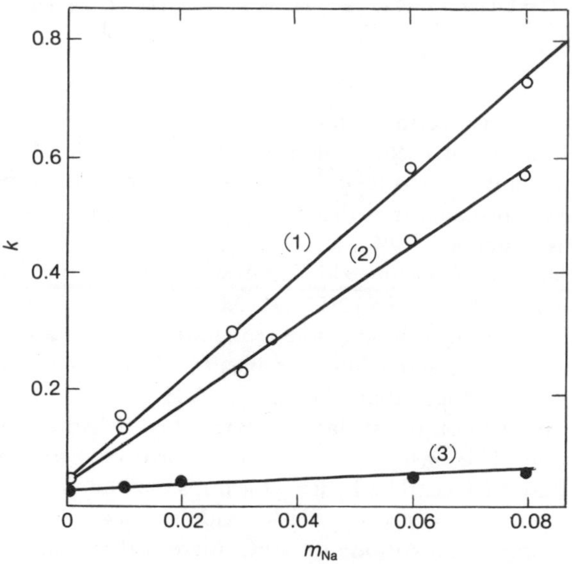

Fig. 4.37 Dependence of apparent rate coefficient, k (sec^{-1}), on sodium content, m_{Na} (mol Na per 100 g cat), in sillica gel catalysts for the vapor phase condensation of formaldehyde with (1) acetaldehyde, (2) acetone, (3) acetonitrile, at 548 K[3)]

The catalytic activities of different alkali hydroxides on SiO_2 were in the following order;[8)] NaOH < KOH < CsOH.

Among anion exchange resins, only the strongly basic ion exchangers are active in self-condensation of aldehydes and ketones. The activity decrease of the functional groups in the following order was found to be: CN > OH > Cl.[11)]

Hydrotalcite, an anionic clay mineral with the formula $Mg_6Al_2(OH)_{16}CO_3{\cdot}4H_2O$, shows a high activity for cross aldol condensation of formaldehyde with acetone to form methyl vinyl ketone.[12] The hydrotalcite becomes an active and selective catalyst on heat treatment at 773 K. Because base sites appears on the surface of hydrotalcite calcined at high temperatures[13] the activity of the hydrotalcite for the cross aldol condensation is considered to be due to the base sites.

Concerning the reaction mechanisms, analogy between the homogeneous and heterogeneous reactions is usually assumed. For acetone aldol condensation, the following mechanisms are accounted for in homogeneous systems.

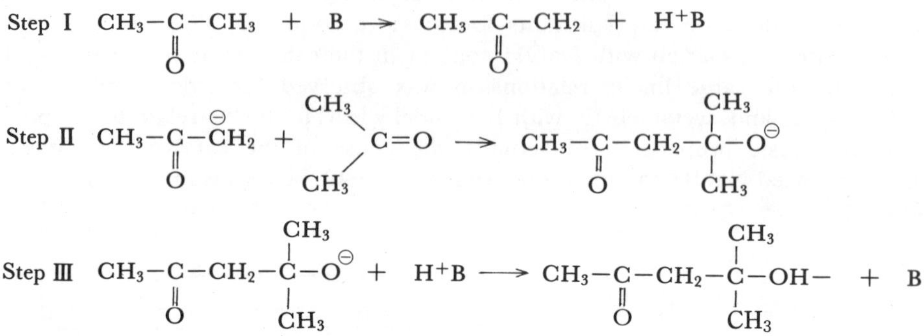

where B represents a base acting catalyst.

Reaction mechanisms of acetone aldol condensation over MgO and La_2O_3 were studied using deuterium as a tracer.[14] Analysis of the isotopic distributions of the product and reactant revealed that the slow step is involved in Step II in accordance with homogeneous systems.

The activities of alkaline earth oxide catalysts on unit surface area bases decrease in the following order: BaO > SrO > CaO > MgO.[15] This order coinsides with the order of the basic strengths of these oxides, suggesting that catalysts possessing strong base sites are efficient. The active sites are basic OH groups on the surfaces, though surface O^{2-} ions are stronger than the surface OH groups. The active surface OH groups are either retained on the surfaces or formed by dehydration of diacetone alcohol to mesityl oxide. The basic properties of the hydroxyl groups reflect the basic properties of the bare surface. The hydroxyl groups may be more strongly basic when water is adsorbed on a more strongly basic oxide surface.

Addition of certain metal cations to MgO increased the catalytic activity. In the case of Cr and Zr ion addition, the catalytic activity reaches maximum at the amount of metal cation of 0.5 – 1.0%.[16] The increases in activity were attributed to the increase in the strength of base sites caused by the addition of proper amounts of the metal cations. The increase in base strength on addition of proper amounts of metal cations was confirmed by TPD experiments for adsorbed carbon dioxide on the catalysts.

One feature which distinguishes acetone aldol condensation from other base-catalyzed reactions is a high resistance to poisoning. The presence of a small amount of water and CO_2 does not significantly retard the conversion rate of acetone.[13] The high resistance to poisoning is quite different from high sensitivity to poisoning of these molecules observed in many base-catalyzed reactions such as butene isomerization, olefin hydrogenation, etc., in which surface O^{2-} ions are believed to be the active sites. This is considered to be due to weak interaction of OH groups with CO_2 and H_2O

in contrast to the O^{2-} ions which strongly interact with these molecules.

The situation for butyraldehyde aldol condensation is different from that for acetone aldol condensation. The active sites for butyraldehyde aldol condensation are not OH groups but O^{2-} ions, and easily poisoned by trace amounts of water and carbon dioxide.[17]

Acid type catalysts catalyze cross aldol reaction of silyl ketene acetals with carbonyl compounds and acetals.[18,19] Aluminum cation and proton exchange montmorillonites are effective catalysts. Although the detailed reaction mechanism is not clear, Brønsted acid sites are considered to be the catalytic sites.

REFERENCES

1. L. Beránek, M. Kraus, in: Comprehensive Chemical Kinetics (C.J. Bamford and C.F.H. Tipper, eds.) Vol. 20, p. 263 *Complex Catalytic Processes,* Elsevier, Amsterdam, 1978.
2. A.T. Nielsen, W.J. Houliham, *Organic Reactions* (A.C. Cope, ed.) Vol. 16, *The Aldol Condensation,* John Wiley and Sons, New York, 1968.
3. S. Malinovski, S. Basinski, S. Szozepanska, W. Kiewlicz, Proc. 3rd Intern. Congr. Catal., Amsterdam, 1964, North-Holland, Amsterdam, 1965, p. 441.
4. S. Malinovski, S. Basinski, *J. Catal.*, **2**, 203 (1963).
5. S. Malinovski, S. Basinski, *Bull Acad. Pol. Sci., Ser. Sci. Chem.*, **11**, 55 (1963).
6. S. Basinski, S. Malinovski, *Rocz. Chim.*, **38**, 635 (1964).
7. S. Basinski, S. Malinovski, *Rocz. Chim.*, **38**, 843 (1964).
8. W. Kiewlicz, S. Malinovski, *Rocz. Chim.*, **44**, 1895 (1970).
9. W. Kiewlicz, S. Malinovski, *Bull. Acad. Pol. Sci., Ser Sci, Chim.*, **17**, 259 (1969).
10. S. Malinowski, S. Basinski, S. Szczepanska, *Rocz. Chem.*, **38**, 1361 (1964).
11. K. Ueno, Y. Yamaguchi, *Kogyo Kagaku Zasshi,* **55**, 234 (1952) (in Japanese).
12. E. Suzuki, Y. Ono, *Bull. Chem. Soc. Jpn.*, **61**, 1008 (1988).
13. S. Miyata, T. Kumura, H. Hattori, K. Tanabe, *Nippon Kagakukaishi,* **92**, 514 (1971) (in Japanese).
14. G. Zhang, H. Hattori, K. Tanabe, *Appl. Catal.*, **40**, 183 (1988).
15. G. Zhang, H. Hattori, K. Tanabe, *Appl. Catal.*, **36**, 189 (1988).
16. K. Tanabe, G. Zhang, H. Hattori, *Appl. Catal.*, **48**, 63 (1989).
17. G. Zhang, H. Hattori, *Bull. Chem. Soc. Jpn.*, **62**, 2070 (1989).
18. M. Kawai, M. Onaka, Y. Izumi, *Chem. Lett.*, **1987** 1581.
19. M. Onaka, R. Ohno, M. Kawai, Y. Izumi, *Bull. Chem. Soc. Jpn.*, **60**, 2689 (1987).

4.18.2 Addition of Amines to Conjugated Dienes

Primary amines and secondary amines added to cojugated dienes over solid base catalysts form unsaturated secondary and tertiary amines, repectively. The general form of the reaction is given below.

$$CH_2{=}CH{-}CH{=}CH_2 + R_1R_2NH \longrightarrow R_1R_2N{-}CH_2{-}CH{=}CH{-}CH_3 \quad (1,4\ \text{adduct})$$

$$\searrow R_1R_2N{-}CH_2{-}CH_2{-}CH{=}CH_2 \quad (1,2\ \text{adduct})$$

As homogeneous catalysts, alkali metals.[1,2] Li – amide,[3] and transition metal complexes such as $Ni[P(OC_2H_5)_3]_4$,[4] Ni acetylacetonate,[5] $PdBr_2(Ph_2PCH_2PPh_2)$, and $(Ph_3P)_3RhCl$[7] have been reported. With alkali metals and Li−amide, the products consist mostly of 1,4-addition products, while a mixture of 1,2 adducts, 1,4 adducts, and telomer was produced with transition metal complex catalysts.

The heterogeneous catalysts active for addition reactions are basic type catalysts such as a series of alkaline earth oxides, La_2O_3, ThO_2.[8,9] Zirconium oxide, which is basic, however, is not active.

Addition of dimethylamine to 1,3-butadiene proceeds at 273 K over MgO, CaO, SrO and La_2O_3, and at 323 K over ThO_2.[8,9] The composition of the products varies with the type of catalyst. One example of time dependence of the composition is shown in Fig. 4.38 for the reaction over CaO. In the initial stage of the reaction, *N,N*-dimethyl-1-2-butenylamine forms by 1,4 addition of an H^+ and a dimethylaminyl group. As the reaction proceeds, *N,N*-dimethyl-2-butenylamine undergoes double bond migration to *N-N*-dimethyl-1-butenylamine (enamine). Relative rate of the addition of amine to diene as compared to the double bond migration of the 1,4 adduct determines the selectivity. The La_2O_3 catalyst shows a high selectivity for *N, N*-dimethyl-2-butenylamine because very little double bond migration occurs. In contrast, *N-N*-dimethyl-2-butenylamine is exclusively formed over SrO catalyst due to a fast double bond migration.

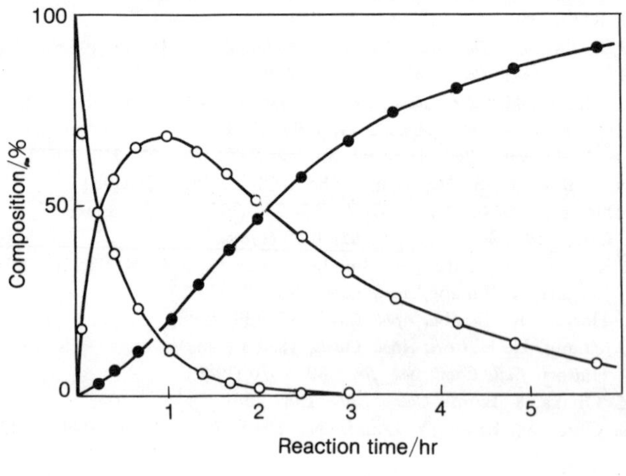

Fig. 4.38 Time dependence of composition in the reaction of 1,3-butadiene with dimethylamine at 273 K over the CaO pretreated at 873 K Catalyst; 100 mg, 1,3-butadiene; 15 Torr, dimethylamine 20 Torr. ○; Dimethylamine, ○; *N, N*-dimethyl-2-butenylamine, ●; *N, N*-dimethyl-1-butenylamine.

The activity of each catalyst is dependent upon the pretreatment temperature. Table 4.33 summarizes the activities following pretreatment at optimum temperature for each catalyst. Calcium oxide shows markedly high activity.

Ethylamine, piperidine, aniline, and trimethylamine are less reactive in the addition to 1,3-butadiene. Ethylamine and piperidine addition reactions proceed at 373 K and 453 K, respectively. Addition of aniline or trimethylamine does not take place at 473 K.

TABLE 4.33 Activities for addition of dimethylamine to 1,3-butadiene

Catalyst	Catalyst weight (mg)	Pretreatment temp. (K)	Reaction temp. (K)	Activity (10^{18} molecules·min^{-1}·g^{-1})
MgO (I)	500	973	273	3.0
MgO (II)	500	773	273	3.9
CaO	100	873	273	173.2
SrO	300	1273	273	16.4
La_2O_3	500	923	273	11.4
ThO_2	500	773	323	1.4
ZrO_2	500	1073	373	0
SiO_2-Al_2O_3	500	773	423	0

The reaction mechanism for the addition of dimethylamine to 1,3-butadiene is shown in Scheme 1.

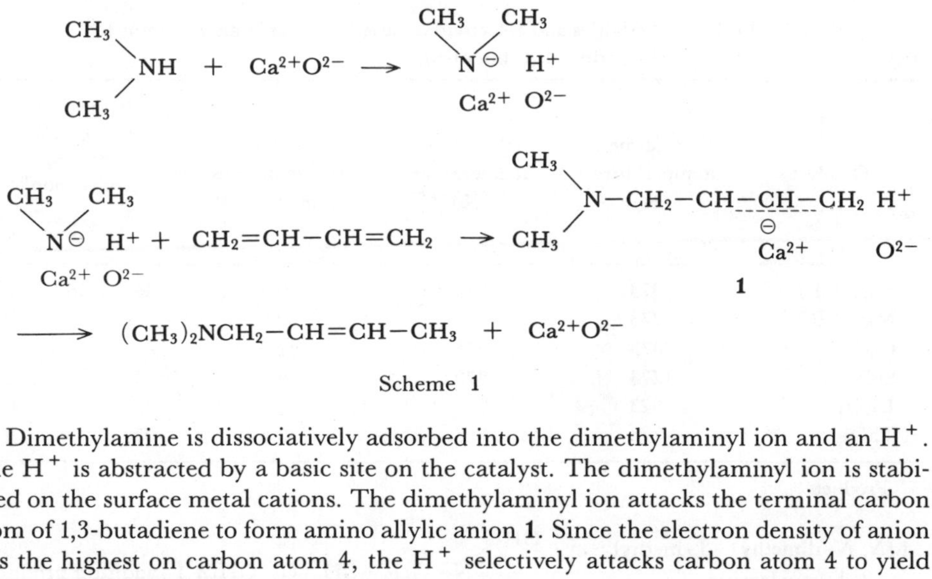

Scheme 1

Dimethylamine is dissociatively adsorbed into the dimethylaminyl ion and an H^+. The H^+ is abstracted by a basic site on the catalyst. The dimethylaminyl ion is stabilized on the surface metal cations. The dimethylaminyl ion attacks the terminal carbon atom of 1,3-butadiene to form amino allylic anion **1**. Since the electron density of anion **1** is the highest on carbon atom 4, the H^+ selectively attacks carbon atom 4 to yield the 1,4 addition product.

The above scheme is analogous to that proposed for 1,3-butadiene hydrogenation over basic catalysts, in which 1,4 addition of an H^+ and an H^- occurs selectively (see Section 4.15)

The activities and selectivities for addition of dimethylamine to 2-methyl-1,3-butadiene are given in Table 4.34. For this reaction too, CaO exhibits the highest activity. The addition occurs in two ways: 1,4 addition and 4,1 addition as illustrated in Scheme 2.

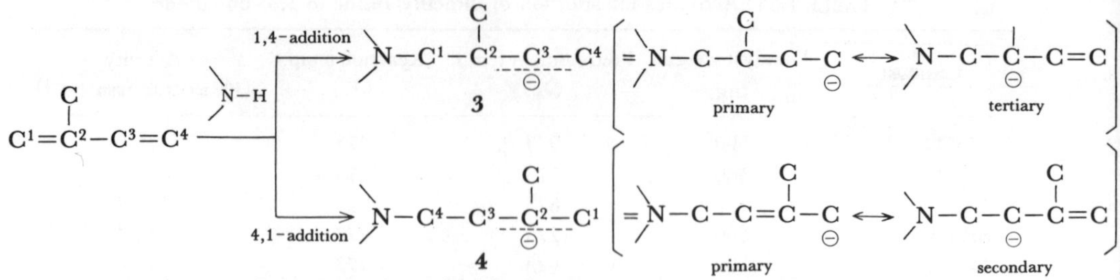

Scheme 2

The anionic mechanism accounts for the selective occurrence of 4,1 rather than 1,4-addition. The allylic anion **3** is a resonance hybrid of a primary anion and a tertiary anion, while allylic anion **4** is a resonance hybrid of a primary anion and a secondary anion. The order of stability is primary > secondary > tertiary for anion. Therefore, allylic anion **4** is more stable than allylic anion **3**. The predominant occurrence of 4,1 addition is mainly due to the difference in the stabilities of allylic anion **4** over allylic

TABLE 4.34 Activities and Selectivities for Addition of Dimethylamine to 2-methyl-1,3-butadiene

Catalyst	Pretreatment temperature (K)	Reaction temperature (K)	Activity (10^{18} molecules·min^{-1}·g^{-1})	Percentage of each product at zero conversion†		
				1	2	3
MgO (I)	973	273	0.3	44	56	0
MgO (II)	773	273	0.8	91	9	0
CaO	873	273	22.0	61	39	0
SrO	1273	273	8.5	75	19	6
La_2O_3	923	273	1.6	73	27	0
ThO_2	773	323	0.6	95	5	0

† Products:

(1) *N*, *N*-dimethyl-3-methyl-2-butenylamine: $(CH_3)_2N-CH_2-CH=C(CH_3)-CH$ (4,1-addition product).

(2) *N*, *N*-dimethyl-2-methyl-2-butenylamine: $(CH_3)_2N-CH_2-C(CH_3)=C-CH$ (1,4-addition product).

(3) *N*, *N*-dimethyl-2-methyl-1-butenylamine: $(CH_3)_2N-CH_2-CH=C(CH_3)-CH$ (enamine).

anion **3**. It should be noted that the $C^1=C^2$ double bond is more sterically hindered and more electron rich than the $C^3=C^4$ double bond. This situation also favors nucleophilic addition of aminyl ion to carbon atom **4**.

The pretreatment temperatures which result in the highest activities are higher for the reaction with primary amine (ethylamine) than those with secondary amines (dimethylamine, piperidine). This is explained by the appearance of stronger basic sites on pretreatment at higher temperatures. The explanation is extended to the hydrogenation activity variation with pretreatment temperature. Dissociation of hydrogen molecule into H^+ and H^- is more difficult than that of amine into H^- and aminyl ion. The maximum activity for the hydrogenation at higher temperatures is explained if the basic strength increases with increase in the pretreatment temperature. Therefore, for both addition of amines and hydrogenation, variation of the activity with pretreatment temperature is explained in terms of capability of dissociating the reacting molecules into H^+ and the residual anions.

Lack of activities of solid base catalysts for the addition of amines to monoenes is due to the instability of intermediate anions; alkyl anions are less stable than allylic anions.

REFERENCES

1. G. T. Martirosyan, E. A. Grigryan, A. T. Babayan, *Izv. Akad. Nauk. Arm. SSR, Khim. Nauki*, **17**, 517 (1964).
2. W. M. Stalic, H. Pines, *J. Orag. Chem.*, **35**, 415 (1970).
3. R. J. Schlott, J. C. Falk, K. W. Narducy, *J. Org. Chem.*, **37**, 4243 (1972).
4. J. Kiji, E. Sasakawa, K. Yamamoto, J. Furukawa, *J. Organomet. Chem.*, **77**, 125 (1974).
5. R. Baker, D. E. Halliday, T. N. Smith, *J. Chem. Soc. Chem. Commun.*, **1971**, 1583.
6. K. Takahashi, A. Miyake, G. Hata, *Bull. Chem. Soc., Jpn.*, **45**, 2773 (1972).
7. R. Baker, D. E. Halliday, *Tetrahedron Lett.*, **27**, 2773 (1972).
8. Y. Kakuno, H. Hattori, K. Tanabe, *Chem. Lett.*, **1982**, 2015.
9. Y. Kakuno, H. Hattori, *J. Catal.*, **85**, 509 (1984).

4.18.3 Reaction of Methanol with Nitriles, Ketones, and Esters

Reactions of methanol with nitriles, ketones and esters to yield α,β-unsaturated compounds were found by Ueda *et al.*[1-7] These reactions proceed by the catalysts possessing both acidic and basic functions.[1)] The general formula of the reactions is

$$RCH_2Z + CH_3OH \longrightarrow CH_2{=}CHRZ + H_2O + H_2$$

where $Z = -CN,\ -\underset{\underset{O}{\|}}{C}R',\ -\underset{\underset{O}{\|}}{C}OR''$ and $R = -H,\ -CH_3$

To complete the above reaction, dehydration, dehydrogenation, and cross-coupling must occur successively. Catalysts active for these reactions are MgO doped with 2 − 15% transition metal ions.

Acetonitrile reacts with methanol to yield acrylonitrile, propionitrile resulting from acrylonitrile hydrogenation being formed as a byproduct.[2)]

$$CH_3CN + CH_3OH \longrightarrow CH_2CHCN + CH_3CH_2CN$$

The activities and selectivities of MgO doped with several transition metal ions are given in Table 4.35.[1)] Among the catalysts examined, MgO doped with Cr(III) shows the highest selectivity.

TABLE 4.35 Reaction of methanol with acetonitrile to acrylonitrile over MgO doped with transition metal ions

Catalyst	Conversion of acetonitrile/%	Selectivity/%	
		Acrylonitrile	Propionitrile
MgO	0.1>	Trace	Trace
Al-MgO	2.5	Trace	Trace
Fe-MgO	11.2	73.2	11.6
Cr-MgO	9.6	94.2	5.4
Ni-MgO	5.5	2.8	33.5
Cu-MgO	2.2	91.0	9.0

Reaction conditions: W/F=20 g h/mol; CH_3OH/CH_3CN=10; Catalyst 1 g; Reaction temp. 623 K. Transition metal ion; 3.1 wt %.

Propionitrile reacts with methanol to yield metacrilonitrile as a main product, and small amounts of isobutylonitrile and crotonitrile are formed as byproducts.[3)] For this reaction too, MgO doped with Cr(III) exhibits high activity and selectivity.

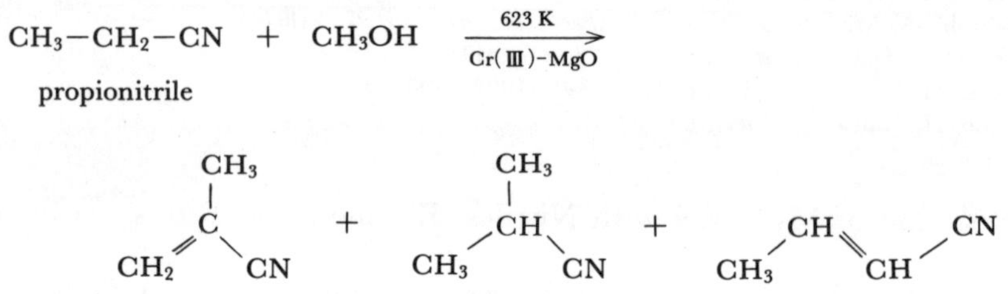

metacrilonitrile (94.1 %) isobutylonitrile (4.7 %) crotonitrile (1.2 %)

Acetone reacts with methanol to yield methyl vinyl ketone as a main product, and methyl ethyl ketone and ketones containing five C atoms are formed as byproducts.[4)] For this reaction, MgO doped with Fe(III), or Cr(III), or Cu(II) exhibits high activity and selectivity. Fe(III)−MgO shows the best result.

$$CH_3-\underset{\underset{O}{\|}}{C}-CH_3 + CH_3OH \xrightarrow[\text{Fe(III)-MgO}]{623\text{ K}} CH_2=CH-\underset{\underset{O}{\|}}{C}-CH_3$$

methyl vinyl ketone

$$+\ CH_3-CH_2-\underset{\underset{O}{\|}}{C}-CH_3\ +\ C_s\text{-ketones}\ +\ CH_3-\underset{\underset{OH}{|}}{CH}-CH_3$$

methyl ethyl ketone

Propionic ester reacts with methanol to yield methyl methacrylate. In addition to methyl methacrylate, methyl isobutyl ketone, and C_5, C_6 ketones are formed as byproducts.[5] The best selectivity for methyl methacrylate, 65%, was observed using MgO doped with Mn(II).

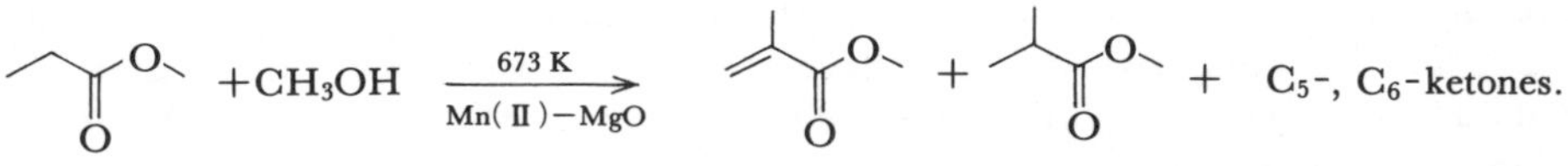

methyl methacrylate   methyl isubutyl ketone

As mentioned above, successive occurrence of dehydration, dehydrogenation, and cross-coupling is required to yield the products. It seems likely that acid sites and base sites participate in the reaction. The role of base sites is particularly important because the conversion rates correlate well with the acidities of the reactants. Acidities of the reactants are compared to the reaction rates in Table 4.36. The p*Ka* values are associated with the dissociation of the H atom which is abstracted in the cross-coupling step.[6] The easier the dissociation of the H atom as an H^+, the faster the reaction rate. This suggests that abstraction of the H^+ from the reactant by base sites is the rate-determining step. Therefore, activity increase is caused by the increase in the basicity of MgO by doping with transition metal ions. The increase in basicity is prominent when the ionic radii of added metal ions are close to magnesium ion radius (see Section 3.2.2)

TABLE 4.36 pK_a values of α-H and reaction rates

Reactant	pK_a for α-H	Rate/-mol min^{-1} g-cat^{-1}	
		Cr (III)-MgO	Fe (III)-MgO
CH_3COPh	19	8.2	34.9
CH_3COCH_3	20	11.9	11.6
CH_3CN	25	3.8	4.5
$CH_3CH_2COOCH_3$	25	0	—
CH_3Ph	37	very slow	very slow

REFERENCES

1. W. Ueda, Y. Moro-oka, T. Ikawa, *Shokubai (Catalyst)*, **28**, 208 (1986) (in Japanese).
2. W. Ueda, Y. Yokoyama, Y. Moro-oka, T. Iwaka, *Ind. Eng. Chem. Prod. Res. Dev.*, **24**, 340 (1985).
3. W. Ueda, T. Yokoyama, Y. Moro-oka, T. Ikawa, *J. Chem. Soc. Chem. Commun.*, **1984**, 39.
4. W. Ueda, H. Kurokawa, Y. Moro-oka, T. Ikawa, *Chem. Lett.*, **1985**, 819.

5. W. Ueda, T. Yokoyama, H. Korokawa, Y. Moro-oka, T. Ikawa, *J. Petrol. Inst.*, **29**, 72 (1986).
6. W. Ueda, H. Kurokawa, Y. Morikawa, Y. Moro-oka, T. Ikawa, Prep. 52nd Spring Meeting, Chem. Soc. Jpn., Tokyo, 1986, 2B46.
7. H. Kurokawa, W. Ueda, Y. Morikawa, Y. Moro-oka, in *Acid-Base Catalysis* (K. Tanabe, H. Hattori, T. Yamaguchi, T. Tanaka, eds.) Kodansha-VCH, 1989, p.93.

4.18.4 Reduction of NO with NH_3

The reduction of NO with NH_3 is an important reaction from the view-point of preventing air from contamination with NO. For this reaction, the oxides of V, Mo, Fe, Cu, and Mn supported on TiO_2 have been extensively studied as catalysts.[1] In the industrial process of denitration, small amounts of SO_3 and SO_2 included in a reacting gas react with O_2 to form SO_4^{2-}, which may be absorbed and combined with catalysts. Titanium oxide itself used in industry contains a few percent SO_4^{2-}, and the sulfate ion cannot be removed unless the catalyst is calcined above 873 K.[2] Thus, the effect of SO_4^{2-} on the catalytic activity is important. Since the acidity of TiO_2 is known to be greatly enhanced by the addition of a small amount of SO_4^{2-}, (see Section 3.9), the acidity of the catalyst is assumed to play an important role for the denitration reaction. In fact, Okazaki *et al.*[3] found that the addition of SO_4^{2-}, which caused increase in catalyst acidity remarkably enhanced the activity for the reduction of NO with NH_3. They also observed that MoO_x-TiO_2 which showed higher activity than TiO_2 for the NO reduction was much more active than TiO_2 for the acid-catalyzed dehydration of isopropyl alcohol. The effect of alkali poisoning of $MoOx-TiO_2$ on the catalytic activities for the NO reduction and the acid-catalyzed isomerization of cyclopropane are shown in Figs. 4.39 and 4.40. Both activities decrease with increase in the amount of NaOH added, indicating the importance of surface acidity in the reaction of NO with NH_3. It is emphasized that the increase of acidity on addition of SO_4^{2-} and on mixing with MoO_x is vitally important to enhance the activity.

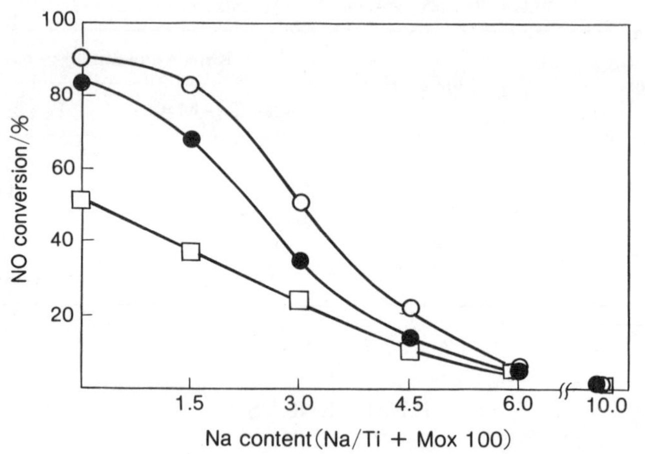

Fig. 4.39 The effect of alkali poisoning of MoO_x-TiO_2 on the catalytic activity in the reduction of NO with NH_3 : MoO_x content; 10 atomic %, reaction temperature, □; 523 K, ●; 548 K, ○; 573 K.

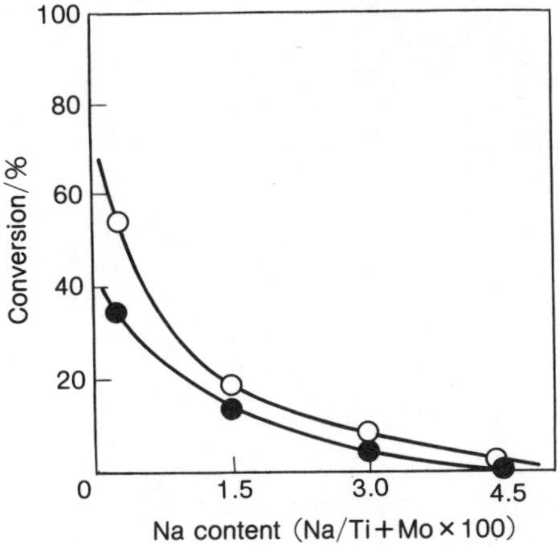

Fig. 4.40 The effect of alkali poisoning of MoO_x-TiO_2 on the catalytic activity in the isomerization of cyclopropane : ○; 3rd pulse, ●; 10th pulse.

REFERENCES

1. Y. Nishimura, *Fuel Combust.*, **45**, 9 (1978); Japanese Patents, Application No., 50-51966, 1975; 50-65467, 1975; 51-58411, 1976 [Mitsubishi Chem. Ind. Ltd.] 50-89264, 1975 [Mitsubishi Petrochemical Co. Ltd.]; 50-89289, 1975; 50-89290, 1975; 50-89291, 1975 [Hitachi Ltd.]
2. S. Okazaki, T. Chinone, A. Kurosaki, *Nippon Kagaku Kaishi*, **1977**, 1282 (in Japanese).
3. S. Okazaki, M. Kumasaka, J. Yoshida, K. Kosaka, K. Tanabe, *Ind. Eng. Chem. Proc. Des. Dev.*, **20**, 301 (1981).

5
Deactivation and Regeneration

5.1 DEACTIVATION

As in the case of other solid catalysts, the deactivation of solid acid and base catalysts occurs during use for catalytic reactions by the following mechanisms.

Adsorption of poisonous compounds Basic molecules are poisons or inhibitors for acidic sites, and acidic molecules deactivate basic sites. Such molecules may be contained in the feed as impurities or formed by reactions. Deposition of metals contained in the feedstock causes serious deactivation, e.g., in hydrotreatment of coal or heavy residual oils. This deposit blocks directly the active sites and/or the entrance of catalyst pores.

Coke deposition Carbonaceous materials which are formed from reactants deposit and block the active site, or they narrow or block the mouths of catalyst pores. This could be included under adsorption of poisonous compounds, but since it is particularly important for the solid acid catalysts used in petroleum refining and the petrochemical industry it will be described in more detail as a separate topic of discussion.

Chemical changes of catalysts Classified under this type of mechanism are: i) sintering which decreases the surface area and pore volume, ii) phase transition of the catalyst components, iii) sublimation or dissolution of active components out of the catalysts, and iv) the chemical reactions between the components of the catalysts or between the catalyst and reactant which transform the active catalyst into more stable and less active substances.

5.2 COKE DEPOSITION

"Coke" is the term used to describe generally the carbonaceous materials which are formed and deposited on the surface of catalysts during catalytic reactions mainly of hydrocarbons. There is a wide range of different types of compounds lumped under the name "coke". According to the review of Wolf and Alfani,[1] it is a mixture of mono- and polycyclic aromatic rings connected by aliphatic and alicyclic fragments. The composition changes depending on the reactants, catalysts, reaction conditions, and reaction time. Hydrogen-to-carbon ratio varies in the range of $0.3-1.5$.[1] The coke formed during the conversion of methanol over H−ZSM-5 has been characterized to be mainly two-ring aromatics such as methylated naphthalenes.[2] It is reported that there are two major types of carbon structure; the largest part consisting of pseudo-graphitic carbon and the rest poorly organized polynuclear aromatic systems.[3)]

The mechanism of coke formation is complex with multi-step reaction sequences and greatly differs by catalyst-reactant system used. Dehydrogenation, isomerization, polymerization (condensation), hydrogen transfer, cyclization and aromatization are the reactions generally involved. A reaction scheme of coke formation from polycyclic model compounds was proposed in an early study.[4)] Hydrogen-transfer reaction and the presence of olefins are believed to be key factors for the acceleration of coke formation. It has been suggested that olefins act as hydrogen acceptor for the hydrogen transfer.[5,6)] Important roles of cycloolefins and cyclodiolefins are seen in the mechanism shown in Fig. 5.1 (schemes in refs. 6 and 7 are combined). Hydrogen plays an important role in suppressing the coke deposit. Coke formation becomes faster and heavier in the following order: olefins > polycyclic, heterocyclic aromatics > paraffins,

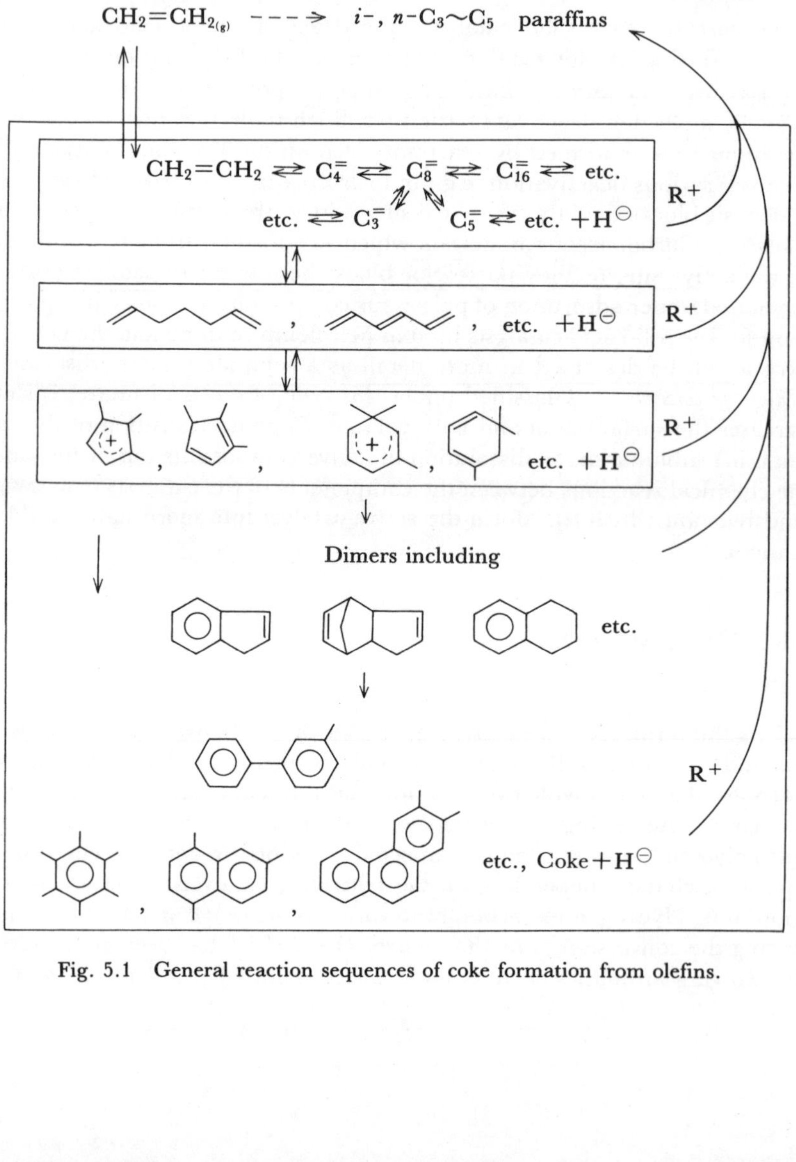

Fig. 5.1 General reaction sequences of coke formation from olefins.

naphthenes. A good correlation between the degree of coke formation and the log of basicity of aromatic hydrocarbons has been demonstrated.[4] This indicates that coke formation is related to the acid-base properties of the catalyst-reactant system.

It has been indicated by several investigators that strong acids rather than weak acids, Lewis acids rather than Brønsted acids favor coke formation, and that the presence of transition metal ions as impurities, e.g., Fe and Ni ions, accelerate the formation of coke. It was reported that coke formation on a nonacidic silica was less than one-twentieth that on acidic silica – alumina.[4] Coke deposition became less serious as the strong acid sites of silica – alumina were weakened by NaOH treatments.[8] Coke formation from hydrocarbons is usually less serious in the case of solid bases. It is reported that the deactivation of MgO and Li/MgO for methane coupling was due to sintering and loss of alkali.[9]

Coking on the surface of reforming catalysts containing noble metals is also of highly industrial importance, although the coking is much less extensive than for cracking catalysts. Hence many studies have been devoted to this subject.[10,11] The rate of coking depends on both metal components and the acidity of supports. Bimetallic catalysts such as Pt – Re and Pt – Ir reduce coke formation and result in longer catalyst life, although the mechanism has not been clarified. The magnitude and balance of the rates of dehydrogenation – hydrogenation on metal *vs.* isomerization – polymerization on acid sites seem to be an important factor. The probable mechanism of coke formation from *n*-praffins is shown in Fig. 5.2.[11] The tendency to coke formation decreases in the order: methylcyclopentane > 3-methylpentane ~ hexane > 2-methylpentane > benzene > cyclohexane.[12]

Deactivation of $MoO_3 - Al_2O_3$ in the hydrotreatment of coal-derived oil and heavy petroleum is mainly due to the deposits of carbonaceous matter and metals coming

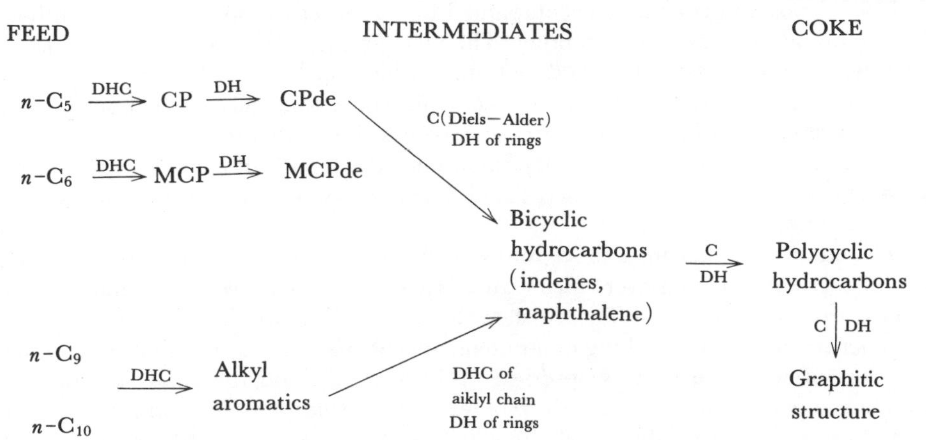

Fig. 5.2 Mechanism of coke formation from *n*-paraffins. DHC; dehydrocyclization (on metal and acid sites); DH; dehydrogenation (on metal sites) or hydrogen transfer (on acid sites); C; condensation on acid sites, CP; cyclopentane, CPde; cyclopentadiene., MCP; methylcyclopentane, MCPde; methylcyclopentadene. (Reproduced with permission by J. M. Parera *et al.*, *Catalyst Deactivation*, 1987, **34**, 145 (1957)).

from feedstocks. It has been shown in the case of coal-derived liquid that nitrogen-containing compounds of basic nature are the precursors for the carbonaceous deposit.[13,14] The addition of promoters such as Ni and Co accelerates not only the hydrodesulfurization (HDS) and hydrodemetallation activity but also the deactivation process. So an unpromoted $MoO_3-Al_2O_3$ exhibited better performance in longterm use for very heavy residue.[15] This is because coke was more uniformly deposited in a catalyst pellet and blocked the entrance of pores more slowly when the catalytic activity was lower. Non-uniform deposition of coke in catalyst is widely observed for cracking and hydrotreating catalysts.[15,16] Catalyst life greatly depends on the pore size supports. Supports having large pores are generally more resistant to deactivation caused by coke and metal deposition.[27]

The rate of coke formation usually follows a phenomenological equation of the type[17,18]

$$C = A \cdot t^n \tag{1}$$

where C is the amount of coke deposited on the catalyst, t is the time on stream, and A and n are constants ($n = 0.4-1.0$).[1]

In addition to coke formation catalyzed by acids, heavy coke formation is observed in the reaction of methanation, Fisher-Tropsch synthesis, and steam reforming over metallic catalysts such as Fe and Ni.[1,19,20]

5.3 Coke Deposition and Deactivation

Coke deposition is usually accompanied by the deactivation of catalysts, although they do not always parallel each other. The mechanisms of the deactivation caused by coke deposit are of two types: i) direct and ii) indirect.[1a]

i) Direct mechanism. Coke or its precursor, which is usually formed on the active sites, irreversibly adsorbs on the sites and blocks the reactants.

ii) Indirect mechanism. Coke deposited near the mouth of catalyst pores narrows the opening or completely blocks the mouth, hindering the diffusion or access of the reactants into the pores.

The latter effect becomes more important when extensive coking proceeds. In the case of ethylene cracking over silica – alumina at 773K there were few changes in the surface area and effective diffusion when the coke deposit was less than 1 wt%.[21] On the other hand, for the cracking of n-butene and phenanthrene over silica – alumina,[4] 33% loss of surface area was observed at 10 wt% coke on the catalyst. The direct mechanism was reported for La – Y zeolites and other large pore solid acids.[22a,b] For large-pore solid acid catalysts, coke tends to deposit in the pores (direct mechanism) and affects acid site distributions, conversion and selectivity.[22b] For small pore catalysts, coke seals off the pore mouth (indirect mechanism) and decrease the number of acid sites, without affecting the acid site distribution and selectivity.[22b] The change in diffusivity brought about by coke deposition sometimes affect shape-selectivity.[22c]

There are few quantitative studies concerning the relationship between the acidic properties of catalyst and the rate of coke formation. The relationship has been investigated for a series of silica – alumina using propylene oligomerization as the model reaction.[8] The acid strength was varied by the treatment of silica-alumina with Na compounds. The rate was expressed by a self-poisoning type equation:

$$r=kp/(1+b\cdot dp) \tag{2}$$

where k is a first-order rate constant, p and dp pressure of propylene and its change, respectively, and b is a constant expressing the degree of deactivation. A good positive correlation, as shown in Fig. 5.3, was noted between the rate of deactivation, b, and the rate constant, k, which increased in accordance with the acid strength and amount measured by amine titration.[23] Furthermore, the H/C ratio of the carbon deposit (a measure of the depth of coking) was low for strong acid catalyst and relatively high for a weak acid catalyst. Based on the close connections among the rate, deactivation, coking and the acid strength, it was concluded that coke formed more rapidly and deeply on the stronger acid sites and deacativated the catalyst more intensively.

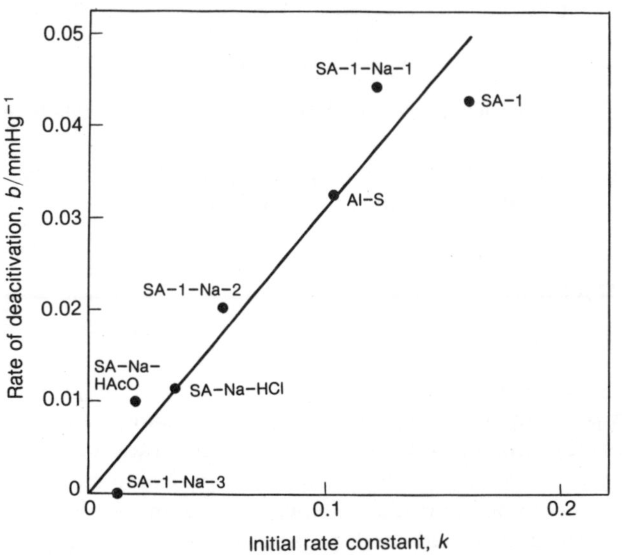

Fig. 5.3 The relationship between the rate constant of propylene polymerization and the rate of deactivation. SA-1 : Silica-alumina (13 % Al_2O_3), SA-1–Na-x : SA-1 was treated with aq. NaOH soln. (Na content increases with x). Some are subsequently washed with acids. A1-S : $Al_2(SO_4)_3$-SiO_2.

The hydrogenation of coal liquid bottom over Ni – Mo/Al_2O_3 catalysts has been investigated in relation to the acid amounts of catalysts.[14] As shown in Fig. 5.4, the catalytic activity was little dependent on the acid amount, while the amount of coke deposit was proportional to the acidity. Based on this correlation and other evidence it was presumed that basic nitrogen-containing compounds contained in the preasphaltheneic fraction are strongly adsorbed on the acid sites and are transformed to coke via polycondensation and dehydrogenation. The coke then covers the surface of the hydrogena-

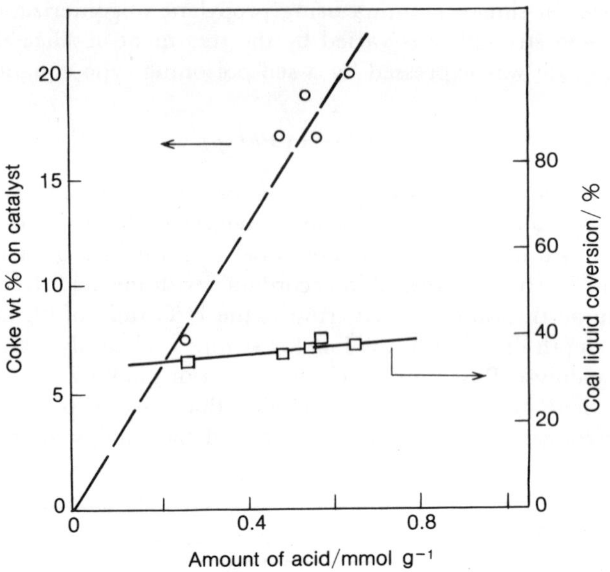

Fig. 5.4 Effect of acidic properties of catalyst on coke formation and activity.

tion sites. The addition of alkali-earth suppresses the acidity and coke deposition, but it also decreases the catalytic activity to a certain degree.[28)]

5.4 REGENERATION

Deactivated catalysts are reused after adequate regeneration processes either on-site or off-site, when possible and economical. Important examples are catalytic cracking (FCC), naphtha-reforming and HDS catalysts. The generation process to be employed differs depending on the kind of catalyst and cause of deactivation. In general carbon deposits are burnt off by air and metal deposits are removed by chemical treatment such as washing with sulfuric acid. Redispersion of sintered metal particles on support is sometimes possible.

FCC catalysts are very rapidly deactivated by coke deposition (the residence time in a so-called riser reactor is a few seconds) and the deactivated catalysts, after stripping of oil, are sent directly to a regenerator, where the coke is burnt off by air. The regenerated catalysts are then returned to the reactor. All these processes are carried out continuously in fluidized reactors. The rate of carbon burning is expressed by

$$dC/dt = -k \cdot C \cdot P(O_2) \tag{3}$$

where C is the amount of coke, and assumed to be independent of the structure of feedstock.[24)] Addition of small amounts of transition metals greatly accelerates the rate of burn-off.

Besides short-term deactivation due to coke, there is long-term aging which is due primarily to permanent damage of active sites. Because of this deactivation, it is necessary to continuously supply fresh catalyst and withdraw a portion of the aged catalyst.

Deactivation of reforming catalysts derives mainly from carbon deposits and sintering of metal particles. So burn-off of the coke and treatments to redisperse the sintered metal are necessary.[25] Temperature-programmed oxidation of coke formed on Pt/Al_2O_3 shows two peaks at 470 and 650K, which are respectively ascribed to cokes

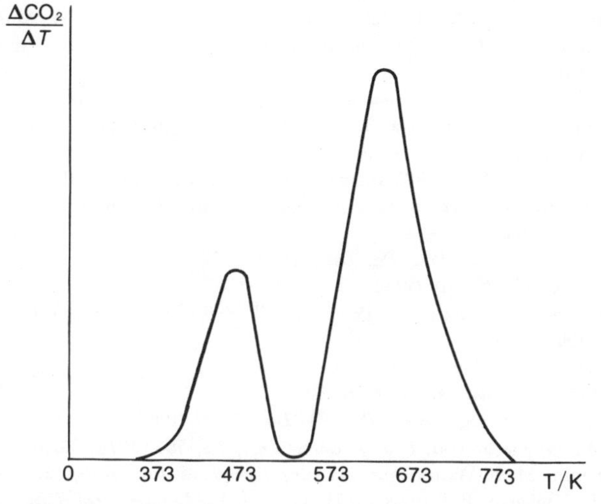

Fig. 5.5 Typical temperature programmed oxidation curve plotting rate of CO_2 formation as a function of temperature.
(Reproduced with permission by J. Barbier *et al.*, *Catalyst Deactivation,* **6**, 55 (1980)).

on metal and on alumina as shown in Fig. 5.5.[26]

The causes of activity decay of HDS catalysts are deposits of carbonaceous materials and heavy metals such as nickel and vanadium. Aggregation or crystal growth of active molybdenum component has been observed as well.[13] Regeneration usually consists of i) removal of heavy metals by washing the catalysts with sulfuric acid after removal of oil, ii) burn-off of coke by air at about 770K, and iii) resulfidation at about 670K. Steps i) and ii) can be interchanged. It has been reported that complete removal of heavy metals is not necessary to restore the activity since they usually only block the entrance of catalyst pores.[25]

REFERENCES

1a) E. E. Wolf, F. Alfani, *Catal. Rec. Sci. Eng.*, **24**, 329 (1982).
b) H. Matsumoto, M. Masai, H. Arai, in: *Shokubai Koza*, Vol. 5, p.224, Kodansha, Tokyo, 1985 (in Japanese).
2. H. Schulz, Z. Siwei, W. Baumgartner, in: *Catalyst Deactivation 1987* (B. Delmon, G. F. Froment, eds.), (Studies in Surface Science and Catalysis 34), Elsevier, Amsterdam, 1987, p.479.

3. R. C. Haldeman, M. C. Botty, *J. Phys. Chem.,* **63**,489 (1959).
4. W. G. Appleby, J. W. Gibson, G. M. Good, *Ind. Eng. Chem.*, *Proc. Des. Dev.*, **1**, 102 (1962).
5. F. E. Shephard, J. J. Rooney, C. Kemball, *J. Catal.*, **1**, 379 (1962).
6. P. B. Venuto, E. T. Habib, Jr., *Fluid Catalytic Cracking with Zeolite Catalysts*, Marcel Dekker, New York, 1979.
7. B. Mercier de Rochettes, C. Marcilly, C. Gueguen, J. Bousquet, in: *Catalyst Deactivation 1987*, (B. Delmon, G. F. Froment, eds.) Elsevier, Amsterdam, 1987, p.589.
8. M. Misono, Y. Yoneda, *Bull. Chem. Soc. Jpn.,* **40**, 42 (1967).
9. C. Mirodatos, V. Perrichon, M. C. Durupty, P. Moral, in: *Catalyst Deactivation 1987* (B. Delmon, G. F. Froment, eds.) Elsevier, Amsterdam, 1987, p.183.
10. J. Barbier in: *Catalyst Deactivation 1987*, (B. Delmon, G. F. Froment, eds.) Elsevier, Amsterdam, 1987, p.1.
11. J. M. Parera, R. J. Verderone, C. A. Querini, in: *Catalyst Deactivation 1987*, (B. Delmon, G. F. Froment, eds.) Elsevier, Amsterdam, 1987, p.135.
12. B. J. Cooper, D. L. Trimm, in: *Catalyst Deactivation 1980*, Elsevier, Amstrerdam, 1980, p.63.
13. A.Nishijima, H. Shimada, Y. Yoshimura, T. Sato, N. Matsubayashi, in: *Catalyst Deactivation 1987*, (B. Delmon, G. F. Froment, eds.) Elsevier, Amsterdam, 1987, p.39.
14. Y. Kageyama, T. Masuyama, Proc. 1985 Intern. Conf. Coal Sci., Sydney, 1985, p.157
15. T. Hisamitsu, K. Komori, H. Ozaki in: *Catalyst Deactivation 1987*, (B. Delmon, G. F. Froment, eds.) Elsevier, Amsterdam, 1987, p.259.
16. J. T. Richardson, *Ind. Eng. Chem., Proc. Des. Dev.,* **11**, 8 (1972).
17. A. Voorhies, *Ind. Eng. Chem.*, **37**, 318 (1945).
18. E.E. Petersen, in: *Catalyst Deactivation* (E. E. Petersen, A.T. Bell eds.), (Chemical Industries/30), Marcel Dekker, New York, 1987, p.39.
19. A. T. Bell, *ibid.*, p.235.
20. Y. Nishiyama, Y. Tamai, *J. Catal.*, **45**, 1 (1976).
21. Y. Ozawa, K. B. Bishoff, *Ind. Eng. Chem., Proc. Des.Dev.*, **7**, 67 (1968).
22a) T. M. John, B. W. Wojciechowski, *Can. J. Chem. Eng.,* **54**, 584 (1976), cited in Ref. (1)
b) J.W.Evans, D.L.Trimm, M. S. Wainwright, *Ind. Eng.Chem. Prod. Res. Develop*, **22**, 242 (1983).
c) S. J. Kulkarni, S.B. Kulkarni, P. Ratnasamy,H. Hattori, K. Tanabe, *Appl. Catal.*, **8**, 43 (1983).
23. J. Take, T. Tsuruya, T. Sato, Y. Yoneda, *Bull. Chem. Soc. Jpn.,* **45**, 3409 (1972).
24. P. B. Weisz, R. D. Goodwin, *J. Catal.*, **2**, 397 (1963); *ibid.* **6**, 227 (1966).
25. M. Tamayama, *Shokubai*, **28**, 571 (1986) (in Japanese).
26. J. Barbier, P. Marecot, N. Martin, L. Elassal, R. Maurel in: *Catalyst Deactivation*, (B. Delmon, G. F. Froment, eds.), Elsevier, Amsterdam, 1980, p.53.
27. H. Shimada, M. Kurita, T. Sato, Y. Yoshimura, Y. Kovayashi, A. Nishijima, *Bull, Chem, Soc, Jpn.,* **59**, 2885 (1986).
28. H. Shimada, T. Sato, Y. Yoshimura, A. Nishijima, M. Matsuda, T. Konakahara, K. Sato, *Sekiyu Gakkaishi,* **31**, 227 (1988) (in Japanese).

6
Related Topics

6.1 GAS SENSORS

Gas sensors are devices which detect or monitor a desired component or components in gases, and quantitatively convert the information to an electric signal. The component(s) may be combustible gases, oxygen, water (humidity), poisonous gases, etc.

For any gas sensor of either a semiconductor-type or catalytic combustion-type, sufficient sensitivity, accuracy, selectivity, reproducibility, life, and stability are required, and the sensing mechanism starts by adsorption or reactive adsorption of the gas component on the surface of sensor materials. Metal oxides such as SnO_2, Al_2O_3,

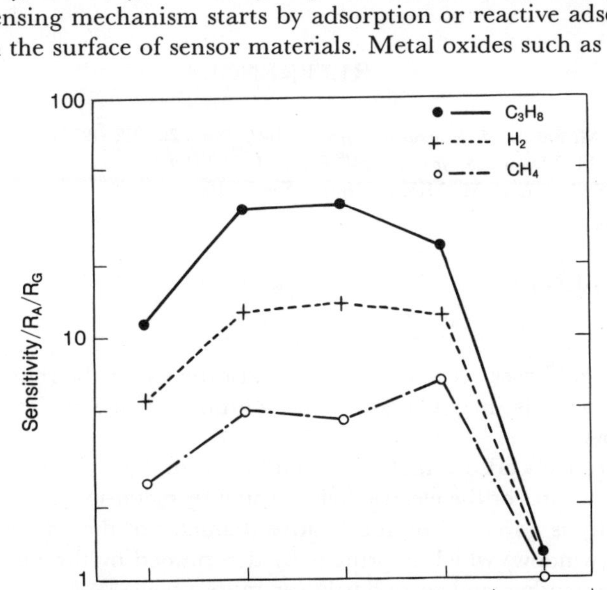

Sample	A	B	C	D	E
Additive	—	$Ti(SO_4)_2$	$ZrOCl_2$	$SnCl_4$	—

Fig. 6.1 Effects of sulfate ion and additives on gas sensitivity.
R_A; resistance in clean air, R_G; resistance in air containing 0.5 vol % of the gas.
Starting Fe salts were ferric sulfate for A—D and ferric nitrate for E.
(Reproduced with permission by Y. Nakatani *et al.*, *Jpn. J. Appl. Phys.*, **22**, 913 (1983)).

Fe_2O_3, ZnO and mixed oxides are widely used as sensor materials. Therefore, the sensitivity and selectivity depend to a considerable extent on the acid-base interactions between the gas component and the surface of metal oxides. However, so far there have been very few studies which explicitly take into account the acid-base properties of the surface of sensor materials.

It has been reported that α-Fe_2O_3 promoted by sulfate ion exhibited higher sensitivity to combustible gases than α-Fe_2O_3 alone.[1] The addition of tetravalent metal ions such as Ti^{4+}, Zr^{4+}, and Sn^{4+} further enhanced the sensitivity (Fig. 6.1). $SO_4^{2-}-Fe_2O_3$ is known to be a very strong acid as mentioned in Section 4.9.1, but the role of sulfate ion in the above case was thought to stabilize fine particles and maintain a high surface area, by suppressing the crystal growth.[2]

Humidity sensors such as $MgCr_2O_4-TiO_2$, $MgO-ZrO_2$, and thin film of Al_2O_3 detect changes of electric properties (conductance or capacitance) caused by hydroxylation, adsorption of water and/or condensation of it in micropores. Therefore, in addition to the pore-size distribution, the affinity of the surface to water vapor (hydrophilic or hydrophobic) also play important roles in the sensing function. For example, the sensitivity, the resistivity-*vs.*-humidity correlation and response of Cs, NH_4, and alkyl ammonium salts of 12-tungstophosphate varied closely corresponding to those properties.[3]

REFERENCES

1. Y. Nakatani, M. Sakai, M. Matsusaka, *Jpn. J. Appl. Phys.*, **22**, 912 (1983).
2. Y. Nakatani, M. Matsusaka, *Jpn. J. Appl. Phys.*, **21**, 758 (1982).
3. N. Mizuno, K. Inumaru, M. Misono, *Hyomen Kagaku,* **10**, 21 (1989).

6.2 ADSORBENTS

Solid acids and bases are widely used as adsorbents for the removal or separattion of certain compounds in gas mixtures or solutions. Some typical examples will be described below.

Zeolites selectively adsorb molecules and ions with the aid of shape-selective micropores and/or acid sites or the electric field created by metal-ions exchanged in the pore. Shape selectivity is governed by the relative diameter of the adsorbate molecule and pore-opening (window) which is principally determined by the number of oxide ions invloved in the window and secondarily (or more precisely) by the metal-ion present in the wall of the window. An example of precise control of shape-selective adsorption is shown in Fig. 6.2.[1] As the extent of exchange of Na^+ with Ca^{2+} increases and slightly widens the window, at a certain level of exchange isobutane starts to adsorb in the pores of A zeolite. The hydrophilic surface of zeolite changes to hydrophobic with increasing silica/alumina ratio. This property can be also used for selective adsorption. Examples of practical applications of zeolites as adsorbents are listed below.

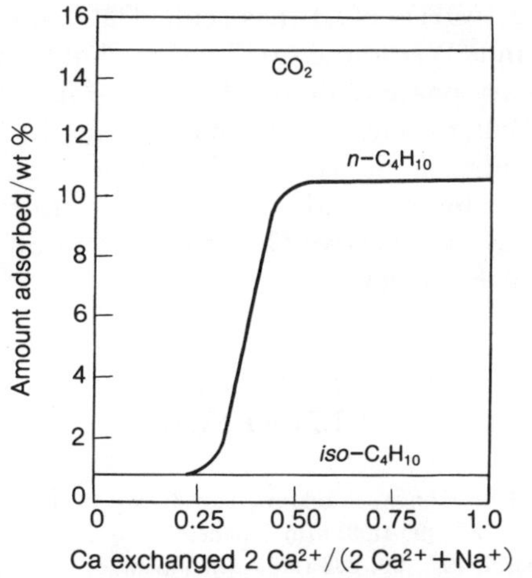

Fig. 6.2 Precise control of the shape-selective adsorption of A zeolite by ion-exchange. (Reproduced with permission by T. L. Thomas, 6th World Petrol. Congr. Frankfurt, Part3, No16(1963)).

Zeolites	Adsorption processes
A	Drying of ethylene and butadiene (3A), drying and removal of CO_2 from natural gas (4A), O_2 separation from air (5A), recovery of *n*-paraffins from naphtha and kerosene (5A), purification of monosilane (NaZnA)
ZSM-5	Dewaxing (selective cracking of *n*-paraffins)
Mordenite	Separation of *p*-xylene from a mixture of xylene isomers and ethylbenzene

Reactive adsorbents which contain alkali-earth metal oxides (MO) are added to the FCC process to remove SO_x by the following reactions.

$$SO_3 + MO \longrightarrow MSO_4 \quad (1)$$

$$MSO_4 + 4\,H_2 \longrightarrow MO + H_2S + 3\,H_2O \quad (2)$$

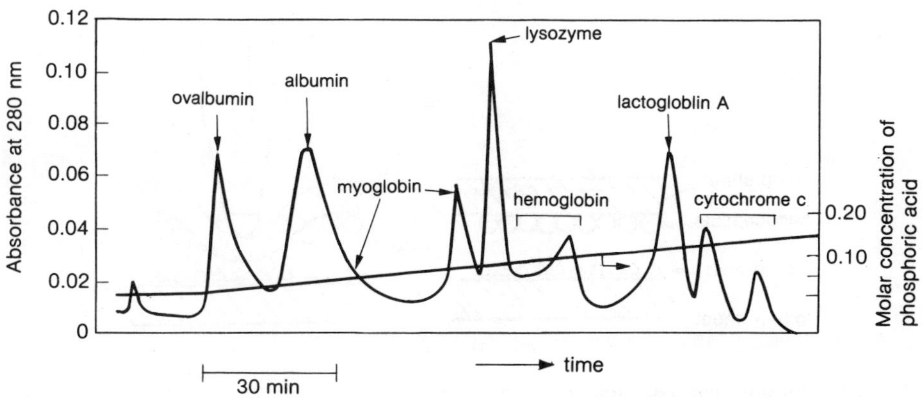

Fig. 6.3 Chromatographic separation of a mixture of proteins with a hydroxyapatite column. (Reproduced with permission by T. Kawasaki, *Ceramics (Japan)*, **20**, 200(1985)).

Hydroxyapaptite, $Ca_{10}(OH)_2(PO_4)_6$ has acidic Ca^{2+} sites (C sites) and basic sites (P sites, one site surrounded by four PO_4^{3-}) of moderate strength on its surface. It can be used as a stationary phase of liquid chromatography for the separation of biopolymers of 10^4-10^9 Daltons such as enzymes, antibody, and other proteins (Fig. 6.3).[2)] C sites interact with anionic parts of biopolymers, carboxylate and phosphate, as well as phosphate ion in buffer solution, and P sites with cations and cationic parts.

TiO_2 is used to remove radioactive Co^{2+} ions[3)] and SiO_2 to remove radioactive RuO_4 in exhausts of nuclear reactor.

REFERENCES

1. T.L. Thomas, 6th World Petr. Congr., Frankfurt, Part 3, No. 16 (1963).
2. T. Kawasaki, *Ceramics Japan*, **20**, 195 (1985) (in Japanese).
3. F. Kawamura, K. Funabashi, *ibid.*, **20**, 203 (1985) (in Japanese).

6.3 PRESSURE SENSITIVE RECORDING PAPER (CARBONLESS PAPER)

6.3.1 Principle

Solid acids are utilized as one component of pressure sensitive recording paper. The principle of the carbonless paper is indicator color change by adsorption on solid acids. The model construction of carbonless paper consists of a set of two types of paper as schematically shown in Fig. 6.4. The top sheet is coated with indicators encapsulated in microcapsules,[1)] and the bottom sheet is coated with solid acids. Applying pressure to the top sheet destroys the microcapsule, and the indicator is released and transferred to the solid acids on the bottom sheet to be adsorbed.

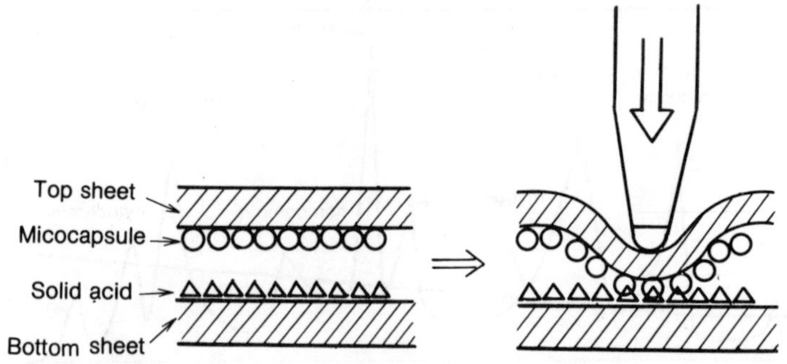

Fig. 6.4 Pressure sensitive paper.
Indicator contained in microcapsules changes color when adsorbed on solid acid.

6.3.2 Types of Solid Acid

The solid acids utilized for carbonless paper must meet the following requirements:
i) Colorless or white
ii) Large adsorption capacity
iii) Fast and strong coloring
iv) High resistance to weather and light
v) Nontoxicity

Both organic and inorganic solid acids are used. As organic acids, Brønsted acids such as condensed compounds of *para*-substituted phenol with formaldehyde,[2-4] and Lewis acids such as zinc salts of the above compounds and those of aromatic carboxyl acids are used. As inorganic acids, montmorillonite clays such as bentonite, fuller's earth, kaoline, chinaclay, and their chemically modified materials are used.

Any kind of indicator can be used provided that it changes colors when adsorbed on solid acids. Selecting the appropriate indicators makes it possible to obtain the desired color. The types of coloring materials in practical use are as follows: phthalides such as crystal violet lactone, acyl leucomethylene blues such as benzoyl leucomethylene blue, and fluorane derivatives. In particular, fluorane derivatives give desired color of yellow, red, and black by changing the substitutional groups.

6.3.3 Preparation of the Paper

The key technique which makes it possible to commercially produce carbonless

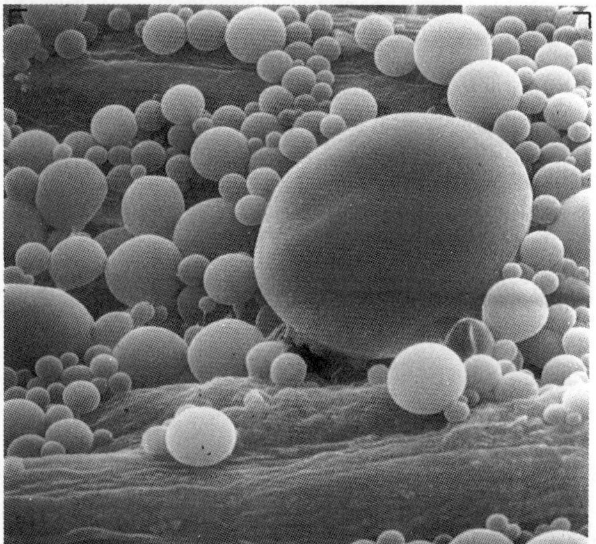

Fig. 6.5 SEM image of microcapsules coated on top sheet, ×2400

Fig. 6.6 SEM image of solid acid coated on bottom sheet, ×2400

paper is microcapsulation of the indicators. The indicators are dissolved in proper solvents, and the solutions are encapsulated into microcapsules made of gelatins and Arabian rubbers. The size of the microcapsules is in the range $1-5$ μm. Aqueous solutions containing the microcapsules are then painted onto the paper.

SEM photographs of the microcapsules coated on the top sheet of paper and the solid acids on the bottom sheet are shown in Figs. 6.5 and 6.6.

REFERENCES

1. US Patent 2,711,375 (1955).
2. Jpn. Patent (Toku-Kou-Shou) 40-9,309
3. Jpn. Patent (Toku-Kou-Shou) 44-9,071
4. Jpn. Patent (Toku-Kou-Shou) 46-22,652

6.4 COSMETIC PIGMENTS

Clays and metal oxides and salts such as talc, kaolinite, mica, zinc oxide, titanium oxide, iron oxide, hydrated chromium oxide, cobalt blue, ultramarine blue, calcium carbonate, barium sulfate, etc. are widely used as pigments for cosmetics. Since these pigments possess acidic and basic surface properties, and hence catalytic activity, cosmetic products containing perfumes, oils, and medicaments as the other components occasionally deteriorate as time elapses by decomposition of the components due to catalytic action of the pigments, markedly depreciating the value of the cosmetic. In the case of perfumes, the isomerization of 2-pinene,[1,2] the reaction of *d*-limonene oxide[3,4] and the dehydrogenation of *d*-limonene to *p*-cymene[5] are known to take place over solid acids and bases. As a result, the odor or the quality of the cosmetic changes.

It is essential to deactivate the catalytic activity of such pigments, and measurement of their catalytic activity is important to maintain the stability of the cosmetics containing them. Fukui *et al.* measured the catalytic activities of various pigments for the dehydration and the dehydrogenation of isopropanol which are known to be catalyzed by acid sites and by both acid and base sites, respectively.[6] As shown in Table 6.1, the pigments used in cosmetics can be classified into four groups by the reaction products of isopropanol: group 1 pigments which mainly produce propylene by dehydration, group 2 pigments which mainly produce acetone by dehydrogenation, group 3 forming both acetone and propylene by dehydrogenation and dehydration, and those of group 4 which do not decompose isopropanol. It was found that the catalytic activities of pigment groups 1−3 decreased markedly by treatment with basic sodium carbonate or acidic acetic acid.[6]

In the reaction between pigments and linalool, which is a common component of perfumes, pigments such as black iron oxide and hydrated chromium oxide which

TABLE 6.1 Decomposition of IPA over various pigments

Pigment	Selectivity to	
	Propylene (%)	Acetone (%)
Group 1		
Talc	8.8	0
Kaolinite	99.5	0
Titanium dioxide (A)	19.2	0
Titanium dioxide (R)	33.8	trace
Ultramarine blue	2.7	0
Cobalt blue	1.0	0
Titanium-coated mica	5.7	0
Group 2		
Zinc oxide	0.9	32.1
Group 3		
Black iron oxide (M)	27.0	33.4
Red iron oxide (H)	3.8	20.9
Hydrated chromium oxide	5.2	17.9
Group 4		
Mica	0	0
Barium sulfate	0	0
Calcium carbonate	0	0
Silica	0	0
Synthesized pigment G	0	0

(Reproduced with permission by H. Fukui, *et al.*, *Cosmetics* 4 *Toiletries*, **96,** 37(1981)).

showed low conversion rates formed only ocimene and myrcene, while limonene and terpinolene were also formed over pigments such as silica and mica with moderate conversion rates.[7] With pigments such as prussian blue and red iron oxide, which showed high conversion rates, alloocimene, terpinene, and *p*-cymene were also formed.[7] *p*-Cymene is the cause of one of the most unpleasant odors in deteriorated cosmetics. The decomposition of perfumes can be controlled by treating pigments with suitable acidic and/or basic materials and the change in odor and hence the deterioration of the quality of cosmetics is prevented.

REFERENCES

1. R. Ohnishi, K. Tanabe, S. Morikawa, T. Nishizaki, *Bull. Chem. Soc. Jpn.*, **47**, 571 (1974).
2. K. Tanabe, R. Ohnishi, K. Arata, *Terpene Chemistry*, (ed. J. Verghese) Tata McGraw, New Delhi, 1981.
3. K. Arata, S. Akutagawa, K. Tanabe, *J. Catal.*, **41**, 173 (1976).
4. K. Arata, K. Tanabe, *Catal. Rev.*, **25**, 365 (1983).
5. H. Pines, J.A. Vesely, V. N. Ipatieff, *J. Am. Chem. Soc.*, **77**, 347 (1955).
6. H. Fukui, T. Saito, M. Tanaka, S. Ohta, *Cosmetics & Toiletries*, **96**, 37 (1981).
7. H. Fukui, R. Namba, M. Tanaka, M. Nakano, S. Fukushima, *J. Soc. Cosmet. Chem.*, **38**, 385 (1987).

Subject Index

Index to Catalysts

Studies in Surface Science and Catalysis

Volume 1 **Preparation of Catalysts I.** Scientific Bases for the Preparation of Heterogeneous Catalysts. Proceedings of the First International Symposium held at the Solvay Research Centre, Brussels, October 14-17, 1975
edited by **B. Delmon, P.A. Jacobs and G. Poncelet**

Volume 2 **The Control of the Reactivity of Solids.** A Critical Survey of the Factors that Influence the Reactivity of Solids, with Special Emphasis on the Control of the Chemical Processes in Relation to Practical Applications
by **V.V. Boldyrev, M. Bulens and B. Delmon**

Volume 3 **Preparation of Catalysts II.** Scientific Bases for the Preparation of Heterogeneous Catalysts. Proceedings of the Second International Symposium, Louvain-la-Neuve, September 4-7, 1978
edited by **B. Delmon, P. Grange, P. Jacobs and G. Poncelet**

Volume 4 **Growth and Properties of Metal Clusters.** Applications to Catalysis and the Photographic Process. Proceedings of the 32nd International Meeting of the Société de Chimie Physique, Villeurbanne, September 24-28, 1979
edited by **J. Bourdon**

Volume 5 **Catalysis by Zeolites.** Proceedings of an International Symposium organized by the Institut de Recherche sur la Catalyse - CNRS - Villeurbanne, and sponsored by the Centre National de Recherche Scientifique, Ecully (Lyon), September 9-11, 1980
edited by **B. Imelik, C. Naccache, Y. Ben Taarit, J.C. Vedrine, G. Coudurier and H. Praliaud**

Volume 6 **Catalyst Deactivation.** Proceedings of the International Symposium, Antwerp, October 13-15, 1980
edited by **B. Delmon and G.F. Froment**

Volume 7 **New Horizons in Catalysis.** Proceedings of the 7th International Congress on Catalysis, Tokyo, June 30 - July 4, 1980
edited by **T. Seiyama and K. Tanabe**

Volume 8 **Catalysis by Supported Complexes**
by **Yu.I. Yermakov, B.N. Kuznetsov and V.A. Zakharov**

Volume 9 **Physics of Solid Surfaces.** Proceedings of the Symposium held in Bechyne, Czechoslovakia, September 29 - October 3, 1980
edited by **M. Láznicka**

Volume 10 **Adsorption at the Gas-Solid and Liquid-Solid Interface.** Proceedings of an International Symposium held in Aix-en-Provence, September 21-23, 1981
edited by **J. Rouquerol and K.S.W. Sing**

Volume 11 **Metal-Support and Metal-Additive Effects in Catalysis.** Proceedings of an International Symposium organized by the Institut de Recherches sur la Catalyse - CNRS - Villeurbanne, and sponsored by the Centre National de la Recherche Scientifique, Ecully (Lyon), September 14-16, 1982
edited by **B. Imelik, C. Naccache, G. Courdurier, H. Praliaud, P. Meriaudeau, P. Gallezot, G.A. Martin and J.C. Vedrine**

Volume 12 **Metal Microstructures in Zeolites.** Preparation - Properties - Applications. Proceedings of a Workshop, Bremen, September 22-24, 1982
edited by **P.A. Jacobs, N.I. Jaeger, P. Jiru and G. Schulz-Ekloff**

Volume 13 **Adsorption on Metal Surfaces.** An Integrated Approach
edited by **J. Bénard**

Volume 14 **Vibration at Surfaces.** Proceedings of the Third International Conference, Asilomar, California, U.S.A., September 1-4, 1982
edited by **C.R. Brundle and H. Morawitz**

Volume 15 **Heterogeneous Catalytic Reactions Involving Molecular Oxygen**
by **G.I. Golodets**

Volume 16 **Preparation of Catalysts III.** Scientific Bases for the Preparation of Heterogeneous Catalysts. Proceedings of the Third International Symposium, Louvain-la-Neuve, September 6-9, 1982
edited by **G. Poncelet, P. Grange and P.A. Jacobs**

Volume 17 **Spillover of Adsorbed Species.** Proceedings of the International Symposium, Lyon-Villeurbanne, September 12-16, 1983
edited by **G.M. Pajonk, S.J. Teichner and J.E. Germain**

Volume 18 **Structure and Reactivity of Modified Zeolites.** Proceedings of an International Conference, Prague, July 9-13, 1984
edited by **P.A. Jacobs, N.I. Jaeger, P. Jiru, V.B. Kazansky and G. Schulz-Ekloff**

Volume 19 **Catalysis on the Energy Scene.** Proceedings of the 9th Canadian Symposium, Québec, P.Q., September 30 - October 3, 1984
edited by **S. Kaliaguine and A. Mahay**

Volume 20 **Catalysis by Acids and Bases.** Proceedings of an International Symposium organized by the Institut de Recherches sur la Catalyse - CNRS - Villeurbanne and sponsored by the Centre National de la Recherche Scientifique, Villeurbanne (Lyon), September 25-27, 1984
edited by **B. Imelik, C. Naccache, G.Coudurier, Y. Ben Taarit and J.C. Vedrine**

Volume 21 **Adsorption and Catalysis on Oxide Surfaces.** Proceedings of a Symposium, Brunel University, Uxbridge, June 28-29, 1984
edited by M. Che and G.C. Bond

Volume 22 **Unsteady Processes in Catalytic Reactors**
by **Yu.Sh. Matros**

Volume 23 **Physics of Solid Surfaces 1984**
edited by **J. Koukal**

Volume 24 **Zeolites. Synthesis, Structure, Technology and Application.** Proceedings of the International Symposium, Portoroz-Portorose, September 3-8, 1984
edited by **B. Drzaj, S. Hocevar and S. Pejovnik**

Volume 25 **Catalytic Polymerization of Olefins.** Proceedings of the International Symposium on Future Aspects of Olefin Polymerization, Tokyo, Japan July 4-6, 1985
edited by **T. Keii and K. Soga**

Volume 26 **Vibrations at Surfaces 1985.** Proceedings of the Fourth International Conference, Bowness-on-Windermere, U.K., September 15-19, 1985
edited by **D.A. King, N.V. Richardson and S. Holloway**

Volume 27 **Catalytic Hydrogenation**
by **L. Cerveny**

Volume 28 **New Developments in Zeolite Science and Technology.** Proceedings of the 7th International Zeolite Conference, Tokyo, August 17-22, 1986
edited by **Y. Murakami, A. Iijima and J.W. Ward**

Volume 29 **Metal Clusters in Catalysis**
edited by **B.C. Gates, L. Guczi and H. Knözinger**

Volume 30 **Catalysis and Automotive Pollution Control.** Proceedings of the First International Symposium (CAPOC I), Brussels, September 8-11, 1986
edited by **A. Crucq and A. Frennet**

Volume 31 **Preparation of Catalysts IV.** Scientific Bases for the Preparation of Heterogeneous Catalysts. Proceedings of the Fourth International Symposium, Louvain-la-Neuve, September 1-4, 1986
edited by **B. Delmon, P. Grange, P.A. Jacobs and G. Poncelet**

Volume 32 **Thin Metal Films and Gas Chemisorption**
edited by **P. Wissmann**

Volume 33 **Synthesis of High-Silica Aluminosolicate Zeolites**
by **P.A. Jacobs and J.A. Martens**

Volume 34 **Catalyst Deactivation 1987.** Proceedings of the 4th International Symposium, Antwerp, September 29 - October 1, 1987
edited by **B. Delmon and G.F. Froment**

Volume 35 **Keynotes in Energy-Related Catalysis**
edited by **S. Kaliaguine**

Volume 36 **Methane Conversion.** Proceedings of a Symposium on the Productions of Fuels and Chemicals from Natural Gas, Auckland, April 27-30, 1987
edited by **D.M. Bibby, C.D. Chang, R.F. Howe and S. Yurchak**

Volume 37 **Innovation in Zeolite Materials Science.** Proceedings of an International Symposium, Nieuwpoort (Belgium), September 13-17, 1987
edited by **P.J. Grobet, W.J. Mortier, E.F. Vansant and G. Schulz-Ekloff**

Volume 38 **Catalysis 1987.** Proceedings of the 10th North American Meeting of the Catalysis Society, San Diego, CA, May 7-22, 1987
edited by **J.W. Ward**

Volume 39 **Characterization of Porous Solids.** Proceedings of the IUPAC Symposium (COPS I), Bad Soden a.Ts., F.R.G., April 26-29, 1987
edited by **K.K. Unger, J. Rouquerol, K.S.W. Sing and H. Kral**

Volume 40 **Physics of Solid Surfaces 1987**
edited by **J. Koukal**

Volume 41 **Heterogeneous Catalysis of Fine Chemicals**
edited by **M. Guisnet, J. Barrault, C. Bouchoule, D. Duprez, C. Montassier and G. Pérot**

Volume 42 **Laboratory Studies of Heterogeneous Catalytic Processes**
by **E.G. Christoffel and Z. Paál**

Volume 43 **Catalytic Processes under Unsteady-State Conditions**
by **Yu.Sh. Matros**

Volume 44 **Successful Design of Catalysts - Future Requirements and Development**
edited by **T. Inui**

Volume 45 **Transition Metal Oxides. Surface Chemistry and Catalysis**
by **H.H. Kung**

Volume 46 **Zeolites as Catalysts, Sorbents and Detergent Builders.** Applications and Innovations. Proceedings of an International Symposium, Würzburg, F.R.G., September 4-8, 1988
edited by **H.G. Karge and J. Weitkamp**

Volume 47 **Photochemistry on Solid Surfaces**
edited by **M. Anpo and T. Matsuura**

Volume 48 **Structure and Reactivity of Surfaces.** Proceedings of a European Conference, Trieste, Italy, September 13-16, 1988
edited by **C. Morterra, A. Zecchina and G. Costa**

Volume 49 **Zeolites: Facts, Figures, Future.** Proceedings of the 8th International Zeolite Conference, Amsterdam, The Netherlands, July 10-14, 1989
edited by **P.A. Jacobs and R.A. van Santen**

Volume 50 **Hydrotreating Catalysts.** Preparation, Characterization and Performance. Proceedings of the Annual International AIChE Meeting, Washington, DC, November 27 - December 2, 1988
edited by **M.L. Occelli and R.G. Anthony**

Volume 51 **New Solid Acids and Bases** - their catalytic properties
by **K. Tanabe, M. Misono, Y. Ono and H. Hattori**